Springer Series in Statistics

Advisors:
P. Bickel, P. Diggle, S. Fienberg K. Krickeberg,
I. Olkin, N. Wermuth, S. Zeger

Springer

New York
Berlin
Heidelberg
Hong Kong
London
Milan
Paris
Tokyo

Springer Series in Statistics

(continued after index)

Geert Verbeke
Geert Molenberghs

Linear Mixed Models for Longitudinal Data

With 128 Illustrations

 Springer

Geert Verbeke
Biostatistical Centre
Katholieke Universiteit Leuven
Kapucijnenvoer 35
B-3000 Leuven
Belgium

Geert Molenberghs
Biostatistics
Center for Statistics
Limburgs Universitair Centrum
Universitaire Campus, Building D
B-3590 Diepenbeek
Belgium

Library of Congress Cataloging-in-Publication Data
Verbeke, Geert.
    Linear mixed models for longitudinal data / Geert Verbeke, Geert Molenberghs.
      p. cm. — (Springer series in statistics)
    Includes bibliographical references and index.
    ISBN 0-387-95027-3 (alk. paper)
    1. Linear models (Statistics).   2. Longitudinal methods. I. Molenberghs, Geert. II. Title.
III. Series.
    QA279 .V458   2000
    519.5′3—dc21   00-026596

Printed on acid-free paper.

Production managed by Jenny Wolkowicki; manufacturing supervised by Jerome Basma.
Photocomposed pages prepared from the authors' LaTeX files.
Printed and bound by Edwards Brothers, Inc., Ann Arbor, MI.
Printed in the United States of America.

9 8 7 6 5

ISBN 0-387-95027-3

Springer-Verlag New York Berlin Heidelberg
A member of BertelsmannSpringer Science+Business Media GmbH

To Godewina, Lien, Noor, and Aart

To Conny, An, and Jasper
To the memory of my father

Preface

The dissemination of the MIXED procedure in SAS and related software have provided a whole class of linear mixed-effects models, some of which with a long history, for routine use. Experience shows that both the ideas behind the techniques and their software implementation are not at all straightforward, and users from various applied backgrounds often encounter difficulties in using the methodology effectively. Courses and consultancy in this domain have been in great demand over the last decade, illustrating the clear need for resource material to aid the user.

As an outgrowth of such courses, Verbeke and Molenberghs (1997) was intended as a contribution to bridging this gap. Since its appearance, it has been the basis for several short and regular courses in academia and industry. In the meantime, many research papers on these and related topics have appeared in the statistical literature. Therefore, it is considered timely to present a second, entirely recast version. Material kept from Verbeke and Molenberghs (1997) has been reworked, and a large range of new topics has been added. The structure of the book reflects not only our own research activity but also our experience in teaching various applied longitudinal modeling courses, such as the Longitudinal Data Analysis course in the Master of Science in Biostatistics Programme of the Limburgs Universitair Centrum, the Repeated Measures course in the International Study Programme in Statistics of the Katholieke Universiteit Leuven, and the Topics in Biostatistics course at the Universiteit Antwerpen.

As with the first version, we hope this book will be of value to a wide audience, including applied statisticians and biomedical researchers, particularly in the pharmaceutical industry, medical and public health research organizations, contract research organizations, and academic departments. This implies that the majority of the chapters is explanatory rather than research oriented and that it emphasizes practice rather than mathematical rigor. In this respect, guidance and advice on practical issues are the main focus of the text. On the other hand, some more advanced topics are included as well, which we believe to be of use to the more demanding modeler.

In the first version, we had placed strong emphasis on the SAS procedure MIXED, without discouraging the non-SAS users. Considerable effort was put in treating data analysis issues in a generic fashion, instead of making them fully software dependent. Therefore, a research question was first translated into a statistical model by means of algebraic notation. In a number of cases, such a model was then implemented using SAS code. This was positively received by many readers and we therefore for most part kept this format. In this version, much of the SAS-related issues are centralized in a single chapter, and we still keep selected examples throughout the text. Additionally, an Appendix is devoted to other software tools (MLwiN, SPlus).

Because SAS Version 7 has not been generally marketed, SAS Version 6.12 was used throughout this book. The Appendix briefly lists the most important changes in Version 7. Selected macros for tools discussed in the text, not otherwise available in commercial software packages, as well as publicly available data sets, can be found at Springer-Verlag's URL: www.springer-ny.com.

Geert Verbeke (Katholieke Universiteit Leuven, Leuven)

Geert Molenberghs (Limburgs Universitair Centrum, Diepenbeek)

Acknowledgments

This book has been accomplished with considerable help from several people. We would like to gratefully acknowledge their support.

A large part of this book is based on joint research. We are grateful to several co-authors: Larry Brant (Gerontology Research Center and The Johns Hopkins University, Baltimore), Luc Bijnens (Janssen Research Foundation, Beerse), Tomasz Burzykowski (Limburgs Universitair Centrum), Marc Buyse (International Institute for Drug Development, Brussels), Desmond Curran (European Organization for Research and Treatment of Cancer, Brussels), Helena Geys (Limburgs Universitair Centrum), Mike Kenward (London School of Hygiene and Tropical Medicine), Emmanuel Lesaffre (Katholieke Universiteit Leuven), Stuart Lipsitz (Medical University of South Carolina, Charleston), Bart Michiels (Janssen Research Foundation, Beerse), Didier Renard (Limburgs Universitair Centrum), Ziv Shkedy (Limburgs Universitair Centrum), Bart Spiessens (Katholieke Universiteit Leuven), Herbert Thijs (Limburgs Universitair Centrum), Tony Vangeneugden (Janssen Research Foundation, Beerse), and Paige Williams (Harvard School of Public Health, Boston).

Russell Wolfinger (SAS Institute, Cary, NC) has been kind enough to provide us with a trial version of SAS Version 7.0 during the development of this text. Bart Spiessens (Katholieke Universiteit Leuven) kindly provided us with technical support. Steffen Fieuws (Katholieke Universiteit Leuven) commented on earlier versions of the text.

We gratefully acknowledge support from Research Project Fonds voor Wetenschappelijk Onderzoek Vlaanderen G.0002.98: "Sensitivity Analysis for Incomplete Data," NATO Collaborative Research Grant CRG950648: "Statistical Research for Environmental Risk Assessment," and from Onderzoeksfonds K.U.Leuven grant PDM/96/105.

It has been a pleasure to work with John Kimmel and Jenny Wolkowicki of Springer-Verlag.

We apologize to our wives, daughters, and son for the time not spent with them during the preparation of this book and we are very grateful for their understanding. The preparation of this book has been a period of close and fruitful collaboration, of which we will keep good memories.

Geert and Geert

Kessel-Lo, December 1999

Contents

Appendix

1

Introduction

In applied sciences, one is often confronted with the collection of *correlated data*. This generic term embraces a multitude of data structures, such as multivariate observations, clustered data, repeated measurements, longitudinal data, and spatially correlated data.

Among those, multivariate data have received most attention in the statistical literature (e.g., Seber 1984, Krzanowski 1988, Johnson and Wichern 1992). Techniques devised for this situation include multivariate regression and multivariate analysis of variance, which have been implemented in the SAS procedure GLM (SAS 1991) for general linear models. In addition, SAS contains a battery of relatively specialized procedures for principal components analysis, canonical correlation analysis, discriminant analysis, factor analysis, cluster analysis, and so forth (SAS 1989).

As an example of a simple multivariate study, assume that a subject's systolic and diastolic blood pressure are measured simultaneously. This is different from a *clustered setting* where, for example, for a number of families, diastolic blood pressure is measured for all of their members. A design where, for each subject, diastolic blood pressure is recorded under several experimental conditions is often termed a *repeated measures* study. In the case that diastolic blood pressure is measured repeatedly over time for each subject, we are dealing with *longitudinal data*. Although one could view all of these data structures as special cases of multivariate designs, we believe there are many fundamental differences, thoroughly affecting the mode of analysis. First, certain multivariate techniques, such as principal components, are hardly useful for the other designs. Second, in a truly multivariate set of outcomes, the variance-covariance structure is usually unstructured, in contrast to, for example, longitudinal data. Therefore, the methodology of the general linear model is too restrictive to perform satisfactory data analyses of these more complex data. In contrast, the *general linear mixed model*, as implemented in the SAS procedure MIXED (Littell *et al.* 1996), is much more flexible.

Replacing the time dimension in a longitudinal setting with one or more spatial dimensions leads naturally to spatial data. While ideas in the longitudinal and spatial areas have developed relatively independently, efforts have been spent in bridging the gap between both disciplines. In 1996, a workshop was devoted to this idea: "The Nantucket Conference on Modeling Longitudinal and Spatially Correlated Data: Methods, Applications, and Future Directions" (Gregoire *et al.* 1997).

Still, restricting attention to the correlated data settings described earlier is too limited to fully grasp the wide applicability of the general linear mixed model. In designed experiments, such as analysis of variance (ANOVA) or nested factorial designs, the variance structure has to reflect the design and thus elaborate structures will be needed. A good mode of analysis should be able to account for various sources of variability. Linear mixed models originated precisely in this area of application. For a review, see Robinson (1991).

Among the clustered data settings, longitudinal data perhaps require the most elaborate modeling of the random variability. Diggle, Liang, and Zeger (1994) distinguish among three components of variability. The first one groups traditional random effects (as in a random-effects ANOVA model) and random coefficients (Longford 1993). It stems from interindividual variability (i.e., heterogeneity between individual profiles). The second component, serial association, is present when residuals close to each other in time are more similar than residuals further apart. This notion is well known in the time-series literature (Ripley 1981, Diggle 1983, Cressie 1991). Finally, in addition to the other two components, there is potentially also measurement error. This results from the fact that, for delicate measurements (e.g., laboratory assays), even immediate replication will not be able to avoid a certain level of variation. In longitudinal data, these three components of variability can be distinguished by virtue of both *replication* as well as a clear *distance* concept (time), one of which is lacking in classical spatial and time-series analysis and in clustered data. This implies that adapting models for longitudinal data to other data structures is in many cases relatively straightforward. For example, clustered data could be analyzed by leaving out all aspects of the model that refer to time.

A very important characteristic of data to be analyzed is the type of outcome. Methods for continuous data form the best developed and most advanced body of research; the same is true for software implementation. This is natural, since the special status and the elegant properties of the normal distribution simplify model building and ease software development. It is in this area that the general linear mixed model and the SAS procedure MIXED, as well as its counterparts in, for example, SPlus and MLwiN, are situated. However, also categorical (nominal, ordinal, and binary) and discrete outcomes are very prominent in statistical practice. For example, quality of life outcomes are often scored on ordinal scales.

Two fairly different views can be adopted. The first one, supported by large-sample results, states that normal theory should be applied as much as possible, even to non-normal data such as ordinal scores and counts. A different view is that each type of outcome should be analyzed using instruments that exploit the nature of the data. We will adopt the second standpoint. In addition, since the statistical community has been familiarized with generalized linear models (GLIM; McCullagh and Nelder 1989), some have taken the view that the normal model for continuous data is but one type of GLIM. Although this is correct in principle, it fails to acknowledge that normal models are much further developed than any other GLIM (e.g., model checks and diagnostic tools) and that it enjoys unique properties (e.g., the existence of closed-form solutions, exact distributions of test statistics, unbiased estimators). Extensions of GLIM to the longitudinal case are discussed in Diggle, Liang, and Zeger (1994), where the main emphasis is on generalized estimating equations (Liang and Zeger 1986). Generalized linear mixed models have been proposed by, for example, Breslow and Clayton (1993). Fahrmeir and Tutz (1994) devote an entire book to GLIM for multivariate settings.

In longitudinal settings, each individual typically has a *vector* Y of responses with a natural (time) ordering among the components. This leads to several, generally nonequivalent, extensions of univariate models. In a *marginal model*, marginal distributions are used to describe the outcome vector Y, given a set X of predictor variables. The correlation among the components of Y can then be captured either by adopting a fully parametric approach or by means of working assumptions, such as in the semiparametric approach of Liang and Zeger (1986). Alternatively, in a *random-effects model*, the predictor variables X are supplemented with a vector b of random effects, conditional upon which the components of Y are usually assumed to be independent. This does not preclude that more elaborate models are possible if residual dependence is detected (Longford 1993). Finally, a *conditional model* describes the distribution of the components of Y, conditional on X but also conditional on (a subset of) the other components of Y. In a longitudinal context, a particular relevant class of conditional models describes a component of Y given the ones recorded earlier in time. Well-known members of this class of *transition models* are *Markov type* models. Several examples are given in Diggle, Liang, and Zeger (1994).

For normally distributed data, marginal models can easily be fitted, for example, with the SAS procedure MIXED, the SPlus function lme, or within the MLwiN package. For such data, integrating a mixed-effects model over the random effects produces a marginal model, in which the regression parameters retain their meaning and the random effects contribute in a simple way to the variance-covariance structure. For example, the marginal model corresponding to a random-intercepts model is a compound symmetry model that can be fitted without explicitly acknowledging the random-intercepts structure. In the same vein, certain types of transition model induce simple marginal covariance structures. For example, some first-order stationary autoregressive models imply an exponential or AR(1) covariance structure. As a consequence, many marginal models derived from random-effects and transition models can be fitted with mixed-models software.

It should be emphasized that the above elegant properties of normal models do not extend to the general GLIM case. For example, opting for a marginal model for longitudinal binary data precludes the researcher from answering conditional and transitional questions in terms of simple model parameters. This implies that each model family requires its own specific software tools. For example, an analysis based on generalized estimating equations can be performed within the GENMOD procedure in SAS, or the SPlus set of functions termed OSWALD (Smith, Robertson, and Diggle 1996). Mixed-effects models for non-Gaussian data can be fitted using the MIXOR program (Hedeker and Gibbons 1994, 1996), MLwiN, or the SAS procedure NLMIXED. The latter procedure is available from Version 7 onward and is the successor of the macros GLIMMIX and NONLINMIX.

Motivated by the above discussion, we have restricted the scope of this book to linear mixed models for continuous outcomes. Fahrmeir and Tutz (1994) discuss generalized linear (mixed) models for multivariate outcomes, while longitudinal versions are treated in Diggle, Liang, and Zeger (1994). Nonlinear models for repeated measurement data are discussed by Davidian and Giltinan (1995).

While research in this area has largely focused on the formulation of linear mixed-effects models, inference, and software implementation, other important aspects, such as exploratory analysis, the investigation of model fit, and the construction of diagnostic tools have received considerably less attention. In addition, longitudinal data are typically very prone to incompleteness, due to dropout or intermediate missing values. This poses particular challenges to methodological development. In this book, we have attempted to give a detailed account of several of these topics. By no means has it been our intention to give a complete or definitive overview. Indeed, given the high research activity, this would be impossible.

Broadly, the structure of the book is as follows. The key examples, used throughout the book, are introduced in Chapter 2. Chapters 3 to 9 provide the core about the linear mixed-effects model, while Chapters 10 to 13 discuss more advanced tools for model exploration, influence diagnostics, as well as extensions of the original model. Chapters 14 to 16 introduce the reader to basic incomplete data concepts. Chapters 17 and 18 discuss strategies to model incomplete longitudinal data, based on the linear mixed model. The sensitivity of such strategies to parametric assumptions is investigated in Chapters 19 and 20. Some additional missing data topics are presented in Chapters 21 and 22. Chapter 23 is devoted to design considerations. Five case studies are treated in detail in Chapter 24. Appendix A reviews a number of software tools for fitting mixed models. Since the book puts relatively more emphasis on SAS than on other packages, this procedure is discussed in detail in Chapter 8, while worked examples can be found throughout the text. Some technical background material from the sensitivity chapters is deferred until Appendix B.

2

Examples

This chapter introduces the longitudinal sets of data which will be used throughout the book. The rat data are presented in Section 2.1. The TDO data, studying toenails, are described in Section 2.2. Section 2.3 is devoted to the Baltimore Longitudinal Study of Aging, with two substudies: prostate-specific antigen data (Section 2.3.1) and data on hearing (Section 2.3.2). Section 2.4 introduces the Vorozole study, focusing on quality of life in breast cancer patients. In Section 2.5, we will introduce data, previously analyzed by Goldstein (1979), on the heights of 20 schoolgirls. Section 2.6 presents the growth data of Potthoff and Roy (1964). Mastitis in dairy cattle is the subject of Section 2.7.

To complement the data introduced in this chapter, five case studies, involving additional sets of data, are presented in Chapter 24.

2.1 The Rat Data

In medical science, there has recently been increased interest in the therapeutic use of hormones. However, such drastic therapies require detailed knowledge about their effect on the different aspects of growth. To this respect, an experiment has been set up at the Department of Orthodontics of the Catholic University of Leuven (KUL) in Belgium (see Verdonck *et al.* 1998). The primary aim was to investigate the effect of the inhibition

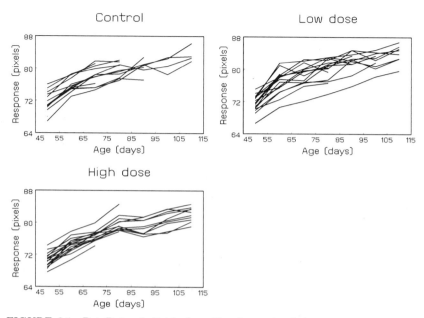

FIGURE 2.1. *Rat Data. Individual profiles for each of the treatment groups in the rat experiment separately.*

of the production of testosterone in male Wistar rats on their craniofacial growth.

A total of 50 male Wistar rats have been randomized to either a control group or one of the two treatment groups where treatment consisted of a low or high dose of the drug Decapeptyl, which is an inhibitor for testosterone production in rats. The treatment started at the age of 45 days, and measurements were taken every 10 days, with the first observation taken at the age of 50 days. The responses of interest are distances (in pixels) between well-defined points on X-ray pictures of the skull of each rat, taken after the rat has been anesthetized. Of primary interest is the estimation of changes over time and testing whether these changes are treatment dependent.

For the purpose of this book, we will consider one of the measurements which can be used to characterize the height of the skull. The individual profiles are shown in Figure 2.1. It is clear that not all rats have measurements up to the age of 110 days. This is due to the fact that many rats do not survive anaesthesia and therefore drop out before the end of the study. Table 2.1 shows the number of rats observed at each occasion. While 50 rats have been randomized at the start of the experiment, only 22 of them survived the 6 first measurements, so measurements on only 22 rats are available in the way anticipated at the design stage. For example, at the

TABLE 2.1. *Rat Data. Summary of the number of observations taken at each occasion in the rat experiment, for each group separately and in total.*

Age (days)	# Observations			Total
	Control	Low	High	
50	15	18	17	50
60	13	17	16	46
70	13	15	15	43
80	10	15	13	38
90	7	12	10	29
100	4	10	10	24
110	4	8	10	22

second occasion (age = 60 days), only 46 rats were available, implying that for 4 rats only 1 measurement could be recorded.

2.2 The Toenail Data (TDO)

The data introduced in this section were obtained from a randomized, double-blind, parallel group, multicenter study for the comparison of two oral treatments (in the sequel coded as A and B) for toenail dermatophyte onychomycosis (TDO), described in full detail by De Backer *et al.* (1996). TDO is a common toenail infection, difficult to treat, affecting more than 2 out of 100 persons (Roberts 1992). Antifungal compounds, classically used for treatment of TDO, need to be taken until the whole nail has grown out healthy. The development of new such compounds, however, has reduced the treatment duration to 3 months. The aim of the present study was to compare the efficacy and safety of 12 weeks of continuous therapy with treatment A or with treatment B.

In total, 2×189 patients were randomized, distributed over 36 centers. Subjects were followed during 12 weeks (3 months) of treatment and followed further, up to a total of 48 weeks (12 months). Measurements were taken at baseline, every month during treatment, and every 3 months afterward, resulting in a maximum of seven measurements per subject. For our purposes, we will only consider one of the secondary endpoints, unaffected nail length, which is measured as follows. At the first occasion, the treating physician indicates one of the affected toenails as the target nail, the nail which will be followed over time. At each occasion, the unaffected nail length (measured from the nail bed to the infected part of the nail,

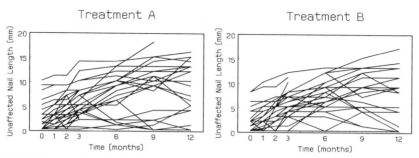

FIGURE 2.2. *Toenail Data. Individual profiles of 30 randomly selected subjects in each of the treatment groups in the toenail experiment.*

which is always at the free end of the nail) of the target nail is measured in millimeters. Obviously, this response will be related to the toe size. Therefore, we will only include here those patients for which the target nail was one of the two big toenails. This reduces our sample under consideration to 150 and 148 subjects, respectively. Figure 2.2 shows the observed profiles of 30 randomly selected subjects from treatment group A and treatment group B, respectively.

Due to a variety of reasons, 72 (24%) out of the 298 participants left the study prematurely. Table 2.2 summarizes the number of subjects still in the study at each occasion, for both treatment groups separately. Although the comparison of the average evolutions in both treatment groups was of primary interest, there was also some interest in studying the relationship between the dropout process and the actual outcome. For example, are patients who drop out doing better or worse than patients who do not drop out from the study ?

2.3 The Baltimore Longitudinal Study of Aging (BLSA)

The Baltimore Longitudinal Study of Aging (BLSA) is an ongoing multidisciplinary observational study, which started in 1958, and with the study of normal human aging as primary objective (Shock *et al.* 1984). Participants in the BLSA are volunteers who return approximately every 2 years for 3 days of biomedical and psychological examinations. They are predominantly white (95%), well educated (over 75% have bachelor's degrees), and financially comfortable (82%). So far, over 1400 men with an average of almost 7 visits and 16 years of follow-up have participated in the study since its inception in 1958. Later on, females have been included in the study as well.

TABLE 2.2. *Toenail Data. Summary of the number of observations taken at each occasion in the TDO study, for each group separately and in total.*

	# Observations		
Time (months)	Treatment A	Treatment B	Total
0	150	148	298
1	149	142	291
2	146	138	284
3	140	131	271
6	131	124	255
9	120	109	229
12	118	108	226

The BLSA (Pearson *et al.* 1994) is a unique resource for rapidly evaluating longitudinal hypotheses because of the availability of data from repeated clinical examinations and a bank of frozen blood samples from the same individuals over 30 years of follow-up (where new studies would require many years to conduct). On the other hand, the observational aspect of the study poses additional complications on the statistical analysis. For example, although repeated visits are scheduled every 2 years, some subjects may have more than one visit within 1 year of time, while others have over 10 years between two successive visits. Also, longitudinal evolutions may be highly influenced by many covariates which may or may not be recorded in the study.

In this book, two of the many responses measured in the BLSA will be used to illustrate the statistical methodology. In Section 2.3.1, it will be discussed how data from the BLSA can be used to study the natural history of prostate disease. Afterward, in Section 2.3.2, the hearing data will be presented.

2.3.1   The Prostate Data

During the last 10 years, many papers have been published on the natural history of prostate disease; see, for example, Carter *et al.* (1992a, 1992b) and Pearson *et al.* (1991, 1994). According to Carter and Coffey (1990), prostate disease is one of the most common and most costly medical problems in the United States, and prostate cancer has become the second leading cause of male cancer deaths. It is therefore very important to look for markers which can detect the disease at an early stage. The prostate-specific antigen (PSA) is such a marker. PSA is an enzyme produced by

TABLE 2.3. *Prostate Data. Description of subjects included in the prostate data set, by diagnostic group. The cancer cases are subdivided into local/regional (L/R) and metastatic (M) cancer cases.*

	Controls	BPH cases	Cancer Cases L/R	Cancer Cases M
Number of participants	16	20	14	4
Age at diagnosis (years)				
Median	66	75.9	73.8	72.1
Range	56.7-80.5	64.6-86.7	63.6-85.4	62.7-82.8
Years of follow-up				
Median	15.1	14.3	17.2	17.4
Range	9.4-16.8	6.9-24.1	10.6-24.9	10-25.3
Time between measurements (years)				
Median	2	2	1.7	1.7
Range	1.1-11.7	0.9-8.3	0.9-10.8	0.9-4.8
Number of measurements per individual				
Median	8	8	11	9.5
Range	4-10	5-11	7-15	7-12

both normal and cancerous prostate cells, and its level is related to the volume of prostate tissue. Still, an elevated PSA level is not necessarily an indicator of prostate cancer because patients with benign prostatic hyperplasia (BPH) also have an enlarged volume of prostate tissue and therefore also an increased PSA level. This overlap of the distribution of PSA values in patients with prostate cancer and BPH has limited the usefulness of a single PSA value as a screening tool since, according to Pearson *et al.* (1991), up to 60% of BPH patients may be falsely identified as potential cancer cases based on a single PSA value.

Based on clinical practice, researchers have hypothesized that the rate of change in PSA level might be a more accurate method of detecting prostate cancer in the early stages of the disease. This has been extensively investigated by Pearson *et al.* (1994), who analyzed repeated PSA measures from the Baltimore Longitudinal Study of Aging (BLSA), using linear mixed models.

A retrospective case-control study was undertaken that utilized frozen serum samples from 18 BLSA participants identified as prostate cancer cases, 20 cases of BPH, and 16 controls with no clinical signs of prostate disease. In order to be eligible for the analyses, men had to meet several criteria:

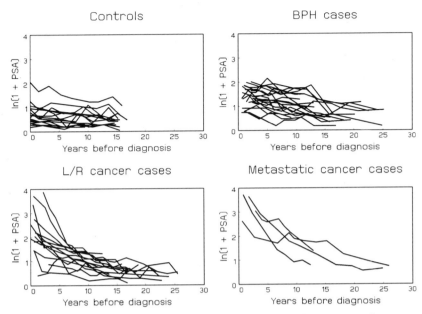

FIGURE 2.3. *Prostate Data. Longitudinal trends in PSA in men with prostate cancer, benign prostatic hyperplasia, or no evidence of prostate disease.*

1. seven or more years of follow-up prior to diagnosis of prostate cancer, simple prostatectomy for BPH, or exclusion of prostate disease by a urologist,

2. confirmation of the pathological diagnosis, and

3. no prostate surgery prior to diagnosis.

To the extent possible, age at diagnosis and years of follow-up were matched for the control, BPH, and cancer groups. However, due to the high prevalence of BPH in men over age 50, it was difficult to find age-matched controls with no evidence of prostate disease. In fact, the control group remained significantly younger at first visit and at diagnosis, compared to the BPH group. For this reason, our analyses of this data set will always correct for age differences at the time of the diagnosis.

A description of the data, differentiating between local/regional (L/R) cancer cases and metastatic cancer cases, is given in Table 2.3. The number of repeated PSA measurements per individual varies between 4 and 15, and the follow-up period ranges from 6.9 to 25.3 years. Since it was anticipated that PSA values would increase exponentially in prostate cancer cases, the responses were transformed to ln(PSA + 1). These transformed individual profiles are shown in Figure 2.3.

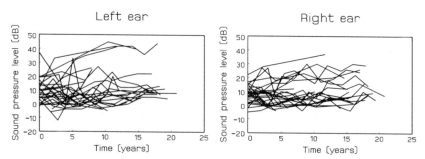

FIGURE 2.4. *Hearing Data. Individual profiles of 30 randomly selected subjects in the hearing data set, for the left and the right ear separately.*

2.3.2 The Hearing Data

Also recorded in the BLSA study are hearing threshold sound pressure levels (SPLs in dB), measured at 11 different frequencies [varying from 125 to 8000 hertz (Hz)] on both ears, yielding a maximum of 22 observations per visit. This was done by means of a sound proof chamber and a Bekesy audiometer. Using these data, Brant and Fozard (1990) have shown that the relationship between hearing threshold level and frequency can be well described by a quadratic function of the logarithm of frequency, the parameters of which depend on age and are highly subject-specific. Morrell and Brant (1991) and Brant and Pearson (1994) considered the data of 268 elderly male participants whose first visit occurred at about 70 years of age or older. They studied how hearing thresholds change over time and how these evolutions depend on age and on the frequency under consideration.

For our purposes, we now consider all available hearing thresholds for 500 Hz, from male BLSA participants only, without otologic disease, unilateral hearing loss, or evidence of noise-induced hearing loss. Individual profiles on the left and right ear separately are shown in Figure 2.4 for 30 randomly selected subjects.

In total, we have 6170 observations (3089 on the left ear and 3081 on the right ear), from 681 males. Their age at the first visit ranged from 17.2 to 90.5 years, with median value equal to 53 years. The number of visits per subject varied from 1 to 15, and some of the participants were followed for over 22 years (median 7.5 years).

TABLE 2.4. *Heights of Schoolgirls. Classification of 20 preadolescent school girls in three groups, according to their mother's height*

	Mothers height	Children numbers
Small mothers	< 155 cm	$1 \to 6$
Medium mothers	[155cm; 164cm]	$7 \to 13$
Tall mothers	> 164 cm	$14 \to 20$

2.4   The Vorozole Study

This study was an open-label, multicenter, parallel group design conducted at 67 North American centers. Patients were randomized to either the new drug Vorozole (2.5 mg taken once daily) or the standard drug megestrol acetate (40 mg four times daily). The patient population consisted of post-menopausal patients with histologically confirmed estrogen-receptor positive metastatic breast carcinoma. All 452 randomized patients were followed until disease progression or death. The main objective was to compare the treatment groups with respect to response rate, whereas secondary objectives included a comparison relative to duration of response, time to progression, survival, safety, pain relief, performance status, and quality of life. Full details of this study are reported in Goss *et al.* (1999). In this book, we will focus on overall quality of life, measured by the total Functional Living Index: Cancer (FLIC; Schipper *et al.* 1984). Precisely, a higher FLIC score is the more desirable outcome. Even though this outcome is, strictly speaking, of the ordinal type, the total number of categories encountered exceeds 70, justifying the use of continuous-outcome methods.

Patients underwent screening and for those deemed eligible, a detailed examination at baseline (occasion 0) took place. Further measurement occasions were months 1, then from months 2 at bimonthly intervals until month 44.

Goss *et al.* (1999) analyzed FLIC using a two-way ANOVA model with effects for treatment, disease status, as well as their interaction. No significant difference was found. Apart from treatment, important covariates are dominant site of the disease as well as clinical stage.

This example will be used, for example, to introduce exploratory tools in Chapter 4.

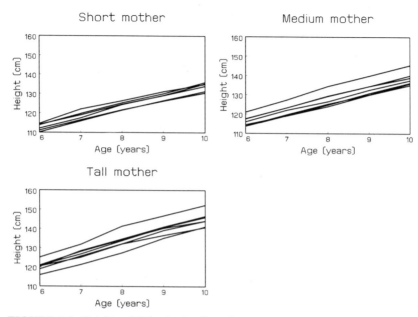

FIGURE 2.5. *Heights of Schoolgirls. Growth curves of 20 school girls from age 6 to 10, for girls with small, medium, or tall mothers.*

2.5 Heights of Schoolgirls

Goldstein (1979, Table 4.3, p. 101) reports growth curves of 20 preadolescent girls, measured on a yearly basis from age 6 to 10. The girls were classified according to the height of their mother, which was discretized as in Table 2.4. The individual profiles are shown in Figure 2.5, for each group separately. The measurements are given at exact years of age, some having been previously adjusted to these. The values Goldstein reports for the fifth girl in the first group are 114.5, 112, 126.4, 131.2, and 135.0. This suggests that the second measurement is incorrect. We therefore replaced it by 122. An extensive analysis of this data set can be found in Section 4.2 of Verbeke and Molenberghs (1997). Of primary interest is to test whether the growth of these schoolgirls is related to the height of their mothers.

2.6 Growth Data

These data, introduced by Potthoff and Roy (1964), contain growth measurements for 11 girls and 16 boys. For each subject, the distance from the center of the pituitary to the maxillary fissure was recorded at ages 8,

TABLE 2.5. *Growth Data for 11 Girls and 16 Boys. Measurements marked with* * *were deleted by Little and Rubin (1987).*

	Age (in years)					Age (in years)			
Girl	8	10	12	14	Boy	8	10	12	14
1	21.0	20.0	21.5	23.0	1	26.0	25.0	29.0	31.0
2	21.0	21.5	24.0	25.5	2	21.5	22.5*	23.0	26.5
3	20.5	24.0*	24.5	26.0	3	23.0	22.5	24.0	27.5
4	23.5	24.5	25.0	26.5	4	25.5	27.5	26.5	27.0
5	21.5	23.0	22.5	23.5	5	20.0	23.5*	22.5	26.0
6	20.0	21.0*	21.0	22.5	6	24.5	25.5	27.0	28.5
7	21.5	22.5	23.0	25.0	7	22.0	22.0	24.5	26.5
8	23.0	23.0	23.5	24.0	8	24.0	21.5	24.5	25.5
9	20.0	21.0*	22.0	21.5	9	23.0	20.5	31.0	26.0
10	16.5	19.0*	19.0	19.5	10	27.5	28.0	31.0	31.5
11	24.5	25.0	28.0	28.0	11	23.0	23.0	23.5	25.0
					12	21.5	23.5*	24.0	28.0
					13	17.0	24.5*	26.0	29.5
					14	22.5	25.5	25.5	26.0
					15	23.0	24.5	26.0	30.0
					16	22.0	21.5*	23.5	25.0

Source: Pothoff and Roy (1964), Jennrich and Schluchter (1986).

10, 12, and 14. The data were used by Jennrich and Schluchter (1986) to illustrate estimation methods for unbalanced data, where unbalancedness is now to be interpreted in the sense of an unequal number of boys and girls.

Little and Rubin (1987) deleted 9 of the $[(11 + 16) \times 4]$ measurements, rendering 9 incomplete subjects. Deletion is confined to the age 10 measurements. Little and Rubin (1987) describe the mechanism to be such that subjects with a low value at age 8 are more likely to have a missing value at age 10. The data are presented in Table 2.5. The measurements that were deleted are marked with an asterisk. In Section 17.4.1, the complete data will be analyzed in some detail. Sections 17.4.2 and 17.4.3 are devoted to frequentist and likelihood-based ignorable analyses of the incomplete version of the data, respectively. Section 17.4.4 is devoted to insight in the missingness mechanism.

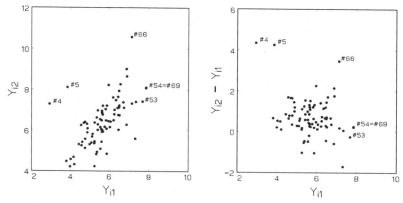

FIGURE 2.6. *Mastitis in Dairy Cattle. The first panel shows a scatter plot of the second measurement versus the first measurement. The second panel shows a scatter plot of the change versus the baseline measurement.*

2.7   Mastitis in Dairy Cattle

This example, concerning the occurrence of the infectious disease mastitis in dairy cows, was introduced in Diggle and Kenward (1994) and reanalyzed in Kenward (1998). Data were available of the milk yields in thousands of liters of 107 dairy cows from a single herd in 2 consecutive years: Y_{ij} $(i = 1, \ldots, 107; j = 1, 2)$. In the first year, all animals were supposedly free of mastitis; in the second year, 27 became infected. Mastitis typically reduces milk yield, and the question of scientific interest is whether the probability of occurrence of mastitis is related to the yield that would have been observed had mastitis not occurred. A graphical representation of the complete data is given in Figure 2.6.

3

A Model for Longitudinal Data

3.1 Introduction

In practice, longitudinal data are often highly unbalanced in the sense that not an equal number of measurements is available for all subjects and/or that measurements are not taken at fixed time points. In the rat data set and the toenail data set, presented in Section 2.1 and in Section 2.2, respectively, a fixed number of measurements was scheduled to be taken on all subjects, at fixed time points. However, during the study, rats died, and patients left the toenail study prematurely, implying unbalance. This is different from the prostate data and the hearing data (Sections 2.3.1 and 2.3.2, respectively), where the unbalance is an immediate result from the fact that the volunteers participating in the BLSA were asked to return *approximately* every 2 years for medical examination.

Due to their unbalanced nature, many longitudinal data sets cannot be analyzed using multivariate regression techniques (see, for example, Seber 1984, Chapters 8 and 9, Hand and Taylor 1987). A natural alternative arises from observing that subject-specific longitudinal profiles can often be well approximated by linear regression functions. One hereby summarizes the vector of repeated measurements for each subject by a vector of a relatively small number of estimated subject-specific regression coefficients. Afterward, in a second stage, multivariate regression techniques can be used to relate these estimates to known covariates such as treatment, disease classification, baseline characteristics, and so forth. This so-called two-stage analysis will be introduced in Section 3.2. Afterward, in Section 3.3, the general linear mixed model will be introduced as a result of combining the two stages into one single statistical model.

3.2   A Two-Stage Analysis

3.2.1   Stage 1

Let the random variable Y_{ij} denote the (possibly transformed) response of interest, for the ith individual, measured at time t_{ij}, $i = 1, \ldots, N$, $j = 1, \ldots, n_i$, and let Y_i be the n_i-dimensional vector of all repeated measurements for the ith subject, that is, $Y_i = (Y_{i1}, Y_{i2}, \ldots, Y_{in_i})'$. The first stage of the two-stage approach assumes that Y_i satisfies the linear regression model

$$Y_i = Z_i \beta_i + \varepsilon_i, \tag{3.1}$$

where Z_i is a $(n_i \times q)$ matrix of known covariates, modeling how the response evolves over time for the ith subject. Further, β_i is a q-dimensional vector of unknown subject-specific regression coefficients, and ε_i is a vector of residual components ε_{ij}, $j = 1, \ldots, n_i$. It is usually assumed that all ε_i are independent and normally distributed with mean vector zero, and covariance matrix $\sigma^2 I_{n_i}$, where I_{n_i} is the n_i-dimensional identity matrix. This latter assumption will be extended in Section 3.3.

Obviously, model (3.1) includes very flexible models for the description of subject-specific profiles. In practice, polynomials will often suffice. However, extensions such as fractional polynomial models (Royston and Altman 1994), or extended spline functions (Pan and Goldstein 1998) can be considered as well. We refer to Lesaffre, Asefa and Verbeke (1999) for an example where subject-specific profiles have been modeled using fractional polynomials.

3.2.2   Stage 2

In a second step, a multivariate regression model of the form

$$\beta_i = K_i \beta + b_i, \tag{3.2}$$

is used to explain the observed variability between the subjects, with respect to their subject-specific regression coefficients β_i. K_i is a $(q \times p)$

matrix of known covariates, and β is a p-dimensional vector of unknown regression parameters. Finally, the b_i are assumed to be independent, following a q-dimensional normal distribution with mean vector zero and general covariance matrix D.

3.2.3 Example: The Rat Data

The rat data presented in Section 2.1 have been analyzed by Verbeke and Lesaffre (1999), who describe the subject-specific profiles shown in Figure 2.1 by straight lines, after transforming the original time scale (age expressed in days) logarithmically (see also Section 4.3.3). The first-stage model (3.1) then becomes

$$Y_{ij} = \beta_{1i} + \beta_{2i}t_{ij} + \varepsilon_{ij}, \quad j = 1, \ldots, n_i, \tag{3.3}$$

where $t_{ij} = \ln[1 + (\text{Age}_{ij} - 45)/10)]$, implying that $t = 0$ corresponds to the start of the treatment. The matrix Z_i has two columns: one containing only ones, and one containing all time points $t_{ij}, j = 1, \ldots, n_i$.

In the second stage, the subject-specific intercepts and time effects are related to the treatment of the rats (low dose, high dose, control). Our second-stage model (3.2) then becomes

$$\begin{cases} \beta_{1i} = \beta_0 + b_{1i}, \\[2mm] \beta_{2i} = \beta_1 L_i + \beta_2 H_i + \beta_3 C_i + b_{2i}, \end{cases} \tag{3.4}$$

in which L_i, H_i, and C_i are indicator variables defined to be one if the rat belongs to the low-dose group, the high-dose group, or the control group, respectively, and zero otherwise. The randomization in combination with the chosen transformation of the original time scale allows us to assume the subject-specific intercepts β_{1i} not to depend on the treatment. The parameter β_0 can be interpreted as the average response at the start of the treatment, whereas the parameters β_1, β_2, and β_3 represent the average time effects for each treatment group separately. Of primary interest is the comparison of these average slopes, since this directly measures the treatment effect on the average growth.

3.2.4 Example: The Prostate Data

Pearson et al. (1994) and Verbeke and Molenberghs (1997, Chapter 3) have previously analyzed the prostate data presented in Section 2.3.1, assuming that each individual profile shown in Figure 2.3 can be well approximated by a quadratic function over time, where time is expressed as years before diagnosis (see also Section 4.3.4). The regression model (3.1) in the first stage is then

$$
\begin{aligned}
Y_{ij} &= \ln(\mathrm{PSA}_{ij} + 1) \\
&= \beta_{1i} + \beta_{2i}t_{ij} + \beta_{3i}t_{ij}^2 + \varepsilon_{ij}, \quad j = 1, \ldots, n_i,
\end{aligned}
\tag{3.5}
$$

and the columns of the covariate matrix Z_i contain only ones, all time points t_{ij}, and all squared time points t_{ij}^2.

In the second stage, the subject-specific intercepts and linear as well as quadratic time effects are related to the diagnostic class of the subject (control, BPH case, local cancer case, or metastatic cancer case). The age at the time of diagnosis is included as a covariate in order to correct for the age differences among the four diagnostic groups. Model (3.2) in the second stage then becomes

$$
\begin{cases}
\beta_{1i} = \beta_1 \mathrm{Age}_i + \beta_2 C_i + \beta_3 B_i + \beta_4 L_i + \beta_5 M_i + b_{1i}, \\[2mm]
\beta_{2i} = \beta_6 \mathrm{Age}_i + \beta_7 C_i + \beta_8 B_i + \beta_9 L_i + \beta_{10} M_i + b_{2i}, \\[2mm]
\beta_{3i} = \beta_{11} \mathrm{Age}_i + \beta_{12} C_i + \beta_{13} B_i + \beta_{14} L_i + \beta_{15} M_i + b_{3i},
\end{cases}
\tag{3.6}
$$

in which Age_i equals the subject's age at diagnosis ($t = 0$), and where C_i, B_i, L_i, and M_i are indicator variables defined to be one if the subject is a control, a BPH case, a local cancer case, or a metastatic cancer case, respectively, and zero otherwise. The parameters β_2, β_3, β_4, and β_5 are the average intercepts for the controls, the BPH cases, the L/R cancer cases, and the metastatic cancer cases, respectively, after correction for age at diagnosis. Similar interpretations hold for the other parameters in (3.6).

3.2.5   Two-Stage Analysis

In practice, the regression parameters in (3.2) are of primary interest. They can be estimated by sequentially fitting the models (3.1) and (3.2). First, all β_i are estimated by fitting model (3.1) to the observed data vector y_i of each subject separately, yielding estimates $\widehat{\beta}_i$. Afterward, model (3.2) is fitted to the estimates $\widehat{\beta}_i$, providing inferences for β.

Fitting the models (3.5) and (3.6) to the prostate data, Verbeke and Molenberghs (1997, Section 3.3) found that the subject-specific regression parameters β_i did not depend on age at diagnosis (at the 5% level of significance), but highly significant differences were found among the diagnostic groups. No significant differences were obtained between the controls and the BPH cases, and the two groups of cancer patients only differed with respect to their intercepts.

Note how this two-stage analysis can be interpreted as the calculation (first stage) and analysis (second stage) of summary statistics. First, the actually observed data vector y_i is summarized by $\widehat{\beta}_i$, for each subject separately. Afterward, regression methods are used to assess the relation between the so-obtained summary statistics and relevant covariates. Other summary statistics frequently used in practice are the area under each individual profile (AUC), the mean response for each individual, the largest observation (peak), the half-time, and so forth (see, for example, Weiner 1981 and Rang and Dale 1990).

As for any analysis of summary statistics, the two-stage analysis obviously suffers from at least two problems. First, information is lost in summarizing the vector y_i of observed measurements for the ith subject by $\widehat{\beta}_i$. Second, random variability is introduced by replacing the β_i in model (3.2) by their estimates $\widehat{\beta}_i$. Moreover, the covariance matrix of $\widehat{\beta}_i$ highly depends on the number of measurements available for the ith subject as well as on the time points at which these measurements were taken, and this has not been taken into account in the second stage of the analysis. In Section 3.3, it will be shown how this can be solved by combining the two stages into one model, the so-called linear mixed-effects model.

3.3   The General Linear Mixed-Effects Model

3.3.1   The Model

In order to combine the models from the two-stage analysis, we replace β_i in (3.1) by expression (3.2), yielding

$$Y_i \;=\; X_i\beta \;+\; Z_ib_i \;+\; \varepsilon_i, \tag{3.7}$$

where $X_i = Z_iK_i$ is the appropriate $(n_i \times p)$ matrix of known covariates, and where all other components are as defined earlier. Model (3.7) is called a linear mixed (-effects) model with fixed effects β and with subject-specific effects b_i. It assumes that the vector of repeated measurements on each subject follows a linear regression model where some of the regression parameters are population-specific (i.e., the same for all subjects), whereas other parameters are subject-specific. As in Section 3.2.2, the b_i are assumed to be random and are therefore often called random effects.

In general, a linear mixed-effects model is any model which satisfies (Laird and Ware 1982)

$$
\begin{cases}
Y_i = X_i\beta + Z_i b_i + \varepsilon_i \\[1ex]
b_i \sim N(0, D), \\[1ex]
\varepsilon_i \sim N(0, \Sigma_i), \\[1ex]
b_1, \ldots, b_N, \varepsilon_1, \ldots, \varepsilon_N \text{ independent,}
\end{cases}
\tag{3.8}
$$

where Y_i is the n_i-dimensional response vector for subject i, $1 \leq i \leq N$, N is the number of subjects, X_i and Z_i are $(n_i \times p)$ and $(n_i \times q)$ dimensional matrices of known covariates, β is a p-dimensional vector containing the fixed effects, b_i is the q-dimensional vector containing the random effects, and ε_i is an n_i-dimensional vector of residual components. Finally, D is a general $(q \times q)$ covariance matrix with (i, j) element $d_{ij} = d_{ji}$ and Σ_i is a $(n_i \times n_i)$ covariance matrix which depends on i only through its dimension n_i, i.e. the set of unknown parameters in Σ_i will not depend upon i. In some cases, one may wish to relax this last assumption. An example of this can be found in Lin, Raz and Harlow (1997).

It follows from (3.8) that, conditional on the random effect b_i, Y_i is normally distributed with mean vector $X_i\beta + Z_i b_i$ and with covariance matrix Σ_i. Further, b_i is assumed to be normally distributed with mean vector 0 and covariance matrix D. Let $f(y_i|b_i)$ and $f(b_i)$ be the corresponding density functions. The marginal density function of Y_i is then given by

$$
f(y_i) \;\; = \;\; \int f(y_i|b_i)\, f(b_i)\, db_i,
$$

which can easily be shown to be the density function of an n_i-dimensional normal distribution with mean vector $X_i\beta$ and with covariance matrix $V_i = Z_i D Z_i' + \Sigma_i$. Hence, the marginal model implied by the two-stage approach makes very specific assumptions about the dependence of the mean structure and the covariance structure on the covariates X_i and Z_i, respectively.

Since model (3.8) is defined through the distributions $f(y_i|b_i)$ and $f(b_i)$, it will be called the hierarchical formulation of the linear mixed model. The corresponding marginal normal distribution with mean $X_i\beta$ and covariance $Z_i D Z_i' + \Sigma_i$ is called the marginal formulation of the model. Note that, although the marginal model naturally follows from the hierarchical one, both models are not equivalent. We refer to Section 5.6.2 for a detailed discussion on the differences between both models.

3.3.2 Example: The Rat Data

Combining models (3.3) and (3.4) previously proposed for a two-stage analysis of the rat data, we obtain

$$Y_{ij} = (\beta_0 + b_{1i}) + (\beta_1 L_i + \beta_2 H_i + \beta_3 C_i + b_{2i})t_{ij} + \varepsilon_{ij}, \quad j = 1, \ldots, n_i,$$

which can be rewritten as

$$Y_{ij} = \begin{cases} \beta_0 + b_{1i} + (\beta_1 + b_{2i})t_{ij} + \varepsilon_{ij}, & \text{if low dose} \\[2mm] \beta_0 + b_{1i} + (\beta_2 + b_{2i})t_{ij} + \varepsilon_{ij}, & \text{if high dose} \\[2mm] \beta_0 + b_{1i} + (\beta_3 + b_{2i})t_{ij} + \varepsilon_{ij}, & \text{if control.} \end{cases} \quad (3.9)$$

The subject-specific profiles are assumed to be linear, with subject-specific intercepts as well as slopes. The average evolution is also linear, with different slopes for the three treatment groups, but with common intercepts. Still assuming the error components ε_{ij} to be independently identically distributed with variance σ^2, we have that the assumed covariance function can be summarized by

$$\begin{aligned} \text{Cov}(Y_i(t_1), Y_i(t_2)) &= \begin{pmatrix} 1 & t_1 \end{pmatrix} D \begin{pmatrix} 1 \\ t_2 \end{pmatrix} + \sigma^2 \\ &= d_{22}t_1 t_2 + d_{12}(t_1 + t_2) + d_{11} + \sigma^2. \end{aligned}$$

Note how the model now implies the variance function of the response to be quadratic over time, with positive curvature d_{22}.

A model which assumes that all variability in subject-specific slopes can be ascribed to treatment differences can be obtained by omitting the random slopes b_{2i} from the above model. This is the so-called random-intercepts model. The subject-specific profiles are then still assumed to be linear, with subject-specific intercepts, but with the same slopes within each treatment group. The implied covariance structure assumes constant variance $d_{11} + \sigma^2$ over time as well as equal positive correlation $\rho_I = d_{11}/(d_{11} + \sigma^2)$ between any two measurements from the same rat. This covariance structure is called compound symmetric, whereas the common correlation ρ_I is often termed the intraclass correlation coefficient (see, e.g., Crowder and Hand 1990, p. 27). Note that ρ_I is large when the intersubject variability d_{11} is large in comparison to the intrasubject variability σ^2.

3.3.3 Example: The Prostate Data

Combining models (3.5) and (3.6) previously proposed for a two-stage analysis of the prostate data, we obtain

$$
\begin{aligned}
Y_{ij} &\equiv \ln(\text{PSA}_{ij} + 1) \\
&= \beta_1 \text{Age}_i + \beta_2 C_i + \beta_3 B_i + \beta_4 L_i + \beta_5 M_i \\
&\quad + (\beta_6 \text{Age}_i + \beta_7 C_i + \beta_8 B_i + \beta_9 L_i + \beta_{10} M_i)\, t_{ij} \\
&\quad + (\beta_{11} \text{Age}_i + \beta_{12} C_i + \beta_{13} B_i + \beta_{14} L_i + \beta_{15} M_i)\, t_{ij}^2 \\
&\quad + b_{1i} + b_{2i} t_{ij} + b_{3i} t_{ij}^2 + \varepsilon_{ij}.
\end{aligned}
\tag{3.10}
$$

The subject-specific profiles are assumed to be quadratic over time, with subject-specific intercepts as well as slopes for the linear as well as quadratic time effect. The average evolution is also quadratic, with different intercepts and slopes for the four diagnostic groups. If we again assume the error components ε_{ij} to be independently identically distributed with variance σ^2, we have that the assumed covariance function is given by

$$
\begin{aligned}
\text{Cov}(\boldsymbol{Y_i}(t_1), \boldsymbol{Y_i}(t_2)) &= \begin{pmatrix} 1 & t_1 & t_1^2 \end{pmatrix} D \begin{pmatrix} 1 \\ t_2 \\ t_2^2 \end{pmatrix} + \sigma^2 \\
&= d_{33} t_1^2\, t_2^2 + d_{23}(t_1^2\, t_2 + t_1\, t_2^2) + d_{22} t_1\, t_2 \\
&\quad + d_{13}(t_1^2 + t_2^2) + d_{12}(t_1 + t_2) + d_{11} + \sigma^2.
\end{aligned}
$$

Consequently, the implied variance function is now a fourth-order polynomial over time.

3.3.4 A Model for the Residual Covariance Structure

Very often, Σ_i is chosen to be equal to $\sigma^2 I_{n_i}$ where I_{n_i} denotes the identity matrix of dimension n_i. We then call model (3.8) the conditional independence model, since it implies that the n_i responses on individual i are independent, conditional on $\boldsymbol{b_i}$ and $\boldsymbol{\beta}$. As shown in Section 3.3.2, the conditional independence model may imply unrealistically simple covariance structures for the response vector $\boldsymbol{Y_i}$, especially for models with few random effects. When there is no evidence for the presence of additional random effects, or when additional random effects have no substantive meaning, the covariance assumptions can often be relaxed by allowing an appropriate, more general, residual covariance structure Σ_i for the vector $\boldsymbol{\varepsilon_i}$ of subject-specific error components.

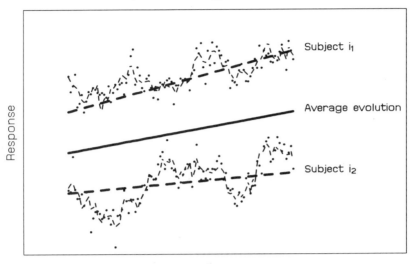

FIGURE 3.1. *Graphical representation of the three stochastic components in the general linear mixed model (3.11). The solid line represents the population-average evolution. The lines with long dashes show subject-specific evolutions for two subjects i_1 and i_2. The residual components of serial correlation and measurement error are indicated by short-dashed lines and dots, respectively.*

A variety of models has been proposed in the statistical literature. See, for example, Mansour, Nordheim, and Rutledge (1985), Diem and Liukkonen (1988), Diggle (1988), Chi and Reinsel (1989), Rochon (1992) , and Núñez-Antón and Woodworth (1994), among many others. Most of these models are special cases of the general model proposed by Diggle, Liang, and Zeger (1994). They assume that ε_i has constant variance and can be decomposed as $\varepsilon_i = \varepsilon_{(1)i} + \varepsilon_{(2)i}$ in which $\varepsilon_{(2)i}$ is a component of serial correlation, suggesting that at least part of an individual's observed profile is a response to time-varying stochastic processes operating within that individual. This type of random variation results in a correlation between serial measurements, which is usually a decreasing function of the time separation between these measurements. Further, $\varepsilon_{(1)i}$ is an extra component of measurement error reflecting variation added by the measurement process itself, and assumed to be independent of $\varepsilon_{(2)i}$. A graphical representation of the three stochastic components (random effects, serial correlation, and measurement error) in the resulting model is shown in Figure 3.1.

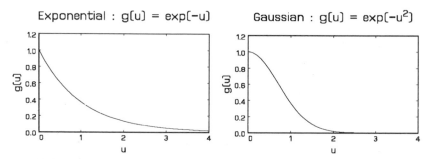

FIGURE 3.2. *Exponential and Gaussian serial correlation functions.*

The resulting linear mixed model can now be written as

$$\begin{cases} Y_i = X_i\beta + Z_i b_i + \varepsilon_{(1)i} + \varepsilon_{(2)i} \\[2mm] b_i \sim N(0, D), \\[2mm] \varepsilon_{(1)i} \sim N(0, \sigma^2 I_{n_i}), \\[2mm] \varepsilon_{(2)i} \sim N(0, \tau^2 H_i), \\[2mm] b_1, \ldots, b_N, \varepsilon_{(1)1}, \ldots, \varepsilon_{(1)N}, \varepsilon_{(2)1}, \ldots, \varepsilon_{(2)N} \text{ independent,} \end{cases} \qquad (3.11)$$

and the model is completed by assuming a specific structure for the $(n_i \times n_i)$ correlation matrix H_i. Such structures are often borrowed from time-series analysis. One usually assumes that the serial effect $\varepsilon_{(2)i}$ is a population phenomenon, independent of the individual. The serial correlation matrix H_i then only depends on i through the number n_i of observations and through the time points t_{ij} at which measurements were taken. Further, it is assumed that the (j, k) element h_{ijk} of H_i is modeled as $h_{ijk} = g(|t_{ij} - t_{ik}|)$ for some decreasing function $g(\cdot)$ with $g(0) = 1$. This means that the correlation between $\varepsilon_{(2)ij}$ and $\varepsilon_{(2)ik}$ only depends on the time interval between the measurements y_{ij} and y_{ik}, and decreases if the length of this interval increases.

Two frequently used functions $g(\cdot)$ are the exponential and Gaussian serial correlation functions defined as $g(u) = \exp(-\phi u)$ and $g(u) = \exp(-\phi u^2)$, respectively ($\phi > 0$), and which are shown in Figure 3.2 for $\phi = 1$. Note that the most important qualitative difference between these functions is their behavior near $u = 0$, although their tail behavior is also different.

Although Diggle, Liang, and Zeger (1994) discuss model (3.11) in full generality, they do not fit any models which simultaneously include serial correlation as well as random effects other than intercepts. They argue that, in applications, the effect of serial correlation is very often dominated by the combination of random effects and measurement error. In practice, this is often reflected in estimation problems for models which include several random effects, serial correlation, as well as measurement error. We refer to Section 9.4 for an example. In Chapter 10, we will discuss how appropriate residual covariance structures can be found in the presence of random effects, other than just intercepts. We also refer to Chapter 4 in the book by Davidian and Giltinan (1995) for a discussion of components of variability in the context of nonlinear mixed models.

4

Exploratory Data Analysis

4.1 Introduction

Most books on longitudinal data discuss exploratory analysis. See, for example, Diggle, Liang, and Zeger (1994). However, most effort is spent to model building and formal aspects of inference. In this section, we present a selected set of techniques to underpin the model building. We distinguish between two modes of display. In Section 4.2, the marginal distribution of the responses in the Vorozole study is explored, that is, we explore the observed profiles averaged over (sub)populations. Three aspects of the data will be looked at in turn: the average evolution, the variance function, and the correlation structure. Afterward, in Section 4.3, we will discuss some procedures for exploring the observed profiles in a subject-specific way.

4.2 Exploring the Marginal Distribution

4.2.1 The Average Evolution

The average evolution describes how the profile for a number of relevant subpopulations (or the population as a whole) evolves over time. The results

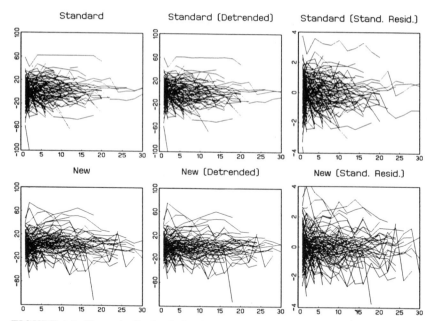

FIGURE 4.1. *Vorozole Study. Individual profiles, raw residuals, and standardized residuals.*

of this exploration will be useful in order to choose a fixed-effects structure for the linear mixed model.

The individual profiles are displayed in Figure 4.1, and the mean profiles, per treatment arm, are plotted in Figure 4.2. The average profiles indicate an increase over time which is slightly stronger for the Vorozole group. In addition, the Vorozole group is, with the exception of month 16, consistently higher than the AGT group. Of course, at this point it is not yet possible to decide on the significance of this difference. It is useful to explore the treatment difference separately since even when both evolutions might be complicated, the treatment difference, which is often of primary interest, could follow a simple model, or vice versa. The treatment difference is plotted in Figure 4.3.

The individual profiles augment the averaged plot with a suggestion of the variability seen within the data. The thinning of the data toward the later study times suggests that trends at later times should be treated with caution. Although these plots also give us some indications about the variability at given times and even about the correlation between measurements of the same individual, it is easier to base such considerations on residual profiles and standardized residual profiles.

Mean Profiles

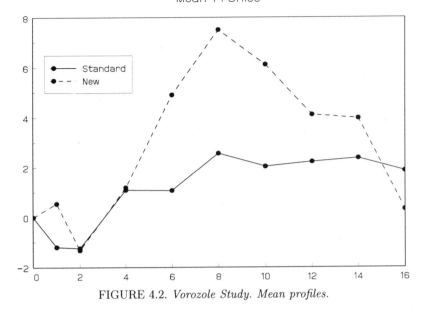

FIGURE 4.2. *Vorozole Study. Mean profiles.*

4.2.2   The Variance Structure

In addition to the average evolution, the evolution of the variance is important to build an appropriate longitudinal model. Clearly, one has to correct the measurements for the fixed-effects structure and hence raw residuals must be used. Again, two plots are of interest. The first one pictures the average evolution of the variance as a function of time; the second one merely produces the individual residual plots.

The detrended profiles are displayed in Figure 4.1, and the corresponding variance function is plotted in Figure 4.4.

The variance function seems to be relatively stable and hence a constant variance model could be a plausible starting point. The individual detrended profiles show subjects' tendency, most clearly in the Vorozole group, to decrease right before they leave the study. In addition, the detrended profiles suggest that the variance would decrease over time. This is in contradiction with the variance function; it is entirely due to considerable attrition. This observation suggests that caution should be used with incomplete data.

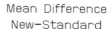

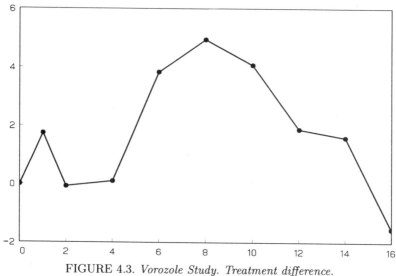

FIGURE 4.3. *Vorozole Study. Treatment difference.*

4.2.3   The Correlation Structure

The correlation structure describes how measurements within a subject correlate. The correlation function depends on a pair of times and only under the assumption of stationarity (see Section 10.4.2 for a formal definition) does this pair of times simplify to the time lag only. This is important since many exploratory and modeling tools are based on this assumption. A plot of standardized residuals is useful in this respect (Figure 4.1). The picture is not radically different from the previous individual plots, which can be explained by the relative flatness of both mean profile and variance functions. If one or both structures is varying with time, the standardized residuals will contribute useful additional information.

A different way of displaying the correlation structure is using a scatter plot matrix, such as in Figure 4.5. The off-diagonal elements picture scatter plots of standardized residuals obtained from pairs of measurement occasions. The decay of correlation with time is studied by considering the evolution of the scatters with increasing distance to the main diagonal. Stationarity on the other hand implies that the scatter plots remain similar within diagonal bands *if measurement occasions are approximately equally spaced*. In addition to the scatter plots, we place histograms on the diagonal, capturing the variance structure, including such features as skewness. If the axes are given the same scales, it is very easy to capture the attrition rate as well.

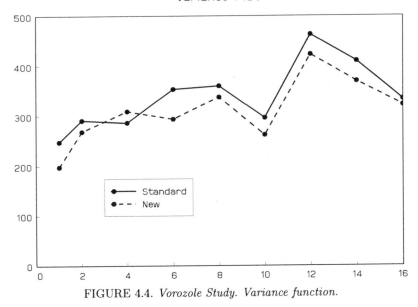

FIGURE 4.4. *Vorozole Study. Variance function.*

4.3   Exploring Subject-Specific Profiles

As shown in the Sections 3.2 and 3.3, linear mixed models can be interpreted as the result from a two-stage approach, where the first stage consists of approximating each observed longitudinal profile by an appropriate linear regression function. In this section, we propose two simple exploratory tools to check to what extent observed longitudinal profiles can be described by a specific linear regression model.

4.3.1   Measuring the Overall Goodness-of-Fit

In practice, one often uses the coefficient of multiple determination R^2 to measure the overall goodness-of-fit of a classical multiple linear regression model. See, for example, Neter, Wasserman and Kutner (1990, Section 7.5). Let

$$\mathbf{Y} \;=\; X\boldsymbol{\beta} + \boldsymbol{\varepsilon} \tag{4.1}$$

be a linear regression model, where $\mathbf{Y}$ is an N-dimensional vector, X is an $(N \times p)$ matrix with known covariate values, and where it is assumed that all elements in $\boldsymbol{\varepsilon}$ are independently normally distributed with mean zero and variance σ^2. The total sum of squares SSTO and the error sum of

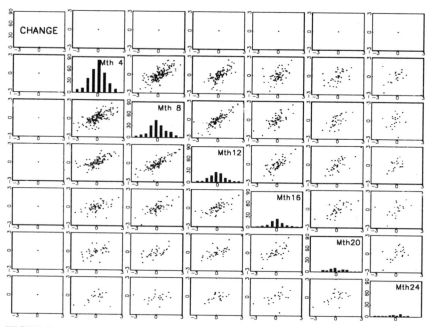

FIGURE 4.5. *Vorozole Study. Scatter plot matrix for selected time points. The same vertical scale is used along the diagonal to display the attrition rate as well.*

squares SSE are then defined as

$$\text{SSTO} \;=\; \left(\boldsymbol{Y} - \mathbf{1}_N \mathbf{1}_N' \boldsymbol{Y}/N\right)' \left(\boldsymbol{Y} - \mathbf{1}_N \mathbf{1}_N' \boldsymbol{Y}/N\right), \tag{4.2}$$

$$\text{SSE} \;=\; \left(\boldsymbol{Y} - X(X'X)^{-1}X'\boldsymbol{Y}\right)' \left(\boldsymbol{Y} - X(X'X)^{-1}X'\boldsymbol{Y}\right), \tag{4.3}$$

respectively, where $\mathbf{1}_N$ is the N-dimensional vector containing only ones. Further, the coefficient of multiple determination is defined as

$$R^2 \;=\; \frac{\text{SSTO} - \text{SSE}}{\text{SSTO}}, \tag{4.4}$$

which expresses what proportion of the total observed variability in the response values can be explained by the covariates in the matrix X. R^2 is always between zero and one. The larger R^2 the better the model describes the observed data.

In order to assess how well a candidate first-stage linear regression model describes observed longitudinal profiles, a coefficient of multiple determination R_i^2 can be calculated for each subject separately. We therefore apply the expressions (4.2), (4.3) and (4.4) to obtain subject-specific total and error sums of squares SSTO_i and SSE_i, as well as subject-specific coefficients of multiple determination R_i^2, respectively. A histogram can now be used to summarize the so-obtained R_i^2. Ideally, all R_i^2 should be large. Typically, only a small or moderate number n_i of repeated measurements is available for (some of) the subjects, which may result in (very) high values R_i^2. This suggests that a fair comparison of the R_i^2 should take into account the numbers of measurements on which they are based. We therefore promote the use of scatter plots of the R_i^2 versus the n_i. Examples will be given in the Sections 4.3.3 and 4.3.4.

Finally, an overall measure for the goodness-of-fit of first-stage linear regression models is

$$R_{\text{meta}}^2 \quad = \quad \frac{\sum_{i=1}^{N}(\text{SSTO}_i - \text{SSE}_i)}{\sum_{i=1}^{N}\text{SSTO}_i},$$

which expresses what proportion of the total within-subject variability can be explained by the first-stage linear regression models.

4.3.2 Testing for the Need of a Model Extension

Another approach toward assessing the adequacy of a linear regression model is to test the assumed model versus an alternative model which is an extended version of the original model. In the context of classical linear regression, we consider again model (4.1). An indirect way of checking whether this model is appropriate is to consider an extended version

$$\mathbf{Y} \quad = \quad X\boldsymbol{\beta} + X^*\boldsymbol{\beta}^* + \boldsymbol{\varepsilon} \tag{4.5}$$

of model (4.1) and to test whether the p^*-dimensional vector $\boldsymbol{\beta}^*$ of regression parameters, corresponding to the additional covariates in X^*, equals zero. Let $\text{SSE}(F)$ and $\text{SSE}(R)$ denote the error sums of squares, as defined in (4.3), for the full model (4.5) and the reduced model (4.1), respectively. The null hypothesis $H_0 : \boldsymbol{\beta}^* = \mathbf{0}$ can then be tested using the test statistic

$$F \quad = \quad \frac{(\text{SSE}(R) - \text{SSE}(F))/p^*}{\text{SSE}(F)/(N - p - p^*)}, \tag{4.6}$$

which follows an F-distribution with p^* and $N - p - p^*$ degrees of freedom, under H_0 and assuming that model (4.5) is correct (see Neter, Wasserman and Kutner 1990, Section 8.6). The null hypothesis is rejected for large observed values of the above F-statistic.

In a longitudinal data context, a similar approach can now be used to test a specific candidate first-stage linear regression model versus an extended version. We therefore first calculate all subject-specific error sums of squares $\text{SSE}_i(F)$ and $\text{SSE}_i(R)$ under the full model and under the reduced model, respectively. Afterward, an overall test statistic can be calculated as

$$
F_{\text{meta}} = \frac{\left\{ \displaystyle\sum_{\{i:n_i \geq p+p^*\}} (\text{SSE}_i(R) - \text{SSE}_i(F)) \right\} \Big/ \left\{ \displaystyle\sum_{\{i:n_i \geq p+p^*\}} p^* \right\}}{\left\{ \displaystyle\sum_{\{i:n_i \geq p+p^*\}} \text{SSE}_i(F) \right\} \Big/ \left\{ \displaystyle\sum_{\{i:n_i \geq p+p^*\}} (n_i - p - p^*) \right\}},
$$

where the sums are taken over all subjects with at least $p + p^*$ measurements. Assuming that the candidate first-stage model was correctly specified, and assuming that all residuals are independently normally distributed with mean zero and with some common variance, we have that the above test statistic follows an F-distribution with $\sum_{\{i:n_i \geq p+p^*\}} p^*$ and $\sum_{\{i:n_i \geq p+p^*\}} (n_i - p - p^*)$ degrees of freedom.

It should be emphasized that the above testing procedure relies on specific distributional assumptions, which are not necessarily satisfied in a longitudinal data context. For example, the within-subjects errors are assumed to be independent, having common variance over the subjects. Hence, referring back to the general linear mixed model discussed in Section 3.3.4, the absence of residual serial correlation within subjects is implicitly assumed. We therefore propose using this approach as a general tool for exploring how a specific first-stage regression model can be improved, rather than a formal testing procedure for the adequacy of the model. In this respect, it is also advisable to always use this procedure in combination with the goodness-of-fit measures discussed in Section 4.3.1. In Sections 4.3.3 and 4.3.4, two examples will be given, for which all calculations have been performed using a SAS macro available from the website.

4.3.3 Example: The Rat Data

As a first example, we explore the adequacy of the first-stage model (3.3) previously used in Section 3.2.3 for the rat data. Recall that it was assumed that, apart from residual variability, the response is a linear function of the transformed timescale $t_{ij} = \ln[1 + (\text{Age}_{ij} - 45)/10)]$. The left panel in Figure 4.6 shows a scatter plot of the subject-specific coefficients R_i^2 of multiple determinations, versus the numbers n_i of repeated measurements. The overall coefficient R_{meta}^2 of multiple determination, represented by the dashed line, equals 0.9294, indicating that the model explains about 93% of the total within-subject variability. All except two coefficients R_i^2 are larger than 0.85, suggesting that our first-stage model fits the observed profiles reasonably well. However, comparing the model with an extended model which assumes quadratic subject-specific evolutions yields $F_{\text{meta}} = 1.5347$,

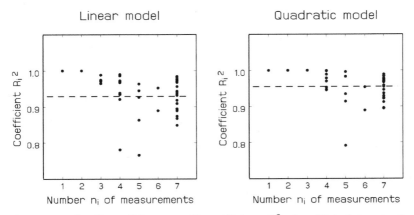

FIGURE 4.6. *Rat Data. Subject-specific coefficients R_i^2 of multiple determination and the overall coefficient R_{meta}^2 of multiple determination (dashed lines), for first-stage models which assume linear (left panel) as well as quadratic (right panel) subject-specific profiles.*

on 43 and 113 degrees of freedom, which is significant on the 5% level ($p = 0.0382$). This suggests that adding quadratic terms to the first-stage model might improve the fit considerably.

The right panel in Figure 4.6 shows a scatter plot of the R_i^2 versus the n_i, for the so-obtained quadratic first-stage model. The overall coefficient R_{meta}^2 now equals 0.9554, which is a rather small improvement when compared to the original model. This is also reflected in the scatter plot. Note also that all rats with at most three measurements have $R_i^2 = 1$. Testing for the need of an additional cubic term in the first-stage model results in $F_{meta} = 1.3039$, which is not significant ($p = 0.1633$) when compared to an F-distribution with 38 and 75 degrees of freedom. Strictly speaking, this is evidence in favor of using the quadratic first-stage model rather than the original linear model. However, in view of the good fits which were already obtained with the linear model, and in order to keep our models as parsimonious as possible, our further analyses of the rat data will be based on the original linear first-stage model (3.3).

4.3.4   Example: The Prostate Data

As a second example, we explore the adequacy of the first-stage model (3.5) previously used in Section 3.2.4 for the prostate data. Recall that it was assumed that, apart from residual variability, the response is a quadratic function of time before diagnosis (expressed in decades). The same checks were performed as for the rat data, described in Section 4.3.3. The results are shown in Figure 4.7.

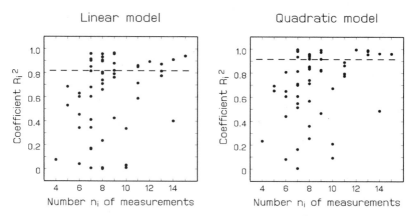

FIGURE 4.7. *Prostate Data. Subject-specific coefficients R_i^2 of multiple determination and the overall coefficient R_{meta}^2 of multiple determination (dashed lines), for first-stage models which assume linear (left panel) as well as quadratic (right panel) subject-specific profiles.*

Although the linear two-stage model explains about 82% of all within-subject variability ($R_{meta}^2 = 0.8188$), many longitudinal profiles are badly described (left panel of Figure 4.7). For example, for 8 out of the 54 profiles, less than 10% of the observed variability could be explained by a linear fit. As can be expected from observing the individual profiles in Figure 2.3, the linear model is strongly rejected when compared to a quadratic first-stage model ($F_{meta} = 6.2181$, 54 and 301 degrees of freedom, $p < 0.0001$).

The right panel in Figure 4.7 shows the coefficients R_i^2 versus the n_i, for this quadratic model. The new model explains about 10% more within-subject variability (R_{meta}^2 increased to 0.9143). Testing for the need of an additional cubic term results in $F_{meta} = 1.2310$, which is not significant ($p = 0.1484$) when compared to an F-distribution with 54 and 247 degrees of freedom. This clearly supports the first-stage model (3.5) proposed in Section 3.2.4. The fact that some individual coefficients R_i^2 are still quite small, although the quadratic model has not been rejected, suggests the presence of a considerable amount of residual variability. This will be confirmed later, in Section 9.5.

5

Estimation of the Marginal Model

5.1 Introduction

As discussed in Section 3.3, the general linear mixed model (3.8) implies the marginal model

$$Y_i \sim N(X_i\beta, Z_i D Z_i' + \Sigma_i). \tag{5.1}$$

Unless the data are analyzed in a Bayesian framework (see, e.g., Gelman *et al.* 1995), inference is based on this marginal distribution for the response Y_i. It should be emphasized that the hierarchical structure of the original model (3.8) is then not taken into account. Indeed, the marginal model (5.1) is not equivalent to the original hierarchical model (3.8). Inferences based on the marginal model do not explicitly assume the presence of random effects representing the natural heterogeneity between subjects. An example of this can be found in Section 5.6.2. In this and the next chapter, we will discuss inference for the parameters in the marginal distribution (5.1). Later, in Chapter 7, it will be shown how the random effects can be estimated under the explicit assumption that Y_i satisfies model (3.8).

Let α denote the vector of all variance and covariance parameters (usually called variance components) found in $V_i = Z_i D Z_i' + \Sigma_i$, that is, α consists of the $q(q+1)/2$ different elements in D and of all parameters in Σ_i. Finally, let $\theta = (\beta', \alpha')'$ be the s-dimensional vector of all parameters in the marginal model for Y_i, and let $\Theta = \Theta_\beta \times \Theta_\alpha$ denote the parameter space for θ, with

Θ_β and Θ_α the parameter spaces for the fixed effects and for the variance components respectively. Note that $\Theta_\beta = I\!\!R^p$, and Θ_α equals the set of values for α such that D and all Σ_i are positive (semi-)definite.

The classical approach to inference is based on estimators obtained from maximizing the marginal likelihood function

$$L_{\mathrm{ML}}(\boldsymbol{\theta}) = \prod_{i=1}^N \left\{ (2\pi)^{-n_i/2} |V_i(\boldsymbol{\alpha})|^{-\frac{1}{2}} \right.$$

$$\left. \times \exp\left(-\frac{1}{2} (\boldsymbol{Y_i} - X_i\boldsymbol{\beta})' V_i^{-1}(\boldsymbol{\alpha}) (\boldsymbol{Y_i} - X_i\boldsymbol{\beta}) \right) \right\} \quad (5.2)$$

with respect to $\boldsymbol{\theta}$. Let us first assume α to be known. The maximum likelihood estimator (MLE) of β, obtained from maximizing (5.2), conditional on α, is then given by (Laird and Ware 1982)

$$\widehat{\beta}(\boldsymbol{\alpha}) = \left(\sum_{i=1}^N X_i'W_iX_i \right)^{-1} \sum_{i=1}^N X_i'W_i\boldsymbol{y_i}, \quad (5.3)$$

where W_i equals V_i^{-1}.

When α is not known, but an estimate $\widehat{\alpha}$ is available, we can set $\widehat{V}_i = V_i(\widehat{\alpha}) = \widehat{W_i}^{-1}$, and estimate β by using the expression (5.3) in which W_i is replaced by $\widehat{W_i}$. Two frequently used methods for estimating α are maximum likelihood estimation and restricted maximum likelihood estimation, which will be discussed and compared in the Sections 5.2 and 5.3, respectively.

5.2 Maximum Likelihood Estimation

The maximum likelihood estimator (MLE) of α is obtained by maximizing (5.2) with respect to α, after β is replaced by (5.3). This approach arises naturally when we consider the estimation of β and α simultaneously by maximizing the joint likelihood (5.2).

5.3   Restricted Maximum Likelihood Estimation

5.3.1   Variance Estimation in Normal Populations

As an introductory example to restricted maximum likelihood estimation, consider the case where the variance of a normal distribution $N(\mu, \sigma^2)$ is to be estimated based on a sample $Y_1, \ldots, Y_N$ of N observations. When the mean μ is known, the MLE for σ^2 equals $\widehat{\sigma}^2 = \sum_i (Y_i - \mu)^2/N$, which is unbiased for σ^2. When μ is not known, we get the same expression for the MLE, but with μ replaced by the sample mean $\overline{Y} = \sum_i Y_i/N$. One can then easily show that

$$E\left(\widehat{\sigma}^2\right) \;\; = \;\; \frac{N-1}{N}\,\sigma^2, \tag{5.4}$$

indicating that the MLE is now biased downward, due to the estimation of μ. Note, however, that an unbiased estimate is easily obtained from expression (5.4), yielding the classical sample variance $S^2 = \sum_i (Y_i - \overline{Y})^2/(N-1)$.

Apparently, directly obtaining an unbiased estimate for σ^2 should be based on a statistical procedure which does not require estimation of μ first. This can be done as follows. Let $Y = (Y_1, \ldots, Y_N)'$ denote the vector of all measurements, and let $\mathbf{1}_N$ be the N-dimensional vector containing only ones. The distribution of Y is then $N(\mu\,\mathbf{1}_N, \sigma^2 I_N)$ where, as before, I_N equals the N-dimensional identity matrix. Let A be any $N \times (N-1)$ matrix with $N-1$ linearly independent columns orthogonal to the vector $\mathbf{1}_N$. We then define the vector U of $N-1$ so-called error contrasts by $U = A'Y$, which now follows a normal distribution with mean vector $\mathbf{0}$ and covariance matrix $\sigma^2 A'A$. Maximizing the corresponding likelihood with respect to the only remaining parameter σ^2 yields $\widehat{\sigma}^2 = Y'A(A'A)^{-1}A'Y/(N-1)$, which can be shown to equal the classical sample variance S^2 previously derived from expression (5.4). Note that any matrix A satisfying the specified conditions leads to the same estimator for σ^2. The resulting estimator for σ^2 is often called the restricted maximum likelihood (REML) estimator since it is restricted to $(N-1)$ error contrasts.

5.3.2   Estimation of Residual Variance in Linear Regression

As a second example, we now consider the estimation of the residual variance σ^2 in a linear regression model $Y = X\beta + \varepsilon$, where Y is an N-dimensional vector, and with X a $(N \times p)$ matrix with known covariate values. It is assumed that all elements in ε are independently normally distributed with mean zero and variance σ^2. The MLE for σ^2 equals

$$\widehat{\sigma}^2 = (\boldsymbol{Y} - X(X'X)^{-1}X'\boldsymbol{Y})'(\boldsymbol{Y} - X(X'X)^{-1}X'\boldsymbol{Y})/N,$$

which can easily be shown to be biased downward by a factor $(N - p)/N$.

Similarly, as in Section 5.3.1, σ^2 can be estimated using a set of error contrasts $\boldsymbol{U} = A'\boldsymbol{Y}$ where A is now any $N \times (N - p)$ matrix with $N - p$ linearly independent columns orthogonal to the columns of the design matrix X. We then have that $\boldsymbol{U}$ follows a normal distribution with mean vector $\boldsymbol{0}$ and covariance matrix $\sigma^2 A'A$, in which σ^2 is again the only unknown parameter. Maximizing the corresponding likelihood with respect to σ^2 yields

$$\widehat{\sigma}^2 = (\boldsymbol{Y} - X(X'X)^{-1}X'\boldsymbol{Y})'(\boldsymbol{Y} - X(X'X)^{-1}X'\boldsymbol{Y})/(N - p),$$

which is the mean squared error, unbiased for σ^2, and classically used as estimator for the residual variance in linear regression analysis (see, for example, Neter, Wasserman, and Kutner 1990, Chapter 7; Seber 1977, Section 3.3). As in Section 5.3.1, we again have that any matrix A satisfying the specified conditions leads to the same estimator for the residual variance, which is again called the REML estimator for σ^2.

5.3.3 REML Estimation for the Linear Mixed Model

In practice, linear mixed models often contain many fixed effects. For example, the linear mixed model (3.10) which immediately followed from the two-stage model proposed in Section 3.2.4 for the prostate data, has a 15-dimensional vector $\boldsymbol{\beta}$ of parameters in the mean structure. In such cases, it may be important to estimate the variance components, explicitly taking into account the loss of the degrees of freedom involved in estimating the fixed effects. In contrast to the simple cases discussed in Sections 5.3.1 and 5.3.2, an unbiased estimator for the vector $\boldsymbol{\alpha}$ of variance components cannot be obtained from appropriately transforming the ML estimator as suggested from the analytic calculation of its bias. However, the error contrasts approach can still be applied as follows. We first combine all N subject-specific regression models (3.8) to one model:

$$\boldsymbol{Y} = X\boldsymbol{\beta} + Z\boldsymbol{b} + \boldsymbol{\varepsilon}, \tag{5.5}$$

where the vectors $\boldsymbol{Y}$, $\boldsymbol{b}$, and $\boldsymbol{\varepsilon}$, and the matrix X are obtained from stacking the vectors $\boldsymbol{Y}_i$, $\boldsymbol{b}_i$, and $\boldsymbol{\varepsilon}_i$, and the matrices X_i respectively, underneath each other, and where Z is the block-diagonal matrix with blocks Z_i on the main diagonal and zeros elsewhere. The dimension of $\boldsymbol{Y}$ equals $\sum_{i=1}^{N} n_i$ and will be denoted by n.

The marginal distribution for Y is normal with mean vector $X\beta$ and with covariance matrix $V(\alpha)$ equal to the block-diagonal matrix with blocks V_i on the main diagonal and zeros elsewhere. The REML estimator for the variance components α is now obtained from maximizing the likelihood function of a set of error contrasts $U = A'Y$ where A is any $(n \times (n - p))$ full-rank matrix with columns orthogonal to the columns of the X matrix. The vector U then follows a normal distribution with mean vector zero and covariance matrix $A'V(\alpha)A$, which is not dependent on β any longer. Further, Harville (1974) has shown that the likelihood function of the error contrasts can be written as

$$
\begin{aligned}
L(\alpha) \quad = \quad & (2\pi)^{-(n-p)/2} \left| \sum_{i=1}^{N} X_i'X_i \right|^{1/2} \\
& \times \left| \sum_{i=1}^{N} X_i'V_i^{-1}X_i \right|^{-1/2} \prod_{i=1}^{N} |V_i|^{-1/2} \\
& \times \exp\left\{ -\frac{1}{2} \sum_{i=1}^{N} \left(Y_i - X_i\widehat{\beta} \right)' V_i^{-1} \left(Y_i - X_i\widehat{\beta} \right) \right\}, \quad (5.6)
\end{aligned}
$$

where $\widehat{\beta}$ is given by (5.3). Hence, as in the simple examples described in Sections 5.3.1 and 5.3.2, the so-obtained REML estimator $\widehat{\alpha}$ does not depend on the error contrasts (i.e., the choice of A).

Note that the maximum likelihood estimator for the mean of a univariate normal population and for the vector of regression parameters in a linear regression model are independent of the residual variance σ^2. Hence, the estimates for the mean structures of the two examples in Sections 5.3.1 and 5.3.2 do not change if REML estimates are used for the variance components, rather than ML estimates. However, it follows from (5.3) that this no longer holds in the general linear mixed model. Thus, we have that although REML estimation is only with respect to the variance components in the model, the "REML" estimator for the vector of fixed effects is not identical to its ML version. This will be illustrated in Section 5.5, where model (3.10) will be fitted to the prostate cancer data.

Finally, note that the likelihood function in (5.6) equals

$$
L(\alpha) \quad = \quad C \left| \sum_{i=1}^{N} X_i'W_i(\alpha)X_i \right|^{-\frac{1}{2}} L_{\text{ML}}(\widehat{\beta}(\alpha), \alpha), \quad (5.7)
$$

where C is a constant not depending on α, where, as earlier, $W_i(\alpha)$ equals $V_i^{-1}(\alpha)$, and where $L_{\text{ML}}(\beta, \alpha) = L_{\text{ML}}(\theta)$ is the ML likelihood function given by (5.2). Because $\left| \sum_{i=1}^{N} X_i'W_i(\alpha)X_i \right|$ in (5.7) does not depend on β, it follows that the REML estimators for α and for β can also be found by maximizing the so-called REML likelihood function

$$L_{\mathrm{REML}}(\boldsymbol{\theta}) = \left| \sum_{i=1}^{N} X_i' W_i(\boldsymbol{\alpha}) X_i \right|^{-\frac{1}{2}} L_{\mathrm{ML}}(\boldsymbol{\theta}) \qquad (5.8)$$

with respect to all parameters simultaneously ($\boldsymbol{\alpha}$ and $\boldsymbol{\beta}$).

5.3.4   Justification of REML Estimation

The main justification of the REML approach has been given by Patterson and Thompson (1971), who prove that, in the absence of information on $\boldsymbol{\beta}$, no information about $\boldsymbol{\alpha}$ is lost when inference is based on U rather than on Y. More precisely, U is marginally sufficient for $\boldsymbol{\alpha}$ in the sense described by Sprott (1975) (see also Harville 1977). Further, Harville (1974) has shown that, from a Bayesian point of view, using only error contrasts to make inferences on $\boldsymbol{\alpha}$ is equivalent to ignoring any prior information on $\boldsymbol{\beta}$ and using all the data to make those inferences.

5.3.5   Comparison Between ML and REML Estimation

Maximum likelihood estimation and restricted maximum likelihood estimation both have the same merits of being based on the likelihood principle which leads to useful properties such as consistency, asymptotic normality, and efficiency. ML estimation also provides estimators of the fixed effects, whereas REML estimation, in itself, does not. On the other hand, for balanced mixed ANOVA models, the REML estimates for the variance components are identical to classical ANOVA-type estimates obtained from solving the equations which set mean squares equal to their expectations. This implies optimal minimum variance properties, and it shows that REML estimates in that context do not rely on any normality assumption since only moment assumptions are involved (Harville 1977 and Searle, Casella, and McCulloch 1992).

Also with regard to the mean squared error for estimating $\boldsymbol{\alpha}$, there is no indisputable preference for either one of the two estimation procedures, since it depends on the specifics of the underlying model and possibly on the true value of $\boldsymbol{\alpha}$. For ordinary ANOVA or regression models, the ML estimator of the residual variance σ^2 has uniformly smaller mean squared error than the REML estimator when $p = \mathrm{rank}(X) \leq 4$, but the opposite is true when $p > 4$ and $n - p$ is sufficiently large ($n - p > 2$ suffices if $p > 12$). In general, one may expect results from ML and REML estimation to differ more as the number p of fixed effects in the model increases. We hereby refer to Section 13.5 for an example with extremely many covariates in the mean structure, leading to severe differences between ML and REML estimates. More details on this and related topics can be found in Harville (1977).

5.4   Model-Fitting Procedures

In the literature, several methods for the actual calculation of the ML or REML estimates have been described. Dempster, Laird and Rubin (1977), for example, have introduced the EM algorithm for the calculation of MLEs based on incomplete data and have illustrated how it can be used for the estimation of variance components in mixed-model analysis of variance. Laird and Ware (1982) have shown how this EM algorithm not only can be applied to obtain MLEs, but also to calculate the REML estimates through an empirical Bayesian approach. Note that, strictly speaking, no data are missing: The EM algorithm is only used to "estimate" the unobservable parameters (i.e., the random effects b_i). The main advantage of the EM algorithm is that the general theory (Dempster, Laird, and Rubin 1977) assures that each iteration increases the likelihood. However, Laird and Ware (1982) report slow convergence of the estimators of the variance components, especially when the maximum likelihood is on or near the boundary of the parameter space. We refer to Chapter 22 for more details on the EM algorithm.

Therefore, nowadays, one usually uses Newton-Raphson-based procedures to estimate all parameters in the model. Details about the implementation of such algorithms, together with expressions for all first- and second-order derivatives of L_{ML} and L_{REML} with respect to all parameters in $\boldsymbol{\theta}$ can be found in Lindstrom and Bates (1988).

Note that fitting the general linear mixed model (3.8) requires maximization of L_{ML} and L_{REML} over the parameter space Θ, which consists of all vectors $\boldsymbol{\theta}$ which yield positive (semi-)definite matrices D and Σ_i. On the other hand, the marginal model (5.1) only requires all $V_i = Z_i D Z_i' + \Sigma_i$ to be positive (semi-)definite. This is why some statistical packages maximize the likelihood functions over a parameter space which is larger than Θ. For example, the SAS procedure MIXED (Version 6.12), by default, only requires the diagonal elements of D and all Σ_i to be positive, which probably stems from classical variance-components models, where the random effects are assumed to be independent of each other (see, for example, Searle, Casella and McCulloch 1992, Chapter 6). An example of this will be given in Section 5.6.2.

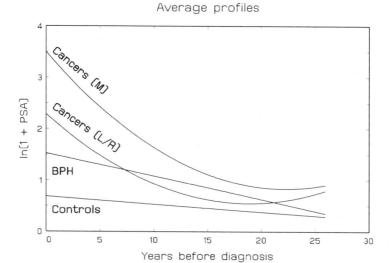

FIGURE 5.1. *Prostate Data. Fitted average profiles for males of median ages at diagnosis, based on the model (3.10), where the parameters are replaced by their REML estimates.*

5.5 Example: The Prostate Data

Table 5.1 shows the maximum likelihood as well as the restricted maximum likelihood estimates for all parameters in the marginal model corresponding to model (3.10), where time is expressed in decades prior to diagnosis rather than years prior to diagnosis (for reasons which will be explained further in Section 5.6.1). Recall that the residual variability in this model is pure measurement error, that is, $\varepsilon_i = \varepsilon_{(1)i}$, with notation as in Section 3.3.4.

As can be expected from the theory in Section 5.3, the ML estimates deviate most from the REML estimates for the variance components in the model. In fact, all REML estimates are larger in absolute value than the ML estimates. Note that the same is true for the REML estimates for the residual variance in normal populations or in linear regression models when compared to the ML estimates, as described in Section 5.3.1 and Section 5.3.2, respectively. Further, Table 5.1 illustrates the fact that the REML estimates for the fixed effects are also different from the ML estimates. Figure 5.1 shows, for each diagnostic group separately, the fitted average profile for a male of median age at diagnosis.

TABLE 5.1. *Prostate Data. Maximum likelihood and restricted maximum likelihood estimates (MLE and REMLE) and standard errors (model based;robust) for all fixed effects and all variance components in model (3.10), with time expressed in decades before diagnosis.*

Effect	Parameter	MLE (s.e.)	REMLE (s.e.)
Age effect	β_1	0.026 (0.013)	0.027 (0.014;0.016)
Intercepts:			
Control	β_2	−1.077 (0.919)	−1.098 (0.976;1.037)
BPH	β_3	−0.493 (1.026)	−0.523 (1.090;1.190)
L/R cancer	β_4	0.314 (0.997)	0.296 (1.059;1.100)
Met. cancer	β_5	1.574 (1.022)	1.549 (1.086;1.213)
Age×time effect	β_6	−0.010 (0.020)	−0.011 (0.021;0.024)
Time effects:			
Control	β_7	0.511 (1.359)	0.568 (1.473;1.640)
BPH	β_8	0.313 (1.511)	0.396 (1.638;1.853)
L/R cancer	β_9	−1.072 (1.469)	−1.036 (1.593;1.646)
Met. cancer	β_{10}	−1.657 (1.499)	−1.605 (1.626;2.038)
Age×time^2 effect	β_{11}	0.002 (0.008)	0.002 (0.009;0.010)
Time^2 effects:			
Control	β_{12}	−0.106 (0.549)	−0.130 (0.610;0.688)
BPH	β_{13}	−0.119 (0.604)	−0.158 (0.672;0.774)
L/R cancer	β_{14}	0.350 (0.590)	0.342 (0.656;0.683)
Met. cancer	β_{15}	0.411 (0.598)	0.395 (0.666;0.844)
Covariance of b_i:			
var(b_{1i})	d_{11}	0.398 (0.083)	0.452 (0.098)
var(b_{2i})	d_{22}	0.768 (0.187)	0.915 (0.230)
var(b_{3i})	d_{33}	0.103 (0.032)	0.131 (0.041)
cov(b_{1i}, b_{2i})	$d_{12} = d_{21}$	−0.443 (0.113)	−0.518 (0.136)
cov(b_{2i}, b_{3i})	$d_{23} = d_{32}$	−0.273 (0.076)	−0.336 (0.095)
cov(b_{3i}, b_{1i})	$d_{13} = d_{31}$	0.133 (0.043)	0.163 (0.053)
Residual variance:			
var(ε_{ij})	σ^2	0.028 (0.002)	0.028 (0.002)
Log-likelihood		−1.788	−31.235

5.6   Estimation Problems

As discussed in Section 5.4, the fitting of linear mixed models is usually done via Newton-Raphson-based procedures. Based on some starting values for the parameters, these procedures iteratively update the estimates until sufficient convergence has been obtained. When fitting complex linear mixed models, the practicing statistician is often faced with nonconverging iteration processes, in the sense that the iterative process does not converge at all, or that it converges to parameter values on or outside the boundary of the parameter space. In some cases, this can be solved by specifying better starting values, or by using other numerical procedures. We refer to Section 9.4 for an example where convergence problems with the Newton-Raphson procedure could be solved by using the Fisher scoring method which uses the expected Hessian matrix (the matrix of second-order derivatives) of the log-likelihood function rather than the observed one. In many cases however, divergence is an indicator of substantial problems with the parameterization of the model or the assumptions implied by the model. Two examples of frequently occurring problems will now be given in Sections 5.6.1 and 5.6.2. It should be emphasized that such numerical problems always arise from estimating the variance components in the model, not from estimating the fixed effects. This can easily be explained from the fact that the classical ordinary least squares estimator for the vector of fixed effects, although completely ignoring the longitudinal structure of the data, is unbiased and consistent and therefore provides good starting values for the fixed effects. This is in contrast to the variance components for which good starting values are often hard to obtain, especially in complex models with many variance components. Also, as explained in Section 5.1, iterative procedures are, strictly speaking, only required for the estimation of the variance components, not for the estimation of the fixed effects.

5.6.1   Estimation Problems due to Small Variance Components

When we first fitted a linear mixed model to the prostate cancer data, time was expressed in decades before diagnosis, rather than years before diagnosis as in the original data set (see Section 5.5). This was done to avoid that the random slopes for the linear and quadratic time effects would show too little variability, which might lead to divergence of the numerical maximization routine. To illustrate this, we refit our mixed model using the SAS procedure MIXED (SAS Version 6.12), but we express time as months before diagnosis. The procedure failed to converge. The estimates for the variance components at the last iteration are shown in Table 5.2.

TABLE 5.2. *Prostate Data. Restricted maximum likelihood estimates (REMLE)
at last iteration for all variance components in model (3.10), with time expressed
in months before diagnosis.*

Effect	Parameter	REMLE
Covariance of b_i:		
$\text{var}(b_{1i})$	d_{11}	0.36893546
$\text{var}(b_{2i})$	d_{22}	0.00003846
$\text{var}(b_{3i})$	d_{33}	0.00000000
$\text{cov}(b_{1i}, b_{2i})$	$d_{12} = d_{21}$	-0.00244046
$\text{cov}(b_{2i}, b_{3i})$	$d_{23} = d_{32}$	-0.00000011
$\text{cov}(b_{3i}, b_{1i})$	$d_{13} = d_{31}$	0.00000449
Residual variance:		
$\text{var}(\varepsilon_{ij})$	σ^2	0.03259207

Note how the reported estimate for the variance of the random slopes for
the quadratic time effect equals $\hat{d}_{33} = 0.00000000$. This suggests that there
is very little variability among the quadratic time effects, requiring the iterative procedure to converge to a point which is very close to the boundary
of the parameter space (since only non-negative variances are allowed), if
not on the boundary of the parameter space. This can produce numerical
difficulties in the maximization process. One way of circumventing this is
by artificially enlarging the true value d_{33}.

Let the model we just fitted for Y_{ij} be written as

$$Y_{ij} = X_i^{[j]}\beta + b_{1i} + b_{2i}t_{ij} + b_{3i}t_{ij}^2 + \varepsilon_{ij},$$

where $X_i^{[j]}$ is the jth row of X_i, where t_{ij} is time expressed as months before
diagnosis, and where the random effects b_{1i}, b_{2i}, and b_{3i} have covariance
matrix

$$\text{var}(b_i) = D = \begin{pmatrix} d_{11} & d_{12} & d_{13} \\ d_{12} & d_{22} & d_{23} \\ d_{31} & d_{32} & d_{33} \end{pmatrix}.$$

We can then reformulate the model as

$$
\begin{aligned}
Y_{ij} &= X_i^{[j]}\beta + b_{1i} + 120\, b_{2i}\left(\frac{t_{ij}}{120}\right) + (120)^2\, b_{3i}\left(\frac{t_{ij}}{120}\right)^2 + \varepsilon_{ij} \\
&= X_i^{[j]}\beta + b_{1i}^* + b_{2i}^* t_{ij}^* + b_{3i}^* t_{ij}^{*\,2} + \varepsilon_{ij},
\end{aligned}
$$

which is a new linear mixed effects model, in which t_{ij}^* is now expressed in
decades before diagnosis, and where the random effects $b_{1i}^* \equiv b_{1i}$, b_{2i}^* and

b_{3i}^* now have covariance matrix

$$\text{var}(\boldsymbol{b_i}^*) = D^* = \begin{pmatrix} (120)^0 \, d_{11} & (120)^1 \, d_{12} & (120)^2 \, d_{13} \\ (120)^1 \, d_{12} & (120)^2 \, d_{22} & (120)^3 \, d_{23} \\ (120)^2 \, d_{31} & (120)^3 \, d_{32} & (120)^4 \, d_{33} \end{pmatrix}.$$

This transformation enlarges the covariance parameters substantially, which implies that the peak of the log-likelihood is well away from the boundary and that the evaluated second-order derivatives are well away from zero. The normal equations, which are the equations to be solved in the maximization algorithm, now form a system which is much more stable and which can easily be solved without any convergence problems. In general, we therefore recommend always rescaling linear mixed models with random effects which are expected to have (very) small variability.

5.6.2 Estimation Problems due to Model Misspecifications

As explained in Section 5.1, inference is based on the marginal model (5.1), rather than on the original, more restrictive, hierarchical model (3.8). In practice, this often causes numerical maximization procedures not to converge to parameter values in the interior of the parameter space implied by the hierarchical model.

As an illustration, we consider the linear mixed model (3.9) proposed in Section 3.3.2 for the rat data introduced in Section 2.1. Table 5.3 shows the restricted maximum likelihood estimates, obtained using the SAS procedure MIXED (version 6.12), for all parameters in the corresponding marginal model. Recall that the residual variability in this model is pure measurement error, that is, $\varepsilon_i = \varepsilon_{(1)i}$, with notation as in Section 3.3.4. Note that the variance of the random slopes is estimated by 0, and no standard error is reported. In contrast to the example given in Section 5.6.1, this cannot be solved by reparameterizing the model. Instead, the zero estimate indicates that the maximum of the REML log-likelihood function is really on the boundary of the parameter space. Indeed, as discussed in Section 5.4, SAS maximizes L_{REML} under the restriction that all diagonal elements d_{ii} in D as well as σ^2 are positive. Our results now suggest that the REML likelihood could be further increased by removing these restrictions and allowing some of the variance components d_{ii} or σ^2 to become negative. The parameter estimates obtained from refitting the model without any restrictions on the parameter space are also reported in Table 5.3. As expected, we now get a negative estimate for the variance d_{22} of the random slopes. Note also that this has further increased the REML log-likelihood value. It should be strongly emphasized, however, that the resulting model does not allow any hierarchical interpretation since no random-effects structure

TABLE 5.3. *Rat Data. Restricted maximum likelihood estimates (REMLE) and standard errors for all fixed effects and all variance components in the marginal model corresponding to model (3.9), for two different parameter restrictions for the variance components α.*

| | | Parameter restrictions for α | |
| | | $d_{ii} \geq 0, \sigma^2 \geq 0$ | $d_{ii} \in \mathbb{R}, \sigma^2 \in \mathbb{R}$ |
Effect	Parameter	REMLE (s.e.)	REMLE (s.e.)
Intercept	β_0	68.606 (0.325)	68.618 (0.313)
Time effects:			
Low dose	β_1	7.503 (0.228)	7.475 (0.198)
High dose	β_2	6.877 (0.231)	6.890 (0.198)
Control	β_3	7.319 (0.285)	7.284 (0.254)
Covariance of b_i:			
var(b_{1i})	d_{11}	3.369 (1.123)	2.921 (1.019)
var(b_{2i})	d_{22}	0.000 (—)	−0.287 (0.169)
cov(b_{1i}, b_{2i})	$d_{12} = d_{21}$	0.090 (0.381)	0.462 (0.357)
Residual variance:			
var(ε_{ij})	σ^2	1.445 (0.145)	1.522 (0.165)
REML log-likelihood		−466.173	−465.193

could ever yield a marginal model as has now been obtained. On the other hand, as long as all covariance matrices $V_i = Z_i D Z_i' + \sigma^2 I_{n_i}$ are positive (semi-) definite, that is, as long as the covariates Z_i take values within a specific range, a valid marginal model is obtained. In our example, the variance function is predicted by

$$
\begin{aligned}
\text{Var}(Y_i(t)) &= \begin{pmatrix} 1 & t \end{pmatrix} \widehat{D} \begin{pmatrix} 1 \\ t \end{pmatrix} + \sigma^2 \\
&= \widehat{d_{22}} t^2 + 2\widehat{d_{12}} t + \widehat{d_{11}} + \widehat{\sigma}^2 \\
&= -0.287 t^2 + 0.924 t + 4.443 \qquad (5.9)
\end{aligned}
$$

and therefore suggests the presence of negative curvature in the variance function. As an informal check, we can calculate the sample variance function of the ordinary least squares (OLS) residuals obtained from fitting a linear regression model with the same mean structure as the marginal model corresponding to (3.9), thereby completely ignoring the correlation structure in the data (see Chapter 4). The obtained variance function, shown in Figure 5.2, indeed supports the negative curvature suggested by our fitted variance function. Note that this is not compatible with the proposed hierarchical model. Hence, although our random-effects model naturally

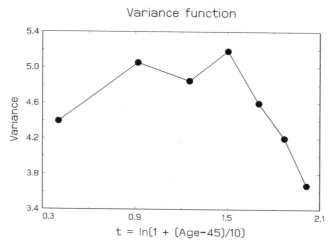

FIGURE 5.2. *Rat Data. Sample variance function for ordinary least squares (OLS) residuals, obtained from fitting a linear regression model with the same mean structure as the marginal model corresponding to (3.9).*

arose from the two-stage approach described in Section 3.2.3, it does not necessarily imply an appropriate marginal model. This again illustrates the need for exploratory data analysis (Chapter 4) prior to fitting linear mixed models. More detailed discussions on negative variance components can be found in Nelder (1954), Thompson (1962), and Searle, Casella and McCulloch (1992, Section 3.5).

6

Inference for the Marginal Model

6.1 Introduction

In practice, the fitting of a model is rarely the ultimate goal of a statistical analysis. Usually, one is primarily interested in drawing inferences on the parameters in a model, in order to generalize results obtained from a specific sample to the general population from which the sample was taken. In Section 6.2, inference for the parameter vector β in the mean structure of model (5.1) is discussed. Afterward, in Section 6.3, inference with respect to the variance components α will be handled.

6.2 Inference for the Fixed Effects

As discussed in Section 5.1, the vector β of fixed effects is estimated by

$$\widehat{\beta}(\alpha) = \left(\sum_{i=1}^{N} X_i' W_i X_i \right)^{-1} \sum_{i=1}^{N} X_i' W_i y_i, \tag{6.1}$$

in which the unknown vector α of variance components is replaced by its ML or REML estimate. Under the marginal model (5.1), and conditionally on α, $\widehat{\beta}(\alpha)$ follows a multivariate normal distribution with mean vector β

and with variance-covariance matrix

$$\text{var}(\widehat{\boldsymbol{\beta}})$$

$$= \left(\sum_{i=1}^{N} X_i' W_i X_i\right)^{-1} \left(\sum_{i=1}^{N} X_i' W_i \text{var}(\boldsymbol{Y_i}) W_i X_i\right) \left(\sum_{i=1}^{N} X_i' W_i X_i\right)^{-1} \quad (6.2)$$

$$= \left(\sum_{i=1}^{N} X_i' W_i X_i\right)^{-1}, \quad (6.3)$$

where W_i equals $V_i^{-1}(\boldsymbol{\alpha})$. In practice, the covariance matrix (6.3) is estimated by replacing $\boldsymbol{\alpha}$ by its ML or REML estimator. For the models previously fitted to the prostate data and to the rat data, the so-obtained standard errors for the fixed effects are also reported in Table 5.1 and Table 5.3, respectively.

6.2.1 Approximate Wald Tests

For each parameter β_j in $\boldsymbol{\beta}$, $j = 1, \ldots, p$, an approximate Wald test (also termed Z-test), as well as an associated confidence interval, is obtained from approximating the distribution of $(\widehat{\beta}_j - \beta_j)/\widehat{\text{s.e.}}(\widehat{\beta}_j)$ by a standard univariate normal distribution. In general, for any known matrix L, a test for the hypothesis

$$H_0 : L\boldsymbol{\beta} = \boldsymbol{0}, \quad \text{versus} \quad H_A : L\boldsymbol{\beta} \neq \boldsymbol{0} \quad (6.4)$$

immediately follows from the fact that the distribution of

$$(\widehat{\boldsymbol{\beta}} - \boldsymbol{\beta})' L' \left[L \left(\sum_{i=1}^{N} X_i' V_i^{-1}(\widehat{\boldsymbol{\alpha}}) X_i\right)^{-1} L'\right]^{-1} L(\widehat{\boldsymbol{\beta}} - \boldsymbol{\beta}) \quad (6.5)$$

asymptotically follows a chi-squared distribution with rank(L) degrees of freedom.

6.2.2 Approximate t-Tests and F-Tests

As noted by Dempster, Rubin and Tsutakawa (1981), the Wald test statistics are based on estimated standard errors which underestimate the true variability in $\widehat{\boldsymbol{\beta}}$ because they do not take into account the variability introduced by estimating $\boldsymbol{\alpha}$. In practice, this downward bias is often resolved by using approximate t- and F-statistics for testing hypotheses about $\boldsymbol{\beta}$.

For each parameter β_j in β, $j = 1, \ldots, p$, an approximate t-test and associated confidence interval can be obtained by approximating the distribution of $(\widehat{\beta}_j - \beta_j)/\widehat{\text{s.e.}}(\widehat{\beta}_j)$ by an appropriate t-distribution. Testing general linear hypotheses of the form (6.4) is now based on an F-approximation to the distribution of

$$F = \frac{(\widehat{\beta} - \beta)' L' \left[L \left(\sum_{i=1}^{N} X_i' V_i^{-1}(\widehat{\alpha}) X_i \right)^{-1} L' \right]^{-1} L(\widehat{\beta} - \beta)}{\text{rank}(L)}. \tag{6.6}$$

The numerator degrees of freedom equals $\text{rank}(L)$. The denominator degrees of freedom needs to be estimated from the data. The same is true for the degrees of freedom needed in the above t-approximation.

In practice, several methods are available for estimating the appropriate number of degrees of freedom needed for a specific t- or F-test. The SAS procedure MIXED (Version 6.12), for example, includes four different estimation methods, one of which is based on a so-called Satterthwaite-type approximation (Satterthwaite 1941). We refer to Section 3.5.2 and Appendix A in Verbeke and Molenberghs (1997) and to SAS (1999) for a detailed discussion on the estimation of the degrees of freedom in SAS. Recently, Kenward and Roger (1997) proposed a scaled Wald statistic, based on an adjusted covariance estimate which accounts for the extra variability introduced by estimating α, and they show that its small sample distribution can be well approximated by an F-distribution with denominator degrees of freedom also obtained via a Satterthwaite-type approximation.

It should be remarked that all these methods usually lead to different results. However, in the analysis of longitudinal data, different subjects contribute independent information, which results in numbers of degrees of freedom which are typically large enough, whatever estimation method is used, to lead to very similar p-values. Only for very small samples, or when linear mixed models are used outside the context of longitudinal analyses, different estimation methods for degrees of freedom may lead to severe differences in the resulting p-values. This will be illustrated in Section 8.3.5.

6.2.3 Example: The Prostate Data

Table 5.1 clearly suggests that the original linear mixed model (3.10), used for describing the prostate data, can be reduced to a more parsimonious model. Classically, this is done in a hierarchical way, starting with the highest-order interaction terms, deleting nonsignificant terms, and combining parameters which do not differ significantly. The so-obtained final model assumes no average evolution over time for the control group, a linear average time trend for the BPH group, the same average quadratic time effects for both cancer groups, and no age dependencies of the average linear as well as quadratic time trends. An overall test for comparing this final model with the original model (3.10) is testing the null hypothesis:

$$
H_0 : \begin{cases}
\beta_6 = 0 & \text{(no age by time interaction)} \\
\beta_7 = 0 & \text{(no linear time effect for controls)} \\
\beta_{11} = 0 & \text{(no age} \times \text{time}^2 \text{ interaction)} \\
\beta_{12} = 0 & \text{(no quadratic time effect for controls)} \\
\beta_{13} = 0 & \text{(no quadratic time effect for BPH)} \\
\beta_{14} = \beta_{15} & \text{(equal quadratic time effect for both cancer groups).}
\end{cases}
$$

The above hypothesis can be rewritten as

$$
H_0 : \begin{pmatrix}
0 & 0 & 0 & 0 & 0 & 1 & 0 & 0 & 0 & 0 & 0 & 0 & 0 & 0 & 0 \\
0 & 0 & 0 & 0 & 0 & 0 & 1 & 0 & 0 & 0 & 0 & 0 & 0 & 0 & 0 \\
0 & 0 & 0 & 0 & 0 & 0 & 0 & 0 & 0 & 0 & 1 & 0 & 0 & 0 & 0 \\
0 & 0 & 0 & 0 & 0 & 0 & 0 & 0 & 0 & 0 & 0 & 1 & 0 & 0 & 0 \\
0 & 0 & 0 & 0 & 0 & 0 & 0 & 0 & 0 & 0 & 0 & 0 & 1 & 0 & 0 \\
0 & 0 & 0 & 0 & 0 & 0 & 0 & 0 & 0 & 0 & 0 & 0 & 0 & 1 & -1
\end{pmatrix} \beta = 0; \tag{6.7}
$$

it is clearly of the form (6.4).

The observed value under the above null hypothesis for the associated Wald statistic (6.5) equals 3.3865, on 6 degrees of freedom. The observed value under H_0 for the associated F-statistic (6.6) equals $3.3865/6 = 0.5664$, on 6 and 46.7 degrees of freedom. The denominator degrees of freedom has been obtained from the Satterthwaite approximation in SAS procedure MIXED (Version 6.12). The corresponding p-values are 0.7587 and 0.7561, respectively, suggesting that no important terms have been left out of the model.

From now on, all further inferences will be based on the reduced final model, which can be written as:

$$
\begin{aligned}
Y_{ij} &= Y_i(t_{ij}) \\
&= \beta_1 \text{Age}_i + \beta_2 C_i + \beta_3 B_i + \beta_4 L_i + \beta_5 M_i \\
&\quad + (\beta_8 B_i + \beta_9 L_i + \beta_{10} M_i) t_{ij} \\
&\quad + \beta_{14} (L_i + M_i) t_{ij}^2 \\
&\quad + b_{1i} + b_{2i} t_{ij} + b_{3i} t_{ij}^2 + \varepsilon_{ij},
\end{aligned} \tag{6.8}
$$

Table 6.1 contains the parameter estimates and estimated standard errors for all fixed effects and variance components in model (6.8). Although the average PSA level for the control patients is not significantly different from

TABLE 6.1. *Prostate Data. Parameter estimates and standard errors (model based;robust) obtained from fitting the final model (6.8) to the prostate cancer data, using restricted maximum likelihood estimation.*

Effect	Parameter	Estimate (s.e.)
Age effect	β_1	0.016 (0.006;0.006)
Intercepts:		
Control	β_2	−0.564 (0.428;0.404)
BPH	β_3	0.275 (0.488;0.486)
L/R cancer	β_4	1.099 (0.486;0.499)
Met. cancer	β_5	2.284 (0.531;0.507)
Time effects:		
BPH	β_8	−0.410 (0.068;0.067)
L/R cancer	β_9	−1.870 (0.233;0.360)
Met. cancer	β_{10}	−2.303 (0.262;0.391)
Time^2 effects:		
Cancer	$\beta_{14} = \beta_{15}$	0.510 (0.088;0.128)
Covariance of b_i:		
var(b_{1i})	d_{11}	0.443 (0.093)
var(b_{2i})	d_{22}	0.842 (0.203)
var(b_{3i})	d_{33}	0.114 (0.035)
cov(b_{1i}, b_{2i})	$d_{12} = d_{21}$	−0.490 (0.124)
cov(b_{2i}, b_{3i})	$d_{23} = d_{32}$	−0.300 (0.082)
cov(b_{3i}, b_{1i})	$d_{13} = d_{31}$	0.148 (0.047)
Residual variance:		
var(ε_{ij})	σ^2	0.028 (0.002)
REML log-likelihood		−20.165

zero ($p = 0.1889$), we will not remove the corresponding effect from the model because a point estimate for the average PSA level in the control group may be of interest.

Figure 6.1 shows the average fitted profiles based on this final model, for a man of median age at diagnosis, for each of the diagnostic groups separately. Note how little difference there is with the average fitted profiles in Figure 5.1, based on the full model (3.10).

When the prostate data were first analyzed, one of the research questions of primary interest was whether early discrimination between cancer cases and BPH cases should be based on the rate of increase of PSA (which can

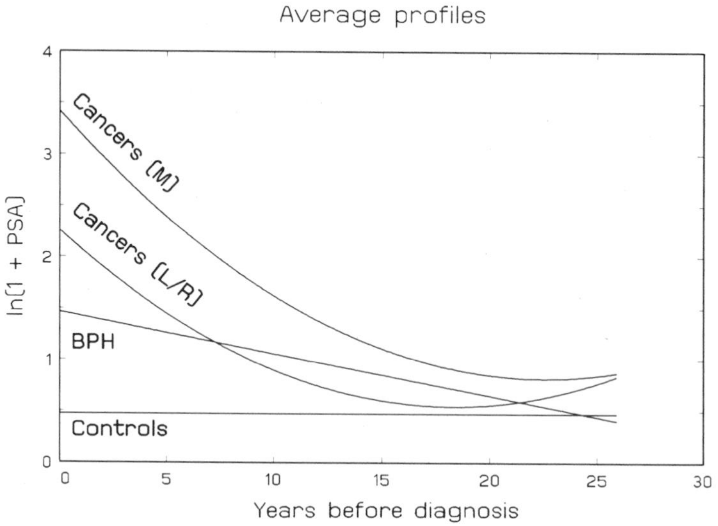

Average profiles

FIGURE 6.1. *Prostate Data. Fitted average profiles for males with median ages at diagnosis, based on the final model (6.8), where the parameters are estimated using restricted maximum likelihood estimation.*

only be estimated when repeated PSA measurements are available) rather than on just one single measurement of PSA (see also Section 2.3.1). In order to assess this, we estimate the average difference in $\ln(1 + \text{PSA})$ between these two groups, as well as the average difference in the rate of increase of $\ln(1+\text{PSA})$ between the two groups, 5 years prior to diagnosis. If we ignore the metastatic cancer cases, this is equivalent to estimating

$$
\begin{aligned}
\text{DIFF}(t = 5 \text{ years}) &= \left.\left(\beta_1 \text{Age} + \beta_4 + \beta_9 t + \beta_{14} t^2\right)\right|_{t=0.5} \\
&\quad - \left.\left(\beta_1 \text{Age} + \beta_3 + \beta_8 t\right)\right|_{t=0.5} \\
&= -\beta_3 + \beta_4 - 0.5\,\beta_8 + 0.5\,\beta_9 + 0.25\,\beta_{14} \quad (6.9)
\end{aligned}
$$

and

$$
\begin{aligned}
\text{DIFFRATE}(t = 5 \text{ years}) &= \left.\frac{\partial}{\partial t}\left(\beta_1 \text{Age} + \beta_4 + \beta_9 t + \beta_{14} t^2\right)\right|_{t=0.5} \\
&\quad - \left.\frac{\partial}{\partial t}\left(\beta_1 \text{Age} + \beta_3 + \beta_8 t\right)\right|_{t=0.5} \\
&= -\beta_8 + \beta_9 + \beta_{14}, \quad\quad\quad (6.10)
\end{aligned}
$$

which are of the form $L\beta$, for specific (1×15) matrices L. Obviously, $L\beta$ will be estimated by $L\widehat{\beta}$. Moreover, the chi-squared approximation for (6.5) as well as the F-approximation for (6.6) can be used to obtain approximate confidence intervals for $L\beta$. The results from the chi-squared approximation are summarized in the top part of Table 6.2.

TABLE 6.2. *Prostate Data. Naive and robust inference for the linear combinations (6.9) and (6.10) of fixed effects in model (6.8), fitted to the prostate cancer data, using restricted maximum likelihood estimation.*

	Naive inference		
			Wald-type approximate
Effect	Estimate	Standard error	95% confidence interval
DIFF	0.221	0.146	$[-0.065, 0.507]$
DIFFRATE	−0.951	0.166	$[-1.276, -0.626]$
	Robust inference		
			Wald-type approximate
Effect	Estimate	Standard error	95% confidence interval
DIFF	0.221	0.159	$[-0.092, 0.533]$
DIFFRATE	−0.951	0.245	$[-1.432, -0.470]$

The average difference in $\ln(1 + \text{PSA})$, 5 years prior to diagnosis, between local cancer cases and BPH cases is estimated by 0.221, with standard error equal to 0.146. The average difference in rate of change of $\ln(1 + \text{PSA})$, 5 years prior to diagnosis, between local cancer cases and BPH cases is estimated by −0.951, with standard error equal to 0.166. The corresponding 95% Wald-type approximate confidence intervals are $[-0.066, 0.508]$ and $[-1.277, -0.624]$, respectively. Hence, there is no significant difference in $\ln(1 + \text{PSA})$ between the local cancer cases and the BPH cases, whereas the rate of increase of PSA differs highly significantly. This illustrates why repeated measures of PSA are needed to discriminate between the different prostate diseases.

6.2.4 Robust Inference

A sufficient condition for $\widehat{\beta}$, given by (6.1), to be unbiased for β is that the mean $E(Y_i)$ is correctly specified as $X_i\beta$. However, the equivalence of (6.2) and (6.3) also assumes the marginal covariance matrix to be correctly specified as $V_i = Z_i D Z_i' + \Sigma_i$. Thus, an analysis based on (6.3) will not be robust with respect to model misspecification of the covariance structure. Liang and Zeger (1986) therefore propose inferential procedures based on the so-called sandwich estimator for $\text{var}(\widehat{\beta})$, obtained by replacing $\text{var}(Y_i)$ in (6.2) by $r_i r_i'$, where $r_i = y_i - X_i\widehat{\beta}$. The resulting estimator, also called robust or empirical variance estimator, can then be shown to be consistent, as long as the mean is correctly specified in the model.

Note that this suggests that as long as interest is only in inferences for average longitudinal evolutions, little effort should be spent in modeling the covariance structure, provided that the data set is sufficiently large. In this respect, an extreme point of view would be to use ordinary least squares regression methods to fit longitudinal models, thereby completely ignoring the presence of any correlation among the repeated measurements, and to use the sandwich estimator to correct for this in the inferential procedures. However, in practice, an appropriate covariance model may be of interest since it helps in interpreting the random variation in the data. For example, it may be of scientific interest to explore the presence of random slopes. Further, efficiency is gained if an appropriate covariance model can be specified (see, for example, Diggle, Liang, and Zeger 1994 Section 4.6). Finally, in the case of missing observations, use of the sandwich estimator only provides valid inferences for the fixed effects under very strict, severe assumptions about the underlying missingness process. This will be extensively discussed in Sections 15.8 and 16.5. Therefore, from now on, all inferences will be based on model-based standard errors, unless explicitly stated otherwise.

As an illustration of robust inference, model-based as well as robust standard errors were reported in the Tables 5.1 and 6.1. For some parameters, the robust standard error is smaller than the naive, model-based one. For other parameters, the opposite is true.

Robust versions of the approximate Wald, t-, and F-test, described in Sections 6.2.1 and 6.2.2, as well as associated confidence intervals, can also be obtained, replacing the naive covariance matrix (6.3) in (6.5) and (6.6) by the robust one given in (6.2). As an illustration, we recalculated the confidence intervals for the linear combinations (6.9) and (6.10) of fixed effects in model (6.8), using robust inference. The results are now shown in the bottom part of Table 6.2. Note that the robust standard errors for both estimates are larger than the naive ones, leading to larger confidence intervals.

6.2.5   Likelihood Ratio Tests

A classical statistical test for the comparison of nested models with different mean structures is the likelihood ratio (LR) test. Suppose that the null hypothesis of interest is given by $H_0 : \boldsymbol{\beta} \in \Theta_{\beta,0}$, for some subspace $\Theta_{\beta,0}$ of the parameter space Θ_β of the fixed effects $\boldsymbol{\beta}$. Let L_{ML} denote again the ML likelihood function (5.2) and let $-2\ln\lambda_N$ be the likelihood ratio test

TABLE 6.3. *Prostate Data. Likelihood ratio test for $H_0 : \beta_1 = 0$, under model (6.8), using ML as well as REML estimation.*

	ML estimation	REML estimation
Under $\beta_1 \in \mathbb{R}$	$L_{\mathrm{ML}} = -3.575$	$L_{\mathrm{REML}} = -20.165$
Under $H_0 : \beta_1 = 0$	$L_{\mathrm{ML}} = -6.876$	$L_{\mathrm{REML}} = -19.003$
$-2\ln\lambda_N$	6.602	-2.324
degrees of freedom	1	——
p-value	0.010	——

statistic defined as

$$-2\ln\lambda_N \quad = \quad -2\ln\left[\frac{L_{\mathrm{ML}}(\widehat{\boldsymbol{\theta}}_{\mathrm{ML},0})}{L_{\mathrm{ML}}(\widehat{\boldsymbol{\theta}}_{\mathrm{ML}})}\right],$$

where $\widehat{\boldsymbol{\theta}}_{\mathrm{ML},0}$ and $\widehat{\boldsymbol{\theta}}_{\mathrm{ML}}$ are the maximum likelihood estimates obtained from maximizing L_{ML} over $\Theta_{\beta,0}$ and Θ_β, respectively. It then follows from classical likelihood theory (see, e.g., Cox and Hinkley 1990, Chapter 9) that, under some regularity conditions, $-2\ln\lambda_N$ follows, asymptotically under H_0, a chi-squared distribution with degrees of freedom equal to the difference between the dimension p of Θ_β and the dimension of $\Theta_{\beta,0}$.

It should be emphasized that the above result is not valid if the models are fitted using REML rather than ML estimation. Indeed, the mean structure of the model fitted under H_0 is not the mean structure $X_i\boldsymbol{\beta}$ of the original model under Θ_β, leading to different error contrasts $\boldsymbol{U} = A'\boldsymbol{Y}$ (see Section 5.3) under both models. Hence, the corresponding REML log-likelihood functions are based on different observations, which makes them no longer comparable. This can be well illustrated in the context of the prostate data. We reconsider the final linear mixed model (6.8), and we use the likelihood ratio test for testing whether correction for the different ages at the time of the diagnosis is really needed. The corresponding null hypothesis equals $H_0 : \beta_1 = 0$. The results obtained under ML as well as under REML estimation are summarized in Table 6.3. Under ML, the observed value for the LR statistic $-2\ln\lambda_N$ equals 6.602, which is significant ($p = 0.010$) when compared to a chi-squared distribution with 1 degree of freedom. Note that a negative observed value for the LR statistic $-2\ln\lambda_N$ is obtained under REML, clearly illustrating the fact that valid classical LR tests for the mean structure can only be obtained in the context of ML inference. We refer to Welham and Thompson (1997) for two alternative LR-type tests, based on profile likelihoods, which do allow comparison of two models with nested mean structures, fitted using the REML estimation method.

6.3   Inference for the Variance Components

In many practical situations, the mean structure rather than the covariance model is of primary interest. However, adequate covariance modeling is useful for the interpretation of the random variation in the data and it is essential to obtain valid model-based inferences for the parameters in the mean structure of the model. Overparameterization of the covariance structure leads to inefficient estimation and potentially poor assessment of standard errors for estimates of the mean response profiles (fixed effects), whereas a too restrictive specification invalidates inferences about the mean response profile when the assumed covariance structure does not hold (Altham 1984). Finally, as will be discussed in Chapters 17, 19, and 21, analyses of longitudinal data subject to dropout often require correct specification of the longitudinal model. In this section, inferential procedures for variance components in linear mixed models will be discussed.

6.3.1   Approximate Wald Tests

It follows from classical likelihood theory (see, e.g., Cox and Hinkley 1990, Chapter 9) that, under some regularity conditions, the distribution of the ML as well as REML estimator $\widehat{\alpha}$ can be well approximated by a normal distribution with mean vector α and with covariance matrix given by the inverse of the Fisher information matrix. Hence, approximate standard errors for the estimates of the variance components in α can be easily calculated from inverting minus the matrix of second-order partial derivatives of the log-likelihood function (ML or REML) with respect to α. These are also the standard errors previously reported in Tables 5.1 and 6.1 and Table 5.3 for the prostate data and the rat data, respectively.

Using the asymptotic normality of the parameter estimates, approximate Wald tests and approximate Wald confidence intervals can now easily be obtained, similarly as described in Section 6.2.1 for the fixed effects. However, the performance of the normal approximation strongly depends on the true value α, with larger samples needed for values of α relatively closer to the boundary of the parameter space Θ_α. In the case that α is a boundary value, the normal approximation completely fails. This has important consequences for significance tests for variance components. Depending on the hypothesis of interest, and depending on whether the marginal or the hierarchical interpretation of the linear mixed model under consideration is used (see discussion in Section 5.6.2), Wald tests may or may not yield valid inferences.

For example, consider the linear mixed model (6.8) previously derived for the prostate data, with REML estimates as reported in Table 6.1. If we assume that the variability between subjects can be explained by random effects b_i, then $H_0 : d_{33} = 0$ cannot be tested using an approximate Wald test. Indeed, given the hierarchical interpretation of the model, D is restricted to be positive (semi-)definite, implying that H_0 is on the boundary of the parameter space Θ_α. Therefore, the asymptotic distribution of $\widehat{d_{33}}/\text{s.e.}(\widehat{d_{33}})$ is not normal under H_0, such that the approximate Wald test is not applicable. On the other hand, if one only assumes that the covariance matrix of each subject's response Y_i can be described by $V_i = Z_i D Z_i' + \sigma^2 I_{n_i}$, not assuming that this covariance matrix results from an underlying random-effects structure, D is no longer restricted to be positive (semi-)definite. Since $d_{33} = 0$ is then interior to Θ_α, $\widehat{d_{33}}/\text{s.e.}(\widehat{d_{33}})$ asymptotically follows a standard normal distribution under $H_0 : d_{33} = 0$, from which a valid approximate Wald test follows. Based on the parameter estimates reported in Table 6.1, we find that the observed value for the test statistic equals $\widehat{d_{33}}/\widehat{\text{s.e.}}(\widehat{d_{33}}) = 0.114/0.035 = 3.257$, which is highly significant when compared to a standard normal distribution ($p = 0.0011$).

Obviously, the above distinction between the hierarchical and marginal interpretation of a linear mixed model is far less crucial for testing significance of covariance parameters in α. For example, the hypothesis $H_0 : d_{23} = 0$ can be tested with an approximate Wald test, even when the variability between subjects is believed to be induced by random effects. However, in order to keep H_0 away from the boundary of Θ_α, one then still has to assume that the variances d_{22} and d_{33} are strictly positive. Hence, based on the parameter estimates reported in Table 6.1, and assuming all diagonal elements in D to be nonzero, we find highly significant correlations between the subject-specific intercepts and slopes in model (6.8), and the only positive correlation is the one between the random intercepts and the random slopes for time[2].

6.3.2 Likelihood Ratio Tests

Similar as for the fixed effects, a LR test can be derived for comparing nested models with different covariance structures. Suppose that the null hypothesis of interest is now given by $H_0 : \alpha \in \Theta_{\alpha,0}$, for some subspace $\Theta_{\alpha,0}$ of the parameter space Θ_α of the variance components α. Let L_{ML} denote again the ML likelihood function (5.2) and let $-2 \ln \lambda_N$ be the likelihood ratio test statistic which is again defined as

$$-2 \ln \lambda_N = -2 \ln \left[\frac{L_{\text{ML}}(\widehat{\theta}_{\text{ML},0})}{L_{\text{ML}}(\widehat{\theta}_{\text{ML}})} \right], \quad (6.11)$$

where $\widehat{\boldsymbol{\theta}}_{\mathrm{ML},0}$ and $\widehat{\boldsymbol{\theta}}_{\mathrm{ML}}$ are now the maximum likelihood estimates obtained from maximizing L_{ML} over $\Theta_{\alpha,0}$ and Θ_α, respectively. It then follows from classical likelihood theory (see, e.g., Cox and Hinkley 1990, Chapter 9) that, under some regularity conditions, $-2\ln\lambda_N$ follows, asymptotically under H_0, a chi-squared distribution with degrees of freedom equal to the difference between the dimension $s - p$ of Θ_α and the dimension of $\Theta_{\alpha,0}$.

One of the regularity conditions under which the chi-squared approximation is valid is that H_0 is not on the boundary of the parameter space Θ_α. Hence, the LR test suffers from exactly the same problems as the approximate Wald test previously described in Section 6.3.1. Further, in contrast to the LR test for fixed effects (see Section 6.2.5), valid LR tests are also obtained under REML estimation. The test statistic $-2\ln\lambda_N$ is then still given by (6.11), but L_{ML} needs to be replaced by the REML likelihood function L_{REML}, given by expression (5.8), and the parameter estimates $\widehat{\boldsymbol{\theta}}_{\mathrm{ML},0}$ and $\widehat{\boldsymbol{\theta}}_{\mathrm{ML}}$ are replaced by their corresponding REML estimates. This is because models with the same mean structure lead to the same error contrasts $U = A'Y$ (see Section 5.3), which makes both REML likelihood functions comparable since they are no longer based on different observations.

6.3.3 Example: The Rat Data

In Section 5.6.2, the marginal model corresponding to model (3.9) was fitted to the rat data, not restricting the parameter space for the variance components. Based on the unrestricted parameter estimates reported in Table 5.3, the variance function was predicted by expression (5.9), which suggested the presence of negative curvature in the variance function. Under the assumed marginal model, an approximate Wald test as well as LR test can be derived to test whether the variance function is significantly different from constant. More specifically, the hypothesis of interest is $H_0 : d_{12} = d_{22} = 0$, which is not on the boundary of the parameter space under the marginal interpretation of the model. The observed value for the Wald statistic equals

$$
\begin{pmatrix} \widehat{d_{12}} & \widehat{d_{22}} \end{pmatrix} \begin{pmatrix} \widehat{\mathrm{Var}(\widehat{d_{12}})} & \widehat{\mathrm{Cov}(\widehat{d_{12}},\widehat{d_{22}})} \\ \widehat{\mathrm{Cov}(\widehat{d_{12}},\widehat{d_{22}})} & \widehat{\mathrm{Var}(\widehat{d_{22}})} \end{pmatrix}^{-1} \begin{pmatrix} \widehat{d_{12}} \\ \widehat{d_{22}} \end{pmatrix}
$$

$$
= \begin{pmatrix} 0.462 & -0.287 \end{pmatrix} \begin{pmatrix} 0.127 & -0.038 \\ -0.038 & 0.029 \end{pmatrix}^{-1} \begin{pmatrix} 0.462 \\ -0.287 \end{pmatrix}
$$

$$
= 2.936,
$$

which is not significant when compared to a chi-squared distribution with 2 degrees of freedom ($p = 0.2304$). The REML estimates of the parameters

TABLE 6.4. *Rat Data. REML estimates and associated estimated standard errors for all parameters in model (3.9), under $H_0 : d_{12} = d_{22} = 0$.*

Effect	Parameter	REMLE (s.e.)
Intercept	β_0	68.607 (0.331)
Time effects:		
Low dose	β_1	7.507 (0.225)
High dose	β_2	6.871 (0.228)
Control	β_3	7.507 (0.225)
Covariance of b_i:		
var(b_{1i})	d_{11}	3.565 (0.808)
Residual variance:		
var(ε_{ij})	σ^2	1.445 (0.145)
REML log-likelihood		-466.202

in the reduced model are shown in Table 6.4. A LR test for the same null hypothesis can now be obtained from comparing the maximized REML log-likelihood values (see Tables 5.3 and 6.4). The observed value for the test statistic equals $-2 \ln \lambda_N = -2(-466.202 + 465.193) = 2.018$, which is also not significant when compared to a chi-squared distribution with 2 degrees of freedom ($p = 0.3646$).

From now on, all further inferences will be based on the reduced model. For each treatment group separately, the predicted average profile based on the estimates reported in Table 6.4 is shown in Figure 6.2. The observed value for the Wald statistic (6.5) for testing the hypothesis $H_0 : \beta_1 = \beta_2 = \beta_3$ of no average treatment effect equals 4.63, which is not significant when compared to a chi-squared distribution with 2 degrees of freedom ($p = 0.0987$).

Since the obtained estimate for d_{11} equals $3.565 > 0$, the fitted model allows a random-effects interpretation. The corresponding hierarchical model is given by

$$
Y_{ij} = \begin{cases} \beta_0 + b_{1i} + \beta_1 t_{ij} + \varepsilon_{ij}, & \text{if low dose} \\ \beta_0 + b_{1i} + \beta_2 t_{ij} + \varepsilon_{ij}, & \text{if high dose} \\ \beta_0 + b_{1i} + \beta_3 t_{ij} + \varepsilon_{ij}, & \text{if control} \end{cases} \quad (6.12)
$$

and is obtained from omitting the subject-specific slopes b_{2i} from the original model (3.9), thereby assuming that all individual profiles have equal

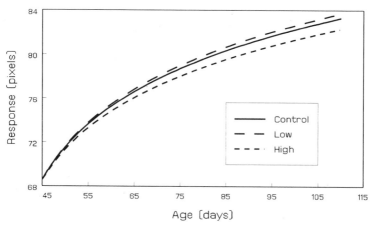

FIGURE 6.2. *Rat Data. Fitted average evolution for each treatment group sepa-rately, obtained from fitting the final model (6.12) using REML estimation.*

slopes, after correction for the treatment. As before, the residual compo-nents ε_{ij} only contain measurement error (i.e., $\varepsilon_i = \varepsilon_{(1)i}$, see Section 3.3.4).

Note that the above random-intercepts model does not only imply the mar-ginal variance function to be constant over time, it also assumes constant correlation between any two repeated measurements from the same rat. The constant correlation is given by

$$\rho_I \;=\; \frac{d_{11}}{d_{11} + \sigma^2}$$

which is the intraclass correlation coefficient previously encountered in Sec-tion 3.3.2. In our example, ρ_I is estimated by $3.565/(3.565+1.445) = 0.712$. The corresponding covariance matrix, with constant variance and constant correlation, is often called compound symmetry. This again illustrates that negative estimates for variance components in a linear mixed model of-ten have meaningful interpretations in the implied marginal model. Here, a nonpositive estimate for the variance d_{11} of the random effects in a random-intercepts model would indicate that the assumption of constant positive correlation between the repeated measurements is not valid for the data set at hand. We refer to Section 5.6.2 for another example in which neg-ative variance estimates indicate misspecifications in the implied marginal model.

6.3.4   Marginal Testing for the Need of Random Effects

As illustrated in Chapter 3, random effects in a linear mixed model represent the variability in subject-specific intercepts and slopes, not explained by the covariates included in the model. Under the hierarchical interpretation of the model, it may therefore be of scientific interest to test for the need of (some of the) random effects in the model. For example, under model (6.8) for the prostate data, it might be of interest to test whether random quadratic time effects are needed. If not, this would suggest that all noncancer patients evolve linearly over time, whereas all cancer patients would have the same quadratic time effect described by the fixed effect $\beta_{14} = \beta_{15}$. Note that, unless a Bayesian approach is followed, this can only indirectly be tested via the induced marginal model. For the prostate example, the corresponding hypothesis of interest is

$$H_0 : d_{13} = d_{23} = d_{33} = 0, \tag{6.13}$$

which is clearly on the boundary of the parameter space Θ_α such that the classical likelihood-based inference cannot be applied (see discussion in Sections 6.3.1 and 6.3.2).

Using results of Self and Liang (1987) on nonstandard testing situations, Stram and Lee (1994, 1995) have been able to show that the asymptotic null distribution for the likelihood ratio test statistic for testing hypotheses of the type (6.13) is often a mixture of chi-squared distributions rather than the classical single chi-squared distribution. This was derived under the assumption of conditional independence, that is, assuming that all residual covariances Σ_i are of the form $\sigma^2 I_{n_i}$. For ANOVA models with independent random effects, this was already briefly discussed by Miller (1977).

Let $-2 \ln \lambda_N$ be the likelihood ratio test statistic as in expression (6.11). Stram and Lee (1994, 1995) then discuss several specific testing situations, which we will briefly summarize. Although their results were derived for the case of maximum likelihood estimation, the same results apply for restricted maximum likelihood estimation, as shown by Morrell (1998). In fact, the REML test statistic performs slightly better than the ML test statistic in the sense that, on average, the rejection proportions are closer to the nominal level for the REML test statistic than for the ML test statistic.

CASE 1: NO RANDOM EFFECTS VERSUS ONE RANDOM EFFECT

For testing $H_0 : D = 0$ versus $H_A : D = d_{11}$, where d_{11} is a non-negative scalar, we have that the asymptotic null distribution of $-2 \ln \lambda_N$ is a mixture of χ_1^2 and χ_0^2 with equal weights 0.5. The χ_0^2 distribution is the distribution which gives probability mass 1 to the value 0. The mixture is shown

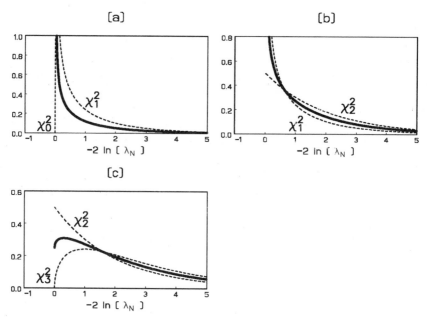

FIGURE 6.3. *Graphical representation of the asymptotic null distribution of the likelihood ratio statistic for testing the significance of random effects in a linear mixed model, for three different types of hypotheses. For each case, the distribution (solid line) is a mixture of two chi-squared distributions (dashed lines), with both weights equal to 0.5:*
(a) Case 1: no random effects versus one random effect.
(b) Case 2: one random effect versus two random effects.
(c) Case 3: two random effects versus three random effects.

in panel (a) of Figure 6.3. Note that if the classical null distribution would be used, all p-values would be overestimated. Therefore, the null hypothesis would be accepted too often, resulting in incorrectly simplifying the covariance structure of the model, which may seriously invalidate inferences, as shown by Altham (1984).

CASE 2: ONE VERSUS TWO RANDOM EFFECTS

In the case that one wishes to test

$$H_0 : D = \begin{pmatrix} d_{11} & 0 \\ 0 & 0 \end{pmatrix},$$

for some strictly positive d_{11}, versus H_A that D is a (2×2) positive semi-definite matrix, we have that the asymptotic null distribution of $-2 \ln \lambda_N$ is a mixture with equal weights 0.5 for χ^2_2 and χ^2_1, shown in Figure 6.3(b). Similar to case 1, we have that ignoring the boundary problems may result in too parsimonious covariance structures.

CASE 3: q VERSUS $q + 1$ RANDOM EFFECTS

For testing the hypothesis

$$H_0 : D = \begin{pmatrix} D_{11} & \mathbf{0} \\ \mathbf{0}' & 0 \end{pmatrix}, \tag{6.14}$$

in which D_{11} is a $(q \times q)$ positive definite matrix, versus H_A that D is a general $((q + 1) \times (q + 1))$ positive semidefinite matrix, the large-sample behavior of the null distribution of $-2 \ln \lambda_N$ is a mixture of χ^2_{q+1} and χ^2_q, again with equal weights 0.5. A graphical representation for the case of testing two random effects $(q = 2)$ versus three random effects is given in the third panel of Figure 6.3. Again, we have that the correction due to the boundary problems reduces the p-values in order to protect against the use of oversimplified covariance structures.

CASE 4: q VERSUS $q + k$ RANDOM EFFECTS

The null distribution of $-2 \ln \lambda_N$ for testing (6.14) versus

$$H_A : D = \begin{pmatrix} D_{11} & D_{12} \\ D'_{12} & D_{22} \end{pmatrix},$$

which is a general $((q + k) \times (q + k))$ positive semidefinite matrix, is a mixture of χ^2 random variables as well as other types of random variables formed by the lengths of projections of multivariate normal random variables upon curved as well as flat surfaces. Apart from very special cases, current statistical knowledge calls for simulation methods to estimate the appropriate null distribution.

Note that the results in the above cases 1 to 4 assume that the likelihood function can be maximized over the space Θ_α of positive *semi*definite matrices D, and that the estimating procedure is able to converge, for example, to values of D which are positive semidefinite but not positive definite. This is software dependent and should be checked when the above results are applied in practice.

For example, according to Stram and Lee (1994), this assumption did not hold for the SAS procedure MIXED when their paper was written, and they therefore discuss how their results had to be corrected. Since the procedure only allowed maximization of the likelihood over a subspace of the required parameter space Θ_α, the likelihood ratio statistics were typically too small. For the third case, for example, the asymptotic null distribution became a mixture of χ^2_{q+1} and χ^2_0 with equal weight 0.5. However, as explained in Section 5.6.2, since release 6.10 of SAS, the only constraint on variance

TABLE 6.5. *Prostate Data. Several random-effects models with the associated value for the log-likelihood value evaluated at the parameter estimates, for maximum as well as restricted maximum likelihood estimation.*

| | $\ln[L(\widehat{\boldsymbol{\theta}})]$ | |
Random effects	ML	REML
Model 1: Intercepts, time, time^2	-3.575	-20.165
Model 2: Intercepts, time	-50.710	-66.563
Model 3: Intercepts	-131.218	-149.430
Model 4: _____	-251.275	-272.367

components estimates is non-negativeness of the variances. In some cases, this can even lead to estimates of D which are not non-negative definite (see, for example, the analysis in Section 5.6.2 of the rat data). Because any symmetric matrix with at least one negative diagonal element is not positive semidefinite, we have that the required parameter space Θ_α is a subspace of the set of all symmetric matrices with non-negative diagonal elements. Hence, we may conclude that, since release 6.10, SAS allows maximizing the likelihood over Θ_α, and therefore that the original results, as described in the cases 1 to 4, are valid even when the procedure MIXED is used. However, since the likelihood is maximized over a parameter space which is larger than Θ_α, one should check the resulting estimate $\widehat{D}$ for positive semidefiniteness. In the next section, the above results will be illustrated in the context of the prostate data. Other examples can be found in Section 17.4 as well as in Section 24.1.

6.3.5   Example: The Prostate Data

For the prostate data, the hypothesis of most interest is that only random intercepts and random slopes for the linear time effect are needed in model (6.8), and hence that the random slopes for the quadratic time effect may be omitted (case 3). However, for illustrative purposes, we tested all hypotheses of deleting one random effect from the model, in a hierarchical way starting from the highest-order time effect. Likelihood ratio tests were used, based on maximum likelihood as well as on restricted maximum likelihood estimation. The models and the associated maximized log-likelihood values are shown in Table 6.5. Further, Table 6.6 shows the likelihood ratio statistics for dropping one random effect at a time, starting from the quadratic time effect. The correct asymptotic null distributions directly follow from the results described in cases 1 to 3. We hereby denote a mixture

TABLE 6.6. *Prostate Data. Likelihood ratio statistics with the correct as well as naive asymptotic null distribution for comparing random-effects models, for maximum as well as restricted maximum likelihood estimation. A mixture of two chi-squared distributions with k_1 and k_2 degrees of freedom and with equal weight for both distributions is denoted by $\chi^2_{k_1:k_2}$.*

Maximum likelihood			
		Asymptotic null distribution	
Hypothesis	$-2\ln(\lambda_N)$	Correct	Naive
Model 2 versus Model 1	94.270	$\chi^2_{2:3}$	χ^2_3
Model 3 versus Model 2	161.016	$\chi^2_{1:2}$	χ^2_2
Model 4 versus Model 3	240.114	$\chi^2_{0:1}$	χ^2_1
Restricted maximum likelihood			
		Asymptotic null distribution	
Hypothesis	$-2\ln(\lambda_N)$	Correct	Naive
Model 2 versus Model 1	92.796	$\chi^2_{2:3}$	χ^2_3
Model 3 versus Model 2	165.734	$\chi^2_{1:2}$	χ^2_2
Model 4 versus Model 3	245.874	$\chi^2_{0:1}$	χ^2_1

ture of two chi-squared distributions with k_1 and k_2 degrees of freedom, with equal weights 0.5, by $\chi^2_{k_1:k_2}$. For example, the p-value obtained under REML estimation for the comparison of Model 2 versus Model 1 can then be calculated as

$$
\begin{aligned}
p &= P(\chi^2_{2:3} > 92.796) \\
&= \frac{1}{2}\, P(\chi^2_2 > 92.796) + \frac{1}{2}\, P(\chi^2_3 > 92.796).
\end{aligned}
$$

The naive asymptotic null distribution is the one which follows from applying the classical likelihood theory, ignoring the boundary problem for the null hypothesis (i.e., a chi-squared distribution with degrees of freedom equal to the number of free parameters which vanish under the null hypothesis). All observed values for $-2\ln(\lambda_N)$ are larger than 90, yielding p-values smaller than 0.0001. We conclude that the covariance structure should not be simplified deleting random effects from the model. We refer to Sections 17.4 and 24.1 for examples where the naive and the corrected p-values show much more difference.

TABLE 6.7. *Overview of frequently used information criteria for comparing linear mixed models. We hereby define n^* equal to the total number $n = \sum_{i=1}^{N} n_i$ of observations or equal to $n - p$, depending on whether ML or REML estimation was used in the calculations.*

Criterion	Definition of $\mathcal{F}(\cdot)$
Akaike (AIC)	$\mathcal{F}(\#\boldsymbol{\theta}) = \#\boldsymbol{\theta}$
Schwarz (SBC)	$\mathcal{F}(\#\boldsymbol{\theta}) = (\#\boldsymbol{\theta} \ \ln n^*)/2$
Hannan and Quinn (HQIC)	$\mathcal{F}(\#\boldsymbol{\theta}) = \#\boldsymbol{\theta} \ \ln(\ln n^*)$
Bozdogan (CAIC)	$\mathcal{F}(\#\boldsymbol{\theta}) = \#\boldsymbol{\theta} \ (\ln n^* + 1)/2$

6.4   Information Criteria

All testing procedures discussed in Sections 6.2 and 6.3 considered the comparison of so-called nested models, in the sense that the model under the null hypothesis could be viewed as a special case from the alternative model. In order to extend this to the case where one wants to discriminate between non-nested models, we take a closer look at the likelihood ratio tests discussed in Sections 6.2.5, 6.3.2, and 6.3.4. Let ℓ_A and ℓ_0 denote the log-likelihood function evaluated at the estimates obtained under the alternative hypothesis and under the null hypothesis, respectively. Further, let $\#\boldsymbol{\theta}_0$ and $\#\boldsymbol{\theta}_A$ denote the number of free parameters under the null hypothesis and under the alternative hypothesis, respectively. The LR test then rejects the null hypothesis if $\ell_A - \ell_0$ is large in comparison to the difference in degrees of freedom between the two models which are to be compared, or, equivalently, if

$$\ell_A - \ell_0 \ > \ \mathcal{F}(\#\boldsymbol{\theta}_A) - \mathcal{F}(\#\boldsymbol{\theta}_0),$$

or, equivalently, if

$$\ell_A - \mathcal{F}(\#\boldsymbol{\theta}_A) \ > \ \ell_0 - \mathcal{F}(\#\boldsymbol{\theta}_0),$$

for an appropriate function $\mathcal{F}(\cdot)$. For example, when tests are performed at the 5% level of significance, for hypotheses of the same form as those described in the third case of Section 6.3.4, $\mathcal{F}$ was such that

$$2 \left[\mathcal{F}(\#\boldsymbol{\theta}_A) \ - \ \mathcal{F}(\#\boldsymbol{\theta}_0)\right] \ = \ \chi^2_{(\#\boldsymbol{\theta}_A \ - \ \#\boldsymbol{\theta}_0 \ - \ 1) \ : \ (\#\boldsymbol{\theta}_A \ - \ \#\boldsymbol{\theta}_0), \ 0.95}$$

where $\chi^2_{k_1 : k_2, 0.95}$ denotes the 95% percentile of the $\chi^2_{k_1 : k_2}$ distribution. This procedure can be interpreted as a formal test of significance only if the model under the null hypothesis is nested within the model under the alternative hypothesis. However, if this is not the case, there is no reason why

TABLE 6.8. *Rat Data. Summary of the results of fitting two different random-intercepts models to the rat data (ML estimation, $n = 252$).*

Mean structure	ℓ_{ML}	$\#\boldsymbol{\theta}$	AIC	SBC
Separate average slopes	-464.326	6	-470.326	-480.914
Common average slope	-466.622	4	-470.622	-477.681

the above procedure could not be used as a rule of thumb, or why no other functions $\mathcal{F}(\cdot)$ could be used to construct empirical rules for discriminating between covariance structures. Some other frequently used functions are shown in Table 6.7, all leading to different discriminating rules, called information criteria. The main idea behind information criteria is to compare models based on their maximized log-likelihood value, but to penalize for the use of too many parameters. The model with the largest AIC, SBC, HQIC, or CAIC is deemed best. Note that, except for the Akaike information criterion (AIC), they all involve the sample size (see Table 6.7), showing that differences in likelihood need to be viewed, not only relative to the differences in numbers of parameters but also relative to the number of observations included in the analysis. As the sample size increases, more severe increases in likelihood are required before a complex model will be preferred over a simple model. Note also that, since REML is based on a set of $n - p$ error contrasts (see Section 5.3), the effective sample size used in the definition of the information criteria is $n^* = n - p$ under REML estimation, while being n under ML estimation. Note also that, as explained in Section 6.2.5, REML log-likelihoods are only fully comparable for models with the same mean structure. Hence, for comparing models with different mean structures, one should only consider information criteria based on ML estimation.

We refer to Akaike (1974), Schwarz (1978), Hannan and Quinn (1979), and Bozdogan (1987) for more information on the information criteria defined in Table 6.7.

It should be strongly emphasized that information criteria only provide rules of thumb to discriminate between several statistical models. They should never be used or interpreted as formal statistical tests of significance. In specific examples, different criteria can even lead to different models. As an illustration, we consider model (6.12) previously derived for the rat data. Using an approximate Wald test, we found evidence in Section 6.3.3 that, under the proposed model, there is no average difference among the three treatment groups ($p = 0.0987$). Hence, we expect that a linear mixed model with equal average slope for all three groups is preferred when compared to the original model (6.12). Table 6.8 shows the AIC and SBC obtained from the ML results for both models, with effective sample size equal to $n = 252$. Note that AIC prefers the model with treatment-specific average slopes, whereas the common-slope model is to be preferred based on SBC. This again illustrates the effect of taking into account the sample size n in the calculation of SBC, which is not the case for AIC.

7

Inference for the Random Effects

/

7.1   Introduction

Although in practice one is usually primarily interested in estimating the parameters in the marginal linear mixed-effects model (the fixed effects β and the variance components in D and in all Σ_i), it is often useful to calculate estimates for the random effects b_i as well, since they reflect how much the subject-specific profiles deviate from the overall average profile. Such estimates can then be interpreted as residuals which may be helpful for detecting special profiles (i.e., outlying individuals) or groups of individuals evolving differently in time. Also, estimates for the random effects are needed whenever interest is in prediction of subject-specific evolutions (see Section 7.5).

As indicated in Section 5.1, it is then no longer sufficient to assume that the marginal distribution of the responses Y_i is given by model (5.1), because it does not imply that the variability in the data can be explained by random effects. In this section, we will therefore explicitly assume that the hierarchical model (3.8) is appropriate. Since random effects represent a natural heterogeneity between the subjects, this assumption will often be justified for data where the between-subjects variability is large in comparison to the within-subject variability.

7.2　Empirical Bayes Inference

Since the random effects in model (3.8) are assumed to be random variables, it is most natural to estimate them using Bayesian techniques (see, for example, Box and Tiao 1992 or Gelman et al. 1995). As discussed in Section 3.3, the distribution of the vector Y_i of responses for the ith individual, conditional on that individual's specific regression coefficients b_i, is multivariate normal with mean vector $X_i\beta + Z_i b_i$ and with covariance matrix Σ_i. Further, the marginal distribution of b_i is multivariate normal with mean vector $\mathbf{0}$ and covariance matrix D. In the Bayesian literature, this last distribution is usually called the prior distribution of the parameters b_i since it does not depend on the data Y_i. Once observed values y_i for Y_i have been collected, the so-called posterior distribution of b_i, defined as the distribution of b_i, conditional on $Y_i = y_i$, can be calculated. If we denote the density function of Y_i conditional on b_i, and the prior density function of b_i by $f(y_i|b_i)$ and $f(b_i)$, respectively, we have that the posterior density function of b_i given $Y_i = y_i$ is given by

$$f(b_i|y_i) \equiv f(b_i|Y_i = y_i) = \frac{f(y_i|b_i)\, f(b_i)}{\int f(y_i|b_i)\, f(b_i)\, db_i}. \qquad (7.1)$$

For the sake of notational convenience, we hereby suppressed the dependence of all above density functions on certain components of θ.

Using the theory on general Bayesian linear models (Smith 1973, Lindley and Smith 1972), it can be shown that (7.1) is the density of a multivariate normal distribution. Very often, b_i is estimated by the mean of this posterior distribution, called the posterior mean of b_i. This estimate is then given by

$$\begin{aligned}
\widehat{b}_i(\theta) &= E\left[b_i \mid Y_i = y_i\right] \\
&= \int b_i\, f(b_i|y_i)\, db_i \\
&= DZ_i'W_i(\alpha)(y_i - X_i\beta), \qquad (7.2)
\end{aligned}$$

and the covariance matrix of the corresponding estimator equals

$$\operatorname{var}(\widehat{b}_i(\theta)) = DZ_i'\left\{ W_i - W_iX_i\left(\sum_{i=1}^{N} X_i'W_iX_i\right)^{-1} X_i'W_i\right\} Z_iD, \qquad (7.3)$$

where, as before, W_i equals V_i^{-1} (Laird and Ware 1982). Note that (7.3) underestimates the variability in $\widehat{b}_i(\theta) - b_i$ since it ignores the variation of b_i. Therefore, inference for b_i is usually based on

$$\operatorname{var}(\widehat{b}_i(\theta) - b_i) = D - \operatorname{var}(\widehat{b}_i(\theta)) \qquad (7.4)$$

as an estimator for the variation in $\widehat{\boldsymbol{b}_i}(\boldsymbol{\theta}) - \boldsymbol{b}_i$ (Laird and Ware 1982).

So far, all calculations were performed conditionally on the vector $\boldsymbol{\theta}$ of parameters in the marginal model. In practice, the unknown parameters $\boldsymbol{\beta}$ and $\boldsymbol{\alpha}$ in (7.2), (7.3), and (7.4) are replaced by their maximum or restricted maximum likelihood estimates. The resulting estimates for the random effects are called "Empirical Bayes" (EB) estimates, which we will denote as $\widehat{\boldsymbol{b}_i}$. Note that (7.3) and (7.4) then underestimate the true variability in the obtained estimate $\widehat{\boldsymbol{b}_i}$ since they do not take into account the variability introduced by replacing the unknown parameter $\boldsymbol{\theta}$ by its estimate. Similar to that for fixed effects (see Section 6.2), inference is therefore often based on approximate t-tests of F-tests, rather than on Wald tests, with similar procedures for the estimation of the denominator degrees of freedom. An example will be given in Section 8.3.6.

In practice, one often uses histograms and scatter plots of components of $\widehat{\boldsymbol{b}_i}$ for diagnostic purposes, such as the detection of outliers which are subjects who seem to evolve differently from the other subjects in the data set. For example, Morrell and Brant (1991) use scattergrams of the EB estimates to pinpoint outlying observations, DeGruttola, Lange, and Dafni (1991) report histograms of the EB estimates, and use a normal quantile plot of standardized estimated random intercepts to check their normality, and Waternaux, Laird, and Ware (1989) use several techniques based on the EB estimates to look for unusual individuals and departures from the model assumptions. However, as will be explained in Section 7.8, results from such procedures should be interpreted with extreme care.

7.3   Henderson's Mixed-Model Equations

In Section 7.2, the estimation of the random effects was approached in a Bayesian way, which was motivated by the assumption that the $\boldsymbol{b}_i$ are random parameters. Henderson has shown that the estimates (7.2) can also be obtained from solving a system of linear equations.

Let the linear mixed model be denoted as in (5.5) in Section 5.3.3; that is, $\boldsymbol{Y} = X\boldsymbol{\beta} + Z\boldsymbol{b} + \boldsymbol{\varepsilon}$, where the vectors $\boldsymbol{Y}$, $\boldsymbol{b}$, and $\boldsymbol{\varepsilon}$, and the matrix X are obtained from stacking the vectors $\boldsymbol{Y}_i$, $\boldsymbol{b}_i$, and $\boldsymbol{\varepsilon}_i$, and the matrices X_i, respectively, underneath each other, and where Z is the block-diagonal matrix with blocks Z_i on the main diagonal and zeros elsewhere. Let $\mathcal{D}$ and Σ be block-diagonal with blocks D and Σ_i on the main diagonal and zeros elsewhere. Henderson *et al.* (1959) showed that, conditional on the vector $\boldsymbol{\alpha}$ of variance components, the estimate (5.3) for $\boldsymbol{\beta}$ and the estimates (7.2) of all random effects $\boldsymbol{b}_i$ can be obtained from solving the so-called mixed-model equations

$$\begin{pmatrix} X'\Sigma^{-1}X & X'\Sigma^{-1}Z \\ Z'\Sigma^{-1}X & Z'\Sigma^{-1}Z + \mathcal{D}^{-1} \end{pmatrix} \begin{pmatrix} \beta \\ b \end{pmatrix} = \begin{pmatrix} X'\Sigma^{-1}y \\ Z'\Sigma^{-1}y \end{pmatrix}$$

with respect to β and b (see also Henderson 1984, Searle, Casella and McCulloch 1992, Section 7.6). Note that, especially with large data sets, this may become computationally very expensive, such that, in practice, it may be (much) more efficient to calculate the estimates directly from the expressions (5.3) and (7.2).

7.4   Best Linear Unbiased Prediction (BLUP)

Suppose interest is in the estimation of a linear combination $u = \lambda'_\beta \beta + \lambda'_b b_i$ of the vector β of fixed effects and the vector b_i of random effects, for some known vectors λ_β and λ_b of dimension p and q, respectively. Conditionally on the variance components in α, an obvious estimator for u is now

$$\widehat{u}(\alpha) = \lambda_\beta'\widehat{\beta}(\alpha) + \lambda_b'\widehat{b_i}(\widehat{\beta}(\alpha), \alpha), \qquad (7.5)$$

where $\widehat{\beta}(\alpha)$ and $\widehat{b_i}(\beta, \alpha) = \widehat{b_i}(\theta)$ are as defined in expressions (5.3) and (7.2), respectively. It can now be shown that $\widehat{u}(\alpha)$ is a best linear unbiased predictor of u, in the sense that it is unbiased for u and has minimum variance among all unbiased estimators of the form $c + \sum_i \lambda_i' Y_i$ (see, for example, Harville 1976, McLean, Sanders and Stroup 1991, Searle, Casella and McCulloch 1992, Chapter 7). Note that, in practice, u is estimated by $\widehat{u}(\widehat{\alpha})$, where $\widehat{\alpha}$ equals the maximum likelihood or restricted maximum likelihood estimator for the vector α of variance components (see Chapter 5).

7.5   Shrinkage

To illustrate the interpretation of the EB estimates, consider the prediction $\widehat{Y_i}$ of the ith profile. It follows from (7.2) that

$$\begin{aligned} \widehat{Y_i} &\equiv X_i\widehat{\beta} + Z_i\widehat{b_i} \\ &= X_i\widehat{\beta} + Z_iDZ_i'V_i^{-1}(y_i - X_i\widehat{\beta}) \\ &= \left(I_{n_i} - Z_iDZ_i'V_i^{-1}\right)X_i\widehat{\beta} + Z_iDZ_i'V_i^{-1}y_i \\ &= \Sigma_iV_i^{-1}X_i\widehat{\beta} + \left(I_{n_i} - \Sigma_iV_i^{-1}\right)y_i, \qquad (7.6) \end{aligned}$$

and therefore can be interpreted as a weighted average of the population-averaged profile $X_i\widehat{\beta}$ and the observed data $\boldsymbol{y_i}$, with weights $\Sigma_i V_i^{-1}$ and $I_{n_i} - \Sigma_i V_i^{-1}$, respectively. Note that the "numerator" of $\Sigma_i V_i^{-1}$ is the residual covariance matrix Σ_i and the "denominator" is the overall covariance matrix V_i. Hence, much weight will be given to the overall average profile if the residual variability is large in comparison to the between-subject variability (modeled by the random effects), whereas much weight will be given to the observed data if the opposite is true.

In the Bayesian literature, one usually refers to phenomena like those exhibited in expression (7.6), as shrinkage (Carlin and Louis 1996, Strenio, Weisberg, and Bryk 1983). The observed data are shrunken toward the prior average profile, which is $X_i\beta$ since the prior mean of the random effects was zero. This is also illustrated in (7.4), which implies that for any linear combination λ of the random effects,

$$\text{var}(\lambda'\widehat{\boldsymbol{b_i}}) \leq \text{var}(\lambda'\boldsymbol{b_i}). \tag{7.7}$$

7.6   Example: The Random-Intercepts Model

As a special case, we consider the random-intercepts model, that is, a linear mixed model where the only subject-specific effects are intercepts. An example was already given in Section 6.3.3, in the context of the rat data. The random-effects covariance matrix D is now a scalar and will be denoted by σ_b^2. Further, all design matrices Z_i are of the form $\boldsymbol{1_{n_i}}$, an n_i-dimensional vector of ones. We will assume here that all residual covariance matrices are of the form $\Sigma_i = \sigma^2 I_{n_i}$; that is, we assume conditional independence (see Section 3.3.4).

Denoting $\boldsymbol{1_{n_i}}\boldsymbol{1_{n_i}}'$ by J_{n_i}, it follows from (7.2) that the EB estimate for the random intercept of subject i is given by

$$
\begin{aligned}
\widehat{b_i} &= \sigma_b^2 \boldsymbol{1_{n_i}}' \left(\sigma_b^2 J_{n_i} - \sigma^2 I_{n_i}\right)^{-1} (\boldsymbol{y_i} - X_i\beta) \\
&= \frac{\sigma_b^2}{\sigma^2} \boldsymbol{1_{n_i}}' \left(I_{n_i} - \frac{\sigma_b^2}{\sigma^2 + n_i\sigma_b^2} J_{n_i}\right) (\boldsymbol{y_i} - X_i\beta) \\
&= \frac{n_i\sigma_b^2}{\sigma^2 + n_i\sigma_b^2} \frac{1}{n_i} \sum_{j=1}^{n_i} (y_{ij} - \boldsymbol{x_{ij}}'\beta),
\end{aligned}
\tag{7.8}
$$

where the vector $\boldsymbol{x_{ij}}$ consists of the jth row in the design matrix X_i. Note that $\bar{r}_{i\cdot} = \sum_{j=1}^{n_i}(y_{ij} - \boldsymbol{x_{ij}}'\beta)/n_i$ is equal to the average residual for subject i.

Expression (7.8) clearly illustrates the shrinkage effect. It immediately follows from $n_i\sigma_b^2/(\sigma^2 + n_i\sigma_b^2) < 1$ that $\hat{b}_i$ is a weighted average of zero (the prior mean of b_i) and the average residual $\bar{r}_{i\cdot}$. The larger the number n_i of measurements available for subject i, the more weight is put on $\bar{r}_{i\cdot}$, yielding less severe shrinkage. Expression (7.8) also shows that more shrinkage is obtained in cases where the within-subject variability is large in comparison to the between-subject variability.

7.7 Example: The Prostate Data

As an illustration of EB estimation, we calculated the EB estimates for the random effects in our final model (6.8). Frequency histograms and scatter plots of these estimates can be found in Figure 7.1. Note how the scatter plots clearly show strong negative correlations between the intercepts and slopes for time, and between the slopes for time and the slopes for time^2. On the other hand, the intercepts are positively correlated with the slopes for the quadratic time effect. This is in agreement with the estimates for the covariance parameters in D (see Table 6.1).

The histograms in Figure 7.1 suggest the presence of outliers. Furthermore, we highlighted subjects #22, #28, #39, and #45, who are the individuals with the highest four slopes for time^2 and the smallest four slopes for time. Hence, these are the subjects with the strongest (quadratic) growth of $\ln(1+\text{PSA})$ over time. Pearson *et al.* (1994) noticed that the local/regional cancer cases #22, #28, and #39 were probably misclassified by the original methods of clinical staging and should have been included in the group of metastatic cancer cases instead. Further, subject #45 is the metastatic cancer case with the highest rate of increase of $\ln(1+\text{PSA})$ over time (see also Figure 2.3).

To illustrate the shrinkage effect, we calculated $\widehat{\boldsymbol{Y}_i}$ and $X_i\widehat{\boldsymbol{\beta}}$ for subjects #15 and #28 in the prostate data set. The resulting predicted profiles and the observed profiles are shown in Figure 7.2. The EB estimates clearly correct the population-average profile toward the observed profile.

The sample covariance matrix of the EB estimates $\widehat{\boldsymbol{b}_i}$ equals

$$\widehat{\text{var}}(\widehat{\boldsymbol{b}_i}) = \begin{pmatrix} 0.4033 & -0.4398 & 0.1311 \\ -0.4398 & 0.7287 & -0.2532 \\ 0.1311 & -0.2532 & 0.0922 \end{pmatrix},$$

which clearly underestimates the variability in the random-effects population (compare to the elements of $\widehat{D}$ in Table 6.1). This again illustrates the shrinkage effect previously obtained in expression (7.7).

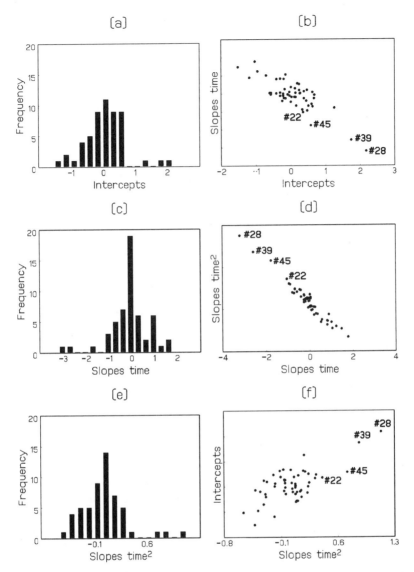

FIGURE 7.1. *Prostate Data. Histograms (panels a, c, and e) and scatter plots (panels b, d, and f) of the empirical Bayes estimates for the random intercepts and slopes in the final model (6.8).*

7.8   The Normality Assumption for Random Effects

7.8.1   Introduction

As described in Section 7.2, EB estimates for subject-specific regression coefficients are often used for diagnostic purposes, such as checking whether

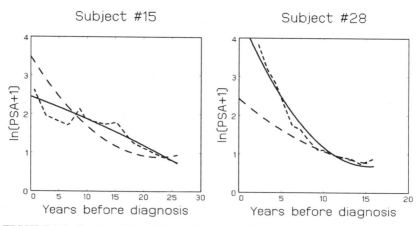

FIGURE 7.2. *Prostate Data. Observed profiles (short dashes), population-average predicted profiles (long dashes), and subject-specific predicted profiles (solid line) for subjects #15 and #28 of the prostate data set.*

the normality assumption for the random effects b_i is appropriate. However, it should be emphasized that, even when the assumed linear mixed model is correctly specified, the EB estimators $\widehat{b_i}$ all have different distributions unless all covariate matrices X_i and Z_i are the same. Hence, it may be questioned to what extent histograms and scatter plots of unstandardized EB estimates are interpretable. This is why DeGruttola, Lange, and Dafni (1991) first standardize the EB estimates prior to constructing normal quantile plots to check the normality assumption for the random effects.

Note also that even when all EB estimates follow the same distribution, it follows from the shrinkage effect (see Section 7.5) that the histogram of the EB estimates shows less variability than actually present in the population of random-effects b_i. This suggests that such histograms do not necessarily reflect the correct random-effects distribution, and hence also that histograms of (standardized) EB estimates might not be suitable for detecting deviations from the normality assumption. Louis (1984) therefore proposes to minimize other well-chosen loss functions than the classical squared-error loss function which corresponds to the posterior mean (7.2). For example, if interest is in the random-effects distribution, one could minimize a distance function between the empirical cumulative distribution function of the estimates and the true parameters.

In Section 7.8.2, it will be shown that, in some cases, a histogram of the classical EB estimates indeed does not reflect the correct random-effects distribution. This indicates that EB estimates are very dependent on their assumed prior distribution. In Section 7.8.3, it will be shown that this is in contrast with the estimation of the vector θ of parameters in the marginal

True random intercepts

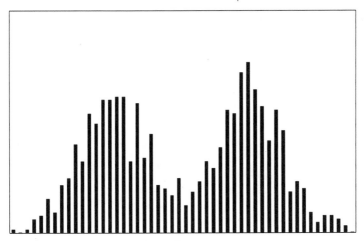

FIGURE 7.3. *Histogram (range* $[-5,5]$*) of 1000 random intercepts drawn from the normal mixture* $0.5N(-2,1) + 0.5N(2,1)$*.*

model (5.1), which is very robust with respect to misspecifications of the random-effects distribution. Finally, Section 7.8.4 briefly discusses how the normality assumption of random effects can be checked in practice.

7.8.2   Impact on EB Estimates

In order to investigate the sensitivity of EB estimates with respect to the assumed underlying random-effects distribution, Verbeke (1995) and Verbeke and Lesaffre (1996a) report results from 1000 simulated longitudinal profiles with 5 repeated measurements each, where univariate random intercepts b_i were drawn from the mixture distribution

$$\tfrac{1}{2}N(-2,1) + \tfrac{1}{2}N(2,1), \tag{7.9}$$

reflecting the presence of heterogeneity in the population. In practice, such a mixture could occur when the population under consideration consists of two subpopulations of equal size, with negative and positive subject-specific intercepts, respectively. A histogram of the realized values is shown in Figure 7.3. They then fitted a linear mixed model, assuming normality for the random effects, and they calculated the EB estimates $\widehat{b_i}$ for the random intercepts in the model. The histogram of these estimates is shown in Figure 7.4. Clearly, the severe amount of shrinkage forces the estimates $\widehat{b_i}$ to satisfy the assumption of a homogeneous, unimodal (normal) population.

Under the conditional independence model (see Section 3.3.4), Verbeke (1995) and Verbeke and Lesaffre (1996a) have shown that the bimodal as-

Empirical Bayes estimates

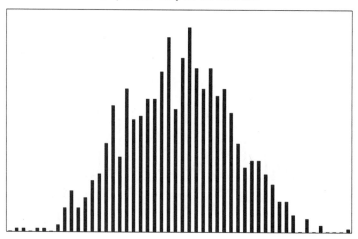

FIGURE 7.4. *Histogram (range $[-5, 5]$) of the Empirical Bayes estimates of the random intercepts shown in Figure 7.3, calculated under the assumption that the random effects are normally distributed.*

pect of the original distribution will not be reflected in the distribution of the EB estimates obtained under the normality assumption, as soon as the eigenvalues of $\sigma^2 (Z_i' Z_i)^{-1}$ are sufficiently large. This means that both the error variance and the covariate structure play an important role in the shape of the distribution of $\widehat{b_i}$. First, if σ^2 is large, it will be difficult to detect heterogeneity in the random-effects population, based on the $\widehat{b_i}$. Thus, if the error variability σ^2 is large compared to the random-effects variability, the $\widehat{b_i}$ may not reflect the correct distributional shape of the random effects. For a linear mixed-effects model with only random intercepts, we previously defined the intraclass correlation ρ_I as $d_{11}/(d_{11} + \sigma^2)$, where d_{11} is the intercept variability (see Section 6.3.3). It represents the correlation between two repeated measurements within the same subject, that is, $\rho_I = \text{corr}(Y_{ik}, Y_{il})$ for all $k \neq l$. We now have that subgroups in the random-effects population will be unrecognized when the within-subject correlation is small. Note that this again illustrates that the degree of shrinkage increases as the residual variability increases (see Section 7.6). The residual variance used in the above simulation by Verbeke and Lesaffre (1996a) equals $\sigma^2 = 30$. Hence, since the variance of the distribution (7.9) equals 5, the implied intraclass correlation equals $\rho_I = 0.14$.

On the other hand, the covariates Z_i are also important. For example, suppose that Z_i is a vector with elements $t_{i1}, t_{i2}, \ldots, t_{in_i}$. We then have that $Z_i' Z_i$ equals $\sum_{j=1}^{n_i} t_{ij}^2$. Hence, heterogeneity will be more likely to be detected when the b_i represent random slopes in a model for measurements taken at large time points t_{ij} than when the b_i are random intercepts (all

t_{ij} equal to 1). If Z_i contains both random intercepts and random slopes for time points t_{ij}, we have that $Z_i'Z_i$ equals

$$
Z_i'Z_i = \begin{pmatrix} n_i & \sum_{j=1}^{n_i} t_{ij} \\ \sum_{j=1}^{n_i} t_{ij} & \sum_{j=1}^{n_i} t_{ij}^2 \end{pmatrix},
$$

which has two positive eigenvalues λ_1 and λ_2, given by

$$
2\lambda_k = \sum_{j=1}^{n_i} t_{ij}^2 + n_i + (-1)^k \sqrt{\left(\sum_{j=1}^{n_i} t_{ij}^2 - n_i\right)^2 + 4\left(\sum_{j=1}^{n_i} t_{ij}\right)^2}.
$$

In a designed experiment with $Z_i = Z$ for all i, and where the time points t_j are centered, the eigenvalues are $\lambda_2 = \sum_{j=1}^{n} t_j^2$ and $\lambda_1 = n_i$, or vice versa. So, if we are interested in detecting subgroups in the random-effects population, we should take as many measurements as possible, at the beginning and at the end of the study (maximal spread of the time points).

In many cases, one is not especially interested in detecting subgroups in the random-effects distribution. However, the above examples clearly illustrate that the EB estimates may be highly affected by their normality assumption. Hence, if there is any interest at all in the distribution of the random effects b_i, one should explore whether the assumed Gaussian distribution is appropriate. This will be discussed in Section 7.8.4. The special case of the detection of heterogeneity in the random-effects population will be handled in detail in Chapter 12.

7.8.3 Impact on the Estimation of the Marginal Model

In contrast to our results in the previous section on the estimation of the random effects, estimation of the vector $\boldsymbol{\theta}$ of parameters in the marginal model is very robust with respect to misspecification of the random-effects distribution. We refer to Section 12.7 for an illustration of this property. Using simulations and the analysis of a real data set, Butler and Louis (1992) have shown that wrongly specifying the random-effects distribution of univariate random effects has little effect on the fixed-effects estimates as well as on the estimates for the residual variance and the variance of the random effects. No evidence was found for any inconsistencies among these estimators. However, it was shown that the standard errors of all parameter estimators need correction in order to get valid inferences. Using theory on maximum likelihood estimation in misspecified models (White 1980, 1982), Verbeke (1995), and Verbeke and Lesaffre (1994, 1996b) extended this to the general model (3.8). Let $N \times A_N(\boldsymbol{\theta})$ be minus the matrix of second-order derivatives of the log-likelihood function with respect to the elements of $\boldsymbol{\theta}$, and let $N \times B_N(\boldsymbol{\theta})$ be the matrix with cross-products of first-order derivatives of the log-likelihood function, also with respect to $\boldsymbol{\theta}$. Their estimated versions, obtained from replacing $\boldsymbol{\theta}$ by its MLE are denoted by $\widehat{A}_N$ and $\widehat{B}_N$, respectively. Verbeke (1995) and Verbeke and Lesaffre (1996b) then prove that, under general regularity conditions, the MLE $\widehat{\boldsymbol{\theta}}$ of $\boldsymbol{\theta}$ is asymptotically normally distributed with mean $\boldsymbol{\theta}$ and with asymptotic covariance matrix

$$\widehat{A}_N^{-1}\,\widehat{B}_N\,\widehat{A}_N^{-1}/N, \tag{7.10}$$

for $N \to \infty$.

It can easily be seen that the covariance matrix obtained from replacing $\boldsymbol{\alpha}$ in (6.3) by its MLE equals

$$\mathrm{var}(\widehat{\boldsymbol{\beta}}) = \left(\sum_{i=1}^{N} X_i' V_i^{-1}(\widehat{\boldsymbol{\alpha}}) X_i \right)^{-1} = \widehat{A}_{N,11}^{-1}/N, \tag{7.11}$$

where $\widehat{A}_{N,11}$ is the leading block in $\widehat{A}_N$, corresponding to the fixed effects. Hence, we have that the asymptotic covariance matrix for $\widehat{\boldsymbol{\beta}}$, obtained from (7.10), adds extra variability to the "naive" estimate (7.11), by taking into account the estimation of the variance components, but it also corrects for possible misspecification of the random-effects distribution. Note also that $\widehat{A}_N^{-1}\widehat{B}_N\widehat{A}_N^{-1}/N$ is of the same form as the so-called "information sandwich" estimator for the asymptotic covariance matrix of fixed effects, estimated with quasilikelihood methods (see Section 6.2.4 and, e.g., Liang and Zeger 1986). However, the above asymptotic result relates to both the fixed effects and the parameters in the "working correlation" model, and the model is incorrectly specified only through the random-effects distribution, whereas the covariance structure is assumed to be correct.

In practice, the asymptotic covariance matrix of MLEs is usually estimated by the inverse Fisher information matrix. However, this is only valid under the assumed model. Verbeke and Lesaffre (1997a) performed extensive simulations to compare this uncorrected covariance matrix with the above sandwich estimator, which corrects for possible non-normality of the random effects. In general, they conclude that, for the fixed effects, the corrected and uncorrected standard errors are very similar. This is in agreement with the results of Sharples and Breslow (1992), who showed that, for correlated binary data, the sandwich estimator for the covariance matrix of fixed effects is almost as efficient as the uncorrected model-based estimator, even under the correct model.

For the random components on the other hand, and more specifically for the elements in D, this is only true under the correct model (normal random effects). When the random effects are not normally distributed, the corrected standard errors are clearly superior to the uncorrected ones. In some cases, the correction enlarges the standard errors to get confidence levels closer to the pursued level. In other cases, the correction results in smaller standard errors protecting against too conservative confidence intervals.

Verbeke and Lesaffre (1997a) calculated the corrected and uncorrected standard errors for all parameters in model (3.10). The ratio of the corrected over the uncorrected standard errors was between 0.52 and 1.72 for all parameters, whereas the same ratio could be shown to be between 0.21 and 2.76 for any linear combination $\lambda'\theta$ of the parameters. Hence, the uncorrected standard errors could be up to five times too large, and almost three times too small, when compared to the standard errors obtained after correcting for possible non-normality of the random effects b_i.

7.8.4   Checking the Normality Assumption

The results presented in the previous section suggest that if interest is only in inference for the marginal model, and especially if interest is only in inference for the fixed effects, valid inferences are obtained even when the random effects have been incorrectly assumed to be normally distributed. This is in strong contrast with the results discussed in Section 7.8.2, showing that the EB estimates may heavily depend on their distributional assumptions. This calls for methods to check these underlying assumptions.

Lange and Ryan (1989) have proposed to check the normality assumption for the random effects based on weighted normal quantile plots of standardized linear combinations

$$v_i = \frac{c'\widehat{b_i}}{\sqrt{c'\mathrm{var}(\widehat{b_i})c}},$$

of the estimates $\widehat{b_i}$. However, since the v_i are functions of the random effects b_i as well as of the error terms ε_i, these normal quantile plots can only indicate that the v_i do not have the distribution one expects under the assumed model, but the plots cannot differentiate a wrong distributional assumption for the random effects or the error terms from a wrong choice of covariates.

This suggests that non-normality of the random effects can only be detected by comparing the results obtained under the normality assumption with results obtained from fitting a linear mixed model with relaxed distributional assumptions for the random effects. Verbeke (1995) and Verbeke and Lesaffre (1996a, 1997b) therefore propose to extend the general linear

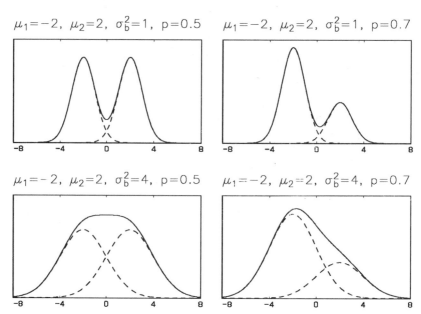

FIGURE 7.5. *Density functions of mixtures $pN(\mu_1, \sigma_b^2) + (1-p)N(\mu_2, \sigma_b^2)$ of two normal distributions, for varying values for p and σ_b^2. The dashed lines represents the densities of the normal components; the solid line represents the density of the mixture.*

mixed model (3.8) by allowing the b_i to be sampled from a mixture of g normal distributions with equal covariance matrix, i.e.,

$$b_i \sim \sum_{j=1}^{g} p_j N(\mu_j, D), \qquad (7.12)$$

with $\sum_{j=1}^{g} p_j = 1$, and such that the mean $\sum_{j=1}^{g} p_j \mu_j$ equals zero. As discussed in Section 7.8.2, this extension naturally arises from assuming that there is unobserved heterogeneity in the random-effects population. Each component in the mixture represents a cluster containing a proportion p_j of the total population. The model is therefore called the heterogeneity model and the linear mixed model discussed so far can then be called the homogeneity model. Also, as shown in Figure 7.5, it extends the assumption about the random-effects distribution to a very broad class of distributions: unimodal as well as multimodal, symmetric as well as very skewed. Note that heterogeneity in the random-effects population may occur very often in practice. Whenever a categorical covariate has been omitted as a fixed effect in a linear mixed-effects model, the random effects will follow a mixture of g normal distributions, where g is the number of categories of the missing covariate.

Verbeke and Lesaffre (1996a) considered the number of components g in (7.12) to be known. In practice, several models can be fitted, with increasing values for g, leading to a series of nested models, and testing procedures such as the likelihood ratio test could be used for the comparison of these models. However, as discussed by Ghosh and Sen (1985), testing for the number of components in a finite mixture is seriously complicated by boundary problems similar to the ones discussed in Section 6.3.4 in the context of testing for the need of random effects. In order to briefly highlight the main problems, we consider testing $H_0 : g = 1$ versus $H_A : g = 2$. The null hypothesis can then be expressed as $H_0 : \mu_1 = \mu_2$. However, the same hypothesis is obtained by setting $H_0 : p_1 = 0$ or $H_0 : p_2 = 0$, which clearly illustrates that H_0 is on the boundary of the parameter space, and hence also that the usual regularity conditions for application of the classical maximum likelihood theory are violated. Therefore, simulations are needed to derive the correct null distribution of the LR test statistic. We refer to Verbeke (1995, Section 4.6) for an example, and to McLachlan and Basford (1988, Section 1.10) for an extensive overview of the literature on the use of the LR test in finite mixture problems. Finally, some informal procedures for checking the goodness-of-fit of heterogeneity models will be discussed in Section 12.5. Obviously, these procedures can also be used to explore the plausibility of the normality assumption for the random effects. In practice however, it may be sufficient to fit several heterogeneity models and to explore how increasing g affects the inference for the parameters of interest. We refer to Chapter 12 for details on the definition and the fitting of the heterogeneity model, and for two examples where the heterogeneity model is used for the classification of longitudinal profiles.

As an example, a two-component heterogeneity model was fitted to the data simulated and analyzed in Section 7.8.2. Figure 7.6 shows the EB estimates of the 1000 simulated random intercepts previously shown in Figure 7.3, and obtained under a two-component heterogeneity model. An expression for these estimates will be derived and discussed in Section 12.3. In comparison to the histogram of the EB estimates under the normality assumption (Figure 7.4), the correct random-effects distribution is (much) better reflected. We do not claim that the $\widehat{b_i}$, calculated under the heterogeneity model, perfectly reflect the correct mixture distribution, but, at least, they suggest that the random effects certainly do not follow a normal distribution, as was suggested by the estimates obtained under the normality assumption.

Magder and Zeger (1996) also considered linear mixed models with mixtures of normal distributions as random-effects distribution, but they treated the number g of components as an unknown parameter, to be estimated from the data. In order to avoid that nonsmooth mixture distributions, with many components, would be obtained, they prespecify a lower boundary h

Empirical Bayes estimates

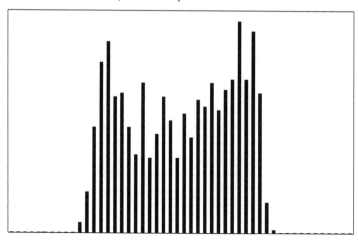

FIGURE 7.6. *Histogram (range* $[-5, 5]$*) of the Empirical Bayes estimates of the random intercepts shown in Figure 7.3, calculated under the assumption that the random effects are drawn from a two-component mixture of normal distributions.*

for the within-component variability measured by $|D|$, the determinant of the covariance matrix of each component in the mixture. In practice, very little difference is expected from the model used by Verbeke and Lesaffre (1996a). Indeed, when a very smooth mixing distribution is required, a large value of h can be specified, which will yield a mixture of a relatively small number of normal distributions.

8
Fitting Linear Mixed Models with SAS

8.1   Introduction

In Chapters 5 and 6, estimation and inference on all parameters in the marginal model (5.1) were discussed. Chapter 7 considered inference for the random effects in the hierarchical model (3.8). At present, among the most flexible commercially available statistical packages is the SAS procedure PROC MIXED (SAS 1992, 1996, 1997). In this chapter, we will therefore explain in full detail how all previously discussed inferences can be obtained with this procedure, using SAS Release 6.12 (SAS 1997). Although this may seem anomalous to many, given the availability of Version 7.0 and higher, it has to be noted that Version 7.0 (SAS 1999) was not available on a commercial basis in 1999, for example, in Europe. For a thorough description of PROC MIXED in SAS Version 7.0, we refer to the on-line manual (SAS 1999). Further, some of the important changes in comparison to Version 6.12 are summarized in Appendix A.

In this chapter, our original model (3.10) for the prostate data will be used as a guiding example. In Section 8.2, the program for fitting the model will be presented, together with some available options. It is by no means our intention to give a full overview of all available statements and options. Instead, we restrict to those statements and options that are, in our experience, most frequently used in the analysis of longitudinal data. When fitting mixed models in other contexts, other statements or options may be more appropriate. We refer to the SAS manuals (SAS 1992, 1996, 1997) and to Littell *et al.* (1996) for more details on the procedure MIXED and for a variety of examples in other contexts.

The SAS output consists of a series of tables, each addressing a specific aspect of the fitted model. These will be discussed in Section 8.3. An alternative SAS procedure, often used in practice for the analysis of longitudinal data, is PROC GLM. In Section 8.6, the most important differences between the procedures GLM and MIXED will be summarized.

8.2   The SAS Program

We now consider fitting the original linear mixed model (3.10) for the prostate data. Let the variable *group* indicate whether a subject is a control (*group* = 1), a BPH case (*group* = 2), a local cancer case (*group* = 3) or a metastatic cancer case (*group* = 4). As in Sections 5.5 and 5.6.1, we express time in decades before diagnosis, rather than years before diagnosis. Further, we define the variable *timeclss* to be equal to *time*. This will enable us to consider time as a continuous covariate and as a classification variable (a factor in the ANOVA terminology) simultaneously. As before, the variable *age* measures the age of the subject at the time of diagnosis. Finally, *id* is a variable containing the subject's identification label, and *lnpsa* is the logarithmic transformation $\ln(1 + x)$ of the original PSA measurements. We can then use the following program to fit model (3.10) and to obtain the inferences described in Chapters 6 and 7:

```
proc mixed data = prostate method = reml asycov asycorr
                         covtest ic;
class id group timeclss;
model lnpsa = group age group*time age*time
            group*time2 age*time2
            / noint solution ddfm = satterth covb chisq;
id id time;
random intercept time time2 / type = un subject = id
                            g gcorr v vcorr solution;
repeated timeclss / type = simple subject = id r rcorr;
contrast 'Final model' age*time 1,
                    group*time  1 0 0 0,
                    age*time2 1,
                    group*time2 1 0 0 0,
                    group*time2 0 1 0 0,
                    group*time2 0  0 1 -1 / chisq;
```

```
estimate 'Diff L/R-BPH, t=5yr' group 0 -4 4 0
                               group*time 0 -2 2 0
                               group*time2 0 0 1 0
                               / cl alpha = 0.05 divisor = 4;
make 'solutionR' out = randeff;
run;
```

Before presenting the results of this analysis, we briefly discuss the statements and options used in the above program.

8.2.1   The PROC MIXED Statement

This statement calls the procedure MIXED and specifies that the data be stored in the SAS data set 'prostate'. If no data set is specified, then the most recently created data set is used. In general, there are two ways of setting up data sets containing repeated measurements. One way is to define a variable for each variable measured and for each time point in the data set at which at least one subject was measured. Each subject then corresponds to exactly one record (one line) in the data set. This setup is convenient when the data are highly balanced, that is, when all measurements are taken at only a few number of time points. However, this approach leads to huge data matrices with many missing values in cases of highly unbalanced data such as the prostate data. The same problem occurs in the presence of missing data or dropout (see Section 17.3). Therefore, the MIXED procedure requires that the data set be structured such that each record corresponds to the measurements available for a subject at only one moment in time. For example, five repeated measurements for individual i are put into five different records. This has the additional advantage that time-varying covariates (such as time) can be easily incorporated into the model. An identification variable id is needed to link measurements to subjects, and a time variable is used to order the repeated measurements within each individual. For example, our prostate cancer data set is set up in the following way:

OBS	ID	LNPSA	TIME	AGE	GROUP
1	1	0.405	1.94	72.4	2
2	1	0.336	1.44	72.4	2
3	1	0.693	1.20	72.4	2
...	..	.....	....	....	.
461	54	0.182	0.46	62.9	1
462	54	0.262	0.25	62.9	1
463	54	0.182	0.00	62.9	1

The option 'method=' specifies the estimation method. In this book, we will always specify 'method=ML' or 'method=REML' requesting ML or REML estimation, respectively. However, it is also possible to use the non-iterative MIVQUE0 method (minimum variance quadratic unbiased estimation), which is used by default to compute starting values for the iterative ML and REML estimation procedures. We refer to the SAS manual (1996) for a treatment of the MIVQUE0 method. If no method is specified, then REML estimation is used by default.

The options 'asycov' and 'asycorr' can be used for printing the asymptotic covariance matrix as well as the associated correlation matrix of the estimators for the variance components in the marginal model. The option 'covtest' requires the printing of the resulting asymptotic standard errors and associated Wald tests for those variance components. Note that these Wald tests are not applicable to all variance components in the model (see discussion in Section 6.3.1). The calculation of the information criteria discussed in Section 6.4 can be requested by adding the option 'ic'.

8.2.2   The CLASS Statement

This statement specifies which variables should be considered as factors. Such classification variables can be either character or numeric. Internally, each of these factors will correspond to a set of dummy variables in the manner described in the SAS manual on linear models (1991, Section 5.5).

8.2.3   The MODEL Statement

The MODEL statement names the response variable (one and only one) and all fixed effects, which determine the X_i matrices. Note that in order to have the same parameterization for the mean structure as model (3.10), no overall intercept (using the 'noint' option) nor overall linear or quadratic time effects should be included into the model, since, otherwise, the mean structure is parameterized using contrasts between the intercepts and slopes of the first three diagnostic groups and those for the last group. Although this would facilitate the testing of group differences (see also Section 8.4), it complicates the interpretation of the parameter estimates.

The 'solution' option is used to request the printing of the estimates for all the fixed effects in the model, together with standard errors, t-statistics, and corresponding p-values for testing their significance (see Section 6.2). When the whole model-based estimated covariance matrix (6.3) for the fixed effects is of interest, it can be obtained by specifying the option 'covb'. The estimation method for the degrees of freedom in the t- and F-approximations needed in tests for fixed-effects estimates (see Section 6.2.2) is specified in the option 'ddfm='. Here, the Satterthwaite approximation was used, but other methods are also available within SAS. When the option 'chisq' is added to the MODEL statement, SAS also provides approximate Wald tests (Section 6.2.1), next to the default t- and F-tests, for all effects specified in the MODEL statement.

8.2.4 The ID Statement

When predicted values are requested with the option 'predmeans' or 'predicted' in the MODEL statement, SAS prints a table with the requested predicted value, the corresponding observed value, and the resulting residual, for each record in the original data set. Although the records are then printed in the same order as they appear in the original data set, it may still be helpful to add columns which help identify the records. This is done via the ID statement. The values of the variables in the ID statement are then printed beside each observed, predicted, and residual value. In our example, we used the identification number of the patient, together with the time point at which a measurement was taken to identify the predictions and residuals which we requested in the MODEL statement.

8.2.5 The RANDOM Statement

This statement is used to define the random effects in the model, that is, the matrices Z_i containing the covariates with subject-specific regression coefficients. Note that when random intercepts are required, this should be specified explicitly, which is in contrast to the MODEL statement where an intercept is included by default.

The 'subject=' option is used to identify the subjects in our data set. Here, 'subject=id' means that all records with the same value for id are assumed to be from the same subject, whereas records with different values for id are assumed to contain independent data. This option also defines the block-diagonality of the matrix Z and of the covariance matrix of b in (5.5). The variable id is permitted to be continuous as well as categorical (specified in the CLASS statement). However, when id is continuous, PROC MIXED considers a record to be from a new subject whenever the value of id is different from the previous record. Hence, one then should first sort the data by the values of id. On the other hand, using a continuous id variable reduces execution times for models with a large number of subjects (manual PROC MIXED).

The 'type=' option specifies the covariance structure D for the random effects b_i. In our example, we specified 'type=un' which corresponds to a general unstructured covariance matrix, i.e., a symmetric positive (semi-) definite $(q \times q)$ matrix D. Many other covariance structures can be specified, some of which are shown in Table 8.1 and Table 8.2. We further refer to the SAS manual (1996) for a complete list of possible structures. Although many structures are available, in longitudinal data analysis, one usually specifies 'type=UN' which does not assume the random-effects covariance matrix to be of any specific form.

Specifying the options 'g' and 'gcorr' requests that the random-effects covariance matrix D as well as its associated correlation matrix are printed, printing blanks for all values that are zero. The options 'v' and 'vcorr' can be used if a printout of the marginal covariance matrix $V_i = Z_i D Z_i' + \Sigma_i$ and the corresponding correlation matrix, respectively, are needed. By default, SAS only prints the covariance and correlation matrices for the first subject in the data set. However, 'v=' and 'vcorr=' can be used to specify the identification numbers of the patients for which the matrix V_i and the associated correlation matrix are needed.

The option 'solution' is needed for the calculation of the empirical Bayes (EB) estimates for the random effects b_i, previously derived and discussed in Chapter 7. The result will be a table containing the EB estimates for the random effects of all subjects included in the analysis. If, for example, scatter plots or histograms of components of these estimates $\widehat{b_i}$ are to be made, then the estimates should be converted to a SAS output data set. This will be discussed in Section 8.2.9.

8.2.6   The REPEATED Statement

The REPEATED statement is used to specify the Σ_i matrices in the mixed model. The repeated effects define the ordering of the repeated measurements within each subject. These effects (in the example, 'timeclss') must be classification variables, which is why we needed two versions of our time variable: a continuous version 'time' needed in the MODEL statement as well as in the RANDOM statement, and a classification version 'timeclss' needed in the REPEATED statement. Usually, one will specify only one repeated effect. Its levels should then be different for each observation within a subject. If not, PROC MIXED constructs identical rows in Σ_i corresponding to the observations with the same level, yielding a singular Σ_i and an infinite likelihood. If the data are ordered similarly for each subject, and any missing data are denoted with missing values, then specifying a repeated effect is not necessary. In this case, the name 'DIAG' appears as the repeated effect in the printed output. Note that this is not the same

TABLE 8.1. *Overview of frequently used covariance structures which can be specified in the RANDOM and REPEATED statements of the SAS procedure MIXED. The σ parameters are used to denote variances and covariances, whereas the ρ parameters are used for correlations.*

Structure	Example
Unstructured type=UN	$\begin{pmatrix} \sigma_1^2 & \sigma_{12} & \sigma_{13} \\ \sigma_{12} & \sigma_2^2 & \sigma_{23} \\ \sigma_{13} & \sigma_{23} & \sigma_3^2 \end{pmatrix}$
Simple Variance Components type=SIMPLE type=VC	$\begin{pmatrix} \sigma^2 & 0 & 0 \\ 0 & \sigma^2 & 0 \\ 0 & 0 & \sigma^2 \end{pmatrix}^{(1)}$ or $\begin{pmatrix} \sigma_1^2 & 0 & 0 \\ 0 & \sigma_2^2 & 0 \\ 0 & 0 & \sigma_3^2 \end{pmatrix}^{(2)}$
Compound symmetry type=CS	$\begin{pmatrix} \sigma_1^2 + \sigma^2 & \sigma_1^2 & \sigma_1^2 \\ \sigma_1^2 & \sigma_1^2 + \sigma^2 & \sigma_1^2 \\ \sigma_1^2 & \sigma_1^2 & \sigma_1^2 + \sigma^2 \end{pmatrix}$
Banded type=UN(2)	$\begin{pmatrix} \sigma_1^2 & \sigma_{12} & 0 \\ \sigma_{12} & \sigma_2^2 & \sigma_{23} \\ 0 & \sigma_{23} & \sigma_3^2 \end{pmatrix}$
First-order autoregressive type=AR(1)	$\begin{pmatrix} \sigma^2 & \rho\sigma^2 & \rho^2\sigma^2 \\ \rho\sigma^2 & \sigma^2 & \rho\sigma^2 \\ \rho^2\sigma^2 & \rho\sigma^2 & \sigma^2 \end{pmatrix}$
Toeplitz type=TOEP	$\begin{pmatrix} \sigma^2 & \sigma_{12} & \sigma_{13} \\ \sigma_{12} & \sigma^2 & \sigma_{12} \\ \sigma_{13} & \sigma_{12} & \sigma^2 \end{pmatrix}$
Toeplitz (1) type=Toep(1)	$\begin{pmatrix} \sigma^2 & 0 & 0 \\ 0 & \sigma^2 & 0 \\ 0 & 0 & \sigma^2 \end{pmatrix}$
Heterogeneous compound symmetry type=CSH	$\begin{pmatrix} \sigma_1^2 & \rho\sigma_1\sigma_2 & \rho\sigma_1\sigma_3 \\ \rho\sigma_1\sigma_2 & \sigma_2^2 & \rho\sigma_2\sigma_3 \\ \rho\sigma_1\sigma_3 & \rho\sigma_2\sigma_3 & \sigma_3^2 \end{pmatrix}$
Heterogeneous first-order autoregressive type=ARH(1)	$\begin{pmatrix} \sigma_1^2 & \rho\sigma_1\sigma_2 & \rho^2\sigma_1\sigma_3 \\ \rho\sigma_1\sigma_2 & \sigma_2^2 & \rho\sigma_2\sigma_3 \\ \rho^2\sigma_1\sigma_3 & \rho\sigma_2\sigma_3 & \sigma_3^2 \end{pmatrix}$
Heterogeneous Toeplitz type=TOEPH	$\begin{pmatrix} \sigma_1^2 & \rho_1\sigma_1\sigma_2 & \rho_2\sigma_1\sigma_3 \\ \rho_1\sigma_1\sigma_2 & \sigma_2^2 & \rho_1\sigma_2\sigma_3 \\ \rho_2\sigma_1\sigma_3 & \rho_1\sigma_2\sigma_3 & \sigma_3^2 \end{pmatrix}$

(1) Example: repeated timeclss / type=simple subject=id;
(2) Example: random intercept time time2 / type=simple subject=id;

TABLE 8.2. *Overview of frequently used (stationary) spatial covariance structures, which can be specified in the RANDOM and REPEATED statements of the SAS procedure MIXED. The correlations are positive decreasing functions of the Euclidean distances d_{ij} between the observations. The coordinates of the observations used to calculate these distances are given by a set of variables, the names of which are specified in the list 'list'. The variance is denoted by σ^2, and ρ defines how fast the correlations decrease as functions of the d_{ij}.*

Structure	Example
Power type=SP(POW)(*list*)	$\sigma^2 \begin{pmatrix} 1 & \rho^{d_{12}} & \rho^{d_{13}} \\ \rho^{d_{12}} & 1 & \rho^{d_{23}} \\ \rho^{d_{13}} & \rho^{d_{23}} & 1 \end{pmatrix}$
Exponential type=SP(EXP)(*list*)	$\sigma^2 \begin{pmatrix} 1 & \exp(-d_{12}/\rho) & \exp(-d_{13}/\rho) \\ \exp(-d_{12}/\rho) & 1 & \exp(-d_{23}/\rho) \\ \exp(-d_{13}/\rho) & \exp(-d_{23}/\rho) & 1 \end{pmatrix}$
Gaussian type=SP(GAU)(*list*)	$\sigma^2 \begin{pmatrix} 1 & \exp(-d_{12}^2/\rho^2) & \exp(-d_{13}^2/\rho^2) \\ \exp(-d_{12}^2/\rho^2) & 1 & \exp(-d_{23}^2/\rho^2) \\ \exp(-d_{13}^2/\rho^2) & \exp(-d_{23}^2/\rho^2) & 1 \end{pmatrix}$

as completely omitting the REPEATED statement, since this would not allow to specify parametric forms for Σ_i other than the simple form $\sigma^2 I_{n_i}$. The options for the REPEATED statement are similar to those for the RANDOM statement.

For example, the option 'subject=' identifies the subjects in the data set, and complete independence is assumed across subjects. It therefore defines the block-diagonality of the covariance matrix of ε in (5.5). With respect to the variable *id*, the same remarks hold as the ones stated in our description of the RANDOM statement. Although this is strictly speaking not required, the RANDOM and REPEATED statement often have the same option 'subject=*id*', as was the case in our example.

Further, the 'type=' option specifies the covariance structure Σ_i for the error components ε_i. All covariance structures previously described for the RANDOM statement can also be specified here. Very often, one selects 'type=simple' which corresponds to the most simple covariance structure $\Sigma_i = \sigma^2 I_{n_i}$. Finally, if no REPEATED statement is used, PROC MIXED automatically fits such a 'simple' covariance structure for the residual components.

Finally, specifying the options 'r' and 'rcorr' requests that the residual co-variance matrix Σ_i as well as its associated correlation matrix be printed. As for the 'v' and 'vcorr' options in the RANDOM statement, SAS only prints by default the covariance and correlation matrices for the first sub-ject in the data set. However 'r=' and 'rcorr=' can be used to specify the identification numbers of the subjects for which the matrix Σ_i and the associated correlation matrix are needed.

8.2.7 The CONTRAST Statement

The contrast statement allows testing general linear hypotheses of the form (6.4). In our program, we have shown how to test whether the original model (3.10) for the prostate data can be reduced to the final model (6.8) [see hypothesis (6.7) obtained in Section 6.2.3].

Since it is possible to test several hypotheses at once (specifying several CONTRAST statements), a label is needed for each contrast in order to identify them in the output. This label can be up to 20 characters long and must be enclosed in single quotes. In the above example, the label was 'Final model'. Following the label, one needs to specify the linear combinations in the hypothesis (i.e., the rows in the matrix L), separated by commas. Each row in L is represented by a list of the effects, specified in the MODEL statement, followed by the appropriate elements in the corresponding row of the matrix L. Effects for which the corresponding parameters only get zero weight in the linear combination may be omitted. For example, the last row in (6.7) only gives nonzero weights to parameters in $\boldsymbol{\beta}$ assigned to the interaction of group by the quadratic time effect. The first two parameters assigned to this effect (β_{12} and β_{13}) get zero weight and the other two parameters (β_{14} and β_{15}) get weights 1 and -1, respectively. This is represented in the last row of the above CONTRAST statement. A similar argument leads to the other rows.

By default, the specified hypothesis is tested based on an F-test (see Sec-tion 6.2.2). The option 'chisq' requests that also an approximate Wald test (see Section 6.2.1) is performed.

8.2.8 The ESTIMATE Statement

The ESTIMATE statement allows estimation and testing of linear combi-nations of the fixed effects. In the above SAS program, we illustrate the estimation of the average difference in $\ln(1 + PSA)$, 5 years prior to diagnosis, between the local cancer cases and the BPH cases, that is, we will estimate the linear combination (6.9), specified in Section 6.2.3.

As for the CONTRAST statement, several ESTIMATE statements can be specified, all having their own label. In our example, the given label equals 'Diff L/R-BPH, t=5yr'. Linear combinations to be estimated are specified in exactly the same way as they would be specified in a CONTRAST statement (see Section 8.2.7). The only difference in output is that an ESTIMATE statement also provides point estimates as well as confidence intervals, whereas a CONTRAST statement only yields approximate F-tests and Wald tests. The options 'cl' and 'alpha=0.05' are used to request that an approximate t-type 95% confidence interval is calculated for the linear combination specified. Finally, the option 'divisor=4' requests division of all coefficients in the specified linear combination by 4.

8.2.9   The MAKE Statement

SAS allows conversion of any table produced by PROC MIXED to a SAS data file. This is done via a MAKE statement. If several tables need to be converted, several MAKE statements can be used. In our example, we are converting the empirical Bayes (EB) estimates which have been requested by the option 'solution' in the RANDOM statement (see Section 8.2.5). This is especially convenient for producing histograms and scatter plots of components of the EB estimates $\widehat{b}_i$, as the ones shown in Figure 7.1. Each table is given a label which can be found in the SAS manuals and which needs to be specified in the MAKE statement. For example, the table with EB estimates is labeled 'solutionR'. The option 'out=randeff' specifies the name of the SAS data set which will contain the requested information. In practice, one will often add the option 'noprint', which avoids printing of the table in the SAS output. This option is particularly useful for large tables such as the one containing the EB estimates.

It should be emphasized that, from Version 7.0 on, the MAKE statement is replaced by the integrated ODS (output delivery system). In Version 7.0, the MAKE statement is still supported, but it is envisaged that this will no longer be the case in later versions. We refer to Appendix A for more details and for an example.

8.2.10   Some Additional Statements and Options

Our program on p. 94 will provide model-based inferences for all parameters in the original linear mixed model (3.10). However, SAS also allows robust inference for all fixed effects in the model (see Section 6.2.4). This can be requested by adding the option 'empirical' to the PROC MIXED statement. All reported standard errors and inferences for the fixed effects will then be based on the robust estimate for the covariance matrix rather than the naive one.

As discussed in Section 5.6, convergence problems may often be avoided by using the Fisher scoring estimation method rather than the default Newton-Raphson-based procedures. This can be required by adding the option 'scoring' to the PROC MIXED statement. The expected Hessian matrix instead of the observed Hessian is then also used to compute approximate standard errors for the covariance parameters. However, as will be discussed in Chapter 21, this may yield invalid inferences when some of the response measurements are missing. In practice, one can start the iteration process using the Fisher scoring algorithm and proceed with Newton-Raphson. This will then still yield inferences based on the observed Hessian rather than the expected one. This can be done in SAS by specifying the option 'scoring=a', where a is the required number of Fisher scoring steps.

Another way of avoiding convergence problems is to specify good starting values for the parameters in the marginal covariance structure. This can be done by adding a PARMS statement to the model. Single values, and also sequences of values can be specified. In the latter case, a grid search is performed and the best point on the grid is used as starting value for the iterative estimation procedure. SAS also allows to fix some of these parameters to known values (option 'eqcons'), which are then no longer included in the iterative estimation procedure.

When predictions are of interest, options 'predmeans' and 'predicted' can be added to the MODEL statement, which require the calculation of predicted means and predicted subject-specific profiles, respectively. The predicted means are given by $X_i\widehat{\beta}$, whereas predicted subject-specific profiles are obtained from calculating $X_i\widehat{\beta}+Z_i\widehat{b_i}$. SAS automatically also calculates residuals defined as the difference between the predicted (mean or subject-specific) and observed values. The options 'predmeans' and 'predicted' are particularly useful when graphs as Figure 7.2 are to be prepared.

Similar to the ESTIMATE statement, one can add the options 'cl' and 'alpha=' to the MODEL statement to request for the construction of t-type confidence limits for each of the fixed-effect parameters.

As discussed in Sections 8.2.5 and 8.2.6, the RANDOM and REPEATED statements specify the structure in the covariance matrices D and Σ_i. The option 'group=' can be added to both statements to specify heterogeneity in these covariance structures. All observations having the same level of the specified effect will have the same covariance parameters. Each new level of the specified effect will produce a new set of covariance parameters with the same structure as specified in the 'type=' option.

In Section 3.3.4, a general but flexible model was presented for the residual covariance Σ_i, which assumed that the residual component ε_i can be decomposed as $\varepsilon_i = \varepsilon_{(1)i} + \varepsilon_{(2)i}$, in which $\varepsilon_{(2)i}$ is a component of serial correlation and where $\varepsilon_{(1)i}$ represents pure measurement error. Such models can also easily be fitted in PROC MIXED. One then has to specify the required covariance structure of $\varepsilon_{(2)i}$ in the 'type=' option of the REPEATED statement, whereas the measurement error component is obtained by adding the option 'local' to the same REPEATED statement. This will be illustrated in Section 9.4.

The CONTRAST and ESTIMATE statements in our program on p. 94 illustrate how linear combinations of the fixed effects in the model can be estimated and tested. However, SAS also allows one to perform subject-specific inferences, which can be obtained by including random effects in the linear combinations. We refer to the SAS manuals and to Littell *et al.* (1996) for examples and more details.

Finally, as discussed in Section 5.4, SAS maximizes likelihood functions under the restriction that all diagonal elements of the random-effects covariance matrix D as well as all diagonal elements of the residual covariance matrices Σ_i are positive. However, an example was given in Section 5.6.2, which shows that, in practice, one may want to remove these constraints on the parameter estimates. This can be done in SAS by adding the option 'nobound' to the PROC MIXED statement or to the PARMS statement.

8.3   The SAS Output

In the following sections, we will discuss the different parts of the SAS output obtained from fitting the model on p. 94 to the prostate data.

8.3.1   Information on the Iteration Procedure

First of all, an 'Estimation Iteration History' table is given, describing the iteration history, that is, the process of maximizing the likelihood function (or equivalently, the log-likelihood function). This table is of the following form:

REML Estimation Iteration History

Iteration	Evaluations	Objective	Criterion
0	1	-259.0577593	
1	2	-753.2423823	0.00962100
2	1	-757.9085275	0.00444385
.	.	. . . .	. . . .
6	1	-760.8988784	0.00000003
7	1	-760.8988902	0.00000000

Convergence criteria met.

The objective function is, apart from a constant which does not depend on the parameters, minus twice the log-likelihood function. In the case of REML estimation, the exact relation between L_{REML} and the objective function $\mathrm{OF}_{\mathrm{REML}}$ is given by

$$-2\ln\left(L_{\mathrm{REML}}(\boldsymbol{\theta})\right) \quad = \quad (n-p)\,\ln(2\pi) + \mathrm{OF}_{\mathrm{REML}}(\boldsymbol{\theta}). \qquad (8.1)$$

For ML estimation, the above equation becomes

$$-2\ln\left(L_{\mathrm{ML}}(\boldsymbol{\theta})\right) \quad = \quad n\,\ln(2\pi) + \mathrm{OF}_{\mathrm{ML}}(\boldsymbol{\theta}).$$

Hence, our final parameter estimates are those which minimize this objective function. The reported number of evaluations is the number of times the objective function has been evaluated during each iteration. In the 'Criterion' column, a measure of convergence is given, where a value equal to zero indicates that the iterative estimation procedure has converged. In practice, the procedure is considered to have converged whenever the convergence criterion is smaller than a so-called tolerance number which is set equal to 10^{-8} by default. Unless specified otherwise, SAS uses the relative Hessian convergence criterion defined as $|g_k'H_k^{-1}g_k|/|f_k|$, where f_k is the value of the objective function at iteration k, g_k is the gradient (vector of first-order derivatives) of f_k, and H_k is the Hessian (matrix of second-order derivatives) of f_k. Other possible choices are the relative function convergence criterion and the relative gradient convergence criterion defined as $|f_k - f_{k-1}|/|f_k|$ and $\max_j |g_{kj}|/|f_k|$, respectively, where g_{kj} is the jth element in g_k.

8.3.2 Information on the Model Fit

The 'Model Fitting Information' table shows the following additional information:

Model Fitting Information for LNPSA

Description	Value
Observations	463.0000
Res Log Likelihood	-31.2350
Akaike's Information Criterion	-38.2350
Schwarz's Bayesian Criterion	-52.6018
-2 Res Log Likelihood	62.4700
Null Model LRT Chi-Square	501.8411
Null Model LRT DF	6.0000
Null Model LRT P-Value	0.0000

In our example, $n = \sum_{i=1}^{N} n_i = 463$ observations were used to calculate the parameter estimates. The value of the REML log-likelihood function evaluated at the REML estimates is reported as 'Res Log Likelihood', and similarly for minus twice the maximized log-likelihood function. Note that the log-likelihood can also easily be calculated from expression (8.1). In our example, this becomes $-[(463 - 15)\ln(2\pi) - 760.8989]/2 = -31.2350$. Note that this value was already reported in Table 5.1. It is also used for the calculation of the reported Akaike and Schwarz information criteria which we defined in Section 6.4. The number of parameters used in the calculation equals the number of variance components in the model. Hence, the criteria reported here should only be used to compare models with the same mean structure, but with different covariance structures, even when maximum likelihood would be used for model fitting. We will come back to this in Section 8.3.3. In our example, we have AIC $= -31.2350 - 7 = -38.2350$ and SBC $= -31.2350 - 3.5\ln(463 - 15) = -52.6018$, respectively.

The 'Null Model LRT Chi-Square' value is -2 times the log-likelihood from the null model minus -2 times the log-likelihood from the fitted model, where the null model is the one with the same fixed effects as the actual model, but without any random effects, and with $\Sigma_i = \sigma^2 I_{n_i}$. This statistic is then compared to a χ^2-distribution with degrees of freedom equal to the number of variance components minus 1, and the reported p-value is the upper tail area from this distribution. It is suggested that this p-value can be used to test whether or not there is any need at all for modeling the covariance structure of the data. However, as discussed in Section 6.3.4, the obtained p-value is not valid since the classical maximum likelihood theory from which it results does not hold due to boundary problems in the parameter space. Hence, it is, in general, not recommended to interpret the reported p-value in any such way.

8.3.3   Information Criteria

Since the option 'ic' was specified in the PROC MIXED statement, all four information criteria previously defined in Section 6.4 were calculated. The results are summarized in the following table:

Information Criteria

Better	Parms	q	p	AIC	HQIC	BIC	CAIC
Larger	7	7	0	-38.2	-43.9	-52.6	-56.1
Larger	22	7	15	-53.2	-71.0	-98.4	-109.4
Smaller	7	7	0	76.5	87.8	105.2	112.2
Smaller	22	7	15	106.5	142.1	196.8	218.8

The first two rows are the information criteria as we defined them in Section 6.4. When comparing different models, the model with the largest value of AIC, SBC, HQIC, or CAIC is deemed best. The last two rows are the versions defined based on minus twice the maximized log-likelihood value rather than on the log-likelihood itself. They equal minus twice the corresponding value in the first or second row. For these versions, small values of AIC, SBC, HQIC, or CAIC are considered as good. The reported parameters p and q in the above output table correspond to the number of fixed effects and the number of variance components that is taken into account in the computation of the information criteria. Note that, since the information criteria reported in the first and third rows do not take into account the number of parameters in the mean structure (p set equal to zero), these criteria should not be used to compare models with different mean structures. Moreover, as explained in Section 6.2.5 and Section 6.4, they are only fully interpretable when models are fitted with the maximum likelihood procedure, rather than the restricted maximum likelihood procedure.

8.3.4   Inference for the Variance Components

First, a table labeled 'Covariance Parameter Estimates (REML)' is given which contains parameter estimates for all variance components in the model. Since we specified the option 'covtest', estimated standard errors and approximate Wald tests are also given:

Covariance Parameter Estimates (REML)

Cov Parm	Subject	Estimate	Std Error	Z	Pr > |Z|
UN(1,1)	ID	0.4517	0.0976	4.63	0.0001
UN(2,1)	ID	-0.5178	0.1355	-3.82	0.0001
UN(2,2)	ID	0.9153	0.2297	3.98	0.0001
UN(3,1)	ID	0.1625	0.0529	3.07	0.0021
UN(3,2)	ID	-0.3356	0.0949	-3.53	0.0004
UN(3,3)	ID	0.1308	0.0409	3.19	0.0014
TIMECLSS	ID	0.0281	0.0022	12.41	0.0001

The entry UN(i,j) corresponds to the element d_{ij} of the random-effects covariance matrix D. The entry TIMECLSS reports the inference results for the residual variance σ^2. Note that the p-values reported for all UN(i,i) entries as well as for the entry TIMECLSS should be ignored since the classical maximum likelihood theory from which they result does not hold due to boundary problems in the parameter space (see our discussion in Section 6.3.1). Note that the estimates and standard errors were already reported in Table 5.1.

Since we specified the options 'asycov' and 'asycorr' in the PROC MIXED statement, we also get a printout of the estimated asymptotic covariance matrix and associated correlation matrix for the above parameter estimates. These are given by

Asymptotic Covariance Matrix of Estimates

Cov Parm	Row	COVP1	COVP2	COVP3	COVP4	COVP5	COVP6	COVP7
UN(1,1)	1	0.0095	-0.0116	0.0145	0.0039	-0.0049	0.0017	-0.0000
UN(2,1)	2	-0.0116	0.0183	-0.0277	-0.0069	0.0103	-0.0038	0.0000
UN(2,2)	3	0.0145	-0.0277	0.0527	0.0112	-0.0212	0.0086	-0.0000
UN(3,1)	4	0.0039	-0.0069	0.0112	0.0027	-0.0044	0.0017	-0.0000
UN(3,2)	5	-0.0049	0.0103	-0.0212	-0.0044	0.0090	-0.0038	0.0000
UN(3,3)	6	0.0017	-0.0038	0.0086	0.0017	-0.0038	0.0016	-0.0000
TIMECLSS	7	-0.0000	0.0000	-0.0000	-0.0000	0.0000	-0.0000	0.0000

and

Asymptotic Correlation Matrix of Estimates

Cov Parm	Row	COVP1	COVP2	COVP3	COVP4	COVP5	COVP6	COVP7
UN(1,1)	1	1.0000	-0.8830	0.6501	0.7594	-0.5383	0.4354	-0.0607
UN(2,1)	2	-0.8830	1.0000	-0.8917	-0.9634	0.8060	-0.7020	0.0983
UN(2,2)	3	0.6501	-0.8917	1.0000	0.9218	-0.9760	0.9168	-0.1325
UN(3,1)	4	0.7594	-0.9634	0.9218	1.0000	-0.8813	0.8038	-0.1291
UN(3,2)	5	-0.5383	0.8060	-0.9760	-0.8813	1.0000	-0.9806	0.1592
UN(3,3)	6	0.4354	-0.7020	0.9168	0.8038	-0.9806	1.0000	-0.1798
TIMECLSS	7	-0.0607	0.0983	-0.1325	-0.1291	0.1592	-0.1798	1.0000

respectively.

For other covariance structures, the table 'Covariance Parameter Estimates (REML)' may look slightly different. Therefore it is useful in practice to print out the complete covariance matrix derived from the estimated variance components. The resulting estimate for the random-effects covariance matrix D and its associated correlation matrix are reported as

G Matrix

Effect	ID	Row	COL1	COL2	COL3
INTERCEPT	1	1	0.4517	-0.5178	0.1625
TIME	1	2	-0.5178	0.9153	-0.3356
TIME2	1	3	0.1625	-0.3356	0.1308

and

G Correlation Matrix

Effect	ID	Row	COL1	COL2	COL3
INTERCEPT	1	1	1.0000	-0.8052	0.6685
TIME	1	2	-0.8052	1.0000	-0.9700
TIME2	1	3	0.6685	-0.9700	1.0000

respectively, and were obtained by specifying the options 'g' and 'gcorr' in the RANDOM statement. The estimate for the residual covariance matrix Σ_1 for the first subject in the data set, as well as the associated correlation matrix are reported as

R Matrix for ID 1

Row	COL1	COL2	COL3	COL4	COL5	COL6	COL7	COL8
1	0.0282							
2		0.0282						
3			0.0282					
4				0.0282				
5					0.0282			
6						0.0282		
7							0.0282	
8								0.0282

and

R Correlation Matrix for ID 1

Row	COL1	COL2	COL3	COL4	COL5	COL6	COL7	COL8
1	1.0000							
2		1.0000						
3			1.0000					
4				1.0000				
5					1.0000			
6						1.0000		
7							1.0000	
8								1.0000

respectively, and were obtained by specifying the options 'r' and 'rcorr' in the REPEATED statement. Note that zero entries are left blank. Finally, combining the estimate for D with the estimate for Σ_1, an estimate is obtained for the marginal covariance matrix $V_1 = Z_1 D Z_1' + \Sigma_1$, as well as for the associated correlation matrix, of the first subject in the data set. These are reported as

V Matrix for ID 1

Row	COL1	COL2	COL3	COL4	COL5	COL6	COL7	COL8
1	0.4133	0.2959	0.2021	0.1460	0.1023	0.0709	0.0501	0.0585
2	0.2959	0.2684	0.1800	0.1427	0.1121	0.0882	0.0691	0.0576
3	0.2021	0.1800	0.1821	0.1357	0.1187	0.1029	0.0864	0.0570
4	0.1460	0.1427	0.1357	0.1566	0.1194	0.1086	0.0942	0.0569
5	0.1023	0.1121	0.1187	0.1194	0.1446	0.1098	0.0977	0.0571
6	0.0709	0.0882	0.1029	0.1086	0.1098	0.1345	0.0968	0.0577
7	0.0501	0.0691	0.0864	0.0942	0.0977	0.0968	0.1186	0.0587
8	0.0585	0.0576	0.0570	0.0569	0.0571	0.0577	0.0587	0.0905

and

V Correlation Matrix for ID 1

Row	COL1	COL2	COL3	COL4	COL5	COL6	COL7	COL8
1	1.0000	0.8885	0.7366	0.5740	0.4186	0.3009	0.2265	0.3025
2	0.8885	1.0000	0.8143	0.6960	0.5689	0.4642	0.3875	0.3696
3	0.7366	0.8143	1.0000	0.8035	0.7316	0.6577	0.5879	0.4440
4	0.5740	0.6960	0.8035	1.0000	0.7934	0.7485	0.6914	0.4781
5	0.4186	0.5689	0.7316	0.7934	1.0000	0.7870	0.7460	0.4996
6	0.3009	0.4642	0.6577	0.7485	0.7870	1.0000	0.7663	0.5231
7	0.2265	0.3875	0.5879	0.6914	0.7460	0.7663	1.0000	0.5672
8	0.3025	0.3696	0.4440	0.4781	0.4996	0.5231	0.5672	1.0000

respectively, and were obtained by specifying the options 'v' and 'vcorr' in the RANDOM statement.

8.3.5 Inference for the Fixed Effects

Due to the specification of the 'solution' option in the MODEL statement, we obtain a table labeled 'Solution for Fixed Effects,' which contains the parameter estimates, estimated standard errors, and approximate t-tests for all fixed effects in the model:

Solution for Fixed Effects

Effect	GROUP	Estimate	Std Error	DF	t	Pr > |t|
GROUP	1	-1.0984	0.9763	47.9	-1.13	0.2662
GROUP	2	-0.5228	1.0895	48	-0.48	0.6335
GROUP	3	0.2964	1.0587	48	0.28	0.7807
GROUP	4	1.5493	1.0856	47.6	1.43	0.1600
AGEDIAG		0.0265	0.0142	47.9	1.87	0.0683
TIME*GROUP	1	0.5680	1.4725	42.7	0.39	0.7016
TIME*GROUP	2	0.3956	1.6376	42.3	0.24	0.8103
TIME*GROUP	3	-1.0359	1.5927	42.4	-0.65	0.5190
TIME*GROUP	4	-1.6049	1.6257	41.6	-0.99	0.3293
AGEDIAG*TIME		-0.0111	0.0214	42.3	-0.52	0.6050
TIME2*GROUP	1	-0.1295	0.6100	32.9	-0.21	0.8332
TIME2*GROUP	2	-0.1584	0.6723	31.9	-0.24	0.8152
TIME2*GROUP	3	0.3419	0.6562	32.1	0.52	0.6060
TIME2*GROUP	4	0.3950	0.6660	31.1	0.59	0.5574
AGEDIAG*TIME2		0.0022	0.0088	32	0.26	0.7997

TABLE 8.3. *Prostate Data. Overview of the hypotheses corresponding to the tests specified in the table labeled 'Tests of Fixed Effects.'*

Source	Null hypothesis
Group	$H_1 : \beta_2 = \beta_3 = \beta_4 = \beta_5 = 0$
Age	$H_2 : \beta_1 = 0$
Time*group	$H_3 : \beta_7 = \beta_8 = \beta_9 = \beta_{10} = 0$
Age*group	$H_4 : \beta_6 = 0$
Time2*group	$H_5 : \beta_{12} = \beta_{13} = \beta_{14} = \beta_{15} = 0$
Age*time2	$H_6 : \beta_{11} = 0$

Note that the estimates and standard errors were already reported in Table 5.1. A printout of the complete estimated covariance matrix for the fixed effects is also obtained (due to the option 'covb' in the MODEL statement), but it is not printed here because of its high dimension (15×15).

By default, SAS provides approximate F-tests (see Section 6.2.2) for all effects specified in the MODEL statement. For continuous covariates, which do not interact with any factors (i.e., with no interaction term included in the MODEL statement), this is equivalent to the t-test reported in the table 'Solution for Fixed Effects.' For each factor specified in the CLASS statement, it is tested whether any of the parameters assigned to this factor is significantly different from zero. The same is true for interactions of factors with other effects. The hypotheses tested in our example are summarized in Table 8.3. Finally, since the option 'chisq' was added to the MODEL statement, approximate Wald tests are also performed (see Section 6.2.1). The output table is given by

Tests of Fixed Effects

Source	NDF	DDF	Type III ChiSq	Type III F	Pr > ChiSq	Pr > F
GROUP	4	47.8	63.60	15.90	0.0001	0.0001
AGEDIAG	1	47.9	3.48	3.48	0.0621	0.0683
TIME*GROUP	4	42.2	31.41	7.85	0.0001	0.0001
AGEDIAG*TIME	1	42.3	0.27	0.27	0.6022	0.6050
TIME2*GROUP	4	31.8	17.78	4.44	0.0014	0.0058
AGEDIAG*TIME2	1	32	0.07	0.07	0.7980	0.7997

Because the option 'ddfm=satterth' has been added to the MODEL statement, a Satterthwaite approximation is used for the calculation of the denominator degrees of freedom needed for the approximate F-tests. As is expected, all p-values obtained from the chi-squared approximation are smaller than those from the F-approximation. However, the difference is rather small, which illustrates the fact that, in a longitudinal context, different estimation methods for the denominator degrees of freedom for the F-test usually lead to very similar results (see also Section 6.2.2).

Two additional tables are also given, labeled 'CONTRAST Statement Results' and 'ESTIMATE Statement Results,' which contain the results from the specified CONTRAST and ESTIMATE statements, respectively. The first one is given by

CONTRAST Statement Results

Source	NDF	DDF	ChiSq	F	Pr > ChiSq	Pr > F
Final model	6	46.7	3.39	0.56	0.7587	0.7561

and provides approximate F- and Wald tests for testing whether the original model (3.10) can be reduced to model (6.8). The results were already reported in Section 6.2.3. The table with the results from the ESTIMATE statement equals

ESTIMATE Statement Results

Parameter	Estimate	Std Error	DF	t	Pr > |t|
Diff L/R-BPH, t=5yr	0.1889	0.2189	71.2	0.86	0.3910

Alpha	Lower	Upper
0.05	-0.2476	0.6255

Note that the results are slightly different from our results given in Table 6.2, which is due to the fact that inferences for the average difference between BPH patients and local cancer cases is now obtained under model (3.10) rather than under model (6.8), as was the case in Table 6.2.

8.3.6 Inference for the Random Effects

Because the option 'solution' was added to the RANDOM statement, a table is printed containing empirical Bayes estimates for all random effects in the model:

Solution for Random Effects

Effect	ID	Estimate	SE Pred	DF	t	Pr > |t|
INTERCEPT	1	0.3470	0.2141	126	1.62	0.1075
TIME	1	-0.8979	0.3822	114	-2.35	0.0205
TIME2	1	0.3138	0.1641	57.3	1.91	0.0609
INTERCEPT	2	-0.3150	0.2781	67.9	-1.13	0.2613
TIME	2	0.9239	0.4218	60.9	2.19	0.0324
TIME2	2	-0.3840	0.1649	38.7	-2.33	0.0252
....	..	......				
INTERCEPT	54	-0.3603	0.2223	93.6	-1.62	0.1083
TIME	54	0.1435	0.4315	108	0.33	0.7400
TIME2	54	-0.0089	0.1896	34.8	-0.05	0.9628

For each subject, the empirical Bayes estimate $\widehat{b}_i$ for its vector b_i of random effects is printed, together with approximate standard errors and t-tests. The standard errors reported here are not based on the covariance matrix (7.3) of $\widehat{b}_i$ but on the covariance matrix (7.4) of $\widehat{b}_i - b_i$.

8.4   Note on the Mean Parameterization

Note that it follows from the way we parameterized the mean structure of our model that the F-tests discussed in Section 8.3.5 cannot be used to test whether the different diagnostic groups have different intercepts or slopes. For example, four parameters are assigned to the effect 'time2 * group', being the slopes for the quadratic time effect for each group separately. The F-test reported for the effect time2 * group was therefore

$$H_5 : \beta_{12} = \beta_{13} = \beta_{14} = \beta_{15} = 0$$

rather than

$$H_0 : \beta_{12} = \beta_{13} = \beta_{14} = \beta_{15}. \qquad (8.2)$$

Note that hypothesis (8.2) is also of the form $H_0 : L\beta = 0$ and can thus also be tested using a CONTRAST statement in PROC MIXED (see Section 8.2.7).

Another possibility, which directly yields a test for group differences, is to reparameterize the mean structure, including an overall intercept, and overall slopes for the linear and quadratic time effects. The MODEL statement of our program on p. 94 then needs to be replaced by

```
model lnpsa = group age time group*time age*time
              time2 group*time2 age*time2
              / solution ddfm = satterth covb chisq;
```

We then get the following output table 'Solution for Fixed Effects' with REML estimates and t-tests for all parameters in the reparameterized mean structure:

Solution for Fixed Effects

Effect	GROUP	Estimate	Std Error	DF	t	Pr > |t|
INTERCEPT		1.5493	1.0856	47.6	1.43	0.1600
GROUP	1	-2.6478	0.3931	48	-6.74	0.0001
GROUP	2	-2.0722	0.3835	48.8	-5.40	0.0001
GROUP	3	-1.2529	0.3932	48.3	-3.19	0.0025
GROUP	4	0.0000		.	.	.
AGEDIAG		0.0265	0.0142	47.9	1.87	0.0683
TIME		-1.6049	1.6257	41.6	-0.99	0.3293
TIME*GROUP	1	2.1729	0.5836	40.9	3.72	0.0006
TIME*GROUP	2	2.0005	0.5678	40.4	3.52	0.0011
TIME*GROUP	3	0.5689	0.5794	39.5	0.98	0.3321
TIME*GROUP	4	0.0000		.	.	.
AGEDIAG*TIME		-0.0111	0.0214	42.3	-0.52	0.6050
TIME2		0.3950	0.6660	31.1	0.59	0.5574
TIME2*GROUP	1	-0.5245	0.2341	29.6	-2.24	0.0327
TIME2*GROUP	2	-0.5535	0.2232	25.9	-2.48	0.0200
TIME2*GROUP	3	-0.0531	0.2267	25.2	-0.23	0.8166
TIME2*GROUP	4	0.0000		.	.	.
AGEDIAG*TIME2		0.0022	0.0088	32	0.26	0.7997

The slope β_{15} for $time^2$ in the last group is now the parameter assigned to the overall $time^2$ effect, and the three parameters assigned to the interaction of $group$ with $time2$ are the contrasts $\beta_{12} - \beta_{15}$, $\beta_{13} - \beta_{15}$, and $\beta_{14} - \beta_{15}$, respectively (see also the original estimates in Table 5.1). Note that this also implies that the CONTRAST statement and the ESTIMATE statement previously used in the program on p. 94 are no longer valid under the new parameterization.

For the reparameterized model, we get the following F-tests for the effects specified in the MODEL statement:

TABLE 8.4. *Prostate Data. Overview of the hypotheses corresponding to the tests specified in the table labeled 'Tests of Fixed Effects' for the model with reparameterized mean structure.*

Source	Null hypothesis
Group	$H_7 : \beta_2 = \beta_3 = \beta_4 = \beta_5$
Age	$H_8 : \beta_1 = 0$
Time	$H_9 : (\beta_7 + \beta_8 + \beta_9 + \beta_{10})/4 = 0$
Time*group	$H_{10} : \beta_7 = \beta_8 = \beta_9 = \beta_{10}$
Age*group	$H_{11} : \beta_6 = 0$
Time2	$H_{12} : (\beta_{12} + \beta_{13} + \beta_{14} + \beta_{15})/4 = 0$
Time2*group	$H_{13} : \beta_{12} = \beta_{13} = \beta_{14} = \beta_{15}$
Age*time2	$H_{14} : \beta_{11} = 0$

Tests of Fixed Effects

Source	NDF	DDF	Type III ChiSq	Type III F	Pr > ChiSq	Pr > F
GROUP	3	47.9	60.39	20.13	0.0001	0.0001
AGEDIAG	1	47.9	3.48	3.48	0.0621	0.0683
TIME	1	42.2	0.07	0.07	0.7874	0.7887
TIME*GROUP	3	41.8	31.24	10.41	0.0001	0.0001
AGEDIAG*TIME	1	42.3	0.27	0.27	0.6022	0.6050
TIME2	1	32	0.03	0.03	0.8609	0.8620
TIME2*GROUP	3	30.8	17.78	5.93	0.0005	0.0026
AGEDIAG*TIME2	1	32	0.07	0.07	0.7980	0.7997

The hypotheses tested in the above output are shown in Table 8.4. Hence, the test for hypothesis (8.2) is now reported as the F-test corresponding to the effect of $time2 * group$. Note also the change in its numerator degrees of freedom due to the fact that we now test for equality of the quadratic time effect in the four diagnostic groups, rather than testing whether there is any quadratic time effect in any of the four diagnostic groups at all. Also, under this parameterization for the mean structure, the F-test reported for $time2$ tests whether there is a quadratic time effect in the overall population and is therefore not equivalent to the t-test reported for $time2$ in the table labeled 'Solution for Fixed Effects,' which was testing for a quadratic time effect for the metastatic cancer cases only. The same remark is true for the F-test reported for $time$.

Although fitting the reparameterized model automatically yields tests for group differences with respect to average intercepts or slopes, it often complicates the interpretation of the parameter estimates since contrasts are estimated rather than the parameters of interest. All further analyses of the prostate data will therefore be based on the original parameterization as in model (3.10) or as in the reduced model (6.8) obtained in Section 6.2.3.

8.5   The RANDOM and REPEATED Statements

In Section 8.2, we introduced the RANDOM statement and the REPEAT-ED statement of PROC MIXED, and both statements were used in our program on p. 94 to fit model (3.10) to the prostate cancer data. However, since the covariance structure for the error components ε_i was taken equal to $\sigma^2 I_{n_i}$, which is the default in PROC MIXED, the same model can be fitted omitting the REPEATED statement. In practice, it is often sufficient to use only a RANDOM statement or only a REPEATED statement. In the first case, a hierarchical model is assumed, in which random effects are used to describe the covariance structure in the data, whereas all remaining variability is assumed to be purely measurement error (the components in ε_i are assumed to be independently, identically distributed). The covariance structure is then assumed to be of the form $V_i = Z_i D Z_i' + \sigma^2 I_{n_i}$. In the other case, no random effects are included, indicating that no part of the observed variability in the data can be ascribed to between-subject variability. The covariance structure for the data is then completely determined by the covariance structure Σ_i for the error components ε_i, which is specified in the REPEATED statement.

It should be noted, however, that although both procedures have different interpretations, they can result in identical marginal models. We hereby also refer to our discussion in Section 5.6.2 on hierarchical and marginal models. As an example, we take the so-called 'random intercepts' or 'compound symmetry' model, which assumes a covariance structure of the form

$$\text{var}(\boldsymbol{Y_i}) = V_i = \begin{pmatrix} \sigma^2 + \sigma_c^2 & \sigma_c^2 & \cdots & \sigma_c^2 \\ \sigma_c^2 & \sigma^2 + \sigma_c^2 & \cdots & \sigma_c^2 \\ \vdots & \vdots & \ddots & \vdots \\ \sigma_c^2 & \sigma_c^2 & \cdots & \sigma^2 + \sigma_c^2 \end{pmatrix}, \tag{8.3}$$

for some non-negative value σ_c^2. Although such an assumption is often not realistic in a longitudinal-data setting (constant variance and all correlations equal), it is frequently used in practice since it immediately follows from random-factor ANOVA models (see, for example, Neter, Wasserman, and Kutner 1990, Section 17.6, Searle 1987, Chapter 13).

Since the covariance matrix in (8.3) can be rewritten as

$$
V_i = \begin{pmatrix} 1 \\ 1 \\ \vdots \\ 1 \end{pmatrix} \sigma_c^2 \begin{pmatrix} 1 & 1 & \cdots & 1 \end{pmatrix} + \sigma^2 \begin{pmatrix} 1 & 0 & \cdots & 0 \\ 0 & 1 & \cdots & 0 \\ \vdots & \vdots & \ddots & \vdots \\ 0 & 0 & \cdots & 1 \end{pmatrix}
$$

$$
= Z_i \mathrm{var}(b_i) Z_i' + \sigma^2 I_{n_i},
$$

it can be interpreted as the marginal covariance structure of a linear mixed model containing only random intercepts with variance $d_{11} = \sigma_c^2$, which can be fitted with PROC MIXED by specifying random intercepts in the RANDOM statement and omitting the REPEATED statement. This hierarchical interpretation of the compound symmetry covariance structure was already encountered in Sections 3.3.2 and 6.3.3. Note also that in this case, no 'type=' option is needed in the RANDOM statement since for univariate random effects b_i, all types result in the same covariance structure.

Further, since the option 'type=CS' in the REPEATED statement results in a covariance structure for ε_i of the same form as (8.3), the same model can be fitted without the RANDOM statement. This shows that there are sometimes several ways of specifying a given model. In such a case, it is recommended to specify the model using the REPEATED statement rather than the RANDOM statement because this can reduce the computing time considerably.

This also implies that one should not conclude that the REPEATED statement is used whenever the data are of the repeated-measures type. Some repeated-measures models are best expressed using the RANDOM statement (see, e.g., the prostate cancer data), whereas there are also random-effects models, which do not fall into the repeated-measures class but where the REPEATED statement is the simplest tool for expressing them in PROC MIXED syntax. For example, suppose 100 exams were randomly assigned for correction to 10 randomly selected teachers. If Y_{ij} then denotes the grade assigned to the jth exam by the ith observer, the following random-factor ANOVA model can be used to analyze the data:

$$
Y_{ij} = \mu + \alpha_i + \varepsilon_{ij}. \tag{8.4}
$$

The parameter μ represents the overall mean, the parameters α_i are the random observer effects, and the ε_{ij} are components of measurement error. It is thereby assumed that all α_i and ε_{ij} are independent of each other, and that they are normally distributed with mean zero and constant variances σ_c^2 and σ^2, respectively. Model (8.4) can then be fitted using the following program:

```
proc mixed;
class observer;
model Y = / solution;
repeated / type = cs subject = observer;
run;
```

In this case, no effects need to be specified in the REPEATED statement since the ordering of observations for each of the observers is of no importance for the estimation of the covariance structure (constant variance and equal correlations).

8.6   PROC MIXED versus PROC GLM

For balanced longitudinal data (i.e., longitudinal data where all subjects have the same number of repeated measures, taken at time points which are also the same for all subjects), one often analyzes the data using the SAS procedure PROC GLM, fitting general multivariate regression models (Seber 1984, Chapters 8 and 9) to the data. Such models can also be fitted with PROC MIXED by omitting the RANDOM statement and including a REPEATED statement with option 'type=UN'. One then fits a linear model with a general unstructured covariance matrix $\Sigma = \Sigma_i$. However, the two procedures do not necessarily yield the same results: PROC GLM only takes into account the data of the completers, that is, only the data of the subjects with all measurements available are used in the calculations. PROC MIXED, on the other hand, uses all available data. Hence, patients for whom not all measurements were recorded will still be taken into account in the analysis. We refer to Section 17.3 for an illustration.

The multivariate approach used in the GLM procedure produces multivariate tests for the fixed effects based on the Wilk's Lambda likelihood ratio test statistic (see, e.g., Rao 1973, Chapter 8, SAS 1989, Chapter 1). The resulting F-tests are based on a better approximation to the actual distribution of the test statistic than that for the F-tests currently given by the MIXED procedure (see Roger and Kenward 1993). Further, apart from the multivariate tests, PROC GLM also provides a univariate analysis for the response at each time point separately. This can also be obtained with PROC MIXED, by specifying a WHERE statement. For example, an analysis of the responses at time $t = 2$ is requested by adding the following line to the main program:

```
where time = 2;
```

Note that this again may yield different results than PROC GLM due to the fact that now all second measurements are analyzed rather than only the measurements from the patients with measurements taken at all time points. Finally, the "split unit" type of analysis provided by the GLM procedure can be obtained using PROC MIXED from fitting a compound symmetry model (see Section 8.5). However, Greenhouse-Geiser and Huynh-Feldt corrections to the F-tests are not available in the MIXED procedure, but they are not really required, as it is very simple to fit and test models with more complex covariance structures.

The main strength of the procedure MIXED is that it does not assume that an equal number of repeated observations is taken from each individual or that all individuals should be measured on the same time points. Hence, the measurements can be viewed as being taken at a continuous rather than discrete time scale. Also, the use of random effects allows us to model covariances as continuous functions of time. Another main advantage in using the MIXED procedure is the fact that all available data (not only the "complete cases") are used in the analysis. Finally, PROC MIXED also allows us to include time-varying covariates in the mean structure, which is not possible in PROC GLM.

For a more elaborate discussion on the comparison between the procedures MIXED and GLM, we refer to Roger and Kenward (1993) and Roger (1993).

9

General Guidelines for Model Building

9.1 Introduction

As discussed in Chapter 8, the SAS procedure PROC MIXED allows the user to fit general linear mixed models, with a large variety of possible covariance structures. Under the linear mixed model (3.11), the data vector Y_i for the ith subject is assumed to be normally distributed with mean vector $X_i\beta$ and covariance matrix of the form $V_i = Z_i D Z_i' + \sigma^2 I_{n_i} + \tau^2 H_i$. Hence, fitting linear mixed models implies that an appropriate mean structure as well as covariance structure needs to be specified. As shown in Figure 9.1, they are not independent of each other.

First, unless robust inference is used (see Section 6.2.4), an appropriate covariance model is essential to obtain valid inferences for the parameters in the mean structure, which is usually of primary interest. This will be especially the case in the presence of missing data, since robust inference then only provides valid results under often unrealistically strict assumptions about the missing data process (see, for example, Section 16.5 as well as Chapters 17, 19, and 21). Too restrictive specifications invalidate inferences when the assumed structure does not hold, whereas overparameterization of the covariance structure leads to inefficient estimation and poor assessment of standard errors (Altham 1984).

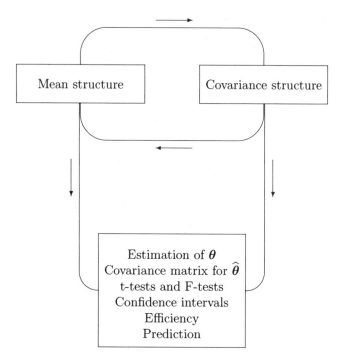

FIGURE 9.1. *Graphical representation of how the mean structure and the covariance structure of a linear mixed model influence one another and how they affect the inference results.*

Further, the covariance structure itself may be of interest for understanding the random variation observed in the data. However, since it only explains the variability not explained by systematic trends, it is highly dependent on the specified mean structure.

Finally, note that an appropriate covariance structure also yields better predictions. For example, the prediction of a future observation y_i^* for individual i, to be taken at time point t_i^*, based on model (3.11) is given by

$$\widehat{y_i^*} = X_i^* \widehat{\beta} + Z_i^* \widehat{b_i} + E(\varepsilon_{(2)i}^* \mid y_i),$$

in which X_i^* and Z_i^* are the fixed-effects and random-effects covariates, respectively, and $\varepsilon_{(2)i}^*$ is the serial error, at time t_i^*. The random effect b_i is estimated as in Section 7.2. Chi and Reinsel (1989) have shown that if the components of $\varepsilon_{(2)i}$ follow an AR(1) process,

$$E(\varepsilon_{(2)i}^* \mid y_i) = \phi^{(t_i^* - t_{i,n_i})} \left[y_i - X_i \widehat{\alpha} - Z_i \widehat{b_i} \right]_{n_i},$$

for ϕ equal to the constant which determines the AR(1) process, $|\phi| < 1$. This means that the inclusion of serial correlation may improve the prediction since it exploits the correlation between the observation to be predicted and the last observed value y_{i,n_i}.

For data sets where most variability in the measurements is due to between-subject variability, one can very often use the two-stage approach to construct an appropriate linear mixed model. This was illustrated for the prostate cancer data in Sections 3.2.4 and 3.3.3. On the other hand, as shown in Section 5.6.2, a two-stage approach does not always automatically yield a valid marginal model for the data. Also, if the intersubject variability is small in comparison to the intrasubject variability, this suggests that the covariance structure cannot be modeled using random effects, but that an appropriate covariance matrix Σ_i for ε_i should be found.

In this chapter, some simple guidelines will be discussed which can help the data analyst to select an appropriate linear mixed model for some specific data set at hand. All steps in this model building process will be illustrated with the prostate cancer data set. It should be emphasized that following the proposed guidelines does not necessarily yield the most appropriate model, nor does it yield a linear mixed model where all distributional assumptions are automatically satisfied. In general, more complex model diagnostics are required to assess the goodness-of-fit of a model. We therefore refer to the exploratory techniques of Chapter 3 and to the more advanced techniques later described in Chapters 10, 11, and 12.

9.2   Selection of a Preliminary Mean Structure

Since the covariance structure models all variability in the data which cannot be explained by the fixed effects, we start by first removing all systematic trends. As proposed by Diggle (1988) and Diggle, Liang, and Zeger (1994) (Sections 4.4 and 5.3), we use an overelaborated model for the mean response profile. When the data are from a designed experiment in which the only relevant explanatory variables are the treatment labels, it is a sensible strategy to use a "saturated model" for the mean structure. This incorporates a separate parameter for the mean response at each time point within each treatment group. For example, when two treatment groups had measurements at four fixed time points, we would use $p = 4 \times 2 = 8$ parameters to model $E(Y_i)$. The X_i matrices would then equal

$$X_i = \begin{pmatrix} 1 & 0 & 0 & 0 & 0 & 0 & 0 & 0 \\ 0 & 1 & 0 & 0 & 0 & 0 & 0 & 0 \\ 0 & 0 & 1 & 0 & 0 & 0 & 0 & 0 \\ 0 & 0 & 0 & 1 & 0 & 0 & 0 & 0 \end{pmatrix} \quad \text{or} \quad X_i = \begin{pmatrix} 0 & 0 & 0 & 0 & 1 & 0 & 0 & 0 \\ 0 & 0 & 0 & 0 & 0 & 1 & 0 & 0 \\ 0 & 0 & 0 & 0 & 0 & 0 & 1 & 0 \\ 0 & 0 & 0 & 0 & 0 & 0 & 0 & 1 \end{pmatrix},$$

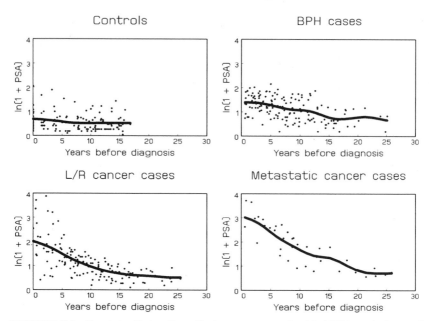

FIGURE 9.2. *Prostate Data. Smoothed average trend of* $\ln(PSA + 1)$ *for each diagnostic group separately.*

depending on whether the ith individual belongs to the first or second treatment group, respectively.

For data in which the times of measurement are not common to all individuals or when there are continuous covariates which are believed to affect the mean response, the concept of a saturated model breaks down and the choice of our most elaborate model becomes less obvious. In such cases, a plot of smoothed average trends or individual profiles often helps to select a candidate mean structure. For the prostate cancer data, Figure 9.2 shows the smoothed average trend in each diagnostic group separately. Note that these trends are not corrected for the age differences between the study participants. Further, the individual profiles (Figure 2.3) and our results from Section 4.3.4 suggest modeling $\ln(1 + \text{PSA})$ as a quadratic function over time. This results in an average intercept and an average linear as well as quadratic time effect within each diagnostic group. Finally, it has been anticipated that age is also an important prognostic covariate. We therefore also include age at time of diagnosis along with its interactions with time and time^2. Our preliminary mean structure therefore contains $4 \times 3 + 3 = 15$ fixed effects, represented by the vector $\boldsymbol{\beta}$. Note that this is the mean structure which was used in our initial model (3.10), and that at this stage, we deliberately favor overparameterized models for $E(\boldsymbol{Y}_i)$ in order to get consistent estimators of the covariance structure in the following steps.

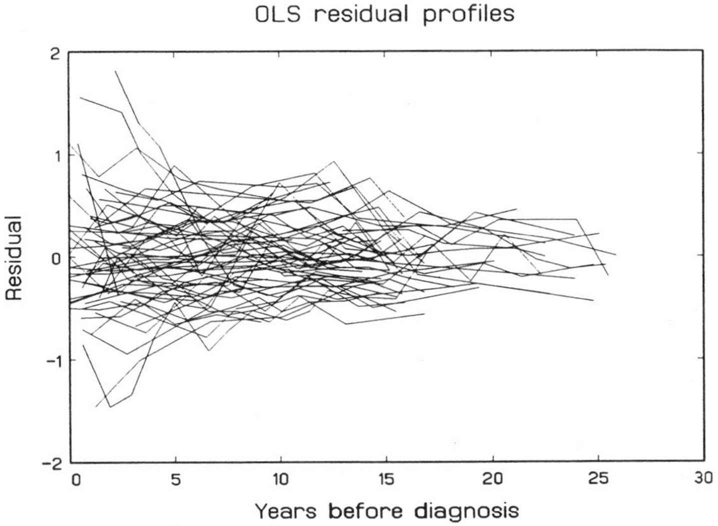

OLS residual profiles

FIGURE 9.3. *Prostate Data. Ordinary least squares (OLS) residual profiles.*

Once an appropriate mean structure $X_i\beta$ for $E(Y_i)$ has been selected, we use the ordinary least squares (OLS) method to estimate β, and we hereby ignore that not all measurements are independent. It follows from the theory of generalized estimating equations (GEE) that this OLS estimator is consistent for β (Liang and Zeger 1986; see also Section 6.2.4). This justifies the use of the OLS residuals $r_i = y_i - X_i\widehat{\beta}_{\text{OLS}}$ for studying the dependence among the repeated measures.

9.3 Selection of a Preliminary Random-Effects Structure

In a second step, we will select a set of random effects to be included in the covariance model. Note that random effects for time-independent covariates can be interpreted as subject-specific corrections to the overall mean structure. This makes them hard to distinguish from random intercepts. Therefore, one often includes random intercepts, and random effects only for time-varying covariates. However, it will be shown in Section 24.1 that in some applications, random effects for time-independent indicators may be useful to model differences in variability between subgroups of subjects or measurements.

A helpful tool for deciding which time-varying covariates should be included in the model is a plot of the OLS residual profiles versus time. For the

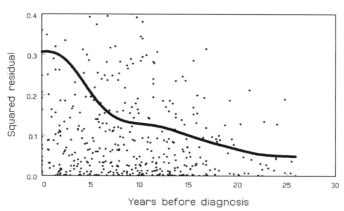

FIGURE 9.4. *Prostate Data. Smoothed average trend of squared OLS residuals. Squared residuals larger than 0.4 are not shown.*

prostate data, this was done in Figure 9.3. A smoothed average trend of the squared OLS residuals is shown in Figure 9.4 and is used to explore the variance function over time. If it is believed that different variance functions should be used for different groups, a plot as in Figure 9.4 could be constructed for each group separately. When this plot shows constant variability over time, we assume stationarity and we do not include random effects other than intercepts. In cases where the variability varies over time and where there is still some remaining systematic structure in the residual profiles (i.e., where the between-subject variability is large in comparison to the overall variation), the following guidelines can be used to select one or more random effects additional to the random intercepts.

- Try to find a regression model for each residual profile in the above plot. Such models contain subject-specific parameters and are therefore perfect candidates as random effects in our general linear mixed model. For example, if the residual profiles can be approximated by straight lines, then only random intercepts and random slopes for time would be included.

- Since our model always assumes the random effects b_i to have mean zero, we only consider covariates Z_i which have already been included as covariates in the fixed part (i.e., in X_i) or which are linear combinations of columns of X_i. Note that this condition was satisfied in model (3.10), which we used to analyze the prostate cancer data. For example, the second column of Z_i, which represents the linear random effect for time, equals the sum of columns 7 to 10 in X_i, which are the columns containing the linear time effects for the controls, the benign prostatic hyperplasia patients, the local cancer cases, and the metastatic cancer cases, respectively.

- Morrell, Pearson, and Brant (1997) have shown that Z_i should not include a polynomial effect if not all hierarchically inferior terms are also included, and similarly for interaction terms. This generalizes the well-known results from linear regression (see, e.g., Peixoto 1987, 1990) to random-effects models. It ensures that the model is invariant to coding transformations and avoids unanticipated covariance structures. This means that if, for example, we want to include quadratic random time effects, then also linear random time effects and random intercepts should be included.

- The choice of a set of random effects for the model automatically implies that the covariance matrix for Y_i is assumed to be of the general form $V_i = Z_i D Z_i' + \Sigma_i$. In the presence of random effects other than intercepts, it is often assumed (see, e.g., Diggle, Liang, and Zeger 1994) that the diagonal elements in Σ_i are all equal such that the variance of $Y_i(t)$ depends on time, only through the component $Z_i(t) D Z_i'(t)$, where it is now explicitly indicated that the covariates Z_i depend on time. As an informal check for the appropriateness of the selected random effects, one can compare the fitted variance function based on a mixed-effects model with $\Sigma_i = \sigma^2 I_{n_i}$ to the smoothed sample variance function of the residuals r_{ij}.

In the example on prostate cancer, Figure 9.4 clearly suggests nonstationarity, and we assume that the remaining structure in the OLS residual profiles in Figure 9.3 can be well described by a quadratic function over time. Hence, random intercepts and linear as well as quadratic random slopes for time are included in the preliminary random-effects structure. Note that, as in Section 9.2, we favor the inclusion of too many random effects rather than omitting some. This ensures that the remaining variability is not due to any missing random effects. However, it also should be emphasized that including high-dimensional random effects b_i with unconstrained covariance matrix D leads to complicated covariance structures and may result in divergence of the maximization procedure.

As an informal check for the variance function, we compared the smoothed average trend of squared OLS residuals, previously shown in Figure 9.4, with the fitted variance function obtained from fitting a linear mixed model with the preliminary mean structure, the preliminary random-effects structure, and measurement error. For the prostate data, this is the original

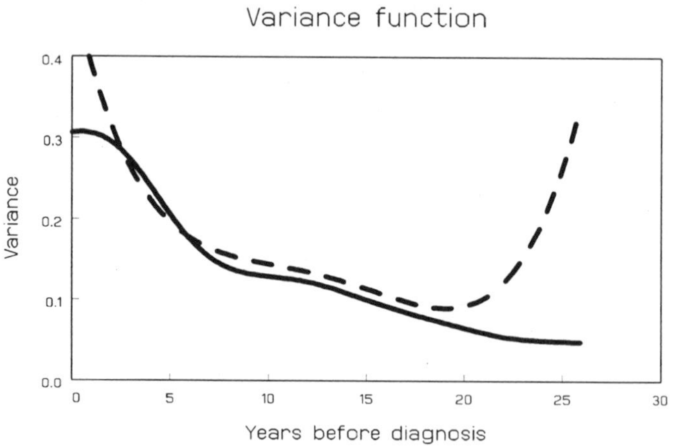

Variance function

FIGURE 9.5. *Prostate Data. Comparison of the smoothed average trend of squared OLS residuals (solid line) with the fitted variance function (dashed line) obtained using the REML estimates in Table 5.1 for the variance components in model (3.10).*

model (3.10), and the fitted variance function is obtained by calculating

$$
\begin{pmatrix} 1 & t & t^2 \end{pmatrix} \widehat{D} \begin{pmatrix} 1 \\ t \\ t^2 \end{pmatrix} + \widehat{\sigma}^2,
$$

where $\widehat{D}$ and $\widehat{\sigma}^2$ are the REML estimates reported in Table 5.1. The result is presented in Figure 9.5. Both variance functions show similar trends, except at the beginning and at the end. The deviation for small time points can be explained by noticing that some subjects have extremely large PSA values close to their time of diagnosis (see individual profiles in Figure 2.3). These correspond to extremely large squared residuals (not shown in Figure 9.4), which may have inflated the fitted variance function much more than the variance function obtained by smoothing the squared OLS residuals. The deviation for large time points can be ascribed to the small amount of data available. Only 24 out of the 463 PSA measurements have been taken more than 20 years prior to the diagnosis.

9.4 Selection of a Residual Covariance Structure

Conditional on our selected set of random effects, we now need to specify the covariance matrix Σ_i for the error components ε_i. Many possible covariance structures are available at this stage. Unfortunately, apart from the information criteria discussed in Section 6.4, there are no general simple techniques available to compare all these models. For highly unbalanced data with many repeated measurements per subject, one usually assumes that random effects can account for most of the variation in the data and that the remaining error components ε_i have a very simple covariance structure, leading to parsimonious models for V_i.

One such model is the model (3.11), introduced in Section 3.3.4. It assumes that ε_i has constant variance and can be decomposed as $\varepsilon_i = \varepsilon_{(1)i} + \varepsilon_{(2)i}$, in which $\varepsilon_{(2)i}$ is a component of serial correlation and where $\varepsilon_{(1)i}$ is a component of measurement error. The model is then completed by specifying a serial correlation function $g(\cdot)$. The most frequently used functions are the exponential and the Gaussian serial correlation functions already shown in Figure 3.2, but other functions can also be specified in the SAS procedure MIXED. We propose to fit a selection of linear mixed models with the same mean and random-effects structure, but with different serial correlation structures, and to use likelihood-based criteria to compare the different models. In some cases, likelihood ratio tests can be applied. In other cases, one might want to use the information criteria introduced in Section 6.4. Alternatively, one can use one of the advanced methods for the exploration of the residual serial correlation, which will be discussed in Chapter 10. However, unless one is especially interested in the serial correlation function, it is usually sufficient to fit and compare a series of serial correlation models.

Using some of the methods to be discussed in Chapter 10, Verbeke, Lesaffre, and Brant (1998) and Verbeke and Lesaffre (1997b) have shown that the residual components in the prostate cancer model indeed contain a serial correlation component $\varepsilon_{(2)i}$, which is probably of the Gaussian type. The linear mixed model with our preliminary mean structure, with random intercepts and random slopes for the linear as well as quadratic time effect, with measurement error, and with Gaussian serial correlation can then be fitted using the following program:

```
proc mixed data = prostate covtest;
class id group timeclss;
model lnpsa = group age group*time age*time
              group*time2 age*time2 / noint solution;
random intercept time time2 / type = un subject = id g;
repeated timeclss / type = sp(gau)(time) local subject = id;
run;
```

The option 'type=sp(gau)(time)' is used to specify the Gaussian serial correlation structure with the SAS variable *time* as the variable which needs to be used to calculate the time differences between the repeated measures.

TABLE 9.1. *Prostate Data. REML estimates and estimated standard errors for all variance components in a linear mixed model with the preliminary mean structure defined in Section 9.2, with random intercepts and random slopes for the linear as well as quadratic time effects, with measurement error, and with Gaussian serial correlation.*

Effect	Parameter	Estimate (s.e.)
Covariance of b_i:		
$\quad$ var(b_{1i})	d_{11}	0.389 (0.096)
$\quad$ var(b_{2i})	d_{22}	0.559 (0.206)
$\quad$ var(b_{3i})	d_{33}	0.059 (0.032)
$\quad$ cov(b_{1i}, b_{2i})	$d_{12} = d_{21}$	-0.382 (0.121)
$\quad$ cov(b_{2i}, b_{3i})	$d_{23} = d_{32}$	-0.175 (0.079)
$\quad$ cov(b_{3i}, b_{1i})	$d_{13} = d_{31}$	0.099 (0.043)
Measurement error variance:		
$\quad$ var$(\varepsilon_{(1)ij})$	σ^2	0.023 (0.002)
Gaussian serial correlation:		
$\quad$ var$(\varepsilon_{(2)ij})$	τ^2	0.032 (0.021)
$\quad$ Rate of exponential decrease	$1/\sqrt{\phi}$	0.619 (0.202)
REML log-likelihood		-24.787
-2 REML log-likelihood		49.574

As explained in Section 8.2.10, the option 'local' requests inclusion of a measurement error component $\varepsilon_{(1)i}$ on top of the serial correlation component $\varepsilon_{(2)i}$. The REML estimates and estimated standard errors of all variance components in this model are shown in Table 9.1. Note how SAS estimates $1/\sqrt{\phi}$ rather than ϕ itself. This is to ensure positiveness of ϕ. In cases where no measurement error is included in the model, this also allows testing whether $\phi = +\infty$, under which H_i becomes equal to the identity matrix I_{n_i}, meaning that no serial correlation would be present in the error components ε_i.

Comparing minus twice the REML log-likelihood of the above model with the value obtained without the serial correlation component (see Table 5.1) yields a difference of 12.896, indicating that adding the serial correlation component really improved the covariance structure of our model. Further, note that the residual variability has now been split up into two components which are about equally important (similar variance). Based on this extended covariance matrix, we repeated our informal check previously presented in Figure 9.5, comparing the smoothed average trend in the squared OLS residuals with the new fitted variance function. The fitted variances

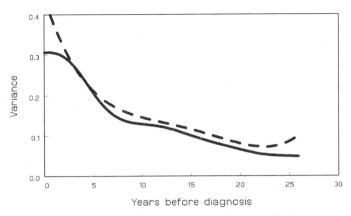

FIGURE 9.6. *Prostate Data. Comparison of the smoothed average trend of squared OLS residuals (solid line) with the fitted variance function (dashed line) obtained using the REML estimates in Table 9.1 for the variance components in model (3.10), extended with a Gaussian serial correlation component.*

are now obtained by calculating

$$
\left(\begin{array}{ccc} 1 & t & t^2 \end{array} \right) \widehat{D} \left(\begin{array}{c} 1 \\ t \\ t^2 \end{array} \right) + \widehat{\sigma}^2 + \widehat{\tau}^2,
$$

where the estimates $\widehat{D}$, $\widehat{\sigma}^2$, and $\widehat{\tau}^2$ are the ones reported in Table 9.1. The result is shown in Figure 9.6. Figures 9.5 and 9.6 are very similar, except for large values of *time*, where the two variance functions are now more alike.

Finally, it should be emphasized that, for fitting complicated covariance structures, with possibly overspecified random-effects structures, one often needs to specify starting values (using the PARMS statement of PROC MIXED, see Section 8.2.10) in order for the iterative procedure to converge. Sometimes it is sufficient to use the Fisher scoring method (option 'scoring' in the PROC MIXED statement, see Section 8.2.10) in the iterative estimating procedure, which uses the expected Hessian matrix instead of the observed one. To illustrate this, we reparameterize the above fitted model by defining the intercept successively at 0 years (= original parameterization), 5 years, 10 years, 15 years, and 20 years prior to diagnosis. The resulting estimates and standard errors for all variance components in the model are shown in Table 9.2.

Obviously, minus twice the maximized REML log-likelihood function is not affected by the reparameterization. The Fisher scoring algorithm was used in two cases in order to attain convergence. It is hereby important that

TABLE 9.2. *Prostate Data. REML estimates and estimated standard errors for all variance components in a linear mixed model with the preliminary mean structure defined in Section 9.2, with random intercepts and random slopes for the linear as well as quadratic time effects, with measurement error, and with Gaussian serial correlation. Each time, another parameterization of the model is used, based on how the intercept has been defined.*

| Parameter | Definition of intercept (time in years before diagnosis) | | | | |
	$t = 0$	$t = 5$	$t = 10^{(1)}$	$t = 15$	$t = 20^{(2)}$
d_{11}	0.389 (0.096)	0.156 (0.039)	0.090 (0.032)	0.061 (0.027)	0.026 (0.025)
d_{22}	0.559 (0.206)	0.267 (0.085)	0.094 (0.027)	0.038 (0.026)	0.099 (0.091)
d_{33}	0.059 (0.032)	0.059 (0.032)	0.059 (0.032)	0.059 (0.032)	0.059 (0.032)
$d_{12} = d_{21}$	-0.382 (0.121)	-0.120 (0.042)	-0.033 (0.019)	-0.033 (0.017)	-0.030 (0.024)
$d_{23} = d_{32}$	-0.175 (0.079)	-0.116 (0.048)	-0.057 (0.021)	0.001 (0.023)	0.060 (0.051)
$d_{13} = d_{31}$	0.099 (0.043)	0.026 (0.023)	-0.018 (0.023)	-0.032 (0.020)	-0.016 (0.016)
σ^2	0.023 (0.002)	0.023 (0.002)	0.023 (0.002)	0.023 (0.002)	0.023 (0.002)
τ^2	0.032 (0.021)	0.032 (0.021)	0.032 (0.021)	0.032 (0.021)	0.032 (0.021)
$1/\sqrt{\phi}$	0.619 (0.202)	0.619 (0.202)	0.619 (0.202)	0.619 (0.202)	0.619 (0.202)
-2 log-lik	49.574	49.574	49.574	49.574	49.574

[1] Five initial steps of Fisher scoring.

[2] One initial step of Fisher scoring.

the final steps in the iterative procedure are based on the default Newton-Raphson method since, otherwise, all reported standard errors are based on the expected rather than observed Hessian matrix, the consequences of which will be discussed in Chapter 21. Note that the variance components in the covariance structure of ε_i as well as the variance d_{33} of the random slopes for the quadratic time effect remain unchanged when the model is reparameterized. This is not the case for the other elements in the random-effects covariance matrix D. As was expected, the random-intercepts variance d_{11} decreases as the intercept moves further away from the time of diagnosis, and the same holds for the overall variance $d_{11} + \sigma^2 + \tau^2$ at the time of the intercept. We therefore recommend defining random intercepts as the response value at the time where the random variation in the data is maximal. This facilitates the discrimination among the three sources of stochastic variability.

9.5 Model Reduction

Based on the residual covariance structure specified in the previous step, we can now investigate whether the random effects which we included in Section 9.3 are really needed in the model. As discussed in Section 9.3, Z_i should not contain a polynomial effect if not all hierarchically inferior terms are also included. Taking into account this hierarchy, one should test the significance of the highest-order random effects first. We reemphasize that the need for random effects cannot be tested using classical likelihood ratio tests, due to the fact that the null hypotheses of interest are on the boundary of the parameter space, which implies that the likelihood ratio statistic does not have the classical asymptotic chi-squared distribution (see our discussion in Section 6.3.4).

Further, now that the final covariance structure for the model has been selected, the tests discussed in Section 6.2 become available for the fixed effects in the preliminary mean structure.

For the prostate cancer data, we then end up with the same mean structure as well as random-effects structure as model (6.8) in Section 6.2.3. This model can now be fitted in SAS with the following program:

```
proc mixed data = prostate covtest;
class id group timeclss;
model lnpsa = group age bph*time loccanc*time metcanc*time
               cancer*time2 / noint solution;
random intercept time time2 / type = un subject = id g;
repeated timeclss / type = sp(gau)(time) local subject = id;
run;
```

The SAS variables *cancer*, *bph*, *loccanc*, and *metcanc* are dummy variables defined to be equal to 1 if the patient has prostate cancer, benign prostatic hyperplasia, local prostate cancer, or metastatic prostate cancer, respectively, and zero otherwise. The variables *id*, *group*, *time*, and *timeclss* are as defined in Section 8.2. The parameter estimates and estimated standard errors are shown in Table 9.3, and they can be compared to the estimates shown in Table 6.1, which were obtained without assuming the presence of residual serial correlation. Note that adding the serial correlation component to the model leads to smaller standard errors for almost all parameters in the marginal model. This illustrates that an adequate covariance model implies efficient model-based inferences for the fixed effects.

Finally, note that the total residual variability is estimated as $\hat{\sigma}^2 + \hat{\tau}^2 = 0.023 + 0.029 = 0.052$, which is as large as the total variability present in the response $\ln(\text{PSA} + 1)$ prior to any development of prostate disease, that is, many years prior to diagnosis (see Figure 9.4). This is in agreement with our findings in Section 4.3.4, where evidence was found in favor of the presence of considerable residual variability not explained by the first-stage model (3.5) used in the two-stage approach proposed in Section 3.2.4.

TABLE 9.3. *Prostate Data. Results from fitting the final model (6.8) to the prostate cancer data, using restricted maximum likelihood estimation. The covariance structure contains three random effects, Gaussian serial correlation, and measurement error.*

Effect	Parameter	Estimate (s.e.)
Age effect	β_1	0.015 (0.006)
Intercepts:		
Control	β_2	-0.496 (0.411)
BPH	β_3	0.320 (0.470)
L/R cancer	β_4	1.216 (0.469)
Met. cancer	β_5	2.353 (0.518)
Time effects:		
BPH	β_8	-0.376 (0.070)
L/R cancer	β_9	-1.877 (0.210)
Met. cancer	β_{10}	-2.274 (0.244)
Time^2 effects:		
Cancer	$\beta_{14} = \beta_{15}$	0.484 (0.073)
Covariance of b_i:		
var(b_{1i})	d_{11}	0.393 (0.093)
var(b_{2i})	d_{22}	0.550 (0.187)
var(b_{3i})	d_{33}	0.056 (0.028)
cov(b_{1i}, b_{2i})	$d_{12} = d_{21}$	-0.382 (0.114)
cov(b_{2i}, b_{3i})	$d_{23} = d_{32}$	-0.170 (0.070)
cov(b_{3i}, b_{1i})	$d_{13} = d_{31}$	0.098 (0.039)
Measurement error variance:		
var($\varepsilon_{(1)ij}$)	σ^2	0.023 (0.002)
Gaussian serial correlation:		
var($\varepsilon_{(2)ij}$)	τ^2	0.029 (0.018)
Rate of exponential decrease	$1/\sqrt{\phi}$	0.599 (0.192)
Observations		463
REML log-likelihood		-13.704
-2 REML log-likelihood		27.407
Akaike's information criterion		-22.704
Schwarz's Bayesian criterion		-41.235

10

Exploring Serial Correlation

10.1 Introduction

As discussed in Sections 3.3.4 and 9.4, the selection of an appropriate residual covariance structure is a nontrivial step in the model selection process, especially in the presence of random effects. This is because the residual variability is in practice very often dominated by the random effects in the model. In this chapter, we will discuss two procedures for exploring the residual covariance, conditionally on a set of random effects already included in the model. This will be done under the assumption that the data can be well described by a general linear mixed model of the form (3.11) introduced in Section 3.3.4; that is, it is assumed that the residual component ε_i has constant variance and can be decomposed as $\varepsilon_i = \varepsilon_{(1)i} + \varepsilon_{(2)i}$, in which $\varepsilon_{(2)i}$ is a component of serial correlation and where $\varepsilon_{(1)i}$ is a component of measurement error. The marginal covariance matrix is then of the form

$$V_i \quad = \quad Z_i D Z_i' + \tau^2 H_i + \sigma^2 I_{n_i}, \tag{10.1}$$

where the (j, k) element of H_i equals $g(|t_{ij} - t_{ik}|)$ for some (usually) decreasing function $g(\cdot)$ with $g(0) = 1$. Exploring the residual covariance structure then reduces to studying the serial correlation function $g(\cdot)$.

We will also assume that all systematic trends have been removed from the data. As explained in Section 9.2, this can be done by calculating ordinary least squares residuals $r_i = y_i - X_i\widehat{\beta}_{\text{OLS}}$, based on some preliminary mean structure, ignoring any dependence among the repeated measurements. The residuals r_i can now be used to study the covariance structure of our data.

In Section 10.2, we will discuss an informal check for the need of a serial correlation component $\tau^2 H_i$ in (10.1). Afterward, in Section 10.3, it will be shown how the residual serial correlation function can be studied by fitting linear mixed models with general flexible parametric functions for $g(\cdot)$. Finally, in Section 10.4, we discuss the use of the so-called variogram to study the serial correlation function nonparametrically, that is, without assuming any parametric form for $g(\cdot)$.

10.2 An Informal Check for Serial Correlation

A simple informal check for the need of a component $\tau^2 H_i$ of serial correlation in (10.1) has been proposed by Verbeke, Lesaffre, and Brant (1998). The main idea is to project the residuals r_i orthogonal to the columns of Z_i, which allows one to directly study the variability in the data not explained by the included random effects. For each i, $i = 1, \ldots, N$, let A_i be an $n_i \times (n_i - q)$ matrix such that $A_i' Z_i = 0$ and such that $A_i' A_i = I_{n_i - q}$. The $(n_i - q)$-dimensional transformed OLS residuals are then defined as $\mathfrak{R}_i = A_i' r_i$. Their covariance matrix equals

$$A_i' V_i A_i = \tau^2 A_i' H_i A_i + \sigma^2 I_{n_i - q}.$$

In the absence of serial correlation, this is equal to $\sigma^2 I_{n_i - q}$, from which it then follows that all $\mathfrak{R}_{ij}$, $i = 1, \ldots, N$ and $j = 1, \ldots, (n_i - q)$, are normally distributed with mean zero and common variance σ^2. Once the transformed residuals $\mathfrak{R}_{ij}$ have been calculated, various techniques are available for checking their normality. Non-normality indicates that the assumed model may not be appropriate, possibly because a component of serial correlation is missing in the covariance structure.

As an illustration, we applied this to the prostate data, introduced in Section 2.3.1. The same preliminary mean structure is used as in Section 9.2, resulting in the OLS residual profiles which have been shown in Figure 9.3. Further, random intercepts as well as slopes for time and time^2 have been included in the model. The Shapiro-Wilk test (Shapiro and Wilk 1965) revealed clear deviations from normality for the transformed residuals $\mathfrak{R}_{ij}$ ($p = 0.0011$). In view of the fact that the assumed random-effects structure describes the variance function quite well (see Figure 9.5), we believe that this suggests that the correlation function is still not adequately modeled. We will therefore use the techniques to be described in Sections 10.3

and 10.4 to explore whether serial correlation should be added to the model, and if so, what function $g(\cdot)$ would be appropriate.

10.3   Flexible Models for Serial Correlation

10.3.1   Introduction

A classical statistical procedure for testing whether a specific statistical model fits some data set at hand is to test the model versus an extended version of that model. A similar idea can now be used to investigate the serial correlation function, conditionally on a prespecified set of random effects. One thereby assumes that $g(\cdot)$ has a parametric form, which is flexible enough to allow various shapes for the function $g(\cdot)$. Lesaffre, Asefa and Verbeke (1999) have used so-called fractional polynomials, which we introduce in the next section. Afterward, in Section 10.3.3, this approach will be applied to the prostate data.

10.3.2   Fractional Polynomials

Royston and Altman (1994) define a fractional polynomial as any function of the form

$$f(u) \;=\; \phi_0 + \sum_{j=1}^{m} \phi_j x^{(p_j)},$$

where the degree m is a positive integer, where $p_1 > \ldots > p_m$ are real-valued prespecified powers, and where ϕ_0 and $\phi_1, \ldots, \phi_m$ are real-valued unknown regression coefficients. Finally, $x^{(p_j)}$ is defined as

$$x^{(p_j)} \;=\; \left\{ \begin{array}{ll} x^{p_j} & \text{if } p_j \neq 0 \\ \ln(x) & \text{if } p_j = 0. \end{array} \right. \tag{10.2}$$

In the context of linear and logistic regression analyses, Royston and Altman (1994) have shown that the family of fractional polynomials is very flexible and that models with degree m larger than 2 are rarely required. In practice, several values for the powers $p_1, \ldots, p_m$ can be tried, and the model with the best fit is then selected.

Lesaffre, Asefa and Verbeke (1999) applied fractional polynomials to model the serial correlation function $g(\cdot)$. Their model is of the form

$$\tau^2 g(u) \;=\; \exp\left\{\phi_0 + \sum_{j=1}^{m} \phi_j u^{(p_j)}\right\}. \tag{10.3}$$

Since u here represents the time lags between repeated measurements, (10.2) needs to be adapted to allow zero time lags. One possibility is to replace (10.2) by

$$x^{(p_j)} \;=\; \begin{cases} x^{p_j} & \text{if } p_j > 0 \\ \ln(x+1) & \text{if } p_j = 0 \\ (x+1)^{p_j} - 1 & \text{if } p_j < 0. \end{cases}$$

Note that $g(0) = \tau^2$ implies that τ^2 is parameterized as $\tau^2 = \exp(\phi_0)$.

A linear mixed model with random effects, measurement error, as well as a serial correlation component with correlation function of the form (10.3) can be fitted using maximum or restricted maximum likelihood estimation. At present, this is not possible in the SAS procedure MIXED, but it can be easily implemented in any software package which allows numerical optimization. However, as discussed in Section 9.4, fitting linear mixed models which contain all of the above three sources of stochastic variation may become quite involved, in the sense that iterative numerical optimization procedures often diverge. This is especially the case when high-degree fractional polynomials are used for modeling the serial correlation function. We therefore propose keeping the degree m of the polynomials relatively small, but fitting several models with a variety of powers $p_1, \ldots, p_m$.

10.3.3 Example: The Prostate Data

As an illustration, we have applied the above approach to the prostate cancer data introduced in Section 2.3.1. The same preliminary mean structure is used as in Section 9.2, resulting in the OLS residual profiles which have been shown in Figure 9.3. Further, random intercepts as well as slopes for time and time^2 have been included in the model. Several models of the form (10.3) have been fitted, with varying degrees m and varying powers $p_1, \ldots, p_m$. For some models, the numerical maximization procedure failed to converge. This should not be surprising, keeping in mind that all models have quite complicated covariance structures, with at least 9 variance components and that the prostate data set contains data from 54 subjects only. Among all models for which estimates for the parameters could be obtained, the best one (largest maximized likelihood value) was the second-

TABLE 10.1. *Prostate Data. ML estimates and estimated standard errors for the variance of the measurement error components as well as the parameters ϕ_j in the serial correlation function (10.4).*

Parameter	Estimate (s.e.)
σ^2	0.018 (0.001)
ϕ_0	-3.173 (0.706)
ϕ_1	-0.800 (3.019)
ϕ_2	-0.907 (3.338)

degree fractional polynomial with powers $p_1 = 2$ and $p_2 = 1$, i.e.,

$$\tau^2 g(u) \quad = \quad \exp\left\{\phi_0 + \phi_1 u^2 + \phi_2 u\right\}. \tag{10.4}$$

Parameter estimates and associated asymptotic standard errors for the parameters ϕ_j as well as for the variance σ^2 of the measurement error components are given in Table 10.1. The fitted function (10.4) is shown in panel (a) of Figure 10.1 (solid line). Note that we report here maximum likelihood (ML) estimates instead of restricted maximum likelihood estimates (REML). Indeed, all parameters reported in the table are based on the OLS residuals r_i, which implies that no mean structure is included in the model. Hence, REML estimation becomes impossible.

The variance of the serial correlation component is estimated as $\tau^2 = \exp(-3.173) = 0.042$, which is more than twice the estimated variance σ^2 of the measurement errors. Note also the relatively large standard error for $\widehat{\phi}_0$ and the relatively small standard error for $\widehat{\sigma}^2$, indicating that we will probably not be able to estimate the serial correlation component very accurately. The expression for the serial correlation function in (10.4) suggests that the "optimal" serial correlation function is a combination of exponential and Gaussian serial correlations. It is therefore interesting to compare the obtained estimate for $\tau^2 g(u)$ with its estimates obtained under exponential and Gaussian serial correlations, respectively. We therefore fit linear mixed models with the same mean structure as the preliminary mean structure and the same random-effects structure as the preliminary random-effects structure.

The parameter estimates for the variance components in the model with Gaussian serial correlation have already been reported in Table 9.1. Those for the model with exponential serial correlation are given in Table 10.2. This latter model can be fitted with the SAS procedure MIXED by replacing the option 'type=sp(gau)(time)' by the option 'type=sp(exp)(time)' in our program on p. 129. Note also that when our exponential serial correlation function is parameterized as $g(u) = \exp(-\phi u)$, SAS provides an

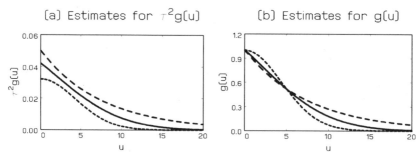

FIGURE 10.1. *Prostate Data. Estimates for the residual serial covariance function* $\tau^2 g(u)$ *and for the residual serial correlation function* $g(u)$. *The solid lines represent the estimate obtained under model (10.4), where the parameter estimates are the ones reported in Table 10.1. The long-dashed lines show the estimates under the exponential serial correlation model* $g(u) = \exp(-\phi u)$, *where the parameter estimates of* τ *and* ϕ *are the ones reported in Table 10.2. The short-dashed lines show the estimates under the Gaussian serial correlation model* $g(u) = \exp(-\phi u^2)$, *where the parameter estimates of* τ *and* ϕ *are the ones reported in Table 9.1.*

estimate for $1/\phi$ rather than for ϕ itself. The so-obtained parametric estimates for $\tau^2 g(u)$ are also included in panel (a) of Figure 10.1. Note that the estimate for $\tau^2 g(u)$ based on model (10.4) is not well approximated by either one of the other two estimates, but this is mainly due to the differences between the estimates for the variance τ^2. For studying the correlation function $g(u)$, independent of the amount of variability explained by the serial correlation component, it is often helpful to plot the corresponding rescaled functions $g(u)$, shown in panel (b) of Figure 10.1. We now find that the exponential as well as Gaussian serial correlation functions are good approximations for the function obtained under model (10.4), with a slightly better performance for the exponential serial correlation model, especially for small time lags. This is also supported by the maximized log-likelihood values which are equal to -24.266 and -24.787 for the exponential and the Gaussian serial correlation model, respectively (see Tables 9.1 and 10.2). The fact that both models are hard to distinguish is also reflected in the high correlations between the estimates for the parameters ϕ_j and the estimate for the variance σ^2, under model (10.4). The estimated correlation matrix is given by

$$\text{Corr}(\widehat{\sigma}^2, \widehat{\phi}_0, \widehat{\phi}_1, \widehat{\phi}_2) = \begin{pmatrix} 1.000 & 0.235 & -0.811 & 0.862 \\ 0.235 & 1.000 & -0.010 & 0.646 \\ -0.811 & -0.010 & 1.000 & -0.706 \\ 0.862 & 0.646 & -0.706 & 1.000 \end{pmatrix},$$

which shows that all parameter estimates are highly correlated with $\widehat{\phi}_2$, including the estimate for the variance σ^2 of the measurement errors.

TABLE 10.2. *Prostate Data. REML estimates and estimated standard errors for all variance components in a linear mixed model with the preliminary mean structure defined in Section 9.2, with random intercepts and random slopes for the linear as well as quadratic time effect, with measurement error, and with exponential serial correlation.*

Effect	Parameter	Estimate (s.e.)
Covariance of b_i:		
$\text{var}(b_{1i})$	d_{11}	0.373 (0.115)
$\text{var}(b_{2i})$	d_{22}	0.557 (0.212)
$\text{var}(b_{3i})$	d_{33}	0.060 (0.032)
$\text{cov}(b_{1i}, b_{2i})$	$d_{12} = d_{21}$	-0.378 (0.120)
$\text{cov}(b_{2i}, b_{3i})$	$d_{23} = d_{32}$	-0.177 (0.078)
$\text{cov}(b_{3i}, b_{1i})$	$d_{13} = d_{31}$	0.099 (0.042)
Measurement error variance:		
$\text{var}(\varepsilon_{(1)ij})$	σ^2	0.015 (0.005)
Gaussian serial correlation:		
$\text{var}(\varepsilon_{(2)ij})$	τ^2	0.050 (0.065)
Rate of exponential decrease	$1/\phi$	0.766 (1.301)
REML log-likelihood		-24.266
-2 REML log-likelihood		48.532

10.4   The Semi-Variogram

10.4.1   Introduction

In Section 10.3, a parametric approach was followed to study the serial correlation function $g(\cdot)$ in linear mixed models. The empirical semi-variogram is an alternative, nonparametric technique which does not require the fitting of linear mixed models. It was first introduced by Diggle (1988) for the case of random-intercepts models (i.e., linear mixed models where the only random effects are intercepts). Later, Verbeke, Lesaffre and Brant (1998) extended the technique to models which may also contain other random effects, additional to the random intercepts (see also Verbeke 1995). In Sections 10.4.2 and 10.4.3, the original empirical semi-variogram for the random-intercepts model will be discussed and illustrated. Afterward, in Sections 10.4.4 and 10.4.5, the extended version will be presented.

10.4.2   The Semi-Variogram for Random-Intercepts Models

Assuming that the only random effects in the model are random intercepts, we have that the marginal covariance matrix is given by

$$\nu^2 \, J_{n_i} \; + \; \tau^2 H_i \; + \; \sigma^2 \, I_{n_i}, \tag{10.5}$$

where J_{n_i} is the $(n_i \times n_i)$ matrix containing only ones and where ν^2 now denotes the random-intercepts variance. This implies that the residuals r_{ij} have constant variance $\nu^2 + \sigma^2 + \tau^2$ and that the correlation between any two residuals r_{ij} and r_{ik} from the same subject i is given by

$$\rho(|t_{ij} - t_{ik}|) \;\; = \;\; \frac{\nu^2 \, + \, \tau^2 \; g(|t_{ij} - t_{ik}|)}{\nu^2 \, + \, \sigma^2 \, + \, \tau^2}.$$

A stochastic process with mean zero, constant variance and a correlation function which only depends on the time lag between the measurements is often called (second-order) stationary (see for example Diggle 1990).

It immediately follows from the stationarity of the random process r_{i1}, $r_{i2}, \ldots,$ that

$$\frac{1}{2} \, E \, (r_{ij} - r_{ik})^2 \;\; = \;\; \sigma^2 \, + \, \tau^2 \, (1 - g(|t_{ij} - t_{ik}|)) \;\; = \;\; v(u_{ijk}),$$

for all $i = 1, \ldots, N$ and for all $j \neq k$. The function $v(u)$ is called the semi-variogram, and it only depends on the time points t_{ij} through the time lags $u_{ijk} = |t_{ij} - t_{ik}|$. Note that decreasing serial correlation functions $g(\cdot)$ yield increasing semi-variograms $v(u)$, with $v(0) = \sigma^2$, which converge to $\sigma^2 + \tau^2$ as u grows to infinity.

This is illustrated in Figure 10.2 for the case of exponential as well as Gaussian serial correlation (see also Figure 3.2). The two graphs show the semi-variogram for a linear random-intercepts model, once with exponential serial correlation and once with Gaussian serial correlation. The variance of the measurement errors was taken equal to $\sigma^2 = 0.7$, the variance of the serial correlation component equal to $\tau^2 = 1.3$, and the random intercepts had variance $\nu^2 = 1$. Thus, the most important source of variability was the serial correlation component $\varepsilon_{(2)i}$. The parameter ϕ was taken equal to 1; that is, the function $g(u)$ was $\exp(-u)$ for the exponential model and $\exp(-u^2)$ for the Gaussian model. Larger values of ϕ would yield semi-variograms which increase much faster, meaning that the function $g(u)$ decays to zero much quicker. The extreme case would be $\phi = +\infty$, which leads to the independence model, assuming no correlation between the components in $\varepsilon_{(2)i}$. On the other hand, values of ϕ smaller than 1 yield semi-variograms which level out much slower, meaning that $g(u)$ does not decay to zero so quickly. The extreme case would be $\phi = 0$, which leads to a model assuming

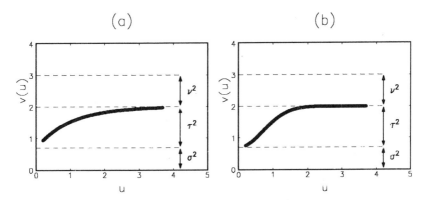

FIGURE 10.2. (a) The semi-variogram for a linear random-intercepts model containing a component with exponential serial correlation. (b) The semi-variogram for a linear random-intercepts model containing a component with Gaussian serial correlation. σ^2, τ^2, and ν^2 represent the variability of the measurement error, the serial correlation component, and the random intercepts respectively.

correlations equal to one between the components of $\varepsilon_{(2)i}$. So, the smaller the value of ϕ, the stronger the serial correlation in the data.

In practice, the function $v(u)$ is estimated by smoothing the scatter plot of the $\sum_{i=1}^{N} n_i(n_i - 1)/2$ half-squared differences $v_{ijk} = (r_{ij} - r_{ik})^2/2$ between pairs of residuals within subjects versus the corresponding time lags $u_{ijk} = |t_{ij} - t_{ik}|$. This estimate will be denoted by $\hat{v}(u)$ and is called the sample semi-variogram. Further, since

$$\frac{1}{2} E[r_{ij} - r_{kl}]^2 \;=\; \sigma^2 + \tau^2 + \nu^2$$

whenever $i \neq k$, we estimate the total process variance by

$$\hat{\sigma}^2 + \hat{\tau}^2 + \hat{\nu}^2 \;=\; \frac{1}{2N^*} \sum_{i \neq k} \sum_{j=1}^{n_i} \sum_{l=1}^{n_l} (r_{ij} - r_{kl})^2,$$

where N^* is the number of terms in the sum. This estimate, together with $\hat{v}(u)$, can now be used for deciding which of the three stochastic components will be included in the model and for selecting an appropriate function $g(u)$ in the case serial correlation will be included. The sample semi-variogram also provides initial values for τ^2, σ^2, and ν^2 when needed for the numerical maximization procedure. Finally, comparing $\hat{v}(u)$ with a fitted semi-variogram yields an informal check on the assumed covariance structure.

More details on this topic can be found in Chapter 5 in the book by Diggle, Liang and Zeger (1994) in which the semi-variogram is discussed for several special cases of the covariance structure (10.5) and the method is illustrated in a covariance analysis for some real data sets.

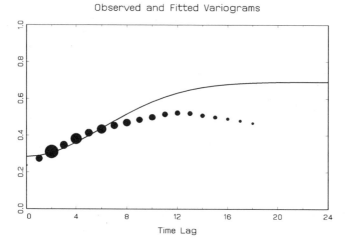

FIGURE 10.3. *Vorozole Study. Observed variogram (bullets with size proportional to the number of pairs on which they are based) and fitted variogram (solid line).*

10.4.3 Example: The Vorozole Study

The Vorozole study was introduced in Section 2.4. In order to construct a variogram, the following fixed effects were considered in the calculation of the residuals r_i: linear, quadratic, and cubic time effects, as well as the interactions between time and the covariates baseline value, treatment, and dominant site.

The variogram is presented in Figure 10.3. Apart from the observed variogram, a fitted version is presented as well. This is based on a linear mixed-effects model with fixed effects: time, time×baseline, time^2, and time^2×baseline. The covariance structure includes a random intercept, a Gaussian serial process, and residual measurement error.

10.4.4 The Semi-Variogram for Random-Effects Models

The semi-variogram described in Section 10.4.2 explicitly assumed that the marginal covariance matrices of the longitudinal vectors Y_i satisfy (10.5), which makes the semi-variogram applicable when the only random effects in the model are intercepts, and therefore only when the variance function can be assumed to be constant. However, in many practical situations, the variance function is clearly not constant, and one may wish to use random effects (other than just intercepts) to take this into account in the model. We will therefore describe now the approach of Verbeke, Lesaffre and Brant (1998), who extended the semi-variogram to the more general linear mixed model with (possibly multivariate) random effects, other than intercepts (see also Verbeke and Lesaffre 1997b).

In the presence of random effects other than intercepts, several authors have reported that the covariance structure (10.1) is often dominated by its first component. It is therefore proposed to first remove all variability, which is explained by the random effects b_i before studying the remaining serial correlation in the data. This can be done by considering again the transformed residuals $\mathfrak{R}_i$, defined in Section 10.2. The $\mathfrak{R}_i$ can now be used to study the covariance structure of $A_i'\varepsilon_{(1)i}+A_i'\varepsilon_{(2)i}$. Another way to remove the effect of the random effects would be to use subject-specific residuals $y_i - X_i\widehat{\beta} - Z_i\widehat{b_i}$, in which $\widehat{b_i} = E(b_i|y_i)$ is the empirical Bayes estimate for the random effect b_i, obtained under a specific linear mixed model. This is the approach followed by Morrell et al. (1995) in the context of piecewise nonlinear mixed-effects models for the prostate data introduced in Section 2.3.1. However, $\widehat{b_i}$ greatly depends on the normality assumption for the random effects [see our discussion in Section 7.8.2, as well as Verbeke and Lesaffre (1996a)], and is also influenced by the assumed covariance structure V_i. This already involves assumptions about the functional form of the serial correlation if present, which is why the subject-specific residuals should not be used to check assumptions about V_i. This is in contrast with the $\mathfrak{R}_i$, which are independent of any distributional assumptions for the b_i and the calculation of which does not require an estimate of the random-effects covariance matrix D. In the context of linear mixed models, we therefore propose to study the serial correlation structure based on the transformed residuals $\mathfrak{R}_i$, rather than on the subject-specific residuals $y_i - X_i\widehat{\beta} - Z_i\widehat{b_i}$.

As mentioned in Section 10.2, all components $\mathfrak{R}_{ij}$ are normally distributed with mean zero and common variance σ^2 if no serial correlation is present. In general, however, we have that

$$\frac{1}{2} E(\mathfrak{R}_{ij} - \mathfrak{R}_{ik})^2$$
$$= \sigma^2 + \tau^2 + \tau^2 \sum_{r<s} (A_{irj} - A_{irk})(A_{isj} - A_{isk}) \, g(u_{irs}), \quad (10.6)$$

where A_{irs} denotes the (r, s) element of the matrix A_i. When all measurements are taken at fixed time points, then only a small number of values u_{irs} can occur, which we denote by $u_0, \ldots, u_M$. Note that (10.6) can then be seen as a multiple regression model with parameters $\sigma^2 + \tau^2$ and $g_j = \tau^2 g(u_j)$, $j = 0, \ldots, M$, and a scatter plot of the OLS estimates $\widehat{g_0}, \ldots, \widehat{g_M}$ versus $u_0, \ldots, u_M$ can be used to select an appropriate parametric form for the function $g(u)$.

Unfortunately, in practice (e.g., in the prostate example, Section 2.3.1), one often has highly unbalanced data, resulting in a very large number of values u_{irs}. Although the above approach is no longer feasible, we can still apply it to estimate $g(u)$ for some set of prespecified values u. Let $u_0 = 0 < u_1 < \ldots < u_M$ be a set of values u for which we wish to estimate $g(u)$, and let $g_0, \ldots, g_M$ be defined as before. We take u_M equal to the largest time lag observed in the data set at hand. For each u_{irs} not equal to any of the prespecified values for u, we apply linear interpolation to approximate $g(u_{irs})$ in (10.6) by a linear combination of $g(u_t)$ and $g(u_{t+1})$, for t such that $u_t < u_{irs} < u_{t+1}$. This yields an approximate linear regression model with parameters $\sigma^2 + \tau^2$ and $g_0 = \tau^2, g_1, \ldots, g_M$. Similar to the balanced case, a scatter plot of the OLS estimates $\widehat{g_0}, \ldots, \widehat{g_M}$ versus $u_0, \ldots, u_M$ can then be used to propose an appropriate structure for the serial correlation in the model. Note that this approach can be seen as an extension of the sample semi-variogram, but with the original OLS residuals r_i replaced by the transformed versions $\Re_i$.

In order to improve the approximation in the linear interpolation, one may be inclined to choose M large, yielding small intervals $[u_t, u_{t+1}]$, $t = 0, \ldots, M - 1$. However, this automatically increases the number of parameters $g_0, \ldots, g_M$ to be estimated. Hence, there is a dilemma between reducing the number of parameters and improving the approximation. One also needs to specify the interpolation points u_t, conditional on the choice of M. We propose to first calculate all distances $u_{ijk} = |t_{ij} - t_{ik}|$ and to take the values u_t equal to the $(1/M)100\%$ percentiles of the distribution of these distances. This usually leads to a set of intervals with increasing length, which has the advantage that most computational effort is spent in the estimation of $g(u)$ for small values of u, where an accurate estimate of g is required for specifying an appropriate parametric structure for the serial correlation in the data (see the difference between exponential and Gaussian serial correlation in Section 3.3.4). Further, it can easily be shown that the covariates in model (10.6) are linearly dependent when the model contains random intercepts. We then set $g_M = 0$, which is equivalent to the assumption that the time span covered by the data is long enough for the serial correlation to have decayed to zero.

Using simulations, Verbeke (1995) has shown that the above approach yields estimates for g_t which are too unstable to be useful for the detection of residual serial correlation. This is caused by the large amount of scatter in the values $(\Re_{ij} - \Re_{ik})^2$ but also by the high degree of multicollinearity (Neter, Wasserman and Kutner, 1990, Section 8.5) in the approximate regression model, induced by the linear interpolation. A classical method to obtain more stable estimates in the presence of multicollinearity is to use ridge regression, which allows a small bias in return for stability (Sen and Srivastava 1990, Section 12.3, Neter, Wasserman and Kutner 1990, Section 1.7). The bias depends on the so-called shrinkage parameter c which is cho-

sen as small as possible (to reduce the bias) and such that the resulting esti-
mates indicate stability and satisfy approximately $\widehat{g}_0 \geq \widehat{g}_1 \geq \ldots \geq \widehat{g}_M = 0$.

Note also that the choice of the matrices A_i is not unique. In the absence
of serial correlation, we have that the distribution of the quantities $(\Re_{ij} - \Re_{ik})^2$ in (10.6) does not depend on A_i. Hence, differences in parameter
estimates due to different transformations A_i then only reflect sampling
variability. This is no longer the case in the presence of serial correlation,
where different choices for A_i not only result in different responses used in
the final regression analysis but also yields different covariates. In that case,
the effect of choosing other transformations A_i is less obvious. However,
Verbeke, Lesaffre and Brant (1998) found empirically that different choices
for the matrices A_i yield slightly different nonparametric estimates for $g(u)$,
but all these estimates lead to the same conclusion with respect to the
presence and type of serial correlation.

Further, the above check for serial correlation acts conditional on the random effects included in the model (i.e., conditional on the covariates Z_i).
However, the resulting semi-variogram is invariant under general reparameterizations of the form $Z_i G_i$. Also, if too many random effects have been
specified, the results will still be valid since the resulting transformation
matrices A_i satisfy $A_i' Z_i^* = 0$ for any matrix Z_i^* consisting of columns of
the overspecified Z_i. This again justifies favoring overspecified models $Z_i b_i$
for the detection of residual serial correlation rather than models which are
too restrictive (see also our discussion in Section 9.3).

10.4.5   Example: The Prostate Data

As an illustration, we have applied the above semi-variogram approach to
the prostate cancer data introduced in Section 2.3.1. The same preliminary
mean structure is used as in Section 9.2, resulting in the OLS residual
profiles which have been shown in Figure 9.3. Further, random intercepts as
well as slopes for time and time^2 have been included in the model. Our aim
is to check whether the lack of fit detected in Section 10.2 can be (partially)
ascribed to the fact that no serial correlation component was included. We
used five intervals, each of which contained 20% of the values u_{ijk}. The
boundaries of these intervals are thus the quintiles of the distribution of
all time lags u_{ijk}, $i = 1, \ldots, N = 54$, $j = 1, \ldots, n_i$, $k = j + 1, \ldots, n_i$, in
the prostate cancer data set. These boundaries equal $u_0 = 0$, $u_1 = 2.7$,
$u_2 = 4.7$, $u_3 = 7.5$, $u_4 = 11.2$, and $u_5 = 25.3$.

The solid line in Figure 10.4 represents the estimate for $\tau^2 g(u)$, obtained
from the methods described in the previous section. It clearly suggests the
presence of serial correlation, which may be appropriately described by a

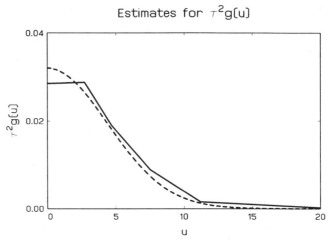

FIGURE 10.4. *Prostate Data. Estimates for the residual serial covariance function* $\tau^2 g(u)$. *The solid line represents the estimate obtained from the extended semi-variogram approach. The dashed line shows the estimated Gaussian serial covariance function* $\tau^2 \exp(-\phi u^2)$, *where the parameter estimates of* τ *and* ϕ *are the ones reported in Table 9.1.*

Gaussian serial correlation function $g(u) = \exp(-\phi u^2)$ for some constant $\phi > 0$. For $u = 0$, we have that $\widehat{g}_0 = \widehat{\tau}^2 = 0.0285$. The intercept $\sigma^2 + \tau^2$ of model (10.6) was estimated as 0.0564, from which σ^2 can be estimated as 0.0279. This suggests that the variability which cannot be explained by the three random effects can be split up into two components which are about equally important (similar variance). We can now compare these results with those obtained from fitting a linear mixed model with the same assumed preliminary mean structure, with the same assumed preliminary random-effects structure, and with Gaussian serial correlation as well as measurement error. The parameter estimates for the variance components in this model have already been reported in Table 9.1. Note that the reported estimates for τ^2 and σ^2 (0.032 and 0.023, respectively) were very close to the values we now obtain nonparametrically. We have also added the parametrically fitted Gaussian serial correlation function to Figure 10.4 (dashed line). Both estimates are very similar, suggesting that a Gaussian serial correlation component might be appropriate here.

10.5   Some Remarks

Comparing parametric models for the serial correlation function, we found in Section 10.3.3, that the exponential serial correlation function fits the prostate data slightly better than the Gaussian serial correlation function, with very similar maximized likelihood values for both models. On the other hand, the nonparametric variogram in Section 10.4.5 seems to suggest the presence of serial correlation of the Gaussian type. This again illustrates the fact that precise characterization of the serial correlation function $g(\cdot)$ is extremely difficult in the presence of several random effects. This was also the conclusion of Lesaffre, Asefa and Verbeke (1999) after the analysis of longitudinal data from more than 1500 children. However, this does not justify ignoring the possible presence of any serial correlation, since this might result in less efficient model-based inferences (see example in Section 9.5). Practical experience suggests that including serial correlation, if present, is more important than correctly specifying the serial correlation function. We therefore propose to use the procedures discussed in this chapter for detecting whether any serial correlation is present, rather than for specifying the actual shape of $g(\cdot)$, which seems to be of minor importance.

Both procedures discussed in this chapter are conditional on a specific random-effects structure. In practice, one often observes strong competition between these two sources of stochastic variation. Chi and Reinsel (1989) report that the inclusion of a sufficient number of random effects in a model with white noise errors may be able to represent the serial correlations among the measurements taken on each individual. Indeed, serial correlation can be replaced by very smooth subject-specific functions. This is also reflected in substantial correlations between the estimates for the variance components in the random-effects covariance matrix D and the estimates for the remaining variance components in the covariance structure. As an example, we consider here the covariance matrix for all variance components in the final linear mixed model (6.8), with Gaussian serial correlation in addition to the three random effects and the measurement error. The REML estimates for all parameters in this model have been reported in Table 9.3. The estimated correlation matrix for the estimates of all variance components is given by

$$\text{Corr}\left(\widehat{d_{11}}, \widehat{d_{12}}, \widehat{d_{22}}, \widehat{d_{13}}, \widehat{d_{23}}, \widehat{d_{33}}, \widehat{\tau}^2, 1/\sqrt{\widehat{\phi}}, \widehat{\sigma}^2\right)$$

$$= \begin{pmatrix}
1.00 & -0.87 & 0.62 & 0.70 & -0.49 & 0.39 & -0.18 & -0.10 & -0.00 \\
-0.87 & 1.00 & -0.85 & -0.94 & 0.75 & -0.63 & 0.21 & 0.08 & -0.03 \\
0.62 & -0.85 & 1.00 & 0.88 & -0.97 & 0.91 & -0.46 & -0.29 & 0.02 \\
0.70 & -0.94 & 0.88 & 1.00 & -0.82 & 0.72 & -0.22 & -0.06 & 0.05 \\
-0.49 & 0.75 & -0.97 & -0.82 & 1.00 & -0.97 & 0.51 & 0.33 & -0.02 \\
0.39 & -0.63 & 0.91 & 0.72 & -0.97 & 1.00 & -0.57 & -0.38 & 0.01 \\
-0.18 & 0.21 & -0.46 & -0.22 & 0.51 & -0.57 & 1.00 & 0.81 & 0.04 \\
-0.10 & 0.08 & -0.29 & -0.06 & 0.33 & -0.38 & 0.81 & 1.00 & 0.32 \\
-0.00 & -0.03 & 0.02 & 0.05 & -0.02 & 0.01 & 0.04 & 0.32 & 1.00
\end{pmatrix}.$$

We indeed get some relatively large correlations between $\hat{\tau}^2$ and the estimates of some of the parameters in D. Note also the small correlations between $\hat{\sigma}^2$ and the other estimates, except for $1/\sqrt{\hat{\phi}}$, which is not completely unexpected since the Gaussian serial correlation component reduces to measurement error for ϕ becoming infinitely large.

11
Local Influence for the Linear Mixed Model

11.1 Introduction

As explained in Chapter 5, the fitting of mixed models is based on likelihood methods (maximum likelihood, restricted maximum likelihood), which are sensitive to peculiar observations. The data analyst should be aware of particular observations that have an unusually large influence on the results of the analysis. Such cases may be found to be completely appropriate and retained in the analysis, or they may represent inappropriate data and may be eliminated from the analysis, or they may suggest that additional data need to be collected or that the current model is inadequate. Of course, an extended investigation of influential cases is only possible once they have been identified.

Many diagnostics have been developed for linear regression models. See, for example, Cook and Weisberg (1982) and Chatterjee and Hadi (1988). Since the linear mixed model can be seen as a concatenation of several subject-specific regression models, it is most obvious to investigate how these diagnostics (residual analysis, leverage, Cook's distance, etc.) can be generalized to the models considered in this book. Unfortunately, such a generalization is far from obvious. First, several kinds of residuals could be defined. For example, the marginal residual $y_i - X_i\widehat{\beta}$ reflects how a specific profile deviates from the overall population mean and can there-fore be interpreted as a residual. Alternatively, the subject-specific residual

$y_i - X_i\widehat{\beta} - Z_i\widehat{b_i}$ measures how much the observed values deviate from the subject's own predicted regression line. Finally, the estimated random effects $\widehat{b_i}$ can also be seen as residuals since they reflect how much specific profiles deviate from the population average profile. Further, the linear mixed model involves two kinds of covariates. The matrix X_i represents the design matrix for the fixed effects, and Z_i is the design matrix for the random effects. Therefore, it is not clear how leverages should be defined, partially because the matrices X_i and Z_i are not necessarily of the same dimension.

The final classification of subjects as influential or not influential can be based on the Cook's distance, first introduced by Cook (1977a, 1977b, 1979), which measures how much the parameter estimates change when a specific individual has been removed from the data set. In ordinary regression analysis, this can easily be calculated due to the availability of closed-form expressions for the parameter estimates, which makes it also possible to ascribe influence to the specific characteristics of the subjects (leverage, outlying). Unfortunately, this is no longer the case in linear mixed models. For exact Cook's distances, the iterative estimation procedure has to be used $N + 1$ times, once to fit the model for all observations and once for each individual that has been excluded from the analysis. This is not only extremely time-consuming, but it also does not give any information on the reason why some individuals are more influential than others.

All these considerations suggest that an influence analysis for the linear mixed model should not be based on the same diagnostic procedures as ordinary least squares regression. DeGruttola, Ware and Louis (1987) describe measures of influence and leverage for a generalized three-step least squares estimator for the regression coefficients in a class of multivariate linear models for repeated measurements. However, their method does not apply to maximum likelihood estimation, and it is also not clear how to extend their diagnostics to the case of unequal covariance matrices V_i.

Christensen, Pearson and Johnson (1992) have noticed that, conditionally on the variance components α, there is an explicit expression for $\widehat{\beta}$ (see Section 5.1), and hence it is possible to extend the Cook's distance to measure influence on the fixed effects in a mixed model. For known α, the so-obtained distances can be compared to a χ_p^2-distribution in order to decide which ones are exceptionally large. For estimated α, they still propose using the χ^2-distribution as approximation. Further, they define Cook's distances, based on one-step estimates, for examining case influence on the estimation of the variance components. These one-step estimates are obtained from one single step in the Newton-Raphson procedure for the maximization of the log-likelihood corresponding to the incomplete data (ith case removed), starting from the estimates obtained for the complete data. Although these procedures seem intuitively appealing, they do not yield any influence diagnostic for the fixed effects and the variance components simultaneously. Further, they do not allow to ascribe global influence to any of the subject's characteristics.

Since case-deletion diagnostics assess the effect of an observation by completely removing it, they fit into the framework of global influence analyses. This contrasts with a local influence analysis, first introduced by Cook (1986). Beckman, Nachtsheim and Cook (1987) used the idea of local influence to develop methods for assessing the effect of perturbations from the usual assumptions in the mixed-models analysis of variance with uncorrelated random components. They investigate how the parameters change under small perturbations of the error variances, the random-effects variances, and the response vector. An alternative perturbation scheme, proposed by Verbeke (1995), Verbeke and Lesaffre (1997b), and Lesaffre and Verbeke (1998), is case-weight perturbation where it is investigated how much the parameter estimates are affected by changes in the weights of the log-likelihood contributions of specific individuals. In Section 11.2, some general theory on local influence will be presented. Afterward, in Section 11.3, this will be applied to case-weight perturbations in the context of linear mixed models. Finally, in Section 11.4, the local influence methodology will be illustrated in an influence analysis for the prostate data.

11.2 Local Influence

In this section, the local influence approach, first introduced by Cook (1986), will be presented. The general idea is to give every individual its own weight in the calculation of the parameter estimates and to investigate how these estimates depend on the weights, locally around the equal-weight case (i.e., all individuals have the same weight 1), which is the ordinary maximum likelihood case. Much of the terminology and most concepts in the sequel of this section are borrowed from differential geometry. More details can be found in any textbook on differential geometry; see, for example, O'Neill (1966).

We know from expression (5.2) that the maximum likelihood log-likelihood function for model (3.8) can be seen as

$$\ell(\boldsymbol{\theta}) \; = \; \sum_{i=1}^{N} \ell_i(\boldsymbol{\theta}), \tag{11.1}$$

in which $\ell_i(\boldsymbol{\theta})$ is the contribution of the ith individual to the log-likelihood. Let $\ell(\boldsymbol{\theta}|\boldsymbol{\omega})$ now denote any perturbed version of $\ell(\boldsymbol{\theta})$, depending on an r-

dimensional vector $\boldsymbol{\omega}$ of weights, which is assumed to belong to an open subset Ω of $I\!\!R^r$, and such that the original log-likelihood $\ell(\boldsymbol{\theta})$ is obtained for $\boldsymbol{\omega} = \boldsymbol{\omega_0}$.

For the detection of influential subjects, our perturbed log-likelihood will be

$$\ell(\boldsymbol{\theta}|\boldsymbol{\omega}) = \sum_{i=1}^{N} w_i \ell_i(\boldsymbol{\theta}), \tag{11.2}$$

which clearly allows different weights for different subjects. The weight vector $\boldsymbol{\omega}$ is then N dimensional, and the original log-likelihood corresponds to $\boldsymbol{\omega} = \boldsymbol{\omega_0} = (1, 1, \ldots, 1)'$. Note also that the log-likelihood with the ith case completely removed corresponds to the vector $\boldsymbol{\omega}$ with $w_i = 0$ and $w_j = 1$ for all $j \neq i$.

Later, in Chapter 19, another perturbation scheme will be used to perform a sensitivity analysis in the context of missing data. For now, we will further consider the case of a general perturbed log-likelihood $\ell(\boldsymbol{\theta}|\boldsymbol{\omega})$ to explain the local influence methodology.

Let $\widehat{\boldsymbol{\theta}}$ be the maximum likelihood estimator for $\boldsymbol{\theta}$, obtained by maximizing $\ell(\boldsymbol{\theta})$, and let $\widehat{\boldsymbol{\theta}}_\omega$ denote the estimator for $\boldsymbol{\theta}$ under $\ell(\boldsymbol{\theta}|\boldsymbol{\omega})$. The local influence approach now compares $\widehat{\boldsymbol{\theta}}_\omega$ with $\widehat{\boldsymbol{\theta}}$. Similar estimates indicate that perturbations have little effect on the parameter estimates. Strongly different estimates suggest that the estimation procedure is highly sensitive to such perturbations. Cook (1986) proposed to measure the distance between $\widehat{\boldsymbol{\theta}}_\omega$ and $\widehat{\boldsymbol{\theta}}$ by the so-called likelihood displacement, defined by

$$\mathrm{LD}(\boldsymbol{\omega}) = 2\left(\ell(\widehat{\boldsymbol{\theta}}) - \ell(\widehat{\boldsymbol{\theta}}_\omega)\right).$$

This way, the variability of $\widehat{\boldsymbol{\theta}}$ is taken into account. $\mathrm{LD}(\boldsymbol{\omega})$ will be large if $\ell(\boldsymbol{\theta})$ is strongly curved at $\widehat{\boldsymbol{\theta}}$ (which means that $\boldsymbol{\theta}$ is estimated with high precision) and $\mathrm{LD}(\boldsymbol{\omega})$ will be small if $\ell(\boldsymbol{\theta})$ is fairly flat at $\widehat{\boldsymbol{\theta}}$ (meaning that $\boldsymbol{\theta}$ is estimated with high variability). From this perspective, a graph of $\mathrm{LD}(\boldsymbol{\omega})$ versus $\boldsymbol{\omega}$ contains essential information on the influence of case-weight perturbations. It is useful to view this graph as the geometric surface formed by the values of the $(r+1)$-dimensional vector

$$\boldsymbol{\xi}(\boldsymbol{\omega}) = \begin{pmatrix} \boldsymbol{\omega} \\ \mathrm{LD}(\boldsymbol{\omega}) \end{pmatrix}$$

as $\boldsymbol{\omega}$ varies throughout Ω. In differential geometry, a surface of this form is frequently called a Monge patch. Following Cook (1986), we will refer to $\boldsymbol{\xi}(\boldsymbol{\omega})$ as an influence graph. It is a surface in $I\!\!R^{r+1}$ and can be used to assess the influence of varying $\boldsymbol{\omega}$ through Ω. A graphical representation, which also illustrates all further developments, is given in Figure 11.1.

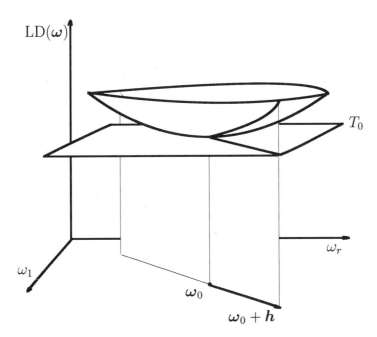

FIGURE 11.1. *Graphical representation of the influence graph and the local influence approach.*

Ideally, we would like a complete influence graph [i.e., a graph of $\boldsymbol{\xi}(\boldsymbol{\omega})$ for varying $\boldsymbol{\omega}$] to assess influence for a particular model and a particular data set. However, this is only possible in cases where the number r of weights in $\boldsymbol{\omega}$ does not exceed 2. Hence, methods are needed for extracting the most relevant information from an influence graph. One possible approach is local influence, which uses normal curvatures of $\boldsymbol{\xi}(\boldsymbol{\omega})$ in $\boldsymbol{\omega}_0$. One proceeds as follows. Let T_0 be the tangent plane to $\boldsymbol{\xi}(\boldsymbol{\omega})$ at $\boldsymbol{\omega}_0$. Since $\mathrm{LD}(\boldsymbol{\omega})$ attains its minimum at $\boldsymbol{\omega}_0$, we have that T_0 is parallel to $\Omega \subset I\!\!R^r$. Each vector $\boldsymbol{h}$ in Ω, of unit length, determines a plane that contains $\boldsymbol{h}$ and which is orthogonal to T_0. The intersection, called a normal section, of this plane with the surface is called a lifted line. It can be graphed by plotting $\mathrm{LD}(\boldsymbol{\omega}_0 + a\boldsymbol{h})$ versus the univariate parameter $a \in I\!\!R$. The normal curvature of the lifted line, denoted by C_h, is now defined as the curvature of the plane curve $(a, \mathrm{LD}(\boldsymbol{\omega}_0 + a\boldsymbol{h}))$ at $a = 0$. It can be visualized by the inverse radius of the best-fitting circle at $a = 0$. The curvature C_h is called the normal curvature of the surface $\boldsymbol{\xi}(\boldsymbol{\omega})$, at $\boldsymbol{\omega}_0$, in the direction $\boldsymbol{h}$. Large values of C_h indicate sensitivity to the induced perturbations in the direction $\boldsymbol{h}$. C_h is called the local influence on the estimation of $\boldsymbol{\theta}$, of perturbing the model corresponding to the log-likelihood (11.1), in the direction $\boldsymbol{h}$.

Let $\boldsymbol{\Delta}_i$ be the s-dimensional vector of second-order derivatives of $\ell(\boldsymbol{\theta}|\boldsymbol{\omega})$, with respect to w_i and all components of $\boldsymbol{\theta}$, and evaluated at $\boldsymbol{\theta} = \widehat{\boldsymbol{\theta}}$ and at $\boldsymbol{\omega} = \boldsymbol{\omega}_0$, and let Δ be the $(s \times r)$ matrix with $\boldsymbol{\Delta}_i$ as the ith column. Further, let $\ddot{L}$ denote the $(s \times s)$ matrix of all second-order derivatives of $\ell(\boldsymbol{\theta})$, also evaluated at $\boldsymbol{\theta} = \widehat{\boldsymbol{\theta}}$. Cook (1986) has then shown that, for any unit vector $\boldsymbol{h}$ in Ω,

$$C_h = 2\left| \boldsymbol{h}'\Delta'\ddot{L}^{-1}\Delta\boldsymbol{h} \right|. \tag{11.3}$$

There are several ways in which (11.3) can be used to study $\boldsymbol{\xi}(\boldsymbol{\omega})$, each corresponding to a specific choice of the unit vector $\boldsymbol{h}$. One evident choice corresponds to the perturbation of the ith weight only (case-weight perturbation). This is obtained by taking $\boldsymbol{h}$ equal to the vector $\boldsymbol{h}_i$ which contains zeros everywhere except on the ith position, where there is a one. The resulting local influence is then given by

$$C_i \equiv C_{h_i} = 2\left| \boldsymbol{\Delta}_i'\ddot{L}^{-1}\boldsymbol{\Delta}_i \right|. \tag{11.4}$$

Another important direction is determined by $\boldsymbol{h}_{\max}$, which corresponds to the maximal normal curvature $C_{\max}$. It follows from Seber (1984, p. 526) that $C_{\max}$ is twice the largest eigenvalue of $-\Delta'\ddot{L}^{-1}\Delta$, and that $\boldsymbol{h}_{\max}$ is the corresponding eigenvector. Hence, the calculation of $C_{\max}$ and $\boldsymbol{h}_{\max}$ involves an eigenvalue analysis of an $(r \times r)$-dimensional matrix and can therefore be both difficult and computationally expensive. Fortunately, $\Delta'\ddot{L}^{-1}\Delta$ is only of rank s, which is often small in comparison to r, and this can be exploited to simplify the computations as follows (see also Beckman, Nachtsheim and Cook 1987). We know from Seber (1984, p. 521) that there exists a nonsingular matrix R such that $\ddot{L}^{-1} = -R'R$, from which it follows that $-\Delta'\ddot{L}^{-1}\Delta$ equals $\Delta'R'R\Delta$. It now follows from Seber (1984, p. 518) that the nonzero eigenvalues of $-\Delta'\ddot{L}^{-1}\Delta$ are the same as those of $R\Delta\Delta'R'$, which is of dimension $(s \times s)$. Hence, $C_{\max}$ can easily be calculated as twice the largest eigenvalue of $R\Delta\Delta'R'$. Let z denote the corresponding eigenvector. The direction of maximal normal curvature can then readily be seen to equal

$$\boldsymbol{h}_{\max} = \frac{\Delta'R'z}{\|\Delta'R'z\|}.$$

The direction $\boldsymbol{h}_{\max}$ is the direction for which the normal curvature is maximal. It shows how to perturb the postulated model (the model for $\boldsymbol{\omega} = \boldsymbol{\omega}_0$) to obtain the largest local changes in the likelihood displacement. If, for example, the ith component of $\boldsymbol{h}_{\max}$ is found to be relatively large, this indicates that perturbations in the weight w_i may lead to substantial changes in the results of the analysis. On the other hand, denoting the s nonzero eigenvalues of $-\Delta'\ddot{L}^{-1}\Delta$ by $C_{\max}/2 \equiv \lambda_1 \geq \ldots \geq \lambda_s > 0$ and the corresponding normalized orthogonal eigenvectors by $\{h_{\max} \equiv v_1, \ldots, v_s\}$, we have that

$$C_i = 2 \sum_{j=1}^{s} \lambda_j v_{ji}^2, \tag{11.5}$$

where v_{ji} is the ith component of the vector v_j. Hence, the local influence C_i of perturbing the ith weight w_i can be large without the ith component of $h_{\max}$ to be large, as long as some of the other eigenvectors of $-\Delta'\ddot{L}^{-1}\Delta$ have large ith components. This will be further illustrated with an example in Section 11.4. Therefore, it is not sufficient to calculate only $h_{\max}$; all C_i should be computed as well. Cook (1986) proposes to inspect $h_{\max}$, regardless of the size of $C_{\max}$, since it may highlight directions which are simultaneously influential. However, since there is usually no analytic expression for $h_{\max}$, it is very difficult to get some insight behind the reasons for such influences. This is not the case for the influence measures C_i. In Section 11.3, it will be shown how expression (11.4) can be used to ascribe local influence to interpretable components such as residuals.

When a subset θ_1 of $\theta = (\theta_1', \theta_2')'$ is of special interest, a similar approach can be used, replacing the log-likelihood by the profile log-likelihood. Let $g(\theta_1)$ be the value of θ_2 that maximizes $\ell(\theta) = \ell(\theta_1, \theta_2)$ for each fixed θ_1. The profile log-likelihood for θ_1 is defined as $\ell(\theta_1, g(\theta_1))$. For $\widehat{\theta}_{1\omega}$ determined from the partition $\widehat{\theta}_\omega' = (\widehat{\theta}_{1\omega}', \widehat{\theta}_{2\omega}')$, the likelihood displacement based on this profile log-likelihood is given by

$$\mathrm{LD}_1(\omega) = 2\left(\ell(\widehat{\theta}) - \ell(\widehat{\theta}_{1\omega}, g(\widehat{\theta}_{1\omega}))\right).$$

The methods discussed above for the full parameter vector now directly carry over to calculate local influences on the geometric surface defined by $\mathrm{LD}_1(\omega)$. We now partition $\ddot{L}$ as

$$\ddot{L} = \begin{pmatrix} \ddot{L}_{11} & \ddot{L}_{12} \\ \ddot{L}_{21} & \ddot{L}_{22} \end{pmatrix},$$

according to the dimensions of θ_1 and θ_2. Cook (1986) has then shown that the local influence on the estimation of θ_1, of perturbing the model in the direction of a normalized vector h, is given by

$$C_h(\theta_1) = 2\left| h'\Delta'\left[\ddot{L}^{-1} - \begin{pmatrix} 0 & 0 \\ 0 & \ddot{L}_{22}^{-1} \end{pmatrix}\right]\Delta h\right|. \tag{11.6}$$

Because all eigenvalues of the matrix

$$\begin{pmatrix} \ddot{L}_{11} & \ddot{L}_{12} \\ \ddot{L}_{21} & \ddot{L}_{22} \end{pmatrix}\begin{pmatrix} 0 & 0 \\ 0 & \ddot{L}_{22}^{-1} \end{pmatrix} = \begin{pmatrix} 0 & \ddot{L}_{12}\ddot{L}_{22}^{-1} \\ 0 & I \end{pmatrix}$$

are either one or zero, we have that (Seber 1984, p. 526) for any vector v

$$0 \geq v' \begin{pmatrix} 0 & 0 \\ 0 & \ddot{L}_{22}^{-1} \end{pmatrix} v \geq v'\ddot{L}^{-1}v$$

and therefore also that

$$\begin{aligned} C_h(\theta_1) &= -2h'\Delta'\ddot{L}^{-1}\Delta h + 2h'\Delta' \begin{pmatrix} 0 & 0 \\ 0 & \ddot{L}_{22}^{-1} \end{pmatrix} \Delta h \\ &= C_h + 2h'\Delta' \begin{pmatrix} 0 & 0 \\ 0 & \ddot{L}_{22}^{-1} \end{pmatrix} \Delta h \qquad (11.7) \\ &\leq C_h. \end{aligned}$$

This means that the normal curvature for θ_1, in the direction h, can never be larger than the normal curvature for θ in that same direction.

Note also that it immediately follows from (11.6) that, for $\ddot{L}_{12} = 0$,

$$\begin{aligned} C_h &= -2h'\Delta' \begin{pmatrix} \ddot{L}_{11}^{-1} & 0 \\ 0 & 0 \end{pmatrix} \Delta h - 2h'\Delta' \begin{pmatrix} 0 & 0 \\ 0 & \ddot{L}_{22}^{-1} \end{pmatrix} \Delta h \\ &= C_h(\theta_1) + C_h(\theta_2). \end{aligned}$$

Hence, we have that for any direction h, the normal curvature for θ in that direction is then the sum of the normal curvatures for θ_1 and θ_2 in the same direction. Intuitively, this can be explained as follows. It follows from the classical maximum likelihood theory that, for sufficiently large samples, and under the correct model specification, $\hat{\theta}$ is asymptotically normally distributed with covariance matrix $-\ddot{L}^{-1}$. So, $\ddot{L}_{12} = 0$ means that $\hat{\theta}_1$ and $\hat{\theta}_2$ are statistically independent. It is then not surprising that $C_h = C_h(\theta_1) + C_h(\theta_2)$ since this expresses the fact that the influence for θ_1 is independent of the influence for θ_2.

Finally, there are again many possible choices for the vector h. For example, the local influence of perturbing the ith weight on the estimation of θ_1 is obtained for $h = h_i$, the vector with zeros everywhere except on the ith position where there is a one. The corresponding normal curvature will be denoted by $C_i(\theta_1)$.

11.3   The Detection of Influential Subjects

We will now show how the local influence approach, introduced in Section 11.2, can be applied to detect subjects which are locally influential for the fitting of a specific linear mixed model. We hereby restrict the discussion to models which assume conditional independence; that is, models of the form (3.8) where all residual covariance matrices Σ_i are equal to $\sigma^2 I_{n_i}$. Our perturbed log-likelihood is defined in (11.2), which allows different subjects to have different weights in the log-likelihood function. The vector Δ_i now equals the s-dimensional vector of first-order derivatives of $\ell_i(\theta)$, with respect to all components of θ and evaluated at $\theta = \widehat{\theta}$. Note that the calculation of Δ_i usually only requires little additional computational effort, since the first-order derivative of the log-likelihood is needed in the iterative Newton-Raphson estimation procedure.

As explained in Section 11.2, local influence measures C_h can be calculated for a variety of unit vectors h. For the detection of influential subjects, the measures C_i will be of particular interest since they represent the local influence of each subject separately on the estimation of θ. In a global case-deletion approach, the maximum likelihood estimate $\widehat{\theta}$ is compared with $\widehat{\theta}_{(i)}$ obtained by maximizing

$$\ell_{(i)}(\theta) = \sum_{j \neq i} \ell_j(\theta).$$

Because this is computationally very expensive, one instead often compares $\widehat{\theta}$ with the approximation of $\widehat{\theta}_{(i)}$ given by

$$\widehat{\theta}^1_{(i)} = \widehat{\theta} - \left(\ddot{L}_{(i)}(\widehat{\theta})\right)^{-1} \sum_{j \neq i} \Delta_j(\widehat{\theta}), \tag{11.8}$$

where $\ddot{L}_{(i)}(\widehat{\theta})$ is the matrix of second-order derivatives of $\ell_{(i)}(\theta)$, evaluated at $\widehat{\theta}$. The vector $\widehat{\theta}^1_{(i)}$ is referred to as the one-step approximation to $\widehat{\theta}_{(i)}$ since it is obtained from a single Newton-Raphson step in the maximization procedure of $\ell_{(i)}(\theta)$, using $\widehat{\theta}$ as the starting value.

A measure of influence, proposed by Pregibon (1981), is then

$$\rho_i = -\left(\widehat{\theta} - \widehat{\theta}^1_{(i)}\right)' \ddot{L} \left(\widehat{\theta} - \widehat{\theta}^1_{(i)}\right).$$

See also Pregibon (1979) (Chapter 5) and Cook and Weisberg (1982) (Chapter 5) for more details. It now follows from (11.8) that

$$\Delta_i = -\sum_{j \neq i} \Delta_j = \ddot{L}_{(i)}(\widehat{\theta}) \left(\widehat{\theta}^1_{(i)} - \widehat{\theta}\right), \tag{11.9}$$

such that expression (11.4) becomes

$$C_i = -2\left(\widehat{\theta} - \widehat{\theta}^1_{(i)}\right)' \ddot{L}_{(i)} \ddot{L}^{-1} \ddot{L}_{(i)} \left(\widehat{\theta} - \widehat{\theta}^1_{(i)}\right). \tag{11.10}$$

Comparison of (11.9) with (11.10) shows that $C_i/2$ and ρ_i are approximately the same for N sufficiently large. Hence, our local influence measures C_i can be interpreted as approximations to the classical global case-deletion diagnostics.

An advantage of C_i, in comparison to, for example, the direction $h_{\max}$ of maximal curvature, is the availability of the analytic expression (11.4) for C_i. Lesaffre and Verbeke (1998) have shown that C_i can be decomposed into five interpretable components. Let $\mathcal{R}_i$, $\mathcal{X}_i$, and $\mathcal{Z}_i$ denote now the "standardized" residuals and covariates for the ith individual, defined by $\mathcal{R}_i = V_i^{-1/2} r_i$, $\mathcal{X}_i = V_i^{-1/2} X_i$, and $\mathcal{Z}_i = V_i^{-1/2} Z_i$, respectively, with r_i equal to $y_i - X_i \widehat{\beta}$. Further, for any matrix A, let $\|A\| = \sqrt{\operatorname{tr}(A'A)}$ be the Frobenius norm of A (see Golub and Van Loan 1989). The interpretable components in C_i are then

$$\|\mathcal{X}_i \mathcal{X}_i'\|, \quad \|\mathcal{R}_i\|, \quad \|\mathcal{Z}_i \mathcal{Z}_i'\|, \quad \|I - \mathcal{R}_i \mathcal{R}_i'\|, \quad \|V_i^{-1}\|. \qquad (11.11)$$

First, $\|\mathcal{X}_i \mathcal{X}_i'\|$ measures the "length" of the standardized covariates in the mean structure and $\|\mathcal{R}_i\|$ is an overall measure for how well the observed data for the ith subject are predicted by the mean structure $X_i \beta$. Second, the components $\|\mathcal{Z}_i \mathcal{Z}_i'\|$ and $\|I - \mathcal{R}_i \mathcal{R}_i'\|$ have a similar meaning, but for the covariance structure. For example, $\|I - \mathcal{R}_i \mathcal{R}_i'\|$ will be zero only if V_i equals $r_i r_i'$. Note that $r_i r_i'$ is an estimate for $\operatorname{var}(y_i)$, which only assumes the mean to be correctly modeled as $X_i \beta$. Therefore, $\|I - \mathcal{R}_i \mathcal{R}_i'\|$ can be interpreted as a residual, measuring how well the covariance structure of the data is modeled by $V_i = Z_i D Z_i' + \sigma^2 I_{n_i}$. Finally, the fifth component $\|V_i^{-1}\|$ will be large if V_i has small eigenvalues, which indicates that the ith subject is assumed to have small variability.

The decomposition of C_i immediately suggests a practical procedure to find an explanation for the influential character of an individual. Namely, when C_i is large, we inspect the diagnostics (11.11). Index plots are useful to graphically inspect the individuals vis-à-vis their influential nature. Hence, we propose to start with an index plot of C_i. In a second step, the index plots of (11.11) can be examined. A recurrent practical difficulty with diagnostics is to establish a threshold above which an individual is defined as "remarkable". It follows from (11.4) that

$$\sum_{i=1}^{N} C_i = -2 \operatorname{tr}\left(\ddot{L}^{-1} \sum_{i=1}^{N} \Delta_i \Delta_i' \right),$$

which converges to $2s$, for N becoming infinitely large. As for leverage in linear regression (see, for example, Neter, Wasserman and Kutner 1990, pp. 395-396), one could classify an individual for which C_i is larger than twice the average value (larger than $4s/N$, for N large) as being influential. However, unlike for the leverage situation, $2s$ is only the approximate sum of the C_i, which will not be accurate if the model is not correctly specified (such that $\ddot{L}^{-1}\sum_i \Delta_i \Delta_i'$ does not converge to I_s) or if N is too small for the asymptotics to yield good approximations. In such cases, we propose to replace $2s$ by the actual sum, and we call the ith subject influential if C_i is larger than the cutoff value $2\sum_{i=1}^N C_i/N$.

It is less evident to find "natural" thresholds for the diagnostics (11.11). Also in the context of linear mixed models, Waternaux, Laird and Ware (1989) proposed to calibrate $\|\mathcal{R}_i\|^2$ with a χ^2-distribution with n_i degrees of freedom to detect outliers. Note that also the other interpretable components depend on the size n_i of the response vector $\boldsymbol{y}_i$. However, it is not immediately clear how to correct them for n_i. We therefore suggest adding an index plot of n_i to the index plots of the interpretable components (11.11). We refer to Section 11.4 for an example.

The general theory on local influence also allows assessing influence on subsets of $\boldsymbol{\theta}$. Here, we will be especially interested in the local influence of subjects on the estimation of the fixed effects $\boldsymbol{\beta}$ or the variance components $\boldsymbol{\alpha}$ separately. The local influence of subject i on the estimation of $\boldsymbol{\beta}$ will be denoted by $C_i(\boldsymbol{\beta})$. For the variance components, this will be denoted by $C_i(\boldsymbol{\alpha})$. It can be shown that $C_i(\boldsymbol{\beta})$ and $C_i(\boldsymbol{\alpha})$ contain the same five interpretable components in their decomposition as C_i. Further, we have that for any variance component α_k from $\boldsymbol{\alpha}$,

$$\frac{\partial^2 \ell(\boldsymbol{\theta})}{\partial\boldsymbol{\beta}\partial\alpha_k} = \sum_{i=1}^N X_i' V_i^{-1} \frac{\partial V_i}{\partial\alpha_k} V_i^{-1}(\boldsymbol{y_i} - X_i\boldsymbol{\beta}),$$

which implies that the maximum likelihood estimates for the fixed effects and for the variance components are asymptotically independent (see Verbeke and Lesaffre 1996b, 1997a, for technical details). It now follows from Section 11.2 that, for N sufficiently large,

$$C_i \approx C_i(\boldsymbol{\beta}) + C_i(\boldsymbol{\alpha}). \tag{11.12}$$

Lesaffre and Verbeke (1998) have shown that this also implies that $C_i(\boldsymbol{\beta})$ can be decomposed using only the first two components $\|\mathcal{X}_i\mathcal{X}_i'\|$ and $\|\mathcal{R}_i\|$, whereas only the last three components $\|\mathcal{Z}_i\mathcal{Z}_i'\|$, $\|I - \mathcal{R}_i\mathcal{R}_i'\|$, and $\|V_i^{-1}\|$ are needed in the decomposition of $C_i(\boldsymbol{\alpha})$. Hence, for sufficiently large data sets, influence for the fixed effects and for the variance components can be further investigated by studying the first two and the last three interpretable components, respectively. This will also be illustrated in our example in Section 11.4.

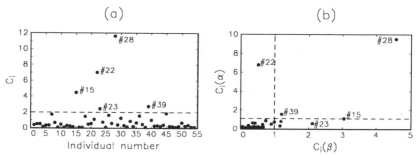

FIGURE 11.2. *Prostate Data. (a) Plot of total local influences C_i versus the identification numbers of the individuals in the BLSA data set. (b) Scatter plot of the local influence measures $C_i(\beta)$ and $C_i(\alpha)$ for the fixed effects and the variance components, respectively. The most influential subjects are indicated by their identification number.*

11.4   Example: The Prostate Data

As an illustration, we now perform a local influence analysis for model (6.8) for the prostate data. The first step is to trace the individuals who have a large impact on the parameter estimates, measured by C_i. In the second step, it is determined which part of the fitted model is affected by the influential cases, the fixed effects, and/or the variance components. Finally, the cause of the influential character has to be established in order to obtain insight in why a case is peculiar. The calculation of the influence measures and the interpretable components can be performed with a SAS macro available from the website.

Figure 11.2(a) is an index plot of the total local influence C_i. The cutoff value used for C_i equals $2\sum_i C_i/N = 1.98$ and has been indicated in the figure by the dashed line. Participants #15, #22, #23, #28, and #39 are found to have a C_i value larger than 1.98 and are therefore considered to be relatively influential for the estimation of the complete parameter vector $\boldsymbol{\theta}$. Their observed and expected profiles are shown in Figure 11.3. Pearson *et al.* (1994) report that subjects #22, #28, and #39, who were classified as local/regional cancer cases, were probably misclassified metastatic cancer cases. It is therefore reassuring that this influence approach flagged these three cases as being special. Subjects #15 and #23 were already in the metastatic cancer group. In Figure 11.2(b), a scatter plot of $C_i(\alpha)$ versus $C_i(\beta)$ is given. Their respective cutoff values are $2\sum_i C_i(\alpha)/N = 1.10$ and $2\sum_i C_i(\beta)/N = 0.99$. Obviously, subject #28, who is the subject with the largest C_i value, is highly influential for both the fixed effects and the variance components. Individuals #15 and #39 are also influential for both parts of the model, but to a much lesser extent. Finally, we have that subject #23 is influential only for the fixed effects β and that, except for subject

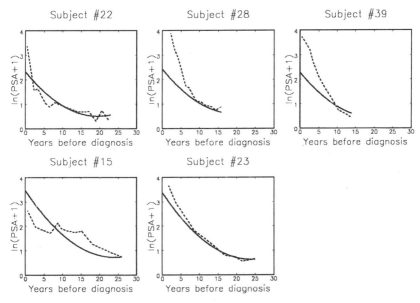

FIGURE 11.3. *Prostate Data. Observed (dashed lines) and fitted (solid lines) profiles for the five most locally influential subjects. All subjects are metastatic cancer cases, but individuals #22, #28, and #39 were wrongly classified as local/regional cancer cases.*

#28, subject #22 has the highest influence for the variance components α, but is not influential for β.

Figure 11.4 shows an index plot of each of the five interpretable components in the decomposition of the local influence measures C_i, $C_i(\beta)$, and $C_i(\alpha)$, as well as of the number n_i of PSA measurements available for each subject. These can now be used to ascribe the influence of the influential subjects to their specific characteristics. As an example, we will illustrate this for subject #22, which has been circled in Figure 11.4. As indicated by Figure 11.2, this subject is highly influential, but only for the estimation of the variance components. If approximation (11.12) is sufficiently accurate, this influence for α can be ascribed to the last three interpretable components only [i.e., the components plotted in the panels (c), (d), and (e) of Figure 11.4]. Hence, although the residual component for the mean structure is the largest for subject #22 [Figure 11.4(b)], it is not the cause of the highly influential character of this subject for the estimation of the variance components, nor did it cause a large influence on the estimation of the fixed effects in the model. Note instead how this subject also has the largest residual for the covariance structure, suggesting that the covariance matrix is poorly predicted by the model-based covariance V_{22}. This is also illustrated in Figure 11.3. Obviously, the large residual for the mean was caused by the poor prediction around the time of diagnosis, but this

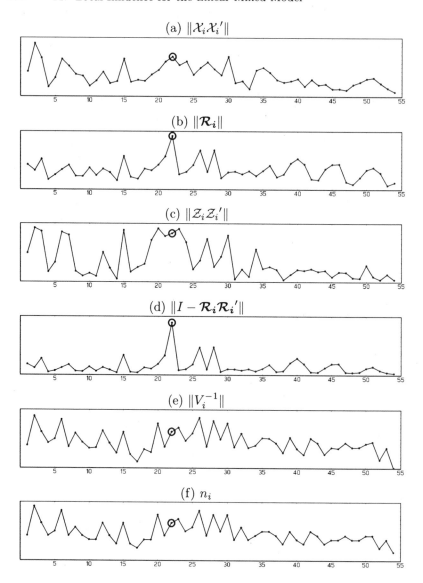

FIGURE 11.4. *Prostate Data. Index plots of the five interpretable components in the decomposition of the total local influence C_i and an index plot of the number n_i of repeated measurements for each subject.*

was not sufficient to make subject #22 influential for the estimation of the complete average profile. Further, a close look at the estimated covariance matrix V_{22} shows that only positive correlations are assumed between the repeated measurements, whereas the positive and negative residuals in Figure 11.3 suggest some negative correlations.

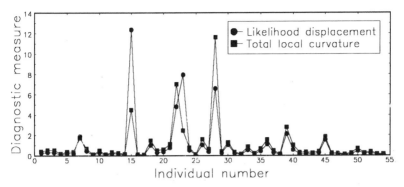

FIGURE 11.5. *Prostate Data. Comparison of the likelihood displacement LD_i with the total local curvature C_i. The individuals are numbered by their identification number.*

As discussed in Section 11.3, the local influence measures C_i can be interpreted as approximations to the classical global case-deletion diagnostics. As an illustration, we also performed such a global influence analysis. Figure 11.5 compares the local influence measures C_i with the likelihood displacements

$$\mathrm{LD}_i \;=\; 2\left[\ell(\widehat{\boldsymbol{\theta}}) - \ell(\widehat{\boldsymbol{\theta}}_{(i)})\right],$$

where $\widehat{\boldsymbol{\theta}}_{(i)}$ denotes the maximum likelihood estimate for $\boldsymbol{\theta}$ after deletion of subject i. From this picture we can observe that for the data set at hand, both approaches highlight the same set of influential observations. However, they do not agree on the ranking of the observations according to their influence.

We also calculated the direction $\boldsymbol{h}_{\max} = \boldsymbol{v_1}$ of maximal curvature $C_{\max}$ for our data set at hand. $C_{\max}$ equals 15.19 and an index plot of the individual components of $\boldsymbol{h}_{\max}$ is shown in Figure 11.6. The vector $\boldsymbol{h}_{\max}$ is pointing toward individuals #15, #28, #39, and #45. These individuals were also found in the index plot of C_i. Indeed, also case #45 is standing out in Figure 11.2, although the C_{45} value is below the threshold. However, subjects #22 and #23 are not highlighted in Figure 11.6. This illustrates the fact that locally influential subjects are not necessarily subjects with a large component in the direction of maximal curvature. Subject #22, for example, has total local influence C_{22} equal to 7.02, but seems to have a small weight in $\boldsymbol{v_1}$ ($v_{1,22} = 0.18$). As already explained in Section 11.2, this occurs when $-\Delta'\ddot{L}^{-1}\Delta$ has large eigenvalues other than λ_1 and for subjects with much weight in one or more eigenvectors corresponding to such eigenvalues. The first five terms in expression (11.5) are shown for subject #22 in Table 11.1. The five largest eigenvalues of $-\Delta'\ddot{L}^{-1}\Delta$ are $\lambda_1 = 7.59$, $\lambda_2 = 5.49$, $\lambda_3 = 2.77$, $\lambda_4 = 1.68$, and $\lambda_5 = 1.53$, and the weight

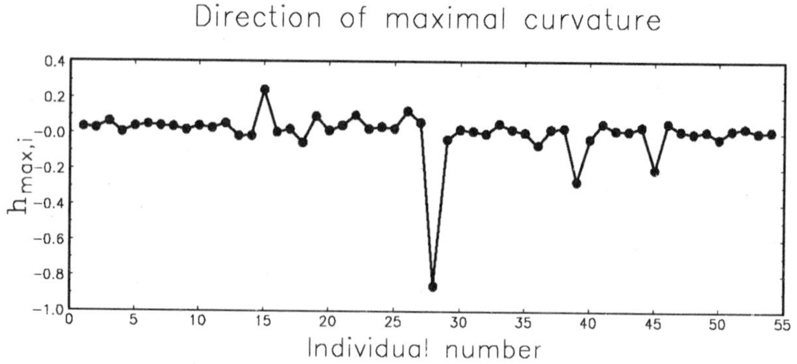

FIGURE 11.6. *Prostate Data. An index plot of the components of* h_{max}.

of this subject in each of the corresponding eigenvectors is $v_{1,22} = 0.18$, $v_{2,22} = -0.74$, $v_{3,22} = -0.20$, $v_{4,22} = 0.11$, and $v_{5,22} = -0.01$. Because of its large weight in v_2, this subject has a large second component in the local influence $(2\lambda_2 v_{2,22}^2 = 6.0413)$, which results in a large C_{22} value. This is why we believe that an influence analysis should not be based on the direction of maximal curvature only.

On the other hand, the index plot in Figure 11.6 offers an extra diagnostic tool. The components of h_{max} have a positive or a negative sign. From this plot, one can observe that case #15 has a different sign than individuals #28, #39, and #45. The impression is given that individual #15 is counterbalancing the effect of these other cases. For the three cases with a negative component of h_{max}, the observed profiles (see also Figure 11.3) are completely above and much steeper than the predicted profiles, whereas for individual #15 the observed response intersects its prediction somewhat halfway in the observation period and is also much less steeper. More detailed similar information could be obtained from h_{max} by deriving it for the fixed effects and the variance components separately. On the other hand, since no analytic expression for h_{max} is available, using h_{max} as a diagnostic tool does not yield much insight into the reasons why some individuals are more influential than others.

Finally, since three subjects were probably misclassified as local cancer cases (participants #22, #28, and #39), we reallocated them to the metastatic cancer group and performed a new influence analysis. Subjects #23 and #39 were not influential anymore, but the local influence of participants #22 and #28 was as high as before, and individual #15 has even become more influential. Hence, the influence of subjects #22 and #28, as found in the first analysis, cannot be entirely ascribed to their incorrect classification as local cancer cases. One possible explanation may be that their Z_i covariate matrix, and hence the assumed model for their covari-

TABLE 11.1. *Prostate Data. Decomposition (11.5) of C_{22} according to the 18 eigenvectors v_j of $-\Delta' \ddot{L}^{-1} \Delta$ with nonzero eigenvalues λ_j. The table shows the first five terms in the decomposition, corresponding to the five largest eigenvalues. The calculations illustrate why subject #22 is influential without having a large contribution in the direction $h_{\max} = v_1$ of maximal curvature.*

$$
\left.
\begin{array}{rcl}
2\,\lambda_1\,v_{1,22}^2 &=& 0.5175 \\[4pt]
2\,\lambda_2\,v_{2,22}^2 &=& 6.0413 \\[4pt]
2\,\lambda_3\,v_{3,22}^2 &=& 0.2311 \\[4pt]
2\,\lambda_4\,v_{4,22}^2 &=& 0.0382 \\[4pt]
2\,\lambda_5\,v_{5,22}^2 &=& 0.0001
\end{array}
\right\} \quad \text{sum: } 6.8283
$$

$$\vdots$$

$$
2\,\sum_{j=1}^{18} \lambda_j\,v_{j,22}^2 \;=\; 7.0214 \;=\; C_{22}
$$

ance matrix V_i, is not changed by reclassifying these patients as metastatic cancer cases.

11.5   Local Influence Under REML Estimation

Our results from Section 11.2 and Section 11.3 assume that the parameters in the marginal linear mixed model are estimated via maximum likelihood. An influence analysis for the REML estimates would also be useful. However, it follows from expression (5.8) that the REML log-likelihood function can no longer be seen as a sum of independent individual contributions and, therefore, it is not obvious how a perturbation scheme, similar to (11.2), should be defined and interpreted. One approach would be to replace $\ell(\theta|\omega)$ in (11.2) by

$$
\ell_{\text{REML}}(\theta \mid \omega) \;=\; -\frac{1}{2}\,\ln\left| \sum_{i=1}^{N} w_i X_i' V_i^{-1} X_i \right| \;+\; \ell(\theta \mid \omega).
$$

The theory of local influence can then also be applied to this new perturbation scheme, but no longer results in simple expressions for the curvature C_i; hence, it becomes much more complicated to ascribe influence to the specific characteristics of the influential subjects. Therefore, we only considered here the maximum likelihood situation.

12

The Heterogeneity Model

12.1 Introduction

In Section 7.8.4, we discussed how Verbeke and Lesaffre (1996a, 1997b) extended the linear mixed model to cases where the random effects are not necessarily normally distributed. Their so-called heterogeneity model assumes the random effects to be sampled from a mixture of normal distributions rather than from just one single normal distribution. This not only extends the assumption about the random-effects distribution to a very broad class of distributions (unimodal as well as multimodal, symmetric as well as highly skewed; see Figure 7.5), it is also perfectly suitable for classification purposes, based on longitudinal profiles.

An example where classification of subjects based on longitudinal profiles is clearly of interest is the prostate data set introduced in Section 2.3.1. Indeed, our analyses of this data set have revealed some significant differences between the control patients, patients with benign prostatic hyperplasia, local cancer cases, and metastatic cancer cases (see, for example, Section 6.2.3), suggesting that repeated measures of PSA might be useful for detecting prostate cancer in an early stage of the disease. Note that such a classification procedure cannot be based on a model for PSA which includes age at diagnosis and time before diagnosis as covariates [as in our final model (6.8) in Section 6.2.3], since these are only available in retrospective studies, such as the Baltimore Longitudinal Study of Aging, where classification of the individuals is superfluous. The indicator variables C_i, B_i, L_i, and M_i are also not available for the same reason. The only possible adjustment of model (6.8) which could be used for our classification purposes is therefore given by

$$
\ln(1 + \text{PSA}_{ij}) = \beta_1 \text{Age}_i + (\beta_2 + b_{1i})
$$
$$
+ (\beta_3 + b_{2i}) \, t_{ij} + (\beta_4 + b_{3i}) \, t_{ij}^2 + \varepsilon_{ij}, \quad (12.1)
$$

where Age_i is now the age of the ith subject at entry in the study (or at the time the first measurement was taken) and where the time points t_{ij} are now expressed as time since entry. The procedure would then be to first fit model (12.1), from which estimates for the random effects can be calculated, as explained in Section 7.2. These estimates could then be used to classify patients in either one of the diagnostic groups.

However, although this approach looks very appealing, it raises many problems with respect to the normality assumption for the random effects, which is automatically made by the linear mixed-effects model. For example, it follows from the results in Section 6.2.3 that the mean quadratic time effect is zero for the noncancer cases and positive for both cancer groups. Hence, the quadratic effects $\beta_4 + b_{3i}$ in model (12.1) should follow a normal distribution with mean zero for the noncancer cases and with positive mean for the cancer cases. This means that the b_{3i} are no longer normally distributed, but follow a mixture of two normal distributions, i.e.,

$$
b_{3i} \sim pN(\mu_1, \sigma_1^2) + (1 - p)N(\mu_2, \sigma_2^2),
$$

in which μ_1, μ_2 and σ_1^2, σ_2^2 denote the means and variances of the b_{3i} in the noncancer and cancer groups, respectively, and where p is the proportion of patients in the data set which belong to the noncancer group. Similar arguments hold for the random intercepts b_{1i} and for the random time slopes b_{2i}, which even may be sampled from mixtures of more than two normal distributions.

It was shown in Section 7.8.2 that for the detection of subgroups in the random-effects population or for the classification of subjects in such subgroups, one should definitely not use empirical Bayes estimates obtained under the normality assumption for the random effects. In this chapter, it will be shown that the heterogeneity model is a natural model for classifying longitudinal profiles. In Sections 12.2 and 12.3, the heterogeneity model will be defined in full detail, and it will be described how the so-called Expectation-Maximization (EM) algorithm can be applied to obtain maximum likelihood estimates for all the parameters in the corresponding marginal model. In Section 12.4, it will be briefly discussed how longitudinal profiles can be classified based on the heterogeneity model. As already mentioned in Section 7.8.4, testing for the number of components in a heterogeneity model is far from straightforward due to boundary problems which make classical likelihood results break down. We will therefore discuss in Section 12.5 some simple informal checks for the goodness-of-fit of heterogeneity models. Finally, two examples will be given in Sections 12.6 and 12.7. Another example can be found in Brant and Verbeke (1997a, 1997b).

12.2   The Heterogeneity Model

As already explained in Section 7.8.4, the heterogeneity model of Verbeke (1995) and Verbeke and Lesaffre (1996a, 1997b) is obtained by replacing the normality assumption for the random effects in the linear mixed model (3.8) by a mixture of g q-dimensional normal distributions with mean vectors $\boldsymbol{\mu_j}$ and covariance matrices D_j, i.e.,

$$\boldsymbol{b_i} \sim \sum_{j=1}^{g} p_j N(\boldsymbol{\mu_j}, D_j), \qquad (12.2)$$

with $\sum_{j=1}^{g} p_j = 1$. We now define $z_{ij} = 1$ if $\boldsymbol{b_i}$ is sampled from the jth component in the mixture, and 0 otherwise, $j = 1, \ldots, g$. We then have that $P(z_{ij} = 1) = E(z_{ij}) = p_j$ and that

$$E(\boldsymbol{b_i}) = E\left(E(\boldsymbol{b_i} \mid z_{i1}, \ldots, z_{ig})\right) = E\left(\sum_{j=1}^{g} \boldsymbol{\mu_j}\, z_{ij}\right) = \sum_{j=1}^{g} p_j\, \boldsymbol{\mu_j}.$$

Therefore, the additional constraint $\sum_{j=1}^{g} p_j \boldsymbol{\mu_j} = 0$ is needed to assure that $E(\boldsymbol{y_i}) = X_i \boldsymbol{\beta}$. Further, we have that the overall covariance matrix of the $\boldsymbol{b_i}$ is given by

$$
\begin{aligned}
D^* &= \operatorname{var}\left(E(\boldsymbol{b_i} \mid z_{i1}, \ldots, z_{ig})\right) + E\left(\operatorname{var}(\boldsymbol{b_i} \mid z_{i1}, \ldots, z_{ig})\right) \\
&= \operatorname{var}\left(\sum_{j=1}^{g} \boldsymbol{\mu_j}\, z_{ij}\right) + E\left(\sum_{j=1}^{g} D_j\, z_{ij}\right) \\
&= \sum_{j=1}^{g} p_j \boldsymbol{\mu_j} \boldsymbol{\mu_j'} + \sum_{j=1}^{g} p_j D_j.
\end{aligned}
\qquad (12.3)
$$

The density function corresponding to (12.2) is given by

$$\sum_{j=1}^{g} p_j\, (2\pi)^{-q/2}\, |D_j|^{-1/2}\, \exp\left\{-\frac{1}{2}\left(\boldsymbol{b_i} - \boldsymbol{\mu_j}\right)' D_j^{-1}\left(\boldsymbol{b_i} - \boldsymbol{\mu_j}\right)\right\}. \qquad (12.4)$$

Note that, for $\boldsymbol{\mu}_1 = \boldsymbol{b_i}$, we have that (12.4) becomes infinitely large if $|D_1| \to 0$. In order to avoid numerical problems in the estimating procedure, which will be described later, we will assume from now on that all covariance matrices D_j are the same (i.e., that $D_j = D$ for all j). Our extended model is then fully determined by

$$
\begin{cases}
\boldsymbol{Y_i} = X_i\boldsymbol{\beta} + Z_i\boldsymbol{b_i} + \boldsymbol{\varepsilon_i} \\[2mm]
\boldsymbol{b_i} \sim \displaystyle\sum_{j=1}^{g} p_j \ N(\boldsymbol{\mu}_j, D), \\[2mm]
\displaystyle\sum_{j=1}^{g} p_j = 1, \\[2mm]
\displaystyle\sum_{j=1}^{g} p_j\boldsymbol{\mu}_j = 0, \\[2mm]
\boldsymbol{\varepsilon_i} \sim N(\boldsymbol{0}, \Sigma_i), \\[2mm]
\boldsymbol{b_1}, \ldots, \boldsymbol{b_N}, \boldsymbol{\varepsilon_1}, \ldots, \boldsymbol{\varepsilon_N} \text{ independent,}
\end{cases}
\tag{12.5}
$$

and it assumes that the random-effects population consists of g subpopulations with mean vectors $\boldsymbol{\mu}_j$ and with common covariance matrix D. The model is therefore called the heterogeneity model. Also, specifying the model as

$$
\begin{cases}
\boldsymbol{Y_i}|\boldsymbol{b_i} \ \sim \ N(X_i\boldsymbol{\beta} + Z_i\boldsymbol{b_i}, \Sigma_i), \\[2mm]
\boldsymbol{b_i}|\boldsymbol{\mu} \ \sim \ N(\boldsymbol{\mu}, D), \\[2mm]
\boldsymbol{\mu} \in \{\boldsymbol{\mu}_1, \ldots, \boldsymbol{\mu}_g\}, \quad \text{with } P\left(\boldsymbol{\mu} = \boldsymbol{\mu}_j\right) = p_j,
\end{cases}
$$

it can be interpreted as a hierarchical Bayes model, but now with an underlying random vector $\boldsymbol{\mu}$ which is no longer identically zero and which therefore represents the heterogeneity for the mean of the random-effects distribution. The classical linear mixed model which assumes $\boldsymbol{b_i} \sim N(\boldsymbol{0}, D)$ does not allow this type of heterogeneity and is therefore called the homogeneity model.

12.3 Estimation of the Heterogeneity Model

The marginal distribution of the measurements Y_i under model (12.5) can easily be seen to be given by

$$Y_i \sim \sum_{j=1}^{g} p_j N(X_i\boldsymbol{\beta} + Z_i\boldsymbol{\mu_j}, V_i), \qquad (12.6)$$

with $V_i = Z_i D Z_i' + \Sigma_i = W_i^{-1}$. Estimation of the parameters $\boldsymbol{\beta}$, $\boldsymbol{\mu_j}$, p_j, and D and the parameters in Σ_i can be done via maximum likelihood estimation. For this, the so-called Expectation-Maximization (EM) algorithm has been advocated; see Laird (1978). The EM algorithm is particularly useful for mixture problems since it often happens that a model is fitted with too many components (g too large), leading to a likelihood which is maximal anywhere on a ridge. As shown by Dempster, Laird and Rubin (1977), the EM algorithm is capable of converging to some particular point on that ridge. Titterington, Smith and Makov (1985, pp. 88-89) compare the EM algorithm with the Newton-Raphson (NR) algorithm. Their conclusions can be summarized as follows:

- EM is usually simple to apply and satisfies the appealing monotonic property in that it increases the objective function at each iteration step. NR is more complicated, and there is no guarantee of monotonicity.

- If NR converges, it is of second order (i.e., fast), whereas EM is often painfully slow. However, if the separation between the components in the mixture is poor, even the numerical performance of NR can be disappointing. Simulations have shown that, in such cases, NR can fail to converge in up to half the simulations, even when the algorithm was started from the true parameter values.

- Convergence is not guaranteed with any of the techniques since EM, even with the monotonicity property, can converge to a local maximum of the likelihood surface.

Böhning and Lindsay (1988) have considered maximization of log-likelihoods for which the quadratic approximation based on the Taylor series is "flatter" than the objective function, thereby sending the solution too far at the next step. They conclude that, in a mixture framework, flat log-likelihoods can lead to problems in convergence and to instabilities for the Newton-Raphson algorithm.

Note also that since the random effects are assumed to follow a mixture of distributions of the same parametric family, the vector of all parameters in model (12.5) is not identifiable. Indeed, the log-likelihood is invariant under the $g!$ possible permutations of the mean vectors and corresponding probabilities of the components in the mixture. Therefore, the likelihood will have at least $g!$ local maxima with the same likelihood value. However, this lack of identifiability is of no concern in practice, as it can easily be overcome by imposing some constraint on the parameters. For example, Aitkin and Rubin (1985) use the constraint that

$$p_1 \geq p_2 \geq \ldots \geq p_g. \tag{12.7}$$

The likelihood is then maximized without the restriction, and the component labels are permuted afterward to achieve (12.7).

The EM algorithm is frequently used for the calculation of maximum likelihood estimates for missing data problems. We have therefore deferred a detailed presentation on this algorithm to the second part of this book, namely to Chapter 22. However, a brief introduction in the context of the heterogeneity model will be given here. We also refer to McLachlan and Basford (1988, Section 1.6) for an application of the EM algorithm in a simpler mixture context, where it is assumed that the available data are all drawn from the same mixture distribution (no different dimensions, no covariates).

Let π be the vector of component probabilities [i.e., $\pi' = (p_1, \ldots, p_g)$] and let γ be the vector containing the remaining parameters β and D, the parameters in all Σ_i, and all μ_j's. Further, $\theta' = (\pi', \gamma')$ denotes the vector of all parameters in the marginal heterogeneity model (12.6), and $f_{ij}(y_i|\gamma)$ is the density function of the normal distribution with mean $X_i\beta + Z_i\mu_j$ and covariance matrix V_i. The likelihood function corresponding to (12.6) is then

$$L(\theta|y) = \prod_{i=1}^{N} \left\{ \sum_{j=1}^{g} p_j \, f_{ij}(y_i \mid \gamma) \right\}, \tag{12.8}$$

where $y' = (y_1', \ldots, y_N')$ is the vector containing all observed response values.

Let z_{ij} be as defined in Section 12.2. The prior probability for an individual to belong to component j is then $P(z_{ij} = 1) = p_j$, the mixture proportion for that component. The log-likelihood function for the observed measurements y and for the vector z of all unobserved z_{ij} is then

$$\ell(\theta|y, z) = \sum_{i=1}^{N} \sum_{j=1}^{g} z_{ij} \left\{ \ln p_j \, + \, \ln f_{ij}(y_i|\gamma) \right\},$$

which is easier to maximize than the log-likelihood function corresponding to the likelihood (12.8) of the observed data vector y only. On the other hand, maximizing $\ell(\boldsymbol{\theta}|\boldsymbol{y}, \boldsymbol{z})$ with respect to $\boldsymbol{\theta}$ yields estimates which depend on the unobserved ("missing") indicators z_{ij}. A compromise is obtained with the EM algorithm, where the expected value of $\ell(\boldsymbol{\theta}|\boldsymbol{y}, \boldsymbol{z})$, rather than $\ell(\boldsymbol{\theta}|\boldsymbol{y}, \boldsymbol{z})$ itself, is maximized with respect to $\boldsymbol{\theta}$, where the expectation is taken over all the unobserved z_{ij}. In the E step (expectation step), the conditional expectation of $\ell(\boldsymbol{\theta}|\boldsymbol{y}, \boldsymbol{z})$, given the observed data vector $\boldsymbol{y}$, is calculated. In the M step (maximization step), the so-obtained expected log-likelihood function is maximized with respect to $\boldsymbol{\theta}$, providing an updated estimate for $\boldsymbol{\theta}$. Finally, one keeps iterating between the E step and the M step until convergence is attained.

More specifically, let $\boldsymbol{\theta}^{(t)}$ be the current estimate for $\boldsymbol{\theta}$, and $\boldsymbol{\theta}^{(t+1)}$ stands for the updated estimate, obtained from one further iteration in the EM algorithm. We then have the following E and M steps in the estimation process for the heterogeneity model.

The E Step. The conditional expectation

$$Q(\boldsymbol{\theta}|\boldsymbol{\theta}^{(t)}) = E\left[\ell(\boldsymbol{\theta}|\boldsymbol{y}, \boldsymbol{z}) \mid \boldsymbol{y}, \boldsymbol{\theta}^{(t)}\right]$$

is given by

$$Q(\boldsymbol{\theta}|\boldsymbol{\theta}^{(t)}) = \sum_{i=1}^{N}\sum_{j=1}^{g} p_{ij}(\boldsymbol{\theta}^{(t)})\left[\ln p_j + \ln f_{ij}(\boldsymbol{y_i}|\boldsymbol{\gamma})\right], \quad (12.9)$$

where only the posterior probability for the ith individual to belong to the jth component of the mixture, given by

$$
\begin{aligned}
p_{ij}(\boldsymbol{\theta}^{(t)}) &= E(z_{ij} \mid \boldsymbol{y_i}, \boldsymbol{\theta}^{(t)}) = P(z_{ij} = 1 \mid \boldsymbol{y_i}, \boldsymbol{\theta}^{(t)}) \\
&= \frac{p_j f_{ij}(\boldsymbol{y_i}|\boldsymbol{\gamma})}{\sum_{k=1}^{g} p_k f_{ik}(\boldsymbol{y_i}|\boldsymbol{\gamma})}\bigg|_{\hat{\pi}^{(t)}, \hat{\gamma}^{(t)}}
\end{aligned}
$$

has to be calculated for each i and j.

The M Step. To get the updated estimate $\boldsymbol{\theta}^{(t+1)}$, we have to maximize expression (12.9) with respect to $\boldsymbol{\theta}$. We first maximize

$$\sum_{i=1}^{N}\sum_{j=1}^{g} p_{ij}(\boldsymbol{\theta}^{(t)}) \ln p_j$$

$$= \sum_{i=1}^{N}\sum_{j=1}^{g-1} p_{ij}(\boldsymbol{\theta}^{(t)}) \ln p_j + \sum_{i=1}^{N} p_{ig}(\boldsymbol{\theta}^{(t)}) \ln\left(1 - \sum_{j=1}^{g-1} p_j\right)$$

with respect to $p_1, \ldots, p_{g-1}$. Setting all first-order derivatives equal to zero yields that the updated estimates satisfy

$$\frac{p_j^{(t+1)}}{p_g^{(t+1)}} = \frac{\sum_{i=1}^{N} p_{ij}(\boldsymbol{\theta}^{(t)})}{\sum_{i=1}^{N} p_{ig}(\boldsymbol{\theta}^{(t)})},$$

for all $j = 1, \ldots, g - 1$. This also implies that

$$1 = \sum_{j=1}^{g} p_j^{(t+1)} = \frac{N\, p_g^{(t+1)}}{\sum_{i=1}^{N} p_{ig}(\boldsymbol{\theta}^{(t)})},$$

from which it follows that all estimates $p_j^{(t+1)}$ satisfy

$$p_j^{(t+1)} = \frac{1}{N} \sum_{i=1}^{N} p_{ij}(\boldsymbol{\theta}^{(t)}).$$

Unfortunately, the second part of (12.9) cannot be maximized analytically, and a numerical maximization procedure such as Newton-Raphson is needed to maximize

$$\sum_{i=1}^{N} \sum_{j=1}^{g} p_{ij}(\boldsymbol{\theta}^{(t)}) \ln f_{ij}(\boldsymbol{y_i}|\boldsymbol{\gamma})$$

with respect to $\boldsymbol{\gamma}$. All necessary derivatives can be obtained from the expressions in a paper by Lindstrom and Bates (1988).

Once all parameters $\boldsymbol{\theta}$ in the marginal heterogeneity model have been estimated, one might be interested in estimating the random effects $\boldsymbol{b_i}$ also. To this end, empirical Bayes (EB) estimates can be calculated, in exactly the same way as EB estimates were obtained under the classical linear mixed model (see Section 7.2). The posterior density of $\boldsymbol{b_i}$ is given by

$$f_i(\boldsymbol{b_i} \mid \boldsymbol{y_i}, \boldsymbol{\theta}) = \sum_{j=1}^{g} p_{ij}(\boldsymbol{\theta}) f_{ij}(\boldsymbol{b_i} \mid \boldsymbol{y_i}, \boldsymbol{\gamma}),$$

where $f_{ij}(\boldsymbol{b_i}|\boldsymbol{y_i}, \boldsymbol{\gamma})$ is the posterior density function of $\boldsymbol{b_i}$, conditional on $z_{ij} = 1$, that is, conditional on the knowledge that $\boldsymbol{b_i}$ was sampled from the jth component in the mixture. Hence, the posterior distribution of $\boldsymbol{b_i}$ is a mixture of the posterior distributions of $\boldsymbol{b_i}$ within each component of the mixture, with the posterior probabilities $p_{ij}(\boldsymbol{\theta})$ as subject-specific mixture proportions. The posterior mean is then

$$\widehat{\boldsymbol{b_i}} = \sum_{j=1}^{g} p_{ij}(\boldsymbol{\theta}) E\left(\boldsymbol{b_i} \mid \boldsymbol{y_i}, \boldsymbol{\gamma}, z_{ij} = 1\right).$$

Conditionally on $\boldsymbol{b_i}$, $\boldsymbol{Y_i}$ is normally distributed with mean $X_i\boldsymbol{\beta} + Z_i\boldsymbol{b_i}$ and variance-covariance matrix Σ_i. The vector $\boldsymbol{b_i}$ is, conditionally on component j, also normally distributed, with mean $\boldsymbol{\mu_j}$ and variance-covariance matrix D. It then follows from Lindley and Smith (1972) that

$$E\left(\boldsymbol{b_i}|\boldsymbol{y_i}, \boldsymbol{\gamma}, z_{ij} = 1\right) = DZ_i'W_i(\boldsymbol{y_i} - X_i\boldsymbol{\beta}) + (I - DZ_i'W_iZ_i)\boldsymbol{\mu_j},$$

from which it follows that

$$\widehat{b_i} = DZ_i'W_i(y_i - X_i\beta) + (I - DZ_i'W_iZ_i)\sum_{j=1}^{g}p_{ij}(\theta)\mu_j. \quad (12.10)$$

Note how the first component of $\widehat{b_i}$ is of exactly the same form as the estimator (7.2) obtained in Section 7.2, assuming normally distributed random effects. However, the overall covariance matrix of the b_i is now replaced by the within-component covariance matrix D. The second component in the expression for $\widehat{b_i}$ can be viewed as a correction term toward the component means μ_j, with most weight on those means, which correspond to components for which the subject has a high posterior probability of belonging. Finally, the unknown parameters in (12.10) are replaced by their maximum likelihood estimates obtained from the EM algorithm. These are the EB estimates shown in Figure 7.6 for the data simulated in Section 7.8.2 and obtained under a two-component heterogeneity model (i.e., g equals 2).

12.4   Classification of Longitudinal Profiles

Interest could also lie in the classification of the subjects into the different mixture components. It is natural in mixture models for such a classification to be based on the estimated posterior probabilities $p_{ij}(\widehat{\theta})$ (McLachlan and Basford 1988, Section 1.4). One then classifies the ith subject into the component for which it has the highest estimated posterior probability to belong to, that is, to the $j(i)$th component, where $j(i)$ is the index for which $p_{i,j(i)}(\widehat{\theta}) = \max_{1 \le j \le g} p_{ij}(\widehat{\theta})$. Note how this technique can be used for cluster analysis within the framework of linear mixed-effects models: If the individual profiles are to be classified into g subgroups, fit a mixture model with g components and use the above rule for classification in either one of the g clusters.

For $g = 2$, the above classification rule implies classification of subject i in the first component if and only if

$$\widehat{p_1}f_{i1}(y_i|\widehat{\gamma}) \ge (1 - \widehat{p_1})f_{i2}(y_i|\widehat{\gamma}).$$

Using some matrix algebra, this can be rewritten as

$$\left[y_i - X_i\widehat{\beta} - Z_i(\widehat{\mu}_1 + \widehat{\mu}_2)/2\right]' \widehat{V}_i^{-1} Z_i(\widehat{\mu}_1 - \widehat{\mu}_2) \ge \ln\left[(1 - \widehat{p_1})/\widehat{p_1}\right],$$

which is the linear discriminant function recently proposed by Tomasko, Helms and Snapinn (1999), also in the context of linear mixed models.

12.5   Goodness-of-Fit Checks

So far, we have not discussed yet how the number g of components in the heterogeneity model should be chosen. One approach is to fit models with increasing numbers of components and to compare them using likelihood ratio tests. However, as explained in Section 7.8.4, this is far from straightforward, due to boundary problems. Also, acceptance of the null hypothesis does not automatically yield a good-fitting model since it was tested against only one specific alternative hypothesis. An alternative approach is to increase g to the level where some of the subpopulations get very small weight (some p_j very small) or where some of the subpopulations coincide (some μ_j approximately the same). Finally, Verbeke and Lesaffre (1996a) proposed some omnibus goodness-of-fit checks for the marginal heterogeneity model (12.6), which can be employed to determine the number of components in the heterogeneity model, but which are also useful for evaluating the final model. These will now be discussed.

Suppose we want to test whether model (12.6) fits our data well, for some specific value of g (possibly 1). The most well-known goodness-of-fit tests are derived for univariate random variables, but, here, all observed vectors y_i possibly have different distributions or different dimensions. Thus, strictly speaking, these tests are not applicable here. However, this problem can be circumvented by considering the stochastic variables $F_i(y_i)$, for F_i equal to the cumulative distribution function of Y_i, under the assumed model. If the assumed model is correct, we have that all $F_i(y_i)$ can be considered sampled from a uniform distribution with support $[0, 1]$. On the other hand, we have that the computation of $F_i(y_i)$ involves the evaluation of multivariate normal distribution functions of dimensions n_i, which may therefore be practically unfeasible for data sets with large numbers of repeated measurements for some of the individuals. We therefore propose to first summarize each vector y_i by some linear combination $a_i' y_i$, and then to calculate $F_i(a_i' y_i)$, where F_i is now the distribution function of $a_i' Y_i$ under the assumed model. We then have that, under model (12.6), the stochastic variables

$$\mathcal{U}_i \; = \; F_i(a_i' Y_i) \; = \; \sum_{j=1}^{g} p_j \, \Phi\left(\frac{a_i'(Y_i - X_i\beta - Z_i\mu_j)}{\sqrt{a_i' V_i a_i}}\right), \quad (12.11)$$

with Φ denoting the cumulative distribution function of a univariate standard normal random variable, are uniformly distributed.

Two procedures can now be followed. First, we can apply the Kolmogorov-Smirnov test (see, for example, Birnbaum 1952, Bickel and Doksum 1977, pp. 378-381) to test whether the observed $\mathcal{U}_i$, calculated by replacing Y_i by y_i, and all parameters in (12.11) by their maximum likelihood estimates obtained from the EM algorithm, indeed follow the uniform distribution as is to be expected under the assumed model. For moderate sample sizes, percentage points are tabulated in Birnbaum (1952) and Neave (1986), whereas Bickel and Doksum (1977, p. 483) give approximations for large sample sizes ($N > 80$). Another possible approach is to test whether or not the values $\Phi^{-1}(\mathcal{U}_i)$ can be assumed to be sampled from a univariate normal distribution. We therefore use the Shapiro-Wilk test, first introduced by Shapiro and Wilk (1965) and since then extensively investigated and compared to other normality tests. See, for example, D'Agostino (1971), Dyer (1974), and Pearson, D'Agostino and Bowman (1977). Percentage points have been tabulated by Shapiro and Wilk (1965), and approximations to the distribution of the test statistic can be found in Shapiro and Wilk (1968) and Leslie, Stephens and Fotopoulos (1986).

Of course, the above goodness-of-fit tests can be performed for any a_i, but a good choice of the linear combination may increase the power of the test. Here, we are specifically interested in exploring whether the number g of components in our heterogeneity model has been taken sufficiently large. Note how this testing for heterogeneity can be viewed as an attempt to break down the total random-effects variability into the within-subgroup variability represented by D, and the between-subgroup variability represented by the component means $\mu_1, \ldots, \mu_g$. However, it is intuitively clear that this will only be successful when the residual variability in the model, represented by the error terms ε_i, is small to moderate, in comparison to the random-effects variability in which we are interested. We therefore recommend choosing a_i such that the variability in $a_i{}'y_i$ due to the random effects is large compared to the variability due to the error terms. Specifically, we choose a_i such that

$$\frac{\mathrm{var}(a_i{}'Z_i b_i)}{\mathrm{var}(a_i{}'\varepsilon_i)} = \frac{a_i{}'Z_i D^* Z_i' a_i}{a_i{}'\Sigma_i a_i},$$

with D^* as defined in (12.3), is maximal. It follows from Seber (1984, p. 526) that this is satisfied for a_i equal to the eigenvector corresponding to the largest eigenvalue of $\Sigma_i^{-1} Z_i D^* Z_i'$. This choice can be further justified as follows. Let $\lambda_{\max}$ be the largest eigenvalue of $\Sigma_i^{-1} Z_i D^* Z_i'$, and a_i be the corresponding eigenvector. We then have that $a_i{}'y_i = a_i' \Sigma_i^{-1} Z_i D^* Z_i' y_i / \lambda_{\max}$ and is therefore a function of $Z_i' y_i$, which can easily be seen to be sufficient for b_i in the conditional distribution of Y_i given b_i (distribution given in Section 3.3.1). So, $a_i{}'y_i$ only uses that part of the information in the sample, which is sufficient for the random effects. In the case that the only random effects in the model are intercepts, we have that $a_i{}'y_i$ is even equivalent to the sufficient statistic $Z_i' y_i$. In practice, all parameters in the above expressions need to be replaced by their maximum likelihood estimates obtained from the EM algorithm.

TABLE 12.1. *Prostate Data. Parameter estimates for all parameters in model (12.1), under a one-, two- and three-component heterogeneity model, and based on the observed data from cancer patients and control patients only.*

Component means $\nu + \mu_j$ Component probabilities		Covariance matrix D			β_1 σ^2
$\begin{pmatrix} 0.1694 \\ 0.0085 \\ 0.0034 \end{pmatrix}$	1.00	$\begin{pmatrix} 0.0760 & 0.0051 & 0.0002 \\ 0.0051 & 0.0023 & -0.0001 \\ 0.0002 & -0.0001 & 0.00003 \end{pmatrix}$			0.008 0.027
$\begin{pmatrix} -0.2373 \\ 0.0260 \\ 0.0012 \end{pmatrix}$	0.79	$\begin{pmatrix} 0.0497 & 0.0089 & -0.0005 \\ 0.0089 & 0.0017 & -0.0001 \\ -0.0005 & -0.0001 & 0.00002 \end{pmatrix}$			0.013 0.026
$\begin{pmatrix} 0.1790 \\ -0.0439 \\ 0.0105 \end{pmatrix}$	0.21				
$\begin{pmatrix} -0.0202 \\ 0.0124 \\ 0.0012 \end{pmatrix}$	0.72	$\begin{pmatrix} 0.0306 & 0.0082 & -0.0003 \\ 0.0082 & 0.0023 & -0.0001 \\ -0.0003 & -0.0001 & 0.00001 \end{pmatrix}$			0.009 0.027
$\begin{pmatrix} 0.5110 \\ -0.0088 \\ 0.0045 \end{pmatrix}$	0.19				
$\begin{pmatrix} 0.2167 \\ -0.0288 \\ 0.0207 \end{pmatrix}$	0.09				

12.6   Example: The Prostate Data

As an example, we will now use several heterogeneity models to analyze the prostate data introduced in Section 2.3.1, and we will hereby ignore any prior information about the disease status of the patients. In order not to complicate the model too much at once, we excluded the benign prostatic hyperplasia patients (BPH) from our analyses, yielding a total of 34 remaining patients. The purpose of our analyses is to investigate (1) whether our mixture approach detects the presence of heterogeneity in the random-effects population, which we know to be present, and (2) whether our classification procedure correctly classifies patients as being

TABLE 12.2. *Prostate Data. Goodness-of-fit checks for model (12.1), under a one-, two- and three-component heterogeneity model, and based on the observed data from cancer patients and control patients only. The reported Kolmogorov-Smirnov statistics are to be compared with the 5% critical value $D_c = 0.2274$. The Shapiro-Wilk statistics are accompanied by their p-values.*

	Kolmogorov-Smirnov	Shapiro-Wilk
1 component	0.2358 ($> D_c$)	0.824 ($p = 0.0001$)
2 components	0.2113 ($< D_c$)	0.848 ($p = 0.0001$)
3 components	0.1061 ($< D_c$)	0.940 ($p = 0.0797$)

controls or cancer cases. As explained in Section 12.1, such an analysis should be based on model (12.1). Several models have been fitted, with one to three components in the random-effects mixture distribution, and all models assume conditional independence (i.e., $\Sigma_i = \sigma^2 I_{n_i}$). The parameter estimates for all fitted models are summarized in Table 12.1, and the results for the goodness-of-fit tests are shown in Table 12.2.

The fixed-effects vector now consists of two parts: β_1 measures the effect of age on PSA, whereas $\boldsymbol{\nu} = (\beta_2, \beta_3, \beta_4)'$ reflects the overall average trend over time, after correction for age differences at the entry in the study. So, $\boldsymbol{\nu}$ contains the overall mean intercept β_2, mean slope β_3 for time, and mean slope β_4 for time^2, which are estimated by the homogeneity model as 0.1694, 0.0085, and 0.0034, respectively. The component means reported in Table 12.1 are the average trends within each component of the mixture (i.e., $\boldsymbol{\nu} + \boldsymbol{\mu}_j$, $j = 1, \ldots, g$).

As can be expected, both goodness-of-fit tests reject the homogeneity model, that is, the one-component model. We therefore extended the model by fitting a two-component heterogeneity model, which yields a first group (79% of the patients) with patients evolving mainly linearly, and a second group (21% of the patients) with patients who clearly evolve quadratically. Although this two-component model is accepted by the Kolmogorov-Smirnov test, it is not by the Shapiro-Wilk test. We therefore also fitted a three-component model, which was accepted by the Kolmogorov-Smirnov test as well as by the Shapiro-Wilk test. The estimated mean profiles, not taking into account the effect of age, are shown in Figure 12.1. Apparently, the first component represents the individuals who evolve mainly linearly: There is a constant increase of PSA over time. This is in contrast with the other two groups in the mixture which contain the subjects who evolve quadratically over time, for the second component after a period of small linear increase, and for the last component immediately after enrollment in the study.

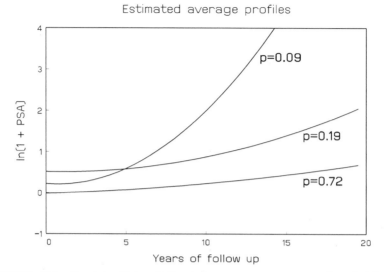

FIGURE 12.1. *Prostate Data. Estimated component means and probabilities, based on a three-component heterogeneity model.*

This model will only be useful for the detection of prostate cancer at an early stage, if one or two of our components can be shown to represent the cancer cases, and the other component(s) then would represent the controls. We therefore compare classification by our mixture approach with the correct classification as control or cancer. The result is shown in Table 12.3. Except for one patient, all controls were classified in the first component, together with 10 cancer cases for which the profiles show hardly any difference from many profiles in the control group (only a moderate, linear increase over time). Three cancer cases were classified in the third component. These are those cases which have entered the study almost simultaneously with the start of the growth of the tumor. The five cancer cases, classified in the second component, are those who were already in the study long before the tumor started to develop and therefore have profiles which hardly change in the beginning, but which start increasing quadratically after some period of time in the study.

Apparently, the detection of the correct diagnostic group is hampered by the different onsets of observation periods. Further, the quadratic mixed-effects model is only a rough approximation to the correct model. For example, Carter *et al.* (1992a) and Pearson *et al.* (1994) have fitted piecewise nonlinear mixed-effects models to estimate the time when rapid increases in PSA were first observable. One could also think of extending the heterogeneity model to the case where the component probabilities p_j are modeled as functions over time. This would take into account the fact that the proportion of cancer cases increases with time. In any case, this exam-

TABLE 12.3. *Prostate Data. Cross-classification of 34 patients according to the three-component mixture model and according to their true disease status.*

		Mixture classification		
		1	2	3
	control	15	1	0
Disease status				
	cancer	10	5	3

ple has shown that the mixture approach does not necessarily model what one might hope. There is no a priori reason why the mixture classification should exactly correspond to some predefined group structure, which may not fully reflect the heterogeneity in growth curves.

12.7   Example: The Heights of Schoolgirls

As a second example of the use of heterogeneity models, we consider the growth curves of 20 preadolescent schoolgirls, introduced in Section 2.5, and previously analyzed by Goldstein (1979), not using linear mixed models. Goldstein found a significant (at the 5% level of significance) group effect as well as a significant interaction of age with group. Note that the individual profiles shown in Figure 2.5 suggest that the variability in the observed heights is mainly due to between-subject variability. That is why we will now reanalyze the data using linear mixed models, which allow us to use subject-specific regression coefficients.

A linear mixed model obtained from a two-stage approach (see Section 3.2) assumes the average evolution within each group to be linear as a function of age and allows for subject-specific intercepts as well as slopes. More specifically, our model is given by

$$\text{Height}_{ij} = \begin{cases} \beta_1 + b_{1i} + (\beta_4 + b_{2i})\text{Age}_{ij} + \varepsilon_{ij}, & \text{if small mother} \\ \beta_2 + b_{1i} + (\beta_5 + b_{2i})\text{Age}_{ij} + \varepsilon_{ij}, & \text{if medium mother} \\ \beta_3 + b_{1i} + (\beta_6 + b_{2i})\text{Age}_{ij} + \varepsilon_{ij}, & \text{if tall mother,} \end{cases}$$

where Height_{ij} and Age_{ij} are the height and the age of the ith girl at the jth measurement, respectively. The model can easily be rewritten as

TABLE 12.4. *Heights of Schoolgirls. REML estimates and associated estimated standard errors for all parameters in model (12.12), under the assumption of conditional independence.*

Effect	Parameter	REMLE (s.e.)
Intercepts:		
Small mothers	β_1	81.300 (1.338)
Medium mothers	β_2	82.974 (1.239)
Tall mothers	β_3	83.123 (1.239)
Age effects:		
Small mothers	β_4	5.270 (0.174)
Medium mothers	β_5	5.567 (0.161)
Tall mothers	β_6	6.249 (0.161)
Covariance of b_i:		
$\text{var}(b_{1i})$	d_{11}	7.603 (3.729)
$\text{var}(b_{2i})$	d_{22}	0.133 (0.063)
$\text{cov}(b_{1i}, b_{2i})$	d_{12}	-0.444 (0.399)
Residual variance:		
$\text{var}(\varepsilon_{ij})$	σ^2	0.476 (0.087)
REML log-likelihood		-157.874

$$
\begin{aligned}
\text{Height}_{ij} = {}& b_{1i} + \beta_1 \text{Small}_i + \beta_2 \text{Medium}_i + \beta_3 \text{Tall}_i \\
& + \{b_{2i} + \beta_4 \text{Small}_i + \beta_5 \text{Medium}_i + \beta_6 \text{Tall}_i\} \, \text{Age}_{ij} \\
& + \varepsilon_{ij}, \quad\quad\quad\quad\quad\quad\quad\quad\quad\quad\quad\quad\quad\quad\quad (12.12)
\end{aligned}
$$

where Small_i, Medium_i, and Tall_i are dummy variables defined to be 1 if the mother of the ith girl is small, medium, or tall, respectively, and defined to be 0 otherwise. So, β_1, β_2, and β_3 represent the average intercepts and β_4, β_5, and β_6 the average slopes in the three groups. The terms b_{1i} and b_{2i} are the random intercepts and random slopes, respectively. REML estimates for all parameters in this model are given in Table 12.4, obtained assuming conditional independence; that is, assuming that all error components ε_{ij} are independent with common variance σ^2.

All of the above analyses are based on the somewhat arbitrary discretization of the heights of the mothers into three different intervals (small, medium, and tall mothers). It is therefore interesting to see how heterogeneity models would classify the children into two, three, or even more groups, ignoring

TABLE 12.5. *Heights of Schoolgirls. Parameter estimates for all parameters in model (12.13), under a one-, two- and three-component heterogeneity model.*

Component means $\beta + \mu_j$		Covariance matrix	Residual variance
Component probabilities		D	σ^2
$\begin{pmatrix} 82.48 \\ 5.72 \end{pmatrix}$	1	$\begin{pmatrix} 6.71 & -0.07 \\ -0.07 & 0.27 \end{pmatrix}$	0.47
$\begin{pmatrix} 82.78 \\ 5.39 \end{pmatrix}$	0.68	$\begin{pmatrix} 6.73 & 0.10 \\ 0.10 & 0.03 \end{pmatrix}$	0.47
$\begin{pmatrix} 82.06 \\ 6.42 \end{pmatrix}$	0.32		
$\begin{pmatrix} 79.46 \\ 5.60 \end{pmatrix}$	0.20	$\begin{pmatrix} 3.64 & 0.32 \\ 0.32 & 0.03 \end{pmatrix}$	0.47
$\begin{pmatrix} 84.21 \\ 5.32 \end{pmatrix}$	0.50		
$\begin{pmatrix} 81.65 \\ 6.47 \end{pmatrix}$	0.30		

the group structure used so far. The corresponding model is then

$$\text{Height}_{ij} = \beta_1 + b_{1i} + (\beta_2 + b_{2i})\text{Age}_{ij} + \varepsilon_{ij}, \qquad (12.13)$$

where β_1 and β_2 denote the overall average intercept and linear age effect, respectively. As before, we will assume all error components ε_{ij} to be independent and normally distributed with mean zero and common variance σ^2.

Three mixture models were fitted: the homogeneous model (one component) and two heterogeneous models (two and three components). For the heterogeneity models, the girls were classified in either one of the mixture components. The parameter estimates and classification rules are summarized in Table 12.5 and Figure 12.2, respectively. The reported component means are the average growth trends within each component of the mixture (i.e., $\beta + \mu_j$, $j = 1, \ldots, g$).

First, it follows from the fit of the homogeneity model that the average intercept and slope are estimated to be 82.48 and 5.72, respectively. Using the goodness-of-fit procedures discussed in Section 12.5, we found that this homogeneity model fits the data sufficiently well; that is, no statistical

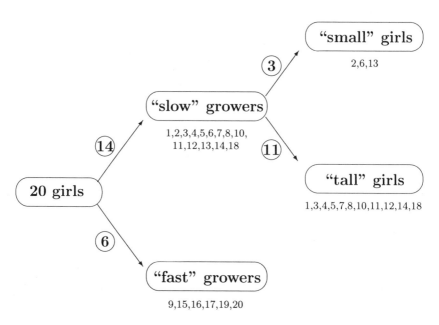

FIGURE 12.2. *Heights of Schoolgirls. Graphical representation of the cluster analysis on the growth curves of 20 preadolescent schoolgirls, based on model (12.13), under a one-, two- and three-component heterogeneity model. The child numbers are given underneath each cluster.*

evidence is found for the presence of heterogeneity in the random-effects population ($p = 0.2970$ for the Shapiro-Wilk test, observed value for the Kolmogorov-Smirnov statistic equal to 0.1505, which is below the 5% critical value $D_c = 0.2941$). Note however, that this is based on a small data set. Also, acceptance of the homogeneity model should not necessarily prevent us from performing cluster analysis. The two-component mixture model subdivides the children into two groups, with similar intercepts but highly different slopes, and therefore can be regarded as discriminating the "slow" growers (68%) from the "fast" growers (32%). Note also that this implies a large reduction in slope variability and no reduction in intercept variability, as indicated by the estimated variances in the matrix D. Finally, the three-component mixture model further subdivides the "slow" growers into "small" and "tall" girls, with an average difference in height of $84.21 - 79.46 = 4.75$ cm.

Although there is no a priori reason why our mixture classification should reconstruct Goldstein's groups, shown in Table 2.4, it may still be interesting to compare his artificial group structure with ours, which naturally arises from the profile structures themselves. Although separation of the children with tall mothers from the rest is achieved fairly well, the children with small mothers could not be well separated from those with medium mothers. However, this is in agreement with results obtained from analyses based on the linear mixed model (12.12), the parameter estimates for which are given in Table 12.4. Indeed, applying the F-tests described in Section 6.2.2 (with the Satterthwaite approximation for the denominator degrees of freedom), we found that the average slopes β_4, β_5, and β_6 are significantly different ($p = 0.0019$), but this can be fully ascribed to differences between the groups of children with small or medium mothers on the one hand and the group of children with tall mothers on the other hand: β_4 and β_6 are significantly different ($p = 0.0007$), also β_5 and β_6 are significantly different ($p = 0.0081$), but β_4 is not significantly different from β_5 ($p = 0.2259$). Since our mixture approach only partially reconstructs the prior group structure of Goldstein (1979), we conclude that the latter does not fully reflect the heterogeneity in the growth curves.

Finally, we note that, under the three-component model, the overall average trend is given by

$$
0.20 \begin{pmatrix} 79.46 \\ 5.60 \end{pmatrix} + 0.50 \begin{pmatrix} 84.21 \\ 5.32 \end{pmatrix} + 0.30 \begin{pmatrix} 81.65 \\ 6.47 \end{pmatrix} = \begin{pmatrix} 82.49 \\ 5.72 \end{pmatrix},
$$

which is very similar to the overall average trend, estimated under the homogeneity model. Further, we also have that, the overall random-effects covariance matrix is given by

$$
\begin{pmatrix} 3.64 & 0.32 \\ 0.32 & 0.03 \end{pmatrix} + 0.20 \begin{pmatrix} -3.03 \\ -0.12 \end{pmatrix} \begin{pmatrix} -3.03 & -0.12 \end{pmatrix}
$$

$$
+ 0.50 \begin{pmatrix} 1.72 \\ -0.40 \end{pmatrix} \begin{pmatrix} 1.72 & -0.40 \end{pmatrix}
$$

$$
+ 0.30 \begin{pmatrix} -0.84 \\ 0.75 \end{pmatrix} \begin{pmatrix} -0.84 & 0.75 \end{pmatrix}
$$

$$
= \begin{pmatrix} 7.17 & -0.14 \\ -0.14 & 0.28 \end{pmatrix},
$$

which is also similar to the overall random-effects covariance matrix obtained under the homogeneity model, but less similar than what we found for the overall average. Finally, the residual variance estimate is the same for the three fitted models. This illustrates the results discussed in Section 7.8.3: Even if the three-component model would be the correct one, the homogeneity model would yield good (consistent) estimators for all parameters in the model.

13

Conditional Linear Mixed Models

13.1 Introduction

As pointed out by Diggle, Liang and Zeger (1994, Section 1.4) and as shown in the examples so far presented in this book, the main advantage of longitudinal studies, when compared to cross-sectional studies, is that they can distinguish changes over time within individuals (longitudinal effects) from differences among people in their baseline values (cross-sectional effects).

Consider a randomized longitudinal clinical trial, where subjects are first randomly assigned to one out of a set of treatments, and then followed for a certain period of time during which measurements are taken at pre-specified time points. Treatment effects are then completely represented by differences in evolutions over time (i.e., by interactions of treatment with time), whereas the randomization assures that, at least in large trials, the treatment groups are completely comparable at baseline with respect to factors which potentially influence change afterward. Hence, a statistical model for such data does not need a cross-sectional model component.

In observational studies however, subjects may be very heterogeneous at baseline such that longitudinal changes need to be studied after correction for potential confounders such as age, gender, and so forth. For example, all our analyses of the prostate data so far have been corrected for age at diagnosis because we knew that, due to the high prevalence of BPH in men over age 50, the control group was significantly younger on average than the BPH cases, at first visit as well as at the time of diagnosis. Brant *et al.* (1992) analyzed repeated measures of systolic blood pressure from 955 healthy males. Their models included cross-sectional effects for age at first visit (linear as well as quadratic effect), obesity, and birth cohort. In a non-linear context, Diggle, Liang and Zeger (1994, Section 9.3) used longitudinal data on 250 children to investigate the evolution of the risk for respiratory infection and its relation to vitamin A deficiency. They adjusted for factors like gender, season, and age at entry in the study.

When analyzing longitudinal data, the longitudinal effects are usually of primary interest, whereas the cross-sectional component of the model is often considered as nuisance, but needed to correct for baseline differences. In this chapter, we will therefore explore how sensitive inference for the longitudinal effects is to model assumptions in the cross-sectional component of the model, and it will be shown how such inferences can be obtained without making any assumptions about this cross-sectional component. Illustrations will be based on the hearing data, introduced in Section 2.3.2, which we will first analyze in the next section.

13.2   A Linear Mixed Model for the Hearing Data

As an example of the effect of the assumed cross-sectional model on the estimation of longitudinal trends, we will now analyze the hearing data introduced in Section 2.3.2, but we restrict our analysis to measurements from the left ear only. Based on the results of Brant and Fozard (1990), Morrell and Brant (1991), and Pearson *et al.* (1995), we propose the following linear mixed model for these data:

$$
\begin{aligned}
Y_{ij} &= (\beta_1 + \beta_2 \text{Age}_{i1} + \beta_3 \text{Age}_{i1}^2 + b_{1i}) \\
&\quad + (\beta_4 + \beta_5 \text{Age}_{i1} + b_{2i}) t_{ij} \\
&\quad + \beta_6 \text{Visit1}_{ij} + \varepsilon_{(1)ij},
\end{aligned} \tag{13.1}
$$

in which t_{ij} is the time point (in decades from entry in the study) at which the jth measurement is taken for the ith subject, and where Age_{i1} is the age (in decades) of the subject at the time of entry in the study. Pearson *et al.* (1995) found evidence for the presence of a learning effect from the first visit to subsequent visits. This is taken into account by the extra time-varying covariate Visit1_{ij}, defined to be one at the first measurement and zero for all other visits. Finally, the b_{1i} are random intercepts, and the b_{2i} are random slopes for time. As before, the $\varepsilon_{(1)ij}$ are measurement error components. Table 13.1 shows the ML estimates and associated standard

TABLE 13.1. *Hearing Data. ML estimates (standard errors) for the parameters in the marginal linear mixed model corresponding to (13.1), with and without inclusion of a cross-sectional quadratic effect of age, for the original data ($\Delta = 0$) as well as for contaminated data ($\Delta = -10$). The last column contains ML estimates (standard errors) obtained from the conditional linear mixed model approach.*

| | Linear mixed model | | | | Conditional |
| | Original data | | Contaminated data | | linear mixed |
Parameter	$\beta_3 \neq 0$	$\beta_3 = 0$	$\beta_3 \neq 0$	$\beta_3 = 0$	model
Fixed effects:					
β_1 (intercept)	3.52 (2.39)	−4.52 (0.86)	3.52 (2.39)	226.39 (3.31)	—
β_2 (age)	−1.63 (1.01)	1.96 (0.15)	−1.63 (1.01)	−101.42 (0.60)	—
β_3 (age^2)	0.35 (0.10)	—	−9.65 (0.10)	—	—
β_4 (time)	−0.20 (0.81)	−0.15 (0.81)	−0.20 (0.81)	−0.58 (0.81)	0.02 (0.81)
β_5 (age×time)	0.86 (0.16)	0.84 (0.16)	0.86 (0.16)	1.03 (0.16)	0.82 (0.17)
β_6 (visit1)	1.85 (0.30)	1.86 (0.30)	1.85 (0.30)	2.07 (0.31)	1.96 (0.31)
Variance components:					
$d_{11} = \mathrm{var}(b_{1i})$	41.81	42.61	41.81	861.93	—
$d_{12} = \mathrm{cov}(b_{1i}, b_{2i})$	3.59	4.11	3.59	−20.75	—
$d_{22} = \mathrm{var}(b_{2i})$	7.61	7.67	7.61	8.06	7.61
$\sigma^2 = \mathrm{var}(\varepsilon_{(1)ij})$	25.16	25.15	25.13	25.16	25.19

errors for all parameters in the marginal model corresponding to the model (13.1).

To study the effect of misspecifying the cross-sectional model on the estimation of the longitudinal model, we refitted model (13.1), not including the cross-sectional quadratic age effect. The results are also shown in Table 13.1. We conclude that removing cross-sectional terms from the model inflates the random-intercepts variance d_{11}, but the estimates of the average longitudinal trends (β_4, β_5, and β_6) are only slightly affected. Intuitively, one might expect the effect of omitting the quadratic age effect to depend on β_3. We checked this by repeatedly fitting the two previous models on contaminated data, obtained from replacing the original response values y_{ij} by $y_{ij} + \Delta \mathrm{Age}_{i1}^2$, for Δ equal to $-10, -9, \ldots, 9$, and 10. For $\Delta = -10$, there is an extremely strong negative quadratic age effect, whereas for $\Delta = 10$, there is an extremely strong positive quadratic age effect. The situation $\Delta = 0$ corresponds to the original data. From now on, model (13.1) will be termed the "correct" model, and the model under $\beta_3 = 0$ will be termed the "incorrect" (misspecified) model.

Figure 13.1 compares the estimates of the average longitudinal effects under both models. The estimates under the correct model are independent of the degree Δ of contamination. Under the incorrect model however, the obtained estimates clearly depend on Δ, and for β_5 and β_6, they differ up

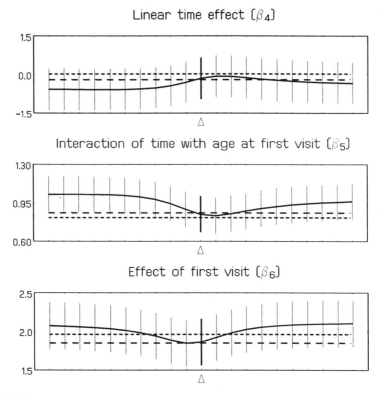

FIGURE 13.1. *Hearing Data. ML estimates for the average longitudinal effects under correct (long dashes) and incorrect (solid) cross-sectional models, as well as for the conditional linear mixed model (short dashes), for several degrees of contamination (Δ). The vertical lines represent one estimated standard deviation under the incorrect model. The bold vertical line corresponds to the original data ($\Delta = 0$).*

to one standard deviation from the estimates obtained under the correct model. The parameter estimates under the correct and under the incorrect model, for the case $\Delta = -10$, are also given in Table 13.1. Under the correct model, we get exactly the same estimates as for the original data, except for the cross-sectional quadratic age effect, for which the difference is exactly $\Delta = -10$. As noticed earlier, the omission of a cross-sectional covariate inflates the random-intercepts variability, but this is now much more pronounced than earlier. Note also that deleting the quadratic age effect for the contaminated data changes the estimated correlation between the random intercepts and random slopes from 0.2013 ($p = 0.0625$, LR test) to -0.2490 ($p = 0.0206$, LR test).

For $\Delta = -10$, we also calculated the empirical Bayes estimates (EB) (see Section 7.2) for the random slopes b_{2i} under the correct and under the in-

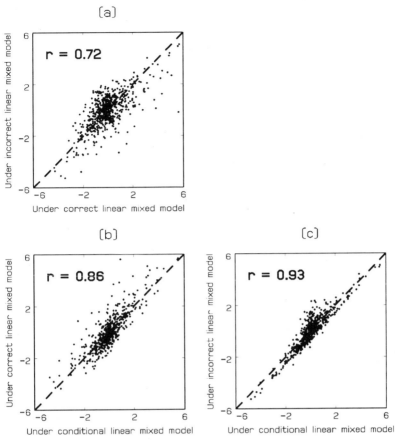

FIGURE 13.2. *Hearing Data. Pairwise scatter plots of the empirical Bayes esti-mates for the random slopes b_{2i} obtained under the correct linear mixed model, the incorrect linear mixed model, and the conditional linear mixed model. All plots are based on contaminated data ($\Delta = -10$), and the Pearson correlation coefficient is denoted by r.*

correct model. A graphical comparison is shown in panel (a) of Figure 13.2. Note that, although the estimates from both procedures are highly corre-lated ($r = 0.72$), many subjects are found to increase faster (slower) than average under the correct model, but slower (faster) than average under the misspecified model, and vice versa.

13.3   Conditional Linear Mixed Models

The results of Section 13.2 illustrate the need for statistical methodology which allows for the study of longitudinal trends in observational data, without having to specify any cross-sectional effects. Verbeke *et al.* (1999) and Verbeke, Spiessens and Lesaffre (2000) propose the use of so-called conditional linear mixed models. In order to simplify notations, we will restrict to discussing this approach in the context of model (13.1) for the hearing data, rather than in full generality, and we refer to the above-mentioned papers for more details.

We first reformulate the linear mixed model (13.1) as

$$
\begin{aligned}
Y_{ij} \;=\; & b_i^* \\
& + (\beta_4 + \beta_5 \text{Age}_{i1} + b_{2i}) t_{ij} \\
& + \beta_6 \text{Visit1}_{ij} + \varepsilon_{(1)ij},
\end{aligned}
\tag{13.2}
$$

where b_i^* represents the cross-sectional component $\beta_1 + \beta_2 \text{Age}_{i1} + \beta_3 \text{Age}_{i1}^2 + b_{1i}$, corresponding to subject i, under the original model. The parameters of interest are the fixed slopes β_4, β_5, and β_6, the subject-specific slopes b_{2i}, and the residual variance σ^2; the cross-sectional component b_i^* is considered as nuisance. Note how model (13.2) is of the form

$$
\boldsymbol{Y_i} \;=\; \mathbf{1}_{n_i} b_i^* + X_i \boldsymbol{\beta} + Z_i \boldsymbol{b_i} + \boldsymbol{\varepsilon_{(1)i}},
\tag{13.3}
$$

where the matrices X_i and Z_i and the vectors $\boldsymbol{\beta}$ and $\boldsymbol{b_i}$ are those submatrices and subvectors of their original counterparts X_i, Z_i, $\boldsymbol{\beta}$, and $\boldsymbol{b_i}$ obtained from deleting the elements which correspond to the cross-sectional component (i.e., the time-independent covariates) of the original model (13.1).

Conditional linear mixed models now proceed in two steps. In a first step, we condition on sufficient statistics for the nuisance parameters b_i^*. In a second step, maximum likelihood or restricted maximum likelihood estimation is used to estimate the remaining parameters in the conditional distribution of the $\boldsymbol{Y_i}$ given these sufficient statistics.

Conditional on the subject-specific parameters b_i^* and $\boldsymbol{b_i}$ in (13.3), we have that $\boldsymbol{Y_i}$ is normally distributed with mean vector $\mathbf{1}_{n_i} b_i^* + X_i \boldsymbol{\beta} + Z_i \boldsymbol{b_i}$ and covariance matrix $\sigma^2 I_{n_i}$, from which it readily follows that $\overline{y}_i = \sum_j y_{ij}/n_i$ is sufficient for b_i^*. Further, the distribution of $\boldsymbol{Y_i}$, conditional on $\overline{y}_i$ and on the remaining subject-specific effects $\boldsymbol{b_i}$, is given by

$$
\begin{aligned}
f_i(\boldsymbol{y_i} | \overline{y}_i, \boldsymbol{b_i}) & \\
&= \frac{f_i(\boldsymbol{y_i} | b_i^*, \boldsymbol{b_i})}{f_i(\overline{y}_i | b_i^*, \boldsymbol{b_i})}
\end{aligned}
$$

$$
= \left(2\pi\sigma^2\right)^{-(n_i-1)/2} \sqrt{n_i}
$$

$$
\times \exp\left\{-\frac{1}{2\sigma^2}\left(\boldsymbol{y_i} - X_i\boldsymbol{\beta} - Z_i\boldsymbol{b_i}\right)'\right.
$$

$$
\left.\times \left(I_{n_i} - \mathbf{1}_{n_i}\left(\mathbf{1}_{n_i}'\mathbf{1}_{n_i}\right)^{-1}\mathbf{1}_{n_i}'\right)\left(\boldsymbol{y_i} - X_i\boldsymbol{\beta} - Z_i\boldsymbol{b_i}\right)\right\}. \quad (13.4)
$$

It now follows directly from some matrix algebra (Seber 1984, property B3.5, p. 536) that (13.4) is proportional to

$$
\left(2\pi\sigma^2\right)^{-(n_i-1)/2} \exp\left\{-\frac{1}{2\sigma^2}\left(A_i'\boldsymbol{y_i} - A_i'X_i\boldsymbol{\beta} - A_i'Z_i\boldsymbol{b_i}\right)'\left(A_i'A_i\right)^{-1}\right.
$$

$$
\left.\times \left(A_i'\boldsymbol{y_i} - A_i'X_i\boldsymbol{\beta} - A_i'Z_i\boldsymbol{b_i}\right)\right\} \quad (13.5)
$$

for any set of $n_i \times (n_i-1)$ matrices A_i of rank n_i-1 which satisfy $A_i'\mathbf{1}_{n_i} = 0$. This shows that the conditional approach is equivalent to transforming each vector $\boldsymbol{Y_i}$ orthogonal to $\mathbf{1}_{n_i}$. If we now also require the A_i to satisfy $A_i'A_i = I_{(n_i-1)}$, we have that the transformed vectors $A_i'\boldsymbol{Y_i}$ satisfy

$$
\begin{aligned}
\boldsymbol{Y_i^*} \equiv A_i'\boldsymbol{Y_i} &= A_i'X_i\boldsymbol{\beta} + A_i'Z_i\boldsymbol{b_i} + A_i'\boldsymbol{\varepsilon}_{(1)i} \\
&= X_i^*\boldsymbol{\beta} + Z_i^*\boldsymbol{b_i} + \boldsymbol{\varepsilon}_{(1)i}^*,
\end{aligned} \quad (13.6)
$$

where $X_i^* = A_i'X_i$ and $Z_i^* = A_i'Z_i$ and where the $\boldsymbol{\varepsilon}_{(1)i}^* = A_i'\boldsymbol{\varepsilon}_{(1)i}$ are normally distributed with mean $\mathbf{0}$ and covariance matrix $\sigma^2 I_{n_i-1}$.

Model (13.6) is now again a linear mixed model, but with transformed data and covariates, and such that the only parameters still in the model are the longitudinal effects and the residual variance. Hence, the second step in fitting conditional linear mixed models is to fit model (13.6) using maximum likelihood or restricted maximum likelihood methods. As earlier, the subject-specific slopes are estimated using empirical Bayes methods (see Section 7.2). Note that once the transformed responses and covariates have been calculated, standard software for fitting linear mixed models (e.g., SAS procedure MIXED) can be used for the estimation of all parameters in model (13.6). A SAS macro for performing the transformation has been provided by Verbeke *et al.* (1999) and is available from the website.

Conditional linear mixed models are based on transformed data, where the transformation is chosen such that a specific set of "nuisance" parameters vanishes from the likelihood. From this respect, the proposed method is very similar to REML estimation in the linear mixed model, where the variance components are estimated after transforming the data such that the fixed effects vanish from the model (see Section 5.3). As shown by Harville (1974, 1977) and by Patterson and Thompson (1971), and as discussed in

Section 5.3.4, the REML estimates for the variance components do not depend on the selected transformation, and no information on the variance components is lost in the absence of information on the fixed effects. It has been shown by Verbeke, Spiessens and Lesaffre (2000) that similar properties hold for inferences obtained from conditional linear mixed models; that is, it was shown that results do not depend on the selected transformation $Y_i \rightarrow A_i' Y_i$ and that no information is lost on the average, nor on the subject-specific longitudinal effects, from conditioning on sufficient statistics for the cross-sectional components b_i^* in the original model.

The simplest example of conditional linear mixed models is obtained for balanced data with only two repeated measurements per subject, where the only time-varying covariate of interest is a binary indicator for the occasion at which the measurement is taken. The proposed approach is then equivalent to analyzing the difference between the first and second measurement of each subject. Hence, conditional linear mixed models can be interpreted as an extension of the well-known paired t-test to longitudinal data sets, possibly unbalanced, with more than two measurements per subject.

In a recent paper, Neuhaus and Kalbfleisch (1998) proposed a similar conditional approach for the analysis of clustered data with generalized linear mixed models. However, since their models do not contain any random effects other than intercepts, they are often too restrictive because they imply unrealistically simple covariance structures. For linear models, the conditional linear mixed models extend this methodology to accommodate models which also allow the presence of subject-specific longitudinal effects (random slopes).

Conditional linear mixed models have many advantages. First, inference becomes available for the parameters of interest, completely disregarding the nuisance parameters, without loss of any information. Further, the fitting of conditional linear mixed models is very straightforward. Note also that the second step in the fitting process does not necessarily require the random longitudinal effects to be normally distributed. Other possibilities include the use of finite mixtures of Gaussian distributions (e.g., Verbeke and Lesaffre 1996a, Magder and Zeger 1996; see Section 7.8.4 and Chapter 12.2). In fact, one could even use nonparametric maximum likelihood estimation, not assuming the mixing distribution to be of any specific parametric form (see, for example, Laird 1978, Aitkin and Francis 1995, Aitkin 1999). The advantage of the first conditional step is then the reduction of the dimensionality of the mixing distribution which seriously reduces the numerical complexity of the fitting algorithms.

A disadvantage of the conditional linear mixed model is that all information is lost on the average as well as on the subject-specific cross-sectional effects. However, it should be emphasized that longitudinal data are collected for studying longitudinal changes, rather than cross-sectional differences among subjects.

13.4 Applied to the Hearing Data

As an illustration, we continue our analysis of Section 13.2 of the hearing data, and we will estimate all longitudinal effects in model (13.1), without having to correct for any baseline differences among the study participants. So-obtained ML estimates for the parameters in the marginal model corresponding to model (13.6) have also been included in Table 13.1 and in Figure 13.1. Note that it is now irrelevant whether or not a quadratic age effect is included in the model, and that exactly the same results are found for the original data as well as for the contaminated data. All parameter estimates are very similar to the ones obtained from fitting model (13.1). This suggests that the baseline differences among the participants in this study can be well described by a quadratic function of the age at entry, which has also been assumed by other authors who studied the relation between hearing thresholds and age (e.g., Morrell and Brant 1991 and Pearson *et al.* 1995).

Panel (b) of Figure 13.2 shows a scatter plot of the EB estimates for the subject-specific slopes b_{2i} in model (13.1), obtained by fitting the associated (correct) linear mixed model versus those obtained from the conditional linear mixed model. Note that the same plot would be obtained for all contaminated data sets. For the contaminated data with $\Delta = -10$, a similar scatter plot has been included for the EB estimates under the incorrect linear mixed model [panel (c) of Figure 13.2]. Surprisingly, the estimates under the conditional model correlate better with the estimates under the incorrect model ($r = 0.93$) than with those obtained from the correct linear mixed model ($r = 0.86$). However, panel (c) in Figure 13.2 reveals the presence of outliers which may inflate the correlation and also suggests that the incorrect model tends to systematically underestimate small (negative) and large (positive) slopes, when compared to the conditional linear mixed model, whereas the opposite is true for the slopes closer to zero.

We also calculated, for all subjects, the difference between their EB estimate obtained under the correct or incorrect model and their EB estimate obtained under the conditional linear mixed model. A plot of these differences versus the subject's age at entry in the study is shown in Figure 13.3. It clearly shows that the omission of the cross-sectional quadratic age-effect results in a systematic bias of the EB estimates, when compared to the estimates from the conditional model: The incorrect model tends to overestimate the subject-specific slope for subjects of low or high age at

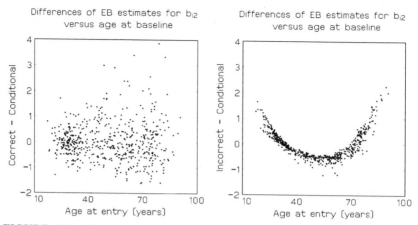

FIGURE 13.3. *Hearing Data. Scatter plots of the differences in empirical Bayes estimates for the random slopes b_{2i} obtained under the correct linear mixed model and the conditional linear mixed model (left panel) as well as under the incorrect model and the conditional linear mixed model (right panel). Both plots are based on contaminated data ($\Delta = -10$).*

entry in the study, whereas the opposite is true for middle-aged subjects. This bias is not present for the EB estimates obtained from the correct linear mixed model. These findings suggest that one way of checking the appropriateness of the cross-sectional component of a classical linear mixed model could be to calculate the difference between the resulting EB estimates for the subject-specific slopes and their EB estimates obtained from the conditional approach, and to plot these differences versus relevant covariates. However, more research is needed in order to fully explore the potentials of such procedures.

13.5   Relation with Fixed-Effects Models

It can be shown that, conditional on the variance components, the same estimates are obtained for longitudinal components in a linear mixed model by applying the conditional linear mixed model approach as obtained by fitting the corresponding model (13.3), thus considering the subject-specific intercepts b_i^* as fixed, rather than random, parameters. So, one can, strictly speaking, perform a conditional linear mixed model analysis without explicit computation of the transformed response vectors y_i^* or of the transformed covariance matrices X_i^* and Z_i^*. In large data sets however, this requires fitting linear mixed models with hundreds, if not thousands, of parameters in the mean structure. In many cases, this will become unfeasible using standard software.

For example, we tried this fixed-effects approach to obtain the results previously derived for the hearing data under a conditional linear mixed model and already reported in Table 13.1. In SAS PROC MIXED, this can be done as follows:

```
proc mixed data = hearing noclprint method = ml;
class id;
model L500 = id time ftime visit1 / solution;
random time / type = un subject = id;
run;
```

The variable *time* contains the time point (in decades from entry in the study) at which the repeated responses were taken, whereas *ftime* represents the interaction term between *time* and the age (in decades) of the subjects at their entry in the study. Further, *id* contains the identification number of each patient, and the variable *visit1* is one at the first visit, and zero for all subsequent visits. Finally, the response variable *L500* contains the hearing thresholds for 500 Hz, taken on the left ear of the study participants. Running the above program requires fitting a linear mixed model with 684 fixed effects and 2 variance components. Unfortunately, this turned out not to be feasible because the 1.12 Gb of free disk space was insufficient for SAS to fit the model.

In order to still be able to illustrate the relationship between the conditional linear mixed models and the fixed-effects approach, we analyzed hearing thresholds from 100 randomly selected subjects in the hearing data set. The results are summarized in Table 13.2. The first two columns of this table show the estimates obtained from fitting a conditional linear mixed model using maximum likelihood (ML) and restricted maximum likelihood (REML), respectively. The third and the fourth columns show the estimates obtained from the fixed-effects approach also under ML and REML estimation, respectively. Note that the ML estimates from both fitting approaches do not coincide and that the fixed-effects approach even fails to detect the presence of random slopes. This is because the equivalence between both procedures acts conditional on the variance components. As discussed in Chapter 5, ML estimates for the variance components do not account for the loss in degrees of freedom due to the estimation of the fixed effects in the mean structure. Here, we are comparing ML estimates from a model with 3 fixed effects (conditional linear mixed model) with those from a model which contains 103 fixed effects (fixed-effects approach), yielding severe differences in the estimates for the variance components, and therefore also in the estimates of the longitudinal components in model (13.1). To illustrate this even better, we refitted the fixed-effects model, using ML estimation, but we restricted the variance components to be equal to those

TABLE 13.2. *Hearing Data (randomly selected subset). ML and REML estimates (standard errors) for the longitudinal effects in (13.1), from a conditional linear mixed model as well as from a fixed-effects approach.*

Parameter	Conditional linear mixed model		Fixed-effects approach	
	ML	REML	ML	REML
Fixed effects:				
β_4 (time)	-1.778 (2.111)	-1.817 (2.146)	-0.483 (1.372)	-1.817 (2.146)
β_5 (age×time)	1.212 (0.430)	1.221 (0.436)	0.918 (0.286)	1.221 (0.436)
β_6 (visit 1)	2.968 (0.697)	2.969 (0.699)	2.939 (0.658)	2.969 (0.699)
Variance components:				
$d_{22} = \mathrm{var}(b_{2i})$	11.466 (4.140)	12.216 (4.381)	0.000 (—)	12.216 (4.381)
$\sigma^2 = \mathrm{var}(\varepsilon_{(1)ij})$	20.148 (1.505)	20.210 (1.513)	19.043 (1.190)	20.210 (1.513)
Random slopes b_{2i}:				
Subject 1	2.597 (2.657)	2.680 (2.712)	0.000 (—)	2.680 (2.712)
Subject 2	-1.847 (2.285)	-1.904 (2.322)	0.000 (—)	-1.904 (2.322)
Subject 3	-1.130 (1.823)	-1.146 (1.844)	0.000 (—)	-1.146 (1.844)

obtained from the conditional linear mixed model (i.e., $d_{22} = 11.466$ and $\sigma^2 = 20.148$). This is done in PROC MIXED using the following program with a PARMS statement:

```
proc mixed data = subset noclprint method = ml;
class id;
model L500 = id time ftime visit1 / solution;
random time / type = un subject = id solution;
parms 11.466  20.148 / eqcons = 1,2;
run;
```

The so-obtained results are now exactly the same as those from the conditional linear mixed model, reported in the first column of Table 13.2.

Finally, since REML estimation does account for the loss in degrees of freedom due to the estimation of the fixed effects, we get the same results from the conditional linear mixed model as from the fixed-effects approach, provided both apply the REML estimation. For our example, this is illustrated in the fact that the second and the last columns in Table 13.2 are identical.

14

Exploring Incomplete Data

In Chapter 4, we introduced several tools, in the context of the Vorozole study, to graphically explore longitudinal data, both from the individual-level standpoint (Figures 4.1 and 4.5) as well as from the population-averaged or group-averaged perspective (Figures 4.2, 4.3, 4.4, and 10.3). These plots are designed to focus on various structural aspects, such as the mean structure, the variance function, and the association structure.

An extra level of complexity is added whenever not all planned measurements are observed. This results in *incompleteness* or *missingness*. Another frequently encountered term is *dropout*, which refers to the case where all observations on a subject are obtained until a certain point in time, after which all measurements are missing.

Several issues arise when data are incomplete. In the remainder of this chapter, we will illustrate some of these issues based on the Vorozole study. To simplify matters, we will focus on dropout. In the next chapter, a formal treatment of missingness, based on the pivotal work of Rubin (1976) and Little and Rubin (1987) will be given. Subsequent chapters present various modeling strategies for incomplete longitudinal data, as well as tools for sensitivity analysis, an area in which interest is strongly rising.

The first issue, resulting from dropout, is evidently a depletion of the study subjects. Of course, a decreasing sample size increases variability which, in turn, decreases precision. In this respect, the Vorozole study is a dramatic example, as can be seen from Figure 14.1 and Table 14.1, which graphi-

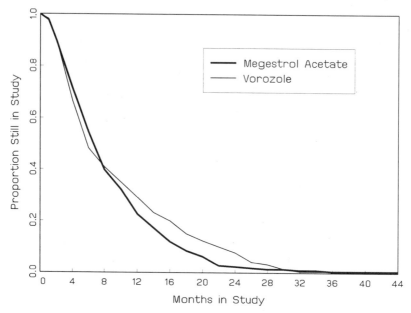

FIGURE 14.1. *Vorozole Study. Representation of dropout.*

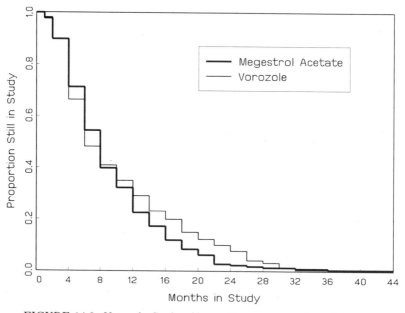

FIGURE 14.2. *Vorozole Study. Alternative representation of dropout.*

cally and numerically present dropout in both treatment arms. Clearly, the dropout rate is high *and* there is a hint of a differential rate between the two arms. This means we have identified one potential factor that could

TABLE 14.1. *Vorozole Study. Evolution of dropout.*

	Standard		Vorozole	
Week	#	(%)	#	(%)
0	226	(100)	220	(100)
1	221	(98)	216	(98)
2	203	(90)	198	(90)
4	161	(71)	146	(66)
6	123	(54)	106	(48)
8	90	(40)	90	(41)
10	73	(32)	77	(35)
12	51	(23)	64	(29)
14	39	(17)	51	(23)
16	27	(12)	44	(20)
18	19	(8)	33	(15)
20	14	(6)	27	(12)
22	6	(3)	22	(10)
24	5	(2)	17	(8)
26	4	(2)	9	(4)
28	3	(1)	7	(3)
30	3	(1)	3	(1)
32	2	(1)	1	(0)
34	2	(1)	1	(0)
36	1	(0)	1	(0)
38	1	(0)	0	(0)
40	1	(0)	0	(0)
42	1	(0)	0	(0)
44	1	(0)	0	(0)

influence a patient's probability of dropping out. Although a large part of the trialist's interest focuses on the treatment effect, we should be aware that it is still a covariate and hence a design factor. Another question that will arise is whether dropout depends on observed or unobserved responses. An equivalent representation is given in Figure 14.2.

A different way of displaying several structural aspects is using a scatter plot matrix, such as in Figure 14.3. The off-diagonal elements picture scatter plots of standardized residuals obtained from pairs of measurement occasions. The decay of correlation with time is studied by considering the evolution of the scatters with increasing distance to the main diagonal. Stationarity, on the other hand, implies that the scatter plots remain similar within diagonal bands *if measurement occasions are approximately equally spaced.* In addition to the scatter plots, we place histograms on the diagonal, capturing the variance structure. Features such as skewness, mul-

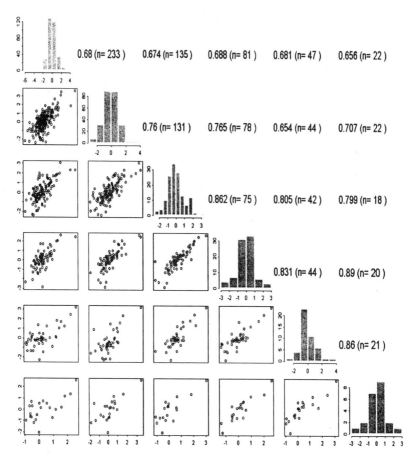

FIGURE 14.3. *Vorozole Study. Scatter plot matrix for selected time points.*

timodality, and so forth, can then be graphically detected. Finally, if the axes are given the same scales, it is very easy to capture the attrition rate as well; see also Figure 4.5.

Another aspect of the impact of dropout is also seen if we consider the average profile in each treatment arm, with pointwise confidence limits added (Figure 14.4). Indeed, near the end of the study, these intervals become extremely wide, as opposed to the relatively narrow intervals at the start of the experiment. Thus, it is clear that dropout leads to efficiency loss. Of course, this effect can be due in part to increasing variability over time. Modeling is needed to obtain more insight into this effect.

To gain further insight into the impact of dropout, it is useful to construct dropout-pattern-specific plots. Figures 14.5 and 14.6 display the individual and averaged profiles per pattern.

Mean Profiles

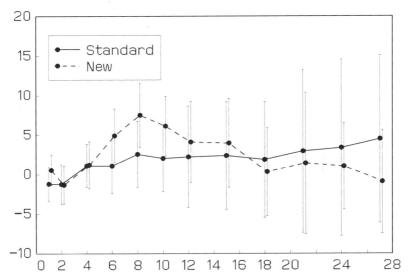

FIGURE 14.4. *Vorozole Study. Mean profiles, with 95% confidence intervals added.*

The individual profiles plot, by definition displaying all available data, has some intrinsic limitations. As is the case with any individual data plot, it tends to be fairly busy. Since there is a lot of early dropout, there are many short sequences and since we decided to use the same time axis for all profiles, also for those that drop out early, very little information can be extracted. Indeed, the evolution over the first few sequences is not clear at all. In addition, the eye assigns more weight to the longer profiles, even though they are considerably less frequent.

Some of these limitations are removed in Figure 14.6, where the pattern-specific average profiles are displayed per treatment arm. Still, care has to be taken for not overinterpreting the longer profiles and neglecting the shorter profiles. Indeed, for this study the latter represent more subjects than the longer profiles.

Several observations can be made. Most profiles clearly show a quadratic trend, which seems to be in contrast with the relatively flat nature of the average profiles in Figure 14.4. This implies that the impression from all patterns together may differ radically from a pattern-specific look. These conclusions seem to be consistent across treatment arms.

Another important observation is that those who drop out rather early seem to decrease from the start, whereas those who remain relatively long in the study exhibit, on average and in turn, a rise, a plateau, and then

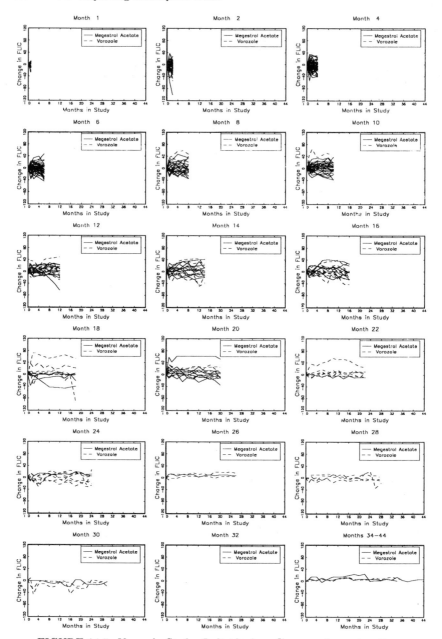

FIGURE 14.5. *Vorozole Study. Individual profiles, per dropout pattern.*

a decrease. Looked upon from the standpoint of dropout, we suggest that there are at least two important characteristics that make dropout increase: (1) a low value of change versus baseline and (2) an unfavorable (downward) evolution.

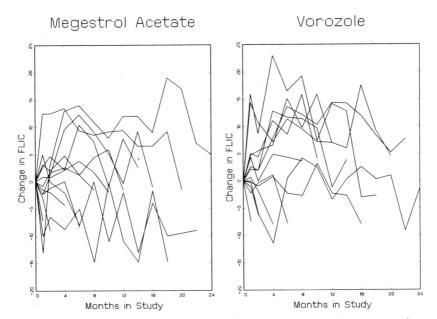

FIGURE 14.6. *Vorozole Study. Mean profiles, per dropout pattern, grouped per treatment arm.*

Arguably, a careful modeling of these data, irrespective of the framework chosen, should reflect these features. In Chapters 17 and 18, we will reconsider these preliminary findings.

15

Joint Modeling of Measurements and Missingness

15.1 Introduction

The problem of dealing with missing values is common throughout statistical work and is almost ever present in the analysis of longitudinal or repeated measurements data. Missing data are indeed common in clinical trials (Piantadosi 1997, Green, Benedetti, and Crowley 1997, Friedman, Furberg, and DeMets 1998), in epidemiologic studies (Kahn and Sempos 1989, Clayton and Hills 1993, Lilienfeld and Stolley 1994, Selvin 1996), and, very prominently, in sample surveys (Fowler 1988, Schafer, Khare and Ezatti-Rice 1993, Rubin 1987, Rubin, Stern, and Vehovar 1995).

Patients who drop out of a clinical trial are usually listed on a separate withdrawal sheet of the case record form with the reasons for withdrawal, entered by the investigator. Reasons typically encountered are adverse events, illness not related to study medication, uncooperative patient, protocol violation, ineffective study medication, and other reasons (with further specification, e.g., lost to follow-up). Based on such a medical typology, Gould (1980) proposed specific methods to handle this type of incompleteness.

Early work on missing values was largely concerned with algorithmic and computational solutions to the induced lack of balance or deviations from the intended study design. See, for example, the reviews by Afifi and Elashoff (1966) and Hartley and Hocking (1971). More recently, general algorithms, such as the Expectation-Maximization (EM; Dempster, Laird, and Rubin 1977), and data imputation and augmentation procedures (Rubin 1987, Tanner and Wong 1987), combined with powerful computing resources have largely provided a solution to this aspect of the problem. There remains the very difficult and important question of assessing the impact of missing data on subsequent statistical inference. Conditions can be formulated and are outlined later (Section 15.8), under which an analysis that proceeds as if the missing data are missing by design (i.e., ignoring the missing value process), can provide valid answers to study questions. The difficulty in practice is that such conditions can rarely be assumed to hold. We will frequently emphasize a key point here: When we undertake such analyses, assumptions will be required that cannot be assessed from the data under analysis. Hence, in this setting, there cannot be anything that could be termed a definitive analysis, and arguably the appropriate statistical framework is one of sensitivity analysis (Chapters 19 and 20).

Much of the treatment in this work will be restricted to dropout (or attrition); that is, to patterns in which missing values are only followed by missing values. There are four related reasons for this restriction. First, the classification of missing value processes has a simpler interpretation with dropout than for patterns with intermediate missing values. Second, it is easier to formulate models for dropout and, third, much of the missing value literature on longitudinal data is restricted to this setting. Finally, dropouts are by far the most common appearance of missingness in longitudinal studies.

15.2   The Impact of Incompleteness

In a strict sense, the conventional justification for the analysis of data from a randomized trial is removed when data are missing for reasons outside the control of the investigator. Before one can address this problem however, it is necessary to clearly establish the purpose of the study (Heyting, Tolboom, and Essers 1992). If one is working within a *pragmatic* setting, the event of dropout, for example, may well be a legitimate component of the response. It may make no sense to ask what response the subject would have shown had they remained in the trial, and the investigator may then require a description of the response *conditional* on a subject remaining in the trial. This, together with the pattern of observed missingness, may then be the appropriate and valid summary of the outcome. We might call this a conditional description. Shih and Quan (1997) argue that such a description will be of more relevance in many clinical trials. On the other hand, from a more *explanatory* perspective, we might be interested in the behavior of the responses that occurred irrespective of whether we were able to record them or not. This might be termed a *marginal* description of the response. For a further discussion of intention-to-treat and explanatory analyses in the context of dropout, we refer to Heyting, Tolboom, and Essers (1992) and Little and Yau (1996). It is commonly suggested (Shih and Quan 1997) that such a marginal representation is not meaningful when the nature of dropout (e.g., death) means that the response cannot subsequently exist, irrespective of whether it is measured.

Although such dropout may in any particular setting imply that a marginal model is not helpful, it does not imply that it necessarily has no meaning. Provided that the underlying model does not attach a probability of 1 to dropout for a particular patient, then non-dropout and subsequent observation is an outcome that is consistent with the model and logically not different from any other event in a probability model. Such distinctions, particularly with respect to the conditional analysis, are complicated by the inevitable mixture of causes behind missing values. The conditional description is a mirror of what has been observed, and so its validity is less of an issue than its interpretation. In contrast, other methods of handling incompleteness make some correction or adjustment to what has been directly observed, and therefore address questions other than those corresponding to the conditional setting. In seeking to understand the validity of these analyses, we need to compare their consequences with their aims.

15.3 Simple ad hoc Methods

Two simple, common approaches to analysis are (1) to discard subjects with incomplete sequences and (2) simple imputation. The first approach has the advantage of simplicity, although the wide availability of more sophisticated methods of analysis minimize the significance of this. It is also an inefficient use of information. In a trivial sense, it provides a description of the response conditional on a subject remaining in the trial. Whether this reflects a question of interest depends entirely on the mechanism(s) generating the missing values. It is not difficult to envisage situations where it can be very misleading, and examples of this exist in the literature (Wang-Clow *et al.* 1995).

There are several forms of simple imputation. For example, a cross-sectional approach replaces a missing observation by the average of available observations at the same time from other subjects with the same covariates and treatment. A simple longitudinal approach carries the last available measurement from a subject forward, replacing the entire sequence of missing values. A more sophisticated version predicts the next missing value using a regression relationship established from available past data. These methods share the same drawbacks, although not all to the same degree. The data set that results will mimic a sample from the population of interest, itself determined by the aims of the analysis, only under particular and potentially unrealistic assumptions. Further, these assumptions depend critically on the missing value mechanism(s). For example, under certain dropout mechanisms, the process of imputation may recover the actual marginal behavior required, whereas under other mechanisms, it may be wildly misleading. It is only under the simplest and most ignorable mechanisms that the relationship between imputation procedure and assumption is easily deduced. Little (1994a) gives two simple examples in which the relationship is clear. A further minor point is that, without further elaboration, the analysis of the completed data set will underestimate the true variability of the data.

In conclusion, we see that when there are missing values, simple methods of analysis do not necessarily imply simple, or even accessible, assumptions, and without understanding properly the assumptions being made in an analysis, we are not in a position to judge its validity or value. It has been argued that although any particular such ad hoc analysis may not represent the true picture behind the data, a collection of such analyses should provide a reasonable envelope within which the truth might lie. This does point to the desirability of a sensitivity analysis, but the main conclusion does not follow. Counterexamples to this can be constructed and, again, without a clear formulation of the assumptions being made, we are not in a position to interpret such an envelope, and we are certainly not justified in assuming that its coverage is, in some practical sense, inclusive. One way to proceed is to consider a formal framework for the missing value problem, and this leads us to Rubin's classification.

A review of simple methods, with their advantages and disadvantages, including complete case analysis and simple forms of imputation, is provided in Chapter 16.

15.4   Modeling Incompleteness

In order to incorporate incompleteness into the modeling process, we need to reflect on the nature of the missing value mechanism and its implications for statistical inference. Rubin (1976) and Little and Rubin (1987, Chapter 6) make important distinctions between different missing values processes. A dropout process is said to be completely random (MCAR) if the dropout is independent of both unobserved and observed data, and random (MAR) if, conditional on the observed data, the dropout is independent of the unobserved measurements; otherwise, the dropout process is termed nonrandom (MNAR). A more formal definition of these concepts is given in Section 15.7. If a dropout process is random, then a valid analysis can be obtained through a likelihood-based analysis that ignores the dropout mechanism, provided the parameters describing the measurement process are functionally independent of the parameters describing the dropout process, the so-called parameter distinctness condition. This situation is termed ignorable by Rubin (1976) and Little and Rubin (1987) and leads to considerable simplification in the analysis (Diggle 1989). See also Section 15.8.

In many examples, however, the reasons for dropout are many and varied and it is therefore difficult to justify on a priori grounds the assumption of random dropout. Arguably, in the presence of nonrandom dropout, a wholly satisfactory analysis of the data is not feasible.

One approach is to estimate from the available data the parameters of a model representing a nonrandom dropout mechanism. It may be difficult to justify the particular choice of dropout model, and it does not necessarily follow that the data contain information on the parameters of the particular model chosen, but where such information exists, the fitted model may provide some insight into the nature of the dropout process and of the sensitivity of the analysis to assumptions about this process. This is the route taken by Diggle and Kenward (1994) in the context of continuous longitudinal data; see also Diggle, Liang and Zeger (1994, Chapter 11) and Section 12.4. Further approaches are proposed by Laird, Lange, and Stram (1987), Wu and Bailey (1988, 1989), Wu and Carroll (1988), and Greenlees, Reece, and Zieschang (1982). An overview of the different modeling approaches is given by Little (1995).

Also, the case of categorical outcomes has received considerable attention. See, for example, Baker and Laird (1988), Stasny (1986), Baker, Rosenberger, and DerSimonian (1992), Conaway (1992, 1993), Park and Brown (1994), and Molenberghs, Kenward, and Lesaffre (1997).

Indeed, one feature in common to all of the more complex approaches is that they rely on untestable assumptions about the relation between the measurement process (often of primary interest) and the dropout process. One should therefore avoid missing data as much as possible, and if dropout occurs, information should be collected on the reasons for this. As an example, consider a clinical trial where outcome and dropout are both strongly related to a specific covariate X and where, conditionally on X, the response Y and the missing data process R are independent. In the selection framework, we then have that $f(Y, R|X) = f(Y|X)f(R|X)$, implying MCAR, whereas omission of X from the model may imply MAR or even MNAR, which has important consequences for selecting valid statistical methods.

Because different models imply different untestable assumptions, thereby possibly affecting the statistical inferences of interest, it is advisable, in practice, to always perform a sensitivity analysis. Various informal and formal ways of performing sensitivity analyses are described in Chapters 19 and 20. Draper (1995) and Copas and Li (1997) provide useful insight in model uncertainty and nonrandomly selected samples.

Section 15.5 develops the necessary terminology and notation and Section 15.6 describes various missing data patterns. The missing data mechanisms, informally introduced in this section, are formalized in Section 15.7. The important case where the missing data mechanism can be excluded from statistical analysis is introduced in Section 15.8. Since much of the subsequent treatment will be confined to dropout, this situation is reviewed in Section 15.9.

15.5 Terminology

In this section, we introduce terminology, building on the standard framework for missing data, which is largely due to Rubin (1976) and Little and Rubin (1987).

In general, we assume that for subject i in the study, a sequence of measurements Y_{ij} is designed to be measured at occasions $j = 1, \ldots, n_i$. As previously, the outcomes are grouped into a vector $\boldsymbol{Y}_i = (Y_{i1}, \ldots, Y_{in_i})'$. In addition, for each occasion j define

$$R_{ij} = \begin{cases} 1 & \text{if } Y_{ij} \text{ is observed} \\ 0 & \text{otherwise.} \end{cases}$$

The *missing data indicators* R_{ij} are grouped into a vector $\boldsymbol{R}_i$ which is, of course, of the same length as $\boldsymbol{Y}_i$.

Partition $\boldsymbol{Y}_i$ into two subvectors such that $\boldsymbol{Y}_i^o$ is the vector containing those Y_{ij} for which $R_{ij} = 1$ and $\boldsymbol{Y}_i^m$ contains the remaining components. These subvectors are referred to as the *observed* and *missing* components, respectively. The following terminology is adopted:

Complete data $\boldsymbol{Y}_i$: the scheduled measurements. This is the outcome vector that would have been recorded if there had been no missing data.

Missing data indicators $\boldsymbol{R}_i$: the process generating $\boldsymbol{R}_i$ is referred to as the missing data process.

Full data $(\boldsymbol{Y}_i, \boldsymbol{R}_i)$: the complete data, together with the missing data indicators. Note that unless all components of $\boldsymbol{R}_i$ equal 1, the full data components are never jointly observed.

Observed data $\boldsymbol{Y}_i^o$.

Missing data $\boldsymbol{Y}_i^m$.

Some confusion might arise between the terms *complete data* introduced here and *complete case analysis* of Sections 15.3 and 16.2. Although the former refers to the (hypothetical) data set that would arise if there were no missing data, "complete case analysis" refers to deletion of all subjects for which at least one component is missing.

Note that one observes the measurements $\boldsymbol{Y}_i^o$ together with the missingness indicators $\boldsymbol{R}_i$.

15.6   Missing Data Patterns

A hierarchy of missing data patterns can be considered. When missingness is due to *attrition*, all measurements for a subject from baseline onward up to a certain measurement time are recorded, whereafter all data are missing. It is then possible to replace the information contained in the vector $\boldsymbol{R}_i$ by a single indicator variable. For example, R_i could indicate the last observed measurement occasion. The sample size decreases over time.

Attrition is a particular *monotone* pattern of missingness. In order to have monotone missingness, there has to exist a permutation of the measurement components such that a measurement earlier in the permuted sequence is observed for at least those subjects that are observed at later measurements. Note that for this definition to be meaningful, we need to have a balanced design in the sense of a common set of measurement occasions. Other patterns are called *nonmonotone*.

15.7   Missing Data Mechanisms

Statistical modeling begins by considering the full data density

$$f(\boldsymbol{y}_i, \boldsymbol{r}_i | X_i, Z_i, \boldsymbol{\theta}, \boldsymbol{\psi}),$$

where X_i and Z_i are the design matrices for fixed and random effects, respectively, and $\boldsymbol{\theta}$ and $\boldsymbol{\psi}$ are vectors that parameterize the joint distribution.

We will use $\boldsymbol{\theta} = (\boldsymbol{\beta}', \boldsymbol{\alpha}')'$ (fixed-effects and covariance parameters) and $\boldsymbol{\psi}$ to describe the measurement and missingness processes, respectively.

The taxonomy, constructed by Rubin (1976), further developed in Little and Rubin (1987), and informally sketched in Section 15.4, is based on the factorization

$$f(\boldsymbol{y}_i, \boldsymbol{r}_i | X_i, Z_i, \boldsymbol{\theta}, \boldsymbol{\psi}) \quad = \quad f(\boldsymbol{y}_i | X_i, Z_i, \boldsymbol{\theta}) f(\boldsymbol{r}_i | \boldsymbol{y}_i, X_i, \boldsymbol{\psi}), \quad (15.1)$$

where the first factor is the marginal density of the measurement process and the second one is the density of the missingness process, conditional on the outcomes. It is possible to have additional covariates in the missingness model, but this is suppressed from notation. Factorization (15.1) forms the basis of *selection modeling*, as the second factor corresponds to the (self-)selection of individuals into "observed" and "missing" groups. Selection modeling is discussed in detail in Chapters 17 and 19. An alternative taxonomy can be built based on so-called *pattern-mixture models* (Little 1993, Little 1994a). These are based on the factorization

$$f(\boldsymbol{y}_i, \boldsymbol{r}_i | X_i, Z_i, \boldsymbol{\theta}, \boldsymbol{\psi}) \quad = \quad f(\boldsymbol{y}_i | \boldsymbol{r}_i, X_i, Z_i, \boldsymbol{\theta}) f(\boldsymbol{r}_i | X_i, \boldsymbol{\psi}). \quad (15.2)$$

Indeed, (15.2) can be seen as a mixture of different populations, characterized by the observed pattern of missingness. Pattern-mixture models are given extensive treatment in Chapters 18 and 20.

The natural parameters of selection models and pattern-mixture models have a different meaning, and transforming a probability model into the other framework is, in general, not straightforward, even not for normal measurement models. When a selection model is used, it is often mentioned that one has to make untestable assumptions about the relationship between dropout and missing data (discussion in Diggle and Kenward 1994, Molenberghs, Kenward, and Lesaffre 1997). In pattern-mixture models, it is explicit which parameters cannot be identified. Little (1993) suggests the use of identifying relationships between identifiable and nonidentifiable parameters. Thus, even though these identifying relationships are also unverifiable (Little 1995), the advantage of pattern-mixture models is that the verifiable and unverifiable assumptions can easily be separated. Note that when interest is confined to describing the *observed* portions of the profiles, no extrapolation and, hence, no restriction is needed. More details are given in Section 18.1 (p. 278).

Further, selection models and pattern-mixture models are not the only possible ways of factorizing the joint distribution of the outcome and missingness processes. Section 17.1 places these models in a broader context.

The selection model taxonomy is based on the second factor of (15.1):

$$f(\boldsymbol{r}_i | \boldsymbol{y}_i, X_i, \boldsymbol{\psi}) \quad = \quad f(\boldsymbol{r}_i | \boldsymbol{y}_i^o, \boldsymbol{y}_i^m, X_i, \boldsymbol{\psi}). \quad (15.3)$$

If (15.3) is independent of the measurements [i.e., when it assumes the form $f(r_i|X_i, Z_i, \psi)$], then the process is termed *missing completely at random* (MCAR).

If (15.3) is independent of the unobserved (missing) measurements y_i^m, but depends on the observed measurements y_i^o, thereby assuming the form $f(r_i|y_i^o, X_i, \psi)$, then the process is referred to as *missing at random* (MAR).

Finally, when (15.3) depends on the missing values y_i^m, the process is referred to as *nonrandom missingness* (MNAR). An MNAR process is allowed to depend on y_i^o.

It is important to note that above terminology is independent of the statistical framework chosen to analyze the data. This is to be contrasted with the terms *ignorable* and *nonignorable* missingness. The latter terms depend crucially on the inferential framework (Rubin 1976).

15.8 Ignorability

Let us decide to use likelihood based estimation. The full data likelihood contribution for subject i assumes the form

$$L^*(\boldsymbol{\theta}, \boldsymbol{\psi}|X_i, Z_i, \boldsymbol{y}_i, \boldsymbol{r}_i) \propto f(\boldsymbol{y}_i, \boldsymbol{r}_i|X_i, Z_i, \boldsymbol{\theta}, \boldsymbol{\psi}).$$

Since inference has to be based on what is observed, the full data likelihood L^* has to be replaced by the observed data likelihood L:

$$L(\boldsymbol{\theta}, \boldsymbol{\psi}|X_i, Z_i, \boldsymbol{y}_i^o, \boldsymbol{r}_i) \propto f(\boldsymbol{y}_i^o, \boldsymbol{r}_i|X_i, Z_i, \boldsymbol{\theta}, \boldsymbol{\psi}) \tag{15.4}$$

with

$$\begin{aligned}
f(\boldsymbol{y}_i^o, \boldsymbol{r}_i|\boldsymbol{\theta}, \boldsymbol{\psi}) &= \int f(\boldsymbol{y}_i, \boldsymbol{r}_i|X_i, Z_i, \boldsymbol{\theta}, \boldsymbol{\psi}) d\boldsymbol{y}_i^m \\
&= \int f(\boldsymbol{y}_i^o, \boldsymbol{y}_i^m|X_i, Z_i, \boldsymbol{\theta}) f(\boldsymbol{r}_i|\boldsymbol{y}_i^o, \boldsymbol{y}_i^m, X_i, \boldsymbol{\psi}) d\boldsymbol{y}_i^m. \tag{15.5}
\end{aligned}$$

Under an MAR process, we obtain

$$\begin{aligned}
f(\boldsymbol{y}_i^o, \boldsymbol{r}_i|\boldsymbol{\theta}, \boldsymbol{\psi}) &= \int f(\boldsymbol{y}_i^o, \boldsymbol{y}_i^m|X_i, Z_i, \boldsymbol{\theta}) f(\boldsymbol{r}_i|\boldsymbol{y}_i^o, X_i, \boldsymbol{\psi}) d\boldsymbol{y}_i^m \\
&= f(\boldsymbol{y}_i^o|X_i, Z_i, \boldsymbol{\theta}) f(\boldsymbol{r}_i|\boldsymbol{y}_i^o, X_i, \boldsymbol{\psi}); \tag{15.6}
\end{aligned}$$

that is, the likelihood factorizes into two components of the same functional form as the general factorization (15.1) of the complete data. If, further,

θ and ψ are disjoint in the sense that the parameter space of the full vector $(\theta', \psi')'$ is the product of the parameter spaces of θ and ψ, then inference can be based on the marginal observed data density only. This technical requirement is referred to as the separability condition. However, still some caution should be used when constructing precision estimators (see Chapter 21).

In conclusion, when the separability condition is satisfied, *within the likelihood framework*, ignorability is equivalent to the union of MAR and MCAR. Hence, nonignorability and MNAR are synonyms in this context. A formal derivation is given in Rubin (1976), where it is also shown that the same requirements hold for Bayesian inference, but that frequentist inference is ignorable only under MCAR. Of course, ignorability is not helpful when at least part of the scientific interest is directed toward the missingness process.

Classical examples of the more stringent condition with frequentist methods are ordinary least squares (see also Sections 16.4 and 17.3) and the generalized estimating equations (GEE) approach of Liang and Zeger (1986). These GEE define an unbiased estimator only under MCAR. Robins, Rotnitzky, and Zhao (1995) and Rotnitzky and Robins (1995) have established that some progress can be made under MAR and that, even under MNAR processes, these methods can be applied (Rotnitzky and Robins 1997, Robins, Rotnitzky, and Scharfstein 1998). Their method is based on including weights that depend on the missingness probability, proving the point that at least some information on the missingness mechanism should be included and, thus, that ignorability does not hold.

15.9   A Special Case: Dropout

Without modifying the notational conventions for the measurement process $\boldsymbol{Y}_i$, we now let the scalar variable D_i be the *dropout indicator*. This is meaningful since, in the case that missingness is restricted to dropout, each vector $\boldsymbol{R}_i$ is of the form $(1, \ldots, 1, 0, \ldots, 0)$ and we can introduce a scalar dropout indicator

$$D_i = 1 + \sum_{j=1}^{n_i} R_{ij}. \tag{15.7}$$

For an incomplete sequence, D_i denotes the occasion at which dropout occurs. For a complete sequence, $D_i = n_i + 1$. In both cases, D_i indicates $1+$ the length of the measurement sequence, whether complete or incomplete.

It will sometimes be convenient to use a different dropout indicator:

$$T_i = \sum_{j=1}^{n_i} R_{ij} = D_i - 1. \tag{15.8}$$

Thus, $T_i = t$ indicates the pattern in which t measurements are obtained. For a complete sequence, $T_i = n_i$. Throughout the text, D_i and T_i will be used in precisely this meaning. Note that, thus far, t_{ij} has been used to indicate the time at which the jth measurement for subject i is taken. Given the difference between single and double subscripts for these different concepts, no confusion should arise.

Selection modeling is now obtained from factorizing the density of the full data $(\boldsymbol{y}_i, d_i)$, $i = 1, \ldots N$, as (suppressing covariate dependence)

$$f(\boldsymbol{y}_i, d_i | \boldsymbol{\theta}, \boldsymbol{\psi}) = f(\boldsymbol{y}_i | \boldsymbol{\theta}) f(d_i | \boldsymbol{y}_i, \boldsymbol{\psi}). \tag{15.9}$$

where the first factor is the marginal density of the measurement process and the second one is the density of the missingness process, conditional on the outcomes.

The observed data likelihood can be expressed as

$$\begin{aligned} f(\boldsymbol{y}_i^o, d_i | \boldsymbol{\theta}, \boldsymbol{\psi}) &= \int f(\boldsymbol{y}_i, d_i | \boldsymbol{\theta}, \boldsymbol{\psi}) d\boldsymbol{y}_i^m \\ &= \int f(\boldsymbol{y}_i^o, \boldsymbol{y}_i^m | \boldsymbol{\theta}) f(d_i | \boldsymbol{y}_i^o, \boldsymbol{y}_i^m, \boldsymbol{\psi}) d\boldsymbol{y}_i^m. \end{aligned} \tag{15.10}$$

If $f(d_i | \boldsymbol{y}_i^o, \boldsymbol{y}_i^m, \boldsymbol{\psi})$ is independent of the measurements [i.e., when it assumes the form $f(d_i | \boldsymbol{\psi})$], then the process is termed missing completely at random (MCAR). If $f(d_i | \boldsymbol{y}_i^o, \boldsymbol{y}_i^m, \boldsymbol{\psi})$ is independent of the unobserved (missing) measurements $\boldsymbol{y}_i^m$, but depends on the observed measurements $\boldsymbol{y}_i^o$, thereby assuming the form $f(d_i | \boldsymbol{y}_i^o, \boldsymbol{\psi})$, then the process is referred to as missing at random (MAR). Finally, when $f(d_i | \boldsymbol{y}_i^o, \boldsymbol{y}_i^m, \boldsymbol{\psi})$ depends on the missing values $\boldsymbol{y}_i^m$, the process is referred to as nonrandom missingness (MNAR). Of course, ignorability is defined in analogy with its definition in Section 15.8.

16

Simple Missing Data Methods

16.1 Introduction

As suggested in Section 15.2, missing data nearly always entail problems for the practicing statistician. First, inference will often be invalidated when the observed measurements do not constitute a simple random subset of the complete set of measurements. Second, even when correct inference follows, it is not always an easy task to trick standard software into operation on a ragged data structure.

Even in the simple case of a one-way ANOVA design (Neter, Wasserman, and Kutner 1990) and under an MCAR mechanism operating, problems occur since missingness destroys the balance between the sizes of the sub-samples. This implies that a slightly more complicated least squares analysis has to be invoked. Of course, a regression module for the latter analysis is included in most statistical software packages. The trouble is that the researcher has to know which tool to choose for particular classes of incomplete data.

Little and Rubin (1987) give an extensive treatment of methods to analyze incomplete data, many of which are intended for continuous, normally distributed data. Some of these methods were proposed more than 50 years ago. Examples are Yates' (1933) iterated ANOVA and Bartlett's (1937) ANCOVA procedures to analyze incomplete ANOVA designs. The former method is an early example of the Expectation-Maximization (EM) algorithm (Dempster, Laird, and Rubin 1977). See also Chapter 22.

We will briefly review a number of techniques that are valid when the measurement and missing data processes are independent and their parameters are separated (MCAR). It is important to realize that many of these methods are used also in situations where the MCAR assumption is not tenable. This should be seen as bad practice since it will often lead to biased estimates and invalid tests and hence to erroneous conclusions. Ample detail and illustrations of several problems are provided in Verbeke and Molenberghs (1997, Chapter 5). Section 16.2 discusses the computationally simplest technique, a *complete case* analysis, in which the analysis is restricted to the subjects for whom all intended measurements have been observed. A complete case analysis is popular because it maps a ragged data matrix into a rectangular one, by deleting incomplete cases. A second family of approaches, with a similar effect on the applicability of complete data software, is based on *imputing* missing values. One distinguishes between single imputation (Section 16.3) and multiple imputation (Section 20.3). In the first case, a single value is substituted for every "hole" in the data set and the resulting data set is analyzed as if it represented the true complete data. Multiple imputation properly acknowledges the uncertainty stemming from filling in missing values rather than observing them.

A third family is based on the principle of analyzing the incomplete data as such. A simple representative is a so-called *available case* analysis. This merely means that every component of a parameter (e.g., made up of mean vector and covariance matrix elements for a multivariate normal sample) is estimated using the maximal amount of information available for that component. This technique is discussed in Section 16.4 and applied in Section 17.4.2 on the growth data that have been introduced in Section 2.6. Although it makes use of more data than a corresponding complete case analysis, it still suffers from some drawbacks. For example, the method requires the missingness process to be MCAR. Section 17.3 describes a simple and convenient analysis, based on the more relaxed MAR assumption, that is consistent with factorization (15.6) and, importantly, can be carried out using PROC MIXED. A popular and very general technique to optimize incomplete data likelihoods under MAR is the EM algorithm (Dempster, Laird, and Rubin 1977). Little and Rubin (1987) used the EM algorithm to analyze their incomplete version of the growth data (Section 17.4.1). The principal ideas behind this method, and its connection to the MAR analysis of Section 17.3 will be given in Chapter 22.

16.2   Complete Case Analysis

A complete case analysis includes only those cases into the analysis, for which all n_i measurements were recorded. This method has obvious advantages. It is very simple to describe and since the data structure is as would have resulted from a complete experiment, standard statistical software can be used. Further, since the complete estimation is done on the same subset of completers, there is a common basis for inference, unlike for the available case methods (see Section 16.4).

Unfortunately, the method suffers from severe drawbacks. First, there is nearly always a substantial loss of information. For example, suppose there are 20 measurements, with 10% of missing data on each measurement. Suppose, further, that missingness on the different measurements is independent; then, the estimated percentage of incomplete observations is as high as 87%. The impact on precision and power is dramatic. Even though the reduction of the number of complete cases will be less dramatic in realistic settings where the missingness indicators R_i are correlated, the effect just sketched will often undermine a lot of complete case analyses. In addition, severe bias can result when the missingness mechanism is MAR but not MCAR. Indeed, should an estimator be consistent in the complete data problem, then the derived complete case analysis is consistent only if the missingness process is MCAR. Unfortunately, the MCAR assumption is much more restrictive than the MAR assumption.

A simple partial check on the MCAR assumption is as follows (Little and Rubin 1987). Divide the observations on measurement j into two groups: (1) those subjects that are also observed on another measurement or set of measurements and (2) those missing on the other measurement(s). Should MCAR hold, then both groups should be random samples of the same population. Failure to reject equality of the distributional parameters of both samples increases the evidence for MCAR, but does not prove it.

16.3   Simple Forms of Imputation

An alternative way to obtain a data set on which complete data methods can be used is filling in the missing values, instead of deleting subjects with incomplete sequences. The principle of imputation is particularly easy. The observed values are used to impute values for the missing observations. There are several ways to use the observed information. First, one can use information on the same subject (e.g., last observation carried forward). Second, information can be borrowed from other subjects (e.g., mean imputation). Finally, both within and between subject information can be used (e.g., conditional mean imputation, hot deck imputation). Standard references are Little and Rubin (1987) and Rubin (1987). Imputation strategies have been very popular in sample survey methods.

However, great care has to be taken with imputation strategies. Dempster and Rubin (1983) write

> The idea of imputation is both seductive and dangerous. It is seductive because it can lull the user into the pleasurable state of believing that the data are complete after all, and it is dangerous because it lumps together situations where the problem is sufficiently minor that it can be legitimately handled in this way and situations where standard estimators applied to the real and imputed data have substantial biases.

For example, Little and Rubin (1987) show that the method could work for a linear model with one fixed effect and one error term, but that it generally does not for hierarchical models, split-plot designs, and repeated measures (with a complicated error structure), random-effects, and mixed-effects models. At the very least, different imputations for different effects would be necessary.

The user of imputation strategies faces several dangers. First, the imputation model could be wrong and, hence, the point estimates would be biased. Second, even for a correct imputation model, the uncertainty resulting from incompleteness is masked. Indeed, even when one is reasonably sure about the mean value the unknown observation would have, the actual stochastic realization, depending on both the mean structure as well as on the error distribution, is still unknown.

In this section, several mean imputation strategies will be described. Applications to real data are discussed in Verbeke and Molenberghs (1997, Chapter 5).

16.3.1 Last Observation Carried Forward

Whenever a value is missing, the last observed value is substituted. The technique can be applied to both monotone and nonmonotone missing data. It is typically applied to settings where incompleteness is due to attrition.

Very strong and often unrealistic assumptions have to be made to ensure validity of this method. First, one has to believe that a subjects' measurement stays at the same level from the moment of dropout onward (or during the period they are unobserved in the case of intermittent missingness). In a clinical trial setting, one might believe that the response profile *changes* as soon as a patient goes off treatment and even that it would flatten. However, the constant profile assumption is even stronger. Further, this method shares with other single imputation methods that it overestimates the precision by treating imputed and actually observed values on equal footing.

16.3.2 *Imputing Unconditional Means*

The idea behind unconditional mean imputation (Little and Rubin 1987) is to replace a missing value with the average of the observed values on the same variable over the other subjects. Thus, the term *unconditional* refers to the fact that one does not use (i.e., condition on) information on the subject for which an imputation is generated.

16.3.3 *Buck's Method: Conditional Mean Imputation*

This approach was suggested by Buck (1960) and reviewed by Little and Rubin (1987). The method is technically hardly more complex than mean imputation. Let us describe it first for a single multivariate normal sample. The first step is to estimate the mean vector μ and the covariance matrix Σ from the complete cases. This step builds on the assumption that $Y \sim N(\mu, \Sigma)$. For a subject with missing components, the regression of the missing components (Y_i^m) on the observed ones (y_i^o) is

$$Y_i^m | y_i^o \sim N(\mu^m + \Sigma^{mo}(\Sigma^{oo})^{-1}(y_i^o - \mu_i^o), \Sigma^{mm} \\ -\Sigma^{mo}(\Sigma^{oo})^{-1}\Sigma^{om}).$$

Superscripts o and m refer to "observed" and "missing" components, respectively. The second step calculates the conditional mean from this regression and substitutes it for the missing values. In this way, "vertical" information (estimates for μ and Σ) is combined with "horizontal" information (y_i^o).

Buck (1960) showed that under mild regularity conditions, the method is valid for MCAR mechanisms. Little and Rubin (1987) added that the method is valid under certain types of MAR mechanism. Even though the distribution of the observed components is allowed to differ between complete and incomplete observations, it is very important that the regression of the missing components on the observed ones is constant across missingness patterns.

Again, this method shares with other single imputation strategies that, although point estimation may be consistent, the precision will be under-estimated. Little and Rubin (1987, p. 46) indicated ways to correct the precision estimation for unconditional mean imputation.

16.3.4   Discussion of Imputation Techniques

The imputation methods reviewed here are clearly not the only ones. Little and Rubin (1987) and Rubin (1987) mention several others. Several methods, such as hot deck imputation, are based on filling in missing values from "matching" subjects, where an appropriate matching criterion is used.

Almost all imputation techniques suffer from the following limitations:

1. The performance of imputation techniques is unreliable. Situations where they do work are difficult to distinguish from situations were they prove misleading. For example, although conditional imputation is considered superior to unconditional imputation, Verbeke and Molenberghs (1997, pp. 217–218) have seen that the latter performs better on the growth data, introduced in Section 2.6.

2. Imputation often requires ad hoc adjustments to yield satisfactory point estimates.

3. The methods fail to provide simple correct precision estimators.

In addition, most methods require the MCAR assumption to hold. Methods such as the last observation carried forward require additional and often unrealistically strong assumptions.

The main advantage, shared with complete case analysis, is that complete data software can be used. Although a complete case analysis is even simpler since one does not need to address the imputation task, the imputation family uses all (and, in fact, too much) of the available information. With the availability of the SAS procedure MIXED, it is no longer necessary to stick to complete data software, since this procedure allows for measurement sequences of unequal length. A discussion of multiple imputation is postponed until Section 20.3.

16.4   Available Case Methods

Consider a single multivariate normal sample, based on $i = 1, \ldots, N$ subjects, for which $j = 1, \ldots, n$ variables are planned. In a longitudinal context, the n variables would refer to n repeated measurements. The data matrix is $Y = (y_{ij})$.

Available case methods (Little and Rubin 1987) use as much of the data as possible. Let us restrict attention to the estimation of the mean vector $\boldsymbol{\mu}$ and the covariance matrix $\boldsymbol{\Sigma}$. The jth component μ_j of the mean vector and the jth diagonal variance element σ_{jj} are estimated using all cases that are observed on the jth variable, disregarding their response status at the other measurement occasions. The (j, k)th element $(j \neq k)$ of the covariance matrix is computed using all cases that are observed on both the jth and the kth variable.

This method is more efficient than the complete case method, since more information is used. The number of components of the outcome vector has no direct effect on the sample available for a particular mean or covariance component.

The method is valid only under MCAR. In this respect, it is no fundamental improvement over a complete case analysis. An added disadvantage is that, although more information is used and a consistent estimator is obtained under MCAR, it is not guaranteed that the covariance matrix is positive (semi-)definite. Of course, this is only a small-sample problem and does not invalidate asymptotic results. However, for samples with a large number of variables and/or with fairly high correlations between pairs of outcomes, this nuisance feature is likely to occur.

Although a complete case analysis is possible for virtually every statistical method and single imputation is also fairly generally applicable, extending an available case analysis beyond multivariate means and covariances can be tedious.

This method will be illustrated in Section 17.4.2, using the growth data introduced in Section 2.6.

16.5   MCAR Analysis of Toenail Data

Let us analyze the toenail data, introduced in Section 2.2, assuming an MCAR process holds. An exploratory graphical tool for studying average evolutions over time is to plot the sample average at each occasion versus

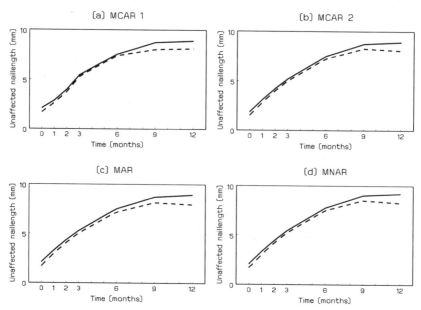

FIGURE 16.1. *Toenail Data. Estimated mean profile under treatment A (solid line) and treatment B (dashed line), obtained under different assumptions for the measurement model and the dropout model. (a) Completely random dropout, without parametric model for the average evolution in both groups. (b) Completely random dropout, assuming quadratic average evolution for both groups. (c) Random dropout, assuming quadratic average evolution for both groups. (d) Nonrandom dropout, assuming quadratic average evolution for both groups.*

time, thereby including all patients still available at that occasion. For the toenail example, this is shown in panel (a) of Figure 16.1. The graph suggests that there is very little difference between both groups, with marginal superiority of treatment A.

Note that the sample averages at a specific occasion are unbiased estimators of the mean responses of those subjects still in the study at that occasion. Hence, the average profiles in panel (a) of Figure 16.1 only reflect the marginal average evolutions if, at each occasion, the mean response of those still in the study equals the mean response of those who already dropped out. Thus, we have to assume that the mean of the response, conditional on dropout status, is independent of the dropout status.

Similar assumptions for variances, correlations, and so forth are needed for drawing valid inferences, based on sample statistics of the observed data. This then comes down to assuming the response Y to be statistically independent of the dropout time D (i.e., MCAR).

Under the assumption of MCAR, valid inferences based on sample statistics can be obtained as follows. First, note that the vector of all 14 sample averages plotted in panel (a) of Figure 16.1 can be interpreted as the ordinary least squares (OLS) estimate obtained from fitting a two-way ANOVA model to all available measurements, thereby ignoring the dependence between repeated measures within subjects. Under MCAR, this provides an unbiased estimator for the marginal average evolution in the population. Further, it follows from the theory on generalized estimating equations (Liang and Zeger 1986) that this OLS estimator is asymptotically normally distributed, and valid standard errors are obtained from the sandwich estimator (see Section 6.2.4). Hence, Wald-type statistics are readily available for testing hypotheses or for the calculation of approximate confidence intervals.

For our toenail example, we used the sample averages displayed in Figure 16.1 [panel (a)] to test for any differences between both treatment groups. The resulting Wald statistic equals $\chi^2 = 4.704$ on 7 degrees of freedom, from which we conclude that there is no evidence in the data for any difference between both groups ($p = 0.696$). Note that the above methodology also applies if we assume the outcome to satisfy a general linear regression model, where the average evolution in both groups may be assumed to be of a specific parametric form. We compared both treatments assuming that the average evolution is quadratic over time, with regression parameters possibly depending on treatment. The resulting OLS profiles are shown in panel (b) of Figure 16.1. The main difference with the profiles obtained from a model with unstructured mean evolutions [panel (a)] is seen during the treatment period (first 3 months). The Wald test statistic, employed for testing treatment differences, now equals $\chi^2 = 2.982$, on 3 degrees of freedom ($p = 0.395$) yielding the same conclusion as before.

In practice, patients often leave the study prematurely due to reasons related to the outcome of interest. The assumption of completely random dropout is then no longer tenable, and statistical methods allowing for less strict assumptions about the relation between the dropout process and the measurement process (MAR or even MNAR) should be investigated. This will be discussed in Sections 17.2 and 18.3.

17

Selection Models

17.1 Introduction

Much of the early development of, and debate about, selection models appeared in the econometrics literature in which the Tobit model (Heckman 1976) played a central role. This combines a marginal Gaussian regression model for the response, as might be used in the absence of missing data, with a Gaussian-based threshold model for the probability of a value being missing. For simplicity, consider a single Gaussian-distributed response variable $Y \sim N(\mu, \sigma^2)$. The probability of Y being missing is assumed to depend on a second Gaussian variable $Y_m \sim N(\mu_m, \sigma_m^2)$, where

$$P(R = 0) = P(Y_m < 0).$$

Dependence of missingness on the response Y is induced by introducing a correlation between Y and Y_m. To avoid some of the complications of direct likelihood maximization, a two-stage estimation procedure was proposed by Heckman (1976) for this type of model. The use of the Tobit model and associated two-stage procedure was the subject of considerable debate in the econometrics literature, much of it focusing on the issues of identifiability and sensitivity (Amemiya 1984, Little 1986).

At first sight, the Tobit model does not appear to have the selection model structure specified in (15.1) in that there is no conditional partition of $f(y, r)$. However, it is simple to show from the joint Gaussian distribution of Y and Y_m that in the Tobit model,

$$P(R = 0 \mid Y = y) \quad = \quad \Phi(\beta_0 + \beta_1 y)$$

for suitably chosen parameters β_0 and β_1 and $\Phi(\cdot)$ the Gaussian cumulative distribution function. This can be seen as a probit regression model for the (binary) missing value process. This basic structure underlies the simplest form of selection model that has been proposed for longitudinal data in the biometric setting. A suitable response model, such as the multivariate Gaussian, is combined with a binary regression model for dropout. At each time point, the occurrence of dropout can be regressed on previous and current values of the response as well as covariates. In this chapter, we explore such models in more detail.

Especially for a continuous response, these models can be constructed in a fairly obvious way, combining the multivariate Gaussian linear model with a suitable dropout model. Diggle and Kenward (1994) used a logistic dropout model in a longitudinal development of the model of Greenlees, Reece, and Zieschang (1982) for nonrandom missingness in a cross-sectional setting. This was subsequently extended to the nonmonotone setting, using an ante-dependence covariance structure for the response with full likelihood and using pseudo-likelihood (Troxel, Harrington, and Lipsitz 1998). For the full likelihood analyses, subject-by-subject integration is required in general, unless MAR is assumed. This makes maximization somewhat cumbersome. The above authors (Diggle and Kenward 1994, Troxel, Harrington, and Lipsitz 1998) used the Nelder and Mead simplex algorithm (Nelder and Mead 1965). However, such an approach lacks flexibility, is inefficient for high-dimensional problems, and does not exploit the well-known routines that are implemented for the two separate components of the model. For some combinations of response and dropout model, the EM algorithm can be used and this does allow separate maximization of response and dropout, hence exploiting the familiar structure, but integration is still required in the expectation step of the algorithm.

Section 17.2 presents an introductory data analysis, based on the toenail data. Both MAR and MNAR analyses are presented in order to appreciate interpretational and computational differences between both. Illustrated with a balanced set of growth data, the validity of an ignorable analysis, using standard statistical software such as the SAS procedure MIXED, is established in Section 17.3. A general MNAR selection model is constructed in Section 17.5 and applied to the Vorozole data in Section 17.6.

17.2   A Selection Model for the Toenail Data

We will familiarize the reader with selection modeling by means of an introductory analysis of the toenail set of data, introduced in Section 2.2 and analyzed in the MCAR context in Section 16.5.

As formally introduced in Chapter 15, under the selection model (15.1), one uses the functional form of $f(d_i|\boldsymbol{y_i})$ to discriminate between different types of dropout processes. Indeed, recall from Section 15.8 that under MCAR or MAR, the joint density of observed measurements and dropout indicator factors as

$$f(\boldsymbol{y_i^o}, d_i) = \begin{cases} f(\boldsymbol{y_i^o})f(d_i) & \text{under MCAR} \\ f(\boldsymbol{y_i^o})f(d_i|\boldsymbol{y_i^o}) & \text{under MAR,} \end{cases}$$

from which it follows that a marginal model for the observed data $\boldsymbol{y_i^o}$ only is required. Moreover, the measurement model $f(\boldsymbol{y_i^o})$ and the dropout model $f(d_i)$ or $f(d_i|\boldsymbol{y_i^o})$ can be fitted separately, provided that the parameters in both models are functionally independent of each other (separability). If interest is in the measurement model only, the dropout model can be completely ignored (Section 15.8). This implies that, under ignorability, MCAR and MAR provide the same fitted measurement model. However, as discussed by Kenward and Molenberghs (1998) and by Verbeke and Molenberghs (1997, Section 5.8), this does not imply that inferences under MCAR and MAR are equivalent.

17.2.1   MAR Analysis

In this section, we will fit a selection model to the toenail dermatophyte onychomycosis (TDO) data, assuming random dropout. Our primary goal is to test for any treatment differences; hence, we do not need to explicitly consider a dropout model, but only need to specify a marginal model for the observed outcomes $\boldsymbol{Y_i^o}$. The measurement model we consider here assumes a quadratic evolution for each subject, possibly with subject-specific intercepts, and we allow the stochastic error components to be correlated within subjects. More formally, we assume that $\boldsymbol{Y_i^o}$ satisfies the following linear mixed-effects model:

$$Y_{ij}^o = \begin{cases} (\beta_{A0} + b_i) + \beta_{A1}t_{ij} + \beta_{A2}t_{ij}^2 + \varepsilon_{(1)ij} + \varepsilon_{(2)ij} & \text{group A} \\ (\beta_{B0} + b_i) + \beta_{B1}t_{ij} + \beta_{B2}t_{ij}^2 + \varepsilon_{(1)ij} + \varepsilon_{(2)ij} & \text{group B.} \end{cases} \quad (17.1)$$

This model is similar to (3.11). All random components have zero mean. The random intercept variance is d_{11}. The variance of the measurement error $\varepsilon_{(1)i}$ is $\sigma^2 I_{n_i}$, whereas the variance of the serial process is $\varepsilon_{(2)i}$ is $\tau^2 H_i$, where H_i follows from the particular serial process considered. The unknown parameters $\beta_{A0}, \beta_{A1}, \beta_{A2}, \beta_{B0}, \beta_{B1}$, and β_{B2} describe the average quadratic evolution of Y_i^o over time.

Let us first assume that $\varepsilon_{(1)ij}$ is absent. The estimated average profiles obtained from fitting model (17.1) to our TDO data are shown in panel (c) of Figure 16.1. Note that there is very little difference from the OLS average profiles shown in panel (b) of the same figure and obtained under the MCAR assumption. The observed likelihood ratio statistic for testing for treatment differences equals $2 \ln \lambda = 4.626$ on 3 degrees of freedom. Hence, under model (17.1) and under the assumption of random dropout, there is little evidence for any average difference between the treatments A and B $(p = 0.201)$.

17.2.2   MNAR analysis

In cases where dropout could be related to the unobserved responses, dropout is no longer ignorable, implying that treatment effects can no longer be tested or estimated without explicitly taking the dropout model $f(d_i | y_i^o, y_i^m)$ into account. Moreover, a marginal model is then required for the complete vector Y_i, rather than for the observed component Y_i^o only.

For the TDO data, we assume that all outcomes Y_{ij} satisfy

$$Y_{ij} = \begin{cases} (\beta_{A0} + b_i) + \beta_{A1} t_{ij} + \beta_{A2} t_{ij}^2 + \varepsilon_{(2)ij} & \text{group A} \\ (\beta_{B0} + b_i) + \beta_{B1} t_{ij} + \beta_{B2} t_{ij}^2 + \varepsilon_{(2)ij} & \text{group B.} \end{cases} \quad (17.2)$$

Note that, for reasons that will become clear later, the measurement error component $\varepsilon_{(1)ij}$ has been removed from the model. Under MAR, model (17.2) reduces to (17.1) with the measurement error removed, but when MAR does not hold, we no longer have that Y_i^o satisfies model (17.1). Further, we assume that the probability for dropout at occasion j $(j = 2, \ldots, n_i)$, given the subject was still in the study at the previous occasion, follows a logistic regression model, in line with Diggle and Kenward (1994),

$$\text{logit}\left[P(D_i = j \mid D_i \geq d, y_i)\right] = \psi_0 + \psi_1 y_{ij} + \psi_2 y_{i,j-1}. \quad (17.3)$$

It is assumed that $d_i \geq 2$. The above model is then used to calculate the dropout probability at each occasion, given the measurements $\boldsymbol{y_i}$:

$$
f(d_i|\boldsymbol{y_i}) = \begin{cases} \begin{aligned} & P(D_i = d_i|D_i \geq d_i, \boldsymbol{y_i}) \\ & \quad \times \prod_{k=2}^{d_i-1} [1 - P(D_i = k|D_i \geq k, \boldsymbol{y_i})] \quad d_i \leq n_i \\[2ex] & \prod_{k=2}^{n_i} [1 - P(D_i = k|D_i \geq k, \boldsymbol{y_i})] \qquad\qquad d_i > n_i. \end{aligned} \end{cases}
\tag{17.4}
$$

Model (17.4) implies that the dropout contribution $f(d_i|\boldsymbol{y_i})$ to the log-likelihood can be written as a product of independent contributions of the form $P(D_i = d_i|D_i \geq d_i, \boldsymbol{y_i})$. They describe a binary outcome, conditional on covariates. For such data, logistic regression methods are a natural choice, as mentioned earlier. Implementations of this will be discussed at a later stage.

At this point, we will relate dropout to the current and previous observation only. No covariates are included, and we do not explicitly take into account the fact that the time points at which measurements have been taken are not evenly spaced. Diggle and Kenward (1994) consider a more general model where dropout at occasion j can depend on the complete history $\{y_{i1}, \ldots, y_{i,j-1}\}$, as well as on external covariates. In addition, one could argue that the dependence of dropout on the measurement history is time dependent. This would imply that the parameters $\boldsymbol{\psi} = (\psi_0, \psi_1, \psi_2)'$ in (17.3) depend on j. However, this level of generality will not be considered here. Note also that, strictly speaking, (15.1) allows dropout at a specific occasion to be related to all future responses. However, this is rather counterintuitive in many cases, especially when it is difficult for the study participants to make projections about the future responses. Moreover, including future outcomes seriously complicates the calculations since computation of the likelihood (15.4) then requires evaluation of a possibly high-dimensional integral. Diggle and Kenward (1994) and Molenberghs, Kenward, and Lesaffre (1997) considered nonrandom versions of this model by including the current, possible unobserved measurement. This requires more elaborate fitting algorithms, given the high-dimensional mentioned earlier. Diggle and Kenward (1994) used the simplex algorithm (Nelder and Mead, 1965), and Molenberghs, Kenward, and Lesaffre fitted their models with the EM algorithm (Dempster, Laird, and Rubin 1977). The algorithm of Diggle and Kenward (1994) is implemented in OSWALD (Smith, Robertson, and Diggle 1996). For further information on OSWALD, please consult Section A.3.2.

We fitted the above model to our TDO data using the PCMID function in the Splus suite of functions called OSWALD (Smith, Robertson and Diggle,

1996). The fitted average profiles are shown in panel (d) of Figure 16.1. Note that we again find very little difference with the estimated average profiles obtained from previous analyses. The observed likelihood ratio statistic for testing for treatment differences equals $2 \ln \lambda = 4.238$ on 3 degrees of freedom. Hence, there is again very little evidence for the presence of any average treatment difference ($p = 0.237$).

The fitted dropout model equals

$$\text{logit}\,[P(D_i = j | D_i \geq j, \boldsymbol{y_i})] \quad = \quad -4.26 + 0.47 y_{ij} - 0.46 y_{i,j-1},$$

which can be rewritten as

$$
\begin{aligned}
&\text{logit}\,[P(D_i = j | D_i \geq j, \boldsymbol{y_i})] \\
&= \quad -4.26 + 0.47(y_{ij} - y_{i,j-1}) + 0.01 y_{i,j-1},
\end{aligned}
$$

showing that dropout is related to the increment $y_{ij} - y_{i,j-1}$, rather than to any of the actual observations y_{ij} or $y_{i,j-1}$, and such that subjects which improve most (large increments) are very likely to drop out from the study. This phenomenon is very common in practice (see, e.g., Diggle and Kenward 1994, Molenberghs, Kenward and Lesaffre 1997). See also Section 19.5.2.

Special cases of model (17.3) are obtained by setting $\psi_1 = 0$ or $\psi_1 = \psi_2 = 0$, respectively. In the first case, dropout is no longer allowed to depend on the current measurement, implying random dropout. In the second case, dropout is independent of the outcome, which corresponds to completely random dropout. Thus, *under the assumed model*, it is possible to test for nonignorable dropout. The likelihood ratio test statistic, comparing the maximized likelihood under model (17.3) with the maximized likelihood under the same model with $\psi_1 = 0$, equals $2 \ln \lambda = 25.386$, which is highly significant ($p < 0.0001$) on 1 degree of freedom. Hence, conditional on the validity of model (17.3), there is a lot of evidence for nonrandom dropout. However, some caution is needed, as will be indicated next.

The main advantage of selection models for nonrandom dropout is that they directly model the quantities which are usually of primary interest: the marginal distribution of the outcome vector $\boldsymbol{Y_i}$ and the distribution of the dropout process conditional on $\boldsymbol{Y_i}$. The former is used for marginal inferences on longitudinal profiles; the latter is used to characterize the dropout process (MCAR, MAR, MNAR). It is important to realize that a model is needed for the complete data vector $\boldsymbol{Y_i}$ rather than for the observed component $\boldsymbol{Y_i^o}$ only. Thus, even if a posited model fits the observed outcomes and the nonresponse data well, one can easily find a model with a similar fit but *different in the predictions for the unobserved outcomes*. Such a model may yield different conclusions about key aspects of the outcome and nonresponse mechanisms. Thus, one is faced with untestable assumptions (see also Section 18.1, p. 278). This clearly indicates a great sensitivity of the conclusions to the stated complete data model. Several authors have pointed to this sensitivity, such as Rubin (1994), Laird (1994), Little (1995), Hogan and Laird (1997), and Molenberghs, Kenward and Lesaffre (1997). A good example is given by Kenward (1998), who reanalyzed data on mastitis in dairy cows, previously analyzed by Diggle and Kenward (1994). He found that all evidence for nonrandom dropout vanishes when the normality assumption for the second, possibly unobserved, outcome conditional on the first, always observed, outcome is replaced by a heavy-tailed t-distribution. See also Section 19.5.1. Further illustrations are Molenberghs, Verbeke, *et al.* (1999) and Kenward and Molenberghs (1999). Thus, clearly, caution is required and, preferably, a sensitivity analysis should be conducted. Formal tools to carry out such an analysis are provided in Chapter 19.

Another reason to formulate a model not only for the complete but also for the observed data is that these components need to be integrated out from the likelihood. Technically, this implies that, at present, specialized software, based on computationally intensive and highly unstable algorithms, is required for fitting nonrandom dropout models.

Another example of the sensitivity of selection models is found in the context of the TDO example. We reanalyzed the data, by allowing a measurement error component in (17.2) to be present:

$$
Y_{ij} = \begin{cases} (\beta_{A0} + b_i) + \beta_{A1}t_{ij} + \beta_{A2}t_{ij}^2 + \varepsilon_{(1)ij} + \varepsilon_{(2)ij} & \text{group A} \\ (\beta_{B0} + b_i) + \beta_{B1}t_{ij} + \beta_{B2}t_{ij}^2 + \varepsilon_{(1)ij} + \varepsilon_{(2)ij} & \text{group B.} \end{cases} \tag{17.5}
$$

Table 17.1 shows a summary of the results from model (17.5) as well as from model (17.2). The estimated amount of variability explained by the $\varepsilon_{(1)ij}$ equals $\hat{\sigma}^2/(\hat{\sigma}^2 + \hat{\tau}^2 + \hat{d}_{11}) = 6\%$. As before, we again find no evidence for any difference in average evolution between both treatment groups ($p = 0.388$). The main difference between the results from both models is found in the LR test for random dropout. The likelihood ratio test statistic reduces from $2 \ln \lambda = 25.386$ under model (17.2) to $2 \ln \lambda = 4.432$ under model (17.5). The reason for this can be found in the estimated correlation $\widehat{\text{corr}}(y_{ij}, y_{ik})$ of outcomes within subjects, also shown in Table 17.1. Under model (17.5), outcomes are much more correlated than under model (17.2), explaining why, under model (17.5), the current observation is less needed for predicting dropout, once the previous measurement is known.

In practice, the covariance structure is often considered a nuisance, and very little effort is spent in finding adequate covariance models. As shown in the above example, this becomes crucial when dropout is present. Statistical packages such as the SAS procedure MIXED (1997) nowadays allow one to fit linear mixed models with a variety of covariance structures. However, as illustrated by Verbeke, Lesaffre and Brant (1998) and by Lesaffre, Asefa and Verbeke (1999), finding appropriate covariance models is often far from

TABLE 17.1. *Toenail Data. Summary of results obtained from fitting model (17.2) and model (17.5), in combination with dropout model (17.3), to the TDO data.*

Model (17.2)	
Variability explained by measurement error:	0%
LR test for treatment differences:	$2\ln\lambda = 4.238$ $p = 0.237$
LR test for MAR:	$2\ln\lambda = 25.386$ $p < 0.0001$

Fitted correlations:
$$\begin{pmatrix} 1 & .87 & .77 & .70 & .58 & .53 & .52 \\ & 1 & .87 & .77 & .61 & .54 & .52 \\ & & 1 & .87 & .65 & .56 & .53 \\ & & & 1 & .70 & .58 & .53 \\ & & & & 1 & .70 & .58 \\ & & & & & 1 & .70 \\ & & & & & & 1 \end{pmatrix}$$

Model (17.5)	
Variability explained by measurement error:	6%
LR test for treatment differences:	$2\ln\lambda = 3.024$ $p = 0.388$
LR test for MAR:	$2\ln\lambda = 4.432$ $p = 0.035$

Fitted correlations:
$$\begin{pmatrix} 1 & .91 & .89 & .86 & .79 & .73 & .70 \\ & 1 & .91 & .89 & .81 & .75 & .70 \\ & & 1 & .91 & .83 & .77 & .71 \\ & & & 1 & .86 & .79 & .73 \\ & & & & 1 & .86 & .79 \\ & & & & & 1 & .86 \\ & & & & & & 1 \end{pmatrix}$$

straightforward. It is therefore advisable to compare results from several models, with varying plausible covariance structures. See also Chapter 10.

17.3   Scope of Ignorability

Let us, as in the toenail data example in Section 17.2.1, assume that MAR holds. Assume, in addition, that the separability condition is satisfied (Section 15.8). In Section 15.8, it was argued, and in Section 17.2.1 it was re-iterated, that likelihood based inference is valid, whenever the mechanism is MAR and provided the technical condition holds that the parameters describing the nonresponse mechanism are distinct from the measurement model parameters. In other words, the missing data process should be ignorable in the likelihood inference sense, since, then, factorization (15.6) applies and the log-likelihood partitions into two functionally independent components.

This implies that a module with likelihood estimation facilities which can handle incompletely observed subjects since its units are measurements rather than subjects, manipulates the correct likelihood and leads to valid likelihood ratios. We will qualify this statement in more detail since, although this is an extremely important feature of PROC MIXED and in fact of any flexibly written linear mixed model likelihood optimization routine, a few cautionary remarks still apply.

1. Ignorability depends on the often implicit assumption that the scientific interest is directed toward the measurement model parameters θ (fixed effects, variance components, or a combination of both) and that the missing data mechanism parameters ψ are nuisance parameters. This is not always true. For instance, when the question of predicting an individual's measurement profile (individual or group averaged) is raised, past the time of dropout and *given that she dropped out*, then *both* parameter vectors θ and ψ need to be estimated. However, due to the ignorability and the resulting partitioning of the likelihood, one can construct a model for nonresponse, separately from the linear mixed measurement model. As a practical consequence, the software module to estimate the missingness model parameters can be chosen independently from PROC MIXED. Often, categorical data analysis methods such as logistic regression will be a sensible choice in this respect.

2. Likelihood inference is often surrounded with references to the sampling distribution (e.g., to construct precision estimators and for statistical hypothesis tests; Kenward and Molenberghs 1998). This issue and its relationship to the PROC MIXED implementation is discussed further in Chapter 21.

3. Even though the assumption of likelihood ignorability encompasses the MAR and not only the more stringent and often implausible

MCAR mechanisms, it is difficult to exclude the option of a more general nonrandom dropout mechanism. One solution is to fit an MNAR model as proposed by Diggle and Kenward (1994). This was done for the toenail data in Section 17.2.2. Diggle and Kenward (1994) fitted models to the full data using the simplex algorithm (Nelder and Mead 1965). Alternatively, the EM algorithm can be used, as proposed by Molenberghs, Kenward, and Lesaffre (1997), for the longitudinal categorical data problem. A module for the linear mixed model with dropout is implemented in the OSWALD software, written for S-Plus (Smith, Robertson, and Diggle 1996). It is based on an extension of the Diggle and Kenward (1994) model. A SAS program for an EM algorithm for the linear model would consist of three parts. First, a macro to carry out the E step has to be written, where the expected value of the observed data likelihood is computed conditional on the current parameter vector and on the observed data. The M step consists of two substeps, where PROC MIXED might be used to maximize the measurement process likelihood and a different routine (e.g., logistic regression) could be called to maximize the nonresponse likelihood. Diggle and Kenward (1994) assumed a logistic model for the dropout process. The EM algorithm will be sketched in Chapter 22.

17.4   Growth Data

The growth data have been introduced in Section 2.6. Section 17.4.1 analyzes the original set of data (i.e., without artificially removed subjects). The incomplete version, generated by Little and Rubin (1987), is studied in Section 17.4.3, and the missingness process is studied in Section 17.4.4.

17.4.1   Analysis of Complete Growth Data

Following guidelines in Chapter 9 and in Diggle, Liang, and Zeger (1994) model building should proceed by constructing an adequate description of the variability on an appropriate set of residuals. These residuals are preferably found by subtracting a saturated sex by time mean model from the measurements. When a satisfactory covariance model is found, attention would then shift to simplification of the mean structure. However, this insight is relatively recent and was certainly not the standard procedure in the mid-eighties. Jennrich and Schluchter (1986) constructed eight models, where the first three concentrate on the mean model, leaving the 4×4

covariance matrix of the repeated measurements completely unstructured. Once an adequate mean model is found, the remaining five models are fit to enable simplification of the covariance structure. Jennrich and Schluchter (1986) primarily wanted to illustrate their estimation procedures and did not envisage a comprehensive data analysis. Moreover, since this procedure can be considered legitimate in small balanced studies and also for reasons of comparability, we will, at first, adopt the same eight models, in the same order. In this section, these models will be fitted to the original data, referred to henceforth as the *complete data set*. The results of Jennrich and Schluchter (1986) will be recovered and additional insight will be given. In Section 17.4.3, these solutions will be compared to the results for the same eight models on the incomplete data. Jennrich and Schluchter (1986) used Newton-Raphson, Fisher scoring, and generalized Expectation-Maximization (EM) algorithms to maximize the log-likelihood. We will show that the data can be analyzed relatively easily using PROC MIXED.

The models of Jennrich and Schluchter (1986) can be expressed in the general linear mixed models family (3.8):

$$\boldsymbol{Y}_i = X_i\boldsymbol{\beta} + Z_i\boldsymbol{b}_i + \boldsymbol{\varepsilon}_i, \tag{17.6}$$

where

$$\boldsymbol{b}_i \sim N(\boldsymbol{0}, D),$$
$$\boldsymbol{\varepsilon}_i \sim N(\boldsymbol{0}, \Sigma),$$

and $\boldsymbol{b}_i$ and $\boldsymbol{\varepsilon}_i$ are statistically independent. As earlier (Section 3.3), $\boldsymbol{Y}_i$ is the (4×1) response vector, X_i is a $(4 \times p)$ design matrix for the fixed effects, $\boldsymbol{\beta}$ is a vector of unknown fixed regression coefficients, Z_i is a $(4 \times q)$ design matrix for the random effects, $\boldsymbol{b}_i$ is a $(q \times 1)$ vector of normally distributed random parameters, with covariance matrix D, and $\boldsymbol{\varepsilon}_i$ is a normally distributed (4×1) random error vector, with covariance matrix Σ. Since every subject contributes exactly four measurements at exactly the same time points, it has been possible to drop the subscript i from the error covariance matrix Σ. The random error $\boldsymbol{\varepsilon}_i$ encompasses both measurement error (as in a cross-sectional study) and serial correlation. In this study, the design will be a function of age, sex, and/or the interaction between both. Let us indicate boys with $x_i = 0$, girls with $x_i = 1$, and age with $t_j = 8, 10, 12, 14$.

MODEL 1

The first model we will consider assumes a separate mean for each of the eight age×sex combinations, together with an unstructured covariance. This is done by assuming that the covariance matrix Σ of the error vector

ε_i is a completely general positive definite matrix and no random effects are included.

This model can be expressed as

$$\begin{array}{rcl}
Y_{i1} & = & \beta_0 + \beta_1(1 - x_i) + \beta_{0,8}(1 - x_i) + \beta_{1,8}x_i + \varepsilon_{i1}, \\
Y_{i2} & = & \beta_0 + \beta_1(1 - x_i) + \beta_{0,10}(1 - x_i) + \beta_{1,10}x_i + \varepsilon_{i2}, \\
Y_{i3} & = & \beta_0 + \beta_1(1 - x_i) + \beta_{0,12}(1 - x_i) + \beta_{1,12}x_i + \varepsilon_{i3}, \\
Y_{i4} & = & \beta_0 + \beta_1(1 - x_i) + \varepsilon_{i4},
\end{array} \qquad (17.7)$$

or, in matrix notation,

$$Y_i \;=\; X_i\beta + \varepsilon_i,$$

with

$$X_i \;=\; \begin{pmatrix}
1 & 1 - x_i & 1 - x_i & 0 & 0 & x_i & 0 & 0 \\
1 & 1 - x_i & 0 & 1 - x_i & 0 & 0 & x_i & 0 \\
1 & 1 - x_i & 0 & 0 & 1 - x_i & 0 & 0 & x_i \\
1 & 1 - x_i & 0 & 0 & 0 & 0 & 0 & 0
\end{pmatrix}$$

and $\beta = (\beta_0, \beta_1, \beta_{0,8}, \beta_{0,10}, \beta_{0,12}, \beta_{1,8}, \beta_{1,10}, \beta_{1,12})'$. With this parameterization, the means for girls are $\beta_0 + \beta_{1,8}$; $\beta_0 + \beta_{1,10}$; $\beta_0 + \beta_{1,12}$; and β_0 at ages 8, 10, 12, and 14, respectively. The corresponding means for boys are $\beta_0 + \beta_1 + \beta_{0,8}$; $\beta_0 + \beta_1 + \beta_{0,10}$; $\beta_0 + \beta_1 + \beta_{0,12}$; and $\beta_0 + \beta_1$, respectively. Of course, there are many equivalent ways to express the set of eight means in terms of eight linearly independent parameters.

This model can, for example, be fitted with the following SAS code:

```
proc mixed data = growth method = ml covtest;
title 'Growth Data, Model 1';
class idnr sex age;
model measure = sex age*sex / s;
repeated / type = un subject = idnr r rcorr;
run;
```

Let us discuss the fit of the model. The deviance (minus twice the log-likelihood at maximum) equals 416.5093, and there are 18 model parameters (8 mean, 4 variance, and 6 covariance parameters). This deviance will serve as a reference to assess the goodness-of-fit of simpler models. Parameter estimates and standard errors are reproduced in Table 17.2. The deviances are listed in Table 17.4.

TABLE 17.2. *Growth Data. Maximum likelihood estimates and standard errors (model based and empirically corrected) for the fixed effects in Model 1 (complete data set).*

Parameter	MLE	(s.e.)$^{(1)}$	(s.e.)$^{(2)}$
β_0	24.0909	(0.6478)	(0.7007)
β_1	3.3778	(0.8415)	(0.8636)
$\beta_{0,8}$	−2.9091	(0.6475)	(0.3793)
$\beta_{0,10}$	−1.8636	(0.4620)	(0.3407)
$\beta_{0,12}$	−1.0000	(0.5174)	(0.2227)
$\beta_{1,8}$	−4.5938	(0.5369)	(0.6468)
$\beta_{1,10}$	−3.6563	(0.3831)	(0.4391)
$\beta_{1,12}$	−1.7500	(0.4290)	(0.5358)

$^{(1)}$ Default s.e.'s under Model 1.

$^{(2)}$ Sandwich s.e.'s, obtained from the "empirical" option;

also obtained under Model 0.

The estimated covariance matrix $\widehat{\Sigma}$ of the error vector, based on this model, equals

$$
\widehat{\Sigma} \;=\; \begin{pmatrix}
5.0143 & 2.5156 & 3.6206 & 2.5095 \\
2.5156 & 3.8748 & 2.7103 & 3.0714 \\
3.6206 & 2.7103 & 5.9775 & 3.8248 \\
2.5095 & 3.0714 & 3.8248 & 4.6164
\end{pmatrix}
\tag{17.8}
$$

with corresponding correlation matrix

$$
\begin{pmatrix}
1.0000 & 0.5707 & 0.6613 & 0.5216 \\
0.5707 & 1.0000 & 0.5632 & 0.7262 \\
0.6613 & 0.5632 & 1.0000 & 0.7281 \\
0.5216 & 0.7262 & 0.7281 & 1.0000
\end{pmatrix}.
\tag{17.9}
$$

These quantities are easily obtained in PROC MIXED by using the options 'r' and 'rcorr' in the REPEATED statement (see Section 8.2.6). Apparently, the variances are close to each other, and so are the correlations.

Even though we opted to follow closely the models discussed in Jennrich and Schluchter (1986), it is instructive to consider a more elaborate model, termed Model 0, where a separate unstructured covariance matrix is assumed for each of the two sex groups. This model has 10 extra parameters and can be fitted to the data using the following SAS code:

TABLE 17.3. *Growth Data. Predicted means.*

Model	Age	Boys Estimate	s.e.	Girls Estimate	s.e.
1	8	22.88	0.56	21.18	0.68
	10	23.81	0.49	22.23	0.59
	12	25.72	0.61	23.09	0.74
	14	27.47	0.54	24.09	0.65
2	8	22.46	0.49	21.24	0.59
	10	24.11	0.45	22.19	0.55
	12	25.76	0.47	23.14	0.57
	14	27.42	0.54	24.10	0.65
3	8	22.82	0.48	20.77	0.57
	10	24.16	0.45	22.12	0.55
	12	25.51	0.47	23.47	0.56
	14	26.86	0.52	24.82	0.60
4	8	22.64	0.53	21.22	0.64
	10	24.23	0.48	22.17	0.57
	12	25.83	0.48	23.12	0.57
	14	27.42	0.53	24.07	0.64
5	8	22.75	0.54	21.19	0.66
	10	24.29	0.44	22.16	0.53
	12	25.83	0.44	23.13	0.53
	14	27.37	0.54	24.09	0.66
6	8	22.62	0.51	21.21	0.61
	10	24.18	0.47	22.17	0.56
	12	25.75	0.48	23.13	0.58
	14	27.32	0.55	24.09	0.67
7	8	22.62	0.52	21.21	0.63
	10	24.18	0.47	22.17	0.57
	12	25.75	0.47	23.13	0.57
	14	27.32	0.52	24.09	0.63
8	8	22.62	0.46	21.21	0.56
	10	24.18	0.30	22.17	0.37
	12	25.75	0.30	23.13	0.37
	14	27.32	0.46	24.09	0.56

Source: Jennrich and Schluchter (1986).

```
proc mixed data = growth method = ml covtest;
title 'Growth Data, Model 0';
class idnr sex  age;
model measure = sex age*sex / s;
repeated / type = un subject = idnr
          r = 1,12 rcorr = 1,12 group = sex;
run;
```

Since this model has individual-specific covariance matrices (although there are only two values, one for each gender), Σ has to be replaced by Σ_i.

These separate covariance matrices are requested by means of the 'group=' option. These matrices and the corresponding correlation matrices are printed using the 'r=' and 'rcorr=' options. The estimated covariance matrix for girls is

$$
\begin{pmatrix}
4.1033 & 3.0496 & 3.9380 & 3.9607 \\
3.0496 & 3.2893 & 3.6612 & 3.7066 \\
3.9380 & 3.6612 & 5.0826 & 4.9690 \\
3.9607 & 3.7066 & 4.9690 & 5.4008
\end{pmatrix}
$$

with corresponding correlation matrix

$$
\begin{pmatrix}
1.0000 & 0.8301 & 0.8623 & 0.8414 \\
0.8301 & 1.0000 & 0.8954 & 0.8794 \\
0.8623 & 0.8954 & 1.0000 & 0.9484 \\
0.8414 & 0.8794 & 0.9484 & 1.0000
\end{pmatrix} .
$$

The corresponding quantities for boys are

$$
\begin{pmatrix}
5.6406 & 2.1484 & 3.4023 & 1.5117 \\
2.1484 & 4.2773 & 2.0566 & 2.6348 \\
3.4023 & 2.0566 & 6.5928 & 3.0381 \\
1.5117 & 2.6348 & 3.0381 & 4.0771
\end{pmatrix}
$$

and

$$
\begin{pmatrix}
1.0000 & 0.4374 & 0.5579 & 0.3152 \\
0.4374 & 1.0000 & 0.3873 & 0.6309 \\
0.5579 & 0.3873 & 1.0000 & 0.5860 \\
0.3152 & 0.6309 & 0.5860 & 1.0000
\end{pmatrix} .
$$

From these, we suspect that there is a non-negligible difference between the covariance structures for boys and girls, with, in particular, a weaker correlation among the boys' measurements. This is indeed supported by a deviance of 23.77 on 10 degrees of freedom ($p = 0.0082$). Nevertheless, the point estimates for the fixed effects coincide exactly with the ones obtained from Model 1 (see Table 17.2). However, even if attention is restricted to fixed-effects inference, one still needs to address the quality of the estimates of precision. To this end, there are, in fact, two solutions. First, the more elaborate Model 0 can be fitted, as was done already. A drawback is that this model has 28 parameters altogether, which is quite a substantial number for such a small data set, implying that the asymptotic behavior of, for example, the deviance statistic becomes questionable. As discussed in Section 6.2.4, an alternative solution consists of retaining Model 1 and estimating the standard errors by means of the so-called robust estimator (equivalently termed "sandwich" or "empirically corrected" estimator; Liang and Zeger 1986). To this end, the following code can be used:

```
proc mixed data = growth method = ml covtest empirical;
title 'Growth Data, Model 1, Empirically Corrected';
class idnr sex  age;
model measure = sex age*sex / s;
repeated / type = un subject = idnr r rcorr;
run;
```

Here, the 'empirical' option is added to the PROC MIXED statement. This method yields a consistent estimator of precision, even if the covariance model is misspecified. In this particular case (a full factorial mean model), both methods (Model 0 on the one hand and the empirically corrected Model 1 on the other hand) lead to *exactly* the same standard errors. This illustrates that the robust method can be advantageous if correct standard errors are required, but finding an adequate covariance model is judged too involved. The robust standard errors are presented in Table 17.2 as the second entry in parentheses. It is seen that the naive standard errors are somewhat smaller than their robust counterparts, except for the parameters $\beta_{1,8}$, $\beta_{1,10}$, and $\beta_{1,12}$, where they are considerably larger. Even though the relation between the standard errors of the "correct model" (here, Model 0) and the empirically corrected "working model" (here, Model 1) will not always be a mathematical identity, the empirically corrected estimator option is a useful tool to compensate for misspecification in the covariance model.

Let us now return to our discussion of Model 1. It is insightful to consider the means for each of the eight categories explicitly. These means are presented in Table 17.3 for Models 1–8. The first panel of Figure 17.1 depicts the eight individual group means, connected to form two profiles, one for each sex group. Clearly, there seems to be a linear trend in both profiles as well as a vague indication for diverging lines, and hence different slopes.

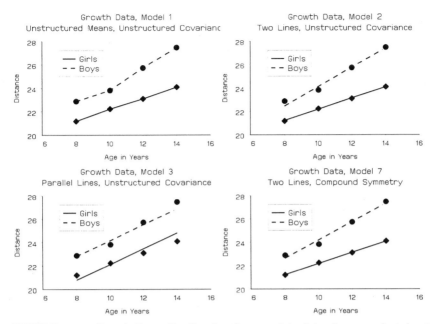

FIGURE 17.1. *Growth Data. Profiles for the complete data, from a selected set of models.*

These hypotheses will be assessed on the basis of likelihood ratio tests, using the simpler Models 2 and 3.

MODEL 2

The first simplification occurs by assuming a linear trend within each sex group. This implies that each profile can be described with two parameters (intercept and slope), instead of with four unstructured means. The error matrix Σ will be left unstructured. The model can be expressed as

$$Y_{ij} = \beta_0 + \beta_{01}(1 - x_i) + \beta_{10}t_j(1 - x_i) + \beta_{11}t_jx_i + \varepsilon_{ij} \quad (17.10)$$

or, in matrix notation,

$$\boldsymbol{Y}_i = X_i\boldsymbol{\beta} + \boldsymbol{\varepsilon}_i,$$

where the design matrix changes to

$$X_i = \begin{pmatrix} 1 & 1 - x_i & 8(1 - x_i) & 8x_i \\ 1 & 1 - x_i & 10(1 - x_i) & 10x_i \\ 1 & 1 - x_i & 12(1 - x_i) & 12x_i \\ 1 & 1 - x_i & 14(1 - x_i) & 14x_i \end{pmatrix}$$

TABLE 17.4. *Growth Data. Complete data set. Model fit summary.*

	Mean	Covar	par	-2ℓ	Ref	G^2	df	p
1	unstr.	unstr.	18	416.509				
2	$\neq$ slopes	unstr.	14	419.477	1	2.968	4	0.5632
3	$=$ slopes	unstr.	13	426.153	2	6.676	1	0.0098
4	$\neq$ slopes	Toepl.	8	424.643	2	5.166	6	0.5227
5	$\neq$ slopes	AR(1)	6	440.681	2	21.204	8	0.0066
					4	16.038	2	0.0003
6	$\neq$ slopes	random	8	427.806	2	8.329	6	0.2150
7	$\neq$ slopes	CS	6	428.639	2	9.162	8	0.3288
					4	3.996	2	0.1356
					6	0.833	2	0.6594
					6	0.833	1:2	0.5104
8	$\neq$ slopes	simple	5	478.242	7	49.603	1	<0.0001
					7	49.603	0:1	<0.0001

and $\boldsymbol{\beta} = (\beta_0, \beta_{01}, \beta_{10}, \beta_{11})'$. Here, $\beta_0 + \beta_{01}$ is the intercept for boys and β_0 is the intercept for girls. The slopes are β_{10} and β_{11}, respectively.

The SAS code for Model 1 can be adapted simply by deleting age from the CLASS statement.

The likelihood ratio test comparing Model 2 to Model 1 does not reject the null hypothesis of linearity. A summary of model fitting information for this and subsequent models as well as for comparisons between models is given in Table 17.4. The first column contains the model number; and a short description of the model is given in the second and third columns, in terms of the mean and covariance structures, respectively. The number of parameters is given next, as well as the deviance (-2ℓ). The column labeled "Ref" displays one (or more) numbers of models to which the current model is compared. The G^2 likelihood ratio statistic is the difference between -2ℓ of the current and the reference model. The final columns contain the number of degrees of freedom, and the p-value corresponds to the likelihood ratio test statistic. Model 2 predicts the following mean growth curves:

$$\text{girls:} \quad \hat{Y}_j = 17.43 + 0.4764 t_j,$$
$$\text{boys:} \quad \hat{Y}_j = 15.84 + 0.8268 t_j.$$

These profiles are visualized in the second panel of Figure 17.1. The observed means are added to the graph. The mean model seems acceptable, consistent with the likelihood ratio test. The estimated covariance and correlation matrices of the measurements are similar to the ones found for Model 1.

MODEL 3

The next step is to investigate whether the two profiles are parallel. Although the plot for Model 2 suggests that the profiles are diverging, the question remains whether this effect is statistically significant. The model can be described as follows:

$$Y_{ij} \quad = \quad \beta_0 + \beta_{01}(1 - x_i) + \beta_1 t_j + \varepsilon_{ij}. \tag{17.11}$$

The design matrix X_i simplifies further:

$$X_i \quad = \quad \begin{pmatrix} 1 & 1 - x_i & 8 \\ 1 & 1 - x_i & 10 \\ 1 & 1 - x_i & 12 \\ 1 & 1 - x_i & 14 \end{pmatrix}$$

and $\beta = (\beta_0, \beta_{01}, \beta_1)'$. The two slopes in Model 2 have been replaced by β_1, a slope common to boys and girls.

Model 3 can be fitted in PROC MIXED by replacing the model statement in Model 2 with

```
model measure = sex age / s;
```

The predicted growth curves are

$$\text{girls:} \quad \hat{Y}_j = 15.37 + 0.6747 t_j,$$
$$\text{boys:} \quad \hat{Y}_j = 17.42 + 0.6747 t_j.$$

Table 17.4 reveals that the likelihood ratio test statistic (comparing Models 2 and 3) rejects the common slope hypothesis ($p = 0.0098$). This is consistent with the systematic deviation between observed and expected means in the third panel of Figure 17.1.

In line with the choice of Jennrich and Schluchter (1986), the mean structure of Model 2 will be kept. We will now turn our attention to simplifying the covariance structure.

GRAPHICAL EXPLORATION

Figure 17.2 presents the 27 individual profiles. The left-hand panel shows the raw profiles, exhibiting the time trend found in the mean model. To

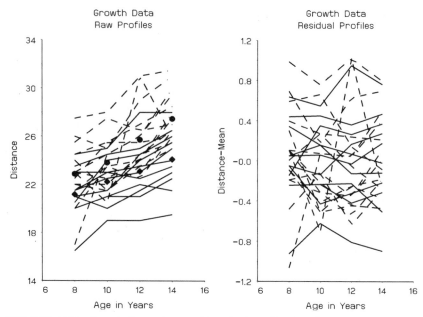

FIGURE 17.2. *Growth Data. Raw and residual profiles for the complete data set.* *(Girls are indicated with solid lines. Boys are indicated with dashed lines.)*

obtain a rough idea about the covariance structure, it is useful to look at the right-hand panel, which gives the profiles after subtracting the means predicted by Model 2. Since these means agree closely with the observed means (see Figure 17.1), the corresponding sets of residuals are equivalent. A noticeable though not fully general trend is that a profile tends to be high or low *as a whole*, which points to a random intercept. Apparently, the variance of the residuals is roughly constant over time, implying that the random-effects structure is probably confined to the intercept. This observation is consistent with correlation matrix (17.9) of the unstructured Model 1. A more formal exploration can be done by means of the variogram (Diggle, Liang, and Zeger 1994, p. 51) or its extensions (Verbeke, Lesaffre, and Brant 1998). See also Section 10.4.4.

Jennrich and Schluchter (1986) considered several covariance structure models, which are all included in PROC MIXED as standard options.

MODEL 4

The first covariance structure model is the so-called *Toeplitz* covariance matrix. Mean model formula (17.10) of Model 2 still applies, but the error vector ε_i is now assumed to follow a $\varepsilon_i \sim N(\mathbf{0}, \Sigma)$ distribution, where Σ is constrained to $\sigma_{ij} = \alpha_{|i-j|}$; that is, the covariance depends on the measurement occasions through the lag between them only. In addition, Σ is assumed to be positive definite. For the growth data, there are only 4 free parameters, α_0, α_1, α_2, and α_3, instead of 10 in the unstructured case. The relationship among the α parameters is left unspecified. In the sequel, such additional constraints will lead to first-order autoregressive (Model 5) or exchangeable (Model 7) covariance structures.

To fit this model with PROC MIXED, the REPEATED statement needs to be changed, leading to the following program:

```
proc mixed data = growth method = ml covtest;
title 'Growth Data, Model 4';
class sex idnr;
model measure = sex age*sex / s;
repeated / type = toep subject = idnr r rcorr;
run;
```

Comparing the likelihood of this model to the one of the reference Model 2 shows that Model 4 is consistent with the data (see Table 17.4). The covariance matrix is

$$\begin{pmatrix} 4.9439 & 3.0507 & 3.4054 & 2.3421 \\ 3.0507 & 4.9439 & 3.0507 & 3.4054 \\ 3.4054 & 3.0507 & 4.9439 & 3.0507 \\ 2.3421 & 3.4054 & 3.0507 & 4.9439 \end{pmatrix}$$

and the derived correlation matrix is

$$\begin{pmatrix} 1.0000 & 0.6171 & 0.6888 & 0.4737 \\ 0.6171 & 1.0000 & 0.6171 & 0.6888 \\ 0.6888 & 0.6171 & 1.0000 & 0.6171 \\ 0.4737 & 0.6888 & 0.6171 & 1.0000 \end{pmatrix}.$$

The lag 2 correlation is slightly higher than the lag 1 correlation, while the lag 3 correlation shows a drop. In light of the standard errors of the covariance parameters (0.9791, 0.9812, and 1.0358, respectively), this observation should not be seen as clear evidence for a particular trend.

Note that this structure constrains the variance to be constant across time. Should this assumption be considered unrealistic, then heterogeneous versions can be fitted instead, combining the correlation matrix from the homogeneous version with variances that are allowed to change over time.

At this point, Model 4 can replace Model 2 as the most parsimonious model, consistent with the data found so far. Whether or not further simplifications are possible will be investigated next.

MODEL 5

A special case of the Toeplitz model is the first-order autoregressive model. This model is based on the assumption that the covariance between two measurements is a decreasing function of the time lag between them:

$$\sigma_{ij} = \sigma^2 \rho^{|i-j|}.$$

In other words, the variance of the measurements equals σ^2, and the covariance decreases with increasing time lag if $\rho > 0$. To fit this model with PROC MIXED, the REPEATED statement should include the option 'type=AR(1)'.

The estimated covariance matrix is

$$\begin{pmatrix} 4.8903 & 2.9687 & 1.8021 & 1.0940 \\ 2.9687 & 4.8903 & 2.9687 & 1.8021 \\ 1.8021 & 2.9687 & 4.8903 & 2.9687 \\ 1.0940 & 1.8021 & 2.9687 & 4.8903 \end{pmatrix}.$$

The correlation matrix is

$$\begin{pmatrix} 1.0000 & 0.6070 & 0.3685 & 0.2237 \\ 0.6070 & 1.0000 & 0.6070 & 0.3685 \\ 0.3685 & 0.6070 & 1.0000 & 0.6070 \\ 0.2237 & 0.3685 & 0.6070 & 1.0000 \end{pmatrix}.$$

Table 17.4 reveals that there is an apparent lack of fit for this model, when compared to Model 2. Jennrich and Schluchter (1986) compared Model 5 to Model 2 as well. Alternatively, we might want to compare Model 5 to Model 4. This more parsimonious test (2 degrees of freedom) yields $p = 0.0003$, strongly rejecting the AR(1) structure.

MODEL 6

An alternative simplification of the unstructured covariance Model 2 is given by allowing the intercept and slope parameters to be random. This is an example of model (17.6) with fixed-effects design matrix X_i as in Model 2 [Eq. (17.10)], random-effects design matrix

$$Z_i = \begin{pmatrix} 1 & 8 \\ 1 & 10 \\ 1 & 12 \\ 1 & 14 \end{pmatrix},$$

as well as measurement error structure $\Sigma = \sigma^2 I_4$.

An unstructured covariance matrix D for the random effects b_i will be assumed. The matrix D (requested by the 'g' option in the RANDOM statement) is estimated to be

$$\hat{D} = \begin{pmatrix} 4.5569 & -0.1983 \\ -0.1983 & 0.0238 \end{pmatrix}. \qquad (17.12)$$

One easily calculates the resulting covariance matrix of Y_i: $V_i = Z_i D Z_i' + \sigma^2 I_4$, which is estimated by

$$\hat{V}_i = Z_i \hat{D} Z_i' + \hat{\sigma}^2 I_4 = \begin{pmatrix} 4.6216 & 2.8891 & 2.8727 & 2.8563 \\ 2.8891 & 4.6839 & 3.0464 & 3.1251 \\ 2.8727 & 3.0464 & 4.9363 & 3.3938 \\ 2.8563 & 3.1251 & 3.3938 & 5.3788 \end{pmatrix}, \qquad (17.13)$$

where $\hat{\sigma}^2 = 1.7162$. Of course, this matrix can be requested by the 'v' option in the REPEATED statement as well. Thus, this covariance matrix is a function of four parameters (three random-effects parameters and one measurement error parameter). The corresponding estimated correlation matrix is

$$\begin{pmatrix} 1.0000 & 0.6209 & 0.6014 & 0.5729 \\ 0.6209 & 1.0000 & 0.6335 & 0.6226 \\ 0.6014 & 0.6335 & 1.0000 & 0.6586 \\ 0.5729 & 0.6226 & 0.6586 & 1.0000 \end{pmatrix}. \qquad (17.14)$$

This model is a submodel of Model 2, but not of Model 4 since the correlations increase within each diagonal, albeit only moderately since the variance of the random slope is very modest. From Table 17.4, we observe that this model is a plausible simplification of Model 2. It has the same number of degrees of freedom as Model 4, although the latter one has a slightly smaller deviance.

Since the variance of the random slope is small, it is natural to explore whether a random intercept model is adequate.

Model 7

A random intercept model is given by $Z_i = (1\ 1\ 1\ 1)'$, with variance of the random intercepts equal to d. The resulting covariance matrix of Y_i is

$$V_i = Z_i d Z_i' + \sigma^2 I_4 = d J_4 + \sigma^2 I_4,$$

where J_4 is a (4×4) matrix of ones. This covariance structure is called exchangeable or compound symmetry. Another term is intraclass correlation (Section 3.3.2). All correlations are equal to $(d + \sigma^2)/\sigma^2$, implying that this model is a submodel of Models 4 and 6, as well as of Model 2. It can be fitted in PROC MIXED with two equivalent programs:

```
proc mixed data = growth method = ml covtest;
title 'Jennrich and Schluchter, Model 7';
class sex idnr;
model measure = sex age*sex / s;
random intercept / type = un subject = idnr g;
run;
```

and

```
proc mixed data = growth method = ml covtest;
title 'Jennrich and Schluchter, Model 7';
class sex idnr;
model measure = sex age*sex / s;
repeated / type = cs subject = idnr r rcorr;
run;
```

These two equivalent views toward the same model have been discussed in Section 3.3.2.

The estimated covariance matrix is

$$
\begin{pmatrix}
4.9052 & 3.0306 & 3.0306 & 3.0306 \\
3.0306 & 4.9052 & 3.0306 & 3.0306 \\
3.0306 & 3.0306 & 4.9052 & 3.0306 \\
3.0306 & 3.0306 & 3.0306 & 4.9052
\end{pmatrix},
$$

with corresponding correlation matrix

$$
\begin{pmatrix}
1.0000 & 0.6178 & 0.6178 & 0.6178 \\
0.6178 & 1.0000 & 0.6178 & 0.6178 \\
0.6178 & 0.6178 & 1.0000 & 0.6178 \\
0.6178 & 0.6178 & 0.6178 & 1.0000
\end{pmatrix}.
$$

Comparing this model to Model 2 yields $p = 0.3288$. Comparisons to Models 4 and 6 lead to the same conclusion. This implies that this model is currently the simplest one consistent with the data. It has to be noted that a comparison of Model 7 with Model 6 is slightly complicated by the fact that the null hypothesis implies that two of the three parameters in the D matrix of Model 6 are zero. For the variance of the random slope, this null value lies on the boundary of the parameter space. As explained in Section 6.3.4, Stram and Lee (1994) show that the corresponding reference distribution is not χ_2^2, but a 50 : 50 mixture of a χ_1^2 and a χ_2^2. Such a mixture is indicated by $\chi_{1:2}^2$, or simply by 1 : 2. As a result, the corrected p-value would be 0.5104, thereby indicating no change in the conclusion. Similarly, comparing Models 2 and 6 as carried out earlier suffers from the same problem. Stram and Lee (1994) indicate that the asymptotic null distribution is

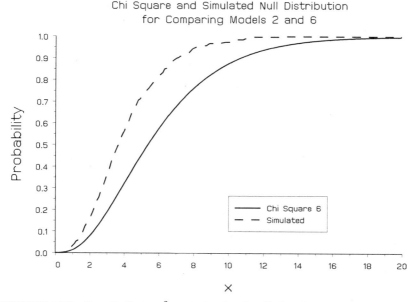

FIGURE 17.3. *Growth Data.* χ_6^2 *and simulated null distributions for comparing Models 2 and 6.*

even more complex and involves projections of random variables on curved surfaces (Stram and Lee 1994). Therefore, the p-value is best determined by means of simulations. A simulation study of 500 samples yields $p = 0.046$, rather than $p = 0.215$, as reported in Table 17.4. To simulate the null distribution, we generated 500 samples of 270 individuals rather than 27 individuals, to reduce small sample effects. Although this choice reflects the desire to perform asymptotic inference, it is debatable since one might rightly argue that generating samples of size 27 would reflect small-sample effects as well. Figure 17.3 shows the simulated as well as the inadequate χ_6^2 null distributions.

Although such a correction is clearly necessary, it is hard to use in general practice in its current form. Additional work in this area is certainly required.

The profiles, predicted by Model 7, are

$$\text{girls:} \quad \hat{Y}_j = 17.37 + 0.4795t_j,$$
$$\text{boys:} \quad \hat{Y}_j = 16.34 + 0.7844t_j.$$

They are shown in the fourth panel of Figure 17.1. Although not exactly the same, they are extremely similar to the profiles of Model 2.

MODEL 8

Finally, the independence model is considered in which the only source of variability is measurement error: $\Sigma = \sigma^2 I_4$. This model can be fitted in PROC MIXED using the 'type=simple' option in the REPEATED statement. Table 17.4 indicates that this model does not fit the data at all. Whether a χ_1^2 is used or a $\chi_{0:1}^2$ does not affect the conclusion.

In summary, among the models presented, Model 7 is preferred to summarize the data. In Sections 17.4.2 and 17.4.3, the trimmed version of the data will be analyzed, using frequentist available data methods and an ignorable likelihood-based analysis.

17.4.2 Frequentist Analysis of Incomplete Growth Data

Let us now turn toward the incomplete version of the data. In this section, we will focus on a straightforward but restrictive *available case* analysis from a frequentist perspective. The method is briefly introduced in Section 16.4. Specifically, the parameters for the unstructured mean and covariance Model 1 will be estimated.

The estimated mean vector for girls is

$$(21.1818, 22.7857, 23.0909, 24.0909),$$

whereas the vector for boys is

$$(22.8750, 24.1364, 25.7188, 27.4688).$$

The mean vector for girls is based on a sample of size 11, except for the second element, which is based on the 7 complete observations. The corresponding sample sizes for boys are 16 and 11, respectively.

The estimated covariance matrix is

$$\begin{pmatrix} 5.4155 & 2.3155 & 3.9102 & 2.7102 \\ 2.3155 & 4.2763 & 2.0420 & 2.5741 \\ 3.9102 & 2.0420 & 6.4557 & 4.1307 \\ 2.7102 & 2.5741 & 4.1307 & 4.9857 \end{pmatrix}, \tag{17.15}$$

with correlation matrix

$$\begin{pmatrix} 1.0000 & 0.4812 & 0.6613 & 0.5216 \\ 0.4812 & 1.0000 & 0.3886 & 0.5575 \\ 0.6613 & 0.3886 & 1.0000 & 0.7281 \\ 0.5216 & 0.5575 & 0.7281 & 1.0000 \end{pmatrix}.$$

The elements of the covariance matrix are computed as

$$\hat{\sigma}_{jk} = \frac{1}{25}\left\{\sum_{i=1}^{11}(y_{ij}^g - \overline{y}_j^g)(y_{ik}^g - \overline{y}_k^g) + \sum_{i=1}^{16}(y_{ij}^b - \overline{y}_j^b)(y_{ik}^b - \overline{y}_k^b)\right\}, \; j, k \neq 2,$$

$$\hat{\sigma}_{j2} = \frac{1}{18}\left\{\sum_{i=1}^{7}(y_{ij}^g - \overline{y}_j^g)(y_{i2}^g - \overline{y}_2^g) + \sum_{i=1}^{11}(y_{ij}^b - \overline{y}_j^b)(y_{i2}^b - \overline{y}_2^b)\right\}.$$

The superscripts g and b refer to girls and boys, respectively. It is assumed that, within each sex subgroup, the ordering is such that completers precede the incompletely measured children.

Looking at the available case procedure from the perspective of the individual observation, one might say that each observation contributes to the subvector of the parameter vector about which it contains information. For example, a complete observation in the growth data set contributes to 4 (sex specific) mean components as well as to all 10 variance-covariance parameters. In an incomplete observation, there is information about three mean components and six variance-covariance parameters (excluding those with a subscript 2).

Whether or not there is nonrandom selection of the incomplete observations does not affect those parameters without a subscript 2. For the ones involving a subscript 2, potential differences between completers and noncompleters are not taken into account and, hence, biased estimation may result when an MAR mechanism is operating. In fact, the estimates for the parameters with at least one subscript 2 equal their complete case analysis counterparts. Thus, MCAR is required. This observation is consistent with the theory in Rubin (1976), since the current available case method is frequentist rather than likelihood based.

17.4.3 Likelihood Analysis of Incomplete Growth Data

As with the available case method of Section 16.4, a complete subject contributes to "more parameters" than an incomplete subject. Whereas these contributions were direct in terms of parameter vector components in Section 16.4, in the current framework subjects contribute information through their factor of the likelihood function. For example, let us consider Model 1. A complete subject contributes by means of a four-dimensional normal log-likelihood term with 4 out of 8 mean components (boys and girls have separate means) and a 4×4 positive definite covariance matrix. An incomplete observation contributes through the three-dimensional marginal density, obtained by integrating over the second component. In practice, this is done by deleting the second component of the mean vector and

TABLE 17.5. *Growth Data. MAR analysis (Little and Rubin). Model fit summary.*

	Mean	Covar	par	-2ℓ	Ref	G^2	df	p
1	unstr.	unstr.	18	386.957				
2	$\neq$ slopes	unstr.	14	393.288	1	6.331	4	0.1758
3	$=$ slopes	unstr.	13	397.400	2	4.112	1	0.0426
4	$\neq$ slopes	Toepl.	8	398.030	2	4.742	6	0.5773
5	$\neq$ slopes	AR(1)	6	409.523	2	16.235	8	0.0391
6	$\neq$ slopes	random	8	400.452	2	7.164	6	0.3059
7	$\neq$ slopes	CS	6	401.313	6	0.861	2	0.6502
					6	0.861	1:2	0.5018
8	$\neq$ slopes	simple	5	441.583	7	40.270	1	<0.0001
					7	40.270	0:1	<0.0001

the second row and the second column of the covariance matrix. In most software packages, such as the SAS procedure MIXED, this is performed automatically, as will be discussed next.

Little and Rubin (1987) fitted the same eight models as Jennrich and Schluchter (1986) to the incomplete growth data set. Whereas Little and Rubin made use of the EM algorithm, we set out to perform our analysis with direct maximization of the observed likelihood (with Fisher scoring or Newton-Raphson) in PROC MIXED. The results ought to coincide. Table 17.5 reproduces the findings of Little and Rubin. We added p-values.

The PROC MIXED programs, constructed in Section 17.3 to analyze the complete data set, will be applied to the artificially incomplete data set. The structure of this data set is given in Table 17.6. Although there would be four records for every subject in the complete data set, now there are nine subjects (e.g., subjects #3 and #27) with only three records.

Applying the programs to the data yields some discrepancies, as seen from the model fit Table 17.7.

Let us take a close look at these discrepancies. Although most of the tests performed lead to the same conclusion, there is one fundamental difference. In Table 17.5, the AR(1) model is rejected whereas it is not in Table 17.7. A puzzling difference is that the maximized log-likelihoods are different for Models 1–5, but not for Models 6–8. The same holds for the mean and covariance parameter estimates. To get a hold on this problem, let us consider the repeated statement (e.g., of Model 1):

```
repeated / type = un subject = idnr r rcorr;
```

TABLE 17.6. *Growth Data. Extract of the incomplete data set.*

OBS	IDNR	AGE	SEX	MEASURE
1	1	8	2	21.0
2	1	10	2	20.0
3	1	12	2	21.5
4	1	14	2	23.0
5	2	8	2	21.0
6	2	10	2	21.5
7	2	12	2	24.0
8	2	14	2	25.5
9	3	8	2	20.5
10	3	12	2	24.5
11	3	14	2	26.0
...				
97	27	8	1	22.0
98	27	12	1	23.5
99	27	14	1	25.0

This statement identifies the subject in terms of IDNR blocks but does not specify the ordering of the observations within a subject. Thus, PROC MIXED assumes the default ordering: 1, 2, 3, 4 for a complete subject and, erroneously, 1, 2, 3 for an incomplete one, whereas the correct incomplete ordering is 1, 2, 4. This means that, by default, dropout is assumed. Since this assumption is inadequate for the growth data, Models 1–5 in Table 17.7 are incorrect. The random-effects Model 6, on the other hand, uses the RANDOM statement

```
random intercept age / type = un subject = idnr g;
```

where the variable AGE conveys the information needed to correctly calculate the random-effects parameters. Indeed, for an incomplete observation, the correct design

$$Z_i = \begin{pmatrix} 1 & 8 \\ 1 & 12 \\ 1 & 14 \end{pmatrix}$$

is generated. Finally, it remains to be discussed why Models 7 and 8 give a correct answer in spite of the fact that they also use the REPEATED statement rather than the RANDOM statement. This is best seen in Model

TABLE 17.7. *Growth Data.* Inadequate *MAR analysis (Little and Rubin).* Model fit summary.

	Mean	Covar	par	-2ℓ	Ref	G^2	df	p
1	unstr.	unstr.	18	394.309				
2	$\neq$ slopes	unstr.	14	397.862	1	3.553	4	0.4699
3	$=$ slopes	unstr.	13	401.935	2	4.073	1	0.0436
4	$\neq$ slopes	banded	8	400.981	2	3.119	6	0.7938
5	$\neq$ slopes	AR(1)	6	408.996	2	11.134	8	0.1942
6	$\neq$ slopes	random	8	400.452	2	2.590	6	0.8583
7	$\neq$ slopes	CS	6	401.312	6	0.860	2	0.6505
					6	0.860	1:2	0.5021
8	$\neq$ slopes	simple	5	441.582	7	40.270	1	<0.0001
					7	40.270	0:1	<0.0001

8, where we assume an independence covariance structure. This covariance structure is equivalent to assuming that the $99 = 108 - 9$ measurements form a simple random sample of size $n = 99$, rather than a longitudinal sample. Consequently, the actual position of a measurement within a subject's sequence is irrelevant. The same holds for the compound symmetry or *exchangeable* model. The only difference with the simple model is that a *common* intraclass correlation between two measurements within the same sequence is assumed. Since this correlation is constant and thus independent of the actual distance between measurements, it can be determined from the full set of pairs of measurements within an individual (six pairs for a complete observation and three pairs for an incomplete observation), with the order being immaterial.

There are two equivalent ways to overcome this problem. The first is to adapt the data set slightly. An example is given in Table 17.8.

The effect of using this data set is, of course, that incomplete records are deleted from the analysis, but that the relative positions are correctly passed on to PROC MIXED. Running Models 1–8 on this data set yields exactly the same results as in Table 17.5.

It is also possible to use the data as presented in Table 17.6. Instead of passing on the position of the missing values through the data set, we have to specify explicitly the ordering by coding it properly into the PROC MIXED program. For Model 1, the following code can be used:

```
proc mixed data = growthav method = ml;
title 'Jennrich and Schluchter (MAR, Altern.), Model 1';
class sex idnr age;
model measure = sex age*sex / s;
repeated age / type = un subject = idnr r rcorr;
run;
```

The REPEATED statement now explicitly includes the ordering by means of the AGE variable. Note that any counter with the correct ordering (e.g., $1, 2, 3, 4$) would be suitable. We consider it good practice to *always* include the (time) ordering variable of the measurements.

The corresponding Model 2 program would be

```
proc mixed data = growthav method = ml;
title 'Jennrich and Schluchter (MAR, Altern.), Model 2';
class sex idnr;
model measure = sex age*sex / s;
repeated age / type = un subject = idnr r rcorr;
run;
```

However, this program generates an error since the variables in the RE-PEATED statement have to be *categorical* variables, termed CLASS variables in PROC MIXED. See Section 8.2.6. One of the tricks to overcome this issue is by using the program

```
data help;
set growthav;
agec = age;
run;

proc mixed data = help method = ml;
title 'Jennrich and Schluchter (MAR, Altern.), Model 2';
class sex idnr agec;
model measure = sex age*sex / s;
repeated agec / type = un subject = idnr r rcorr;
run;
```

Thus, there are two identical copies of the variable AGE, only one of which is treated as a class variable.

Let us now turn attention to the performance of the ignorable method of analysis and compare the results with the ones obtained earlier. First, the model comparisons performed in Tables 17.4 and 17.5 qualitatively

TABLE 17.8. *Growth Data. Extract of the incomplete data set. The missing observations are explicitly indicated.*

OBS	IDNR	AGE	SEX	MEASURE
1	1	8	2	21.0
2	1	10	2	20.0
3	1	12	2	21.5
4	1	14	2	23.0
5	2	8	2	21.0
6	2	10	2	21.5
7	2	12	2	24.0
8	2	14	2	25.5
9	3	8	2	20.5
10	3	10	2	.
11	3	12	2	24.5
12	3	14	2	26.0
...				
105	27	8	1	22.0
106	27	10	1	.
107	27	12	1	23.5
108	27	14	1	25.0

yield the same conclusions. In both cases, linear profiles turn out to be consistent with the data, but parallel profiles do not. A Toeplitz correlation structure (Model 5) is acceptable, as well as a random intercepts and slopes model (Model 6). These models can be simplified further to compound symmetry (Model 7). The assumption of no correlation between repeated measures (Model 8) is untenable. This means that Model 7 is again the most parsimonious description of the data among the eight models considered. It has to be noted that the rejection of Models 3 and 5 is less compelling in the MAR analysis than it was in the complete data set. Of course, this is to be expected due to the reduction in the sample size, or rather in the number of available measurements. The likelihood ratio test statistic for a direct comparison of Model 5 to Model 4 is 11.494 on 2 degrees of freedom ($p = 0.0032$), which is, again, a clear indication of an unacceptable fit.

Figure 17.4 displays the fit of Models 1, 2, 3, and 7. Let us consider the fit of Model 1 first. As mentioned earlier, the complete observations at age 10 are those with a higher measurement at age 8. Due to the within-subject correlation, they are the ones with a higher measurement at age 10 as well. This is seen by comparing the large dot with the corresponding small dot,

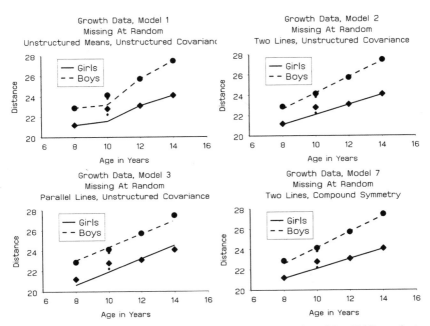

FIGURE 17.4. *Growth Data. Profiles for a selected set of models. MAR analysis. (The small dots are the observed group means for the complete data set. The large dots are the corresponding quantities for the incomplete data.)*

reflecting the means for the complete data set and for those observed at age 10, respectively. Since the average of the observed measurements at age 10 is biased upward, the fitted profiles from the complete case analysis and from unconditional mean imputation were too high. Clearly, the average observed from the data is the same for the complete case analysis, the unconditional mean imputation, the available case analysis, and the present analysis. The most crucial difference is that the current Model 1, although saturated in the sense that there are eight mean parameters (one for each age by sex combination), does *not* let the (biased) observed and fitted averages at age 10 coincide, in contrast to the means at ages 8, 12, and 14. Indeed, if the model specification is correct, then an ignorable likelihood analysis is consistent for the correct complete data mean, rather than for the observed data mean. Of course, this effect might be blurred in relatively small data sets due to small-sample variability.

This discussion touches upon the key distinction between the frequentist available case analysis of Section 16.4, with example in Section 17.4.2, and the present likelihood based available case analysis. The method of Section 16.4 constructs an estimate for the age 10 parameters, irrespective of the (extra) information available for the other parameters. The likelihood

TABLE 17.9. *Growth Data. Means under unstructured Model 1.*

Age	Complete	Incomplete	
		Obs.	Pred.
	Girls		
8	21.18	21.18	21.18
10	22.23	22.79	21.58
12	23.09	23.09	23.09
14	24.09	24.09	24.09
	Boys		
8	22.88	22.88	22.88
10	23.81	24.14	23.17
12	25.72	25.72	25.72
14	27.47	27.47	27.47

approach implicitly constructs a correction, based on (1) the fact that the measurements at ages 8, 12, and 14 differ between the subgroups of complete and incomplete observations and (2) the fairly strong correlation between the measurement at age 10 on the one hand, and the measurements at ages 8, 12, and 14 on the other hand. A detailed treatment of likelihood estimation in incomplete multivariate normal samples is given in Little and Rubin (1987, Chapter 6). Clearly, this correction leads to an overshoot in the fairly small growth data set, whence the predicted mean at age 10 is actually *smaller* than the one of the complete data set. The means are reproduced in Table 17.9. All means coincide for ages 8, 12, and 14. Irrespective of the small-sample behavior encountered here, the validity under MAR and the ease of implementation are good arguments that favor this ignorable analysis over other techniques.

We now present the predicted mean curves for Models 2, 3, and 7:

- Model 2:

$$\text{girls:} \quad \hat{Y}_j = 17.18 + 0.4917t_j,$$
$$\text{boys:} \quad \hat{Y}_j = 16.32 + 0.7886t_j.$$

- Model 3:

$$\text{girls:} \quad \hat{Y}_j = 15.40 + 0.6519t_j,$$
$$\text{boys:} \quad \hat{Y}_j = 17.82 + 0.6519t_j.$$

- Model 7:

$$\text{girls:} \quad \hat{Y}_j = 17.22 + 0.4890 t_j,$$
$$\text{boys:} \quad \hat{Y}_j = 16.30 + 0.7867 t_j.$$

These profiles are fairly similar to their complete data counterparts. This is in contrast to analyses obtained from the simple methods, described in Chapter 16 and applied in Verbeke and Molenberghs (1997, Sections 5.4–5.6).

Let us now study this method in terms of the effect on the estimated covariance structure. The estimated covariance matrix of Model 1 is

$$\hat{\Sigma} = \begin{pmatrix} 5.0142 & 4.8796 & 3.6205 & 2.5095 \\ 4.8796 & 6.6341 & 3.3772 & 3.0621 \\ 3.6205 & 3.3772 & 5.9775 & 3.8248 \\ 2.5095 & 3.0621 & 3.8248 & 4.6164 \end{pmatrix}.$$

The variance at age 10 is inflated compared to its complete data set counterpart (17.8). The dominating reason is that the sample size at age 10 is only two-thirds of the original one, thereby making all estimators involved more variable. In other settings, the variance may increase due to an increased homogeneity in the selected subset. A correct analysis, such as the ignorable one considered here, should acknowledge this additional source of uncertainty. The correlation matrices are as follows:

- Model 1 (unstructured):

$$\begin{pmatrix} 1.0000 & 0.8460 & 0.6613 & 0.5216 \\ 0.8460 & 1.0000 & 0.5363 & 0.5533 \\ 0.6613 & 0.5363 & 1.0000 & 0.7281 \\ 0.5216 & 0.5533 & 0.7281 & 1.0000 \end{pmatrix}. \qquad (17.16)$$

- Model 4 (Toeplitz):

$$\begin{pmatrix} 1.0000 & 0.6248 & 0.6688 & 0.4307 \\ 0.6248 & 1.0000 & 0.6248 & 0.6688 \\ 0.6688 & 0.6248 & 1.0000 & 0.6248 \\ 0.4307 & 0.6688 & 0.6248 & 1.0000 \end{pmatrix}.$$

- Model 5 (AR(1)):

$$\begin{pmatrix} 1.0000 & 0.6265 & 0.3925 & 0.2459 \\ 0.6265 & 1.0000 & 0.6265 & 0.3925 \\ 0.3925 & 0.6265 & 1.0000 & 0.6265 \\ 0.2459 & 0.3925 & 0.6265 & 1.0000 \end{pmatrix}.$$

• Model 6 (random effects):

$$\begin{pmatrix} 1.0000 & 0.6341 & 0.5971 & 0.5465 \\ 0.6341 & 1.0000 & 0.6302 & 0.6041 \\ 0.5971 & 0.6302 & 1.0000 & 0.6461 \\ 0.5465 & 0.6041 & 0.6461 & 1.0000 \end{pmatrix}.$$

• Model 7 (compound symmetry):

$$\begin{pmatrix} 1.0000 & 0.6054 & 0.6054 & 0.6054 \\ 0.6054 & 1.0000 & 0.6054 & 0.6054 \\ 0.6054 & 0.6054 & 1.0000 & 0.6054 \\ 0.6054 & 0.6054 & 0.6054 & 1.0000 \end{pmatrix}.$$

The unstructured model reveals an increased correlation between ages 10 and 8, but a decrease between ages 10 and 12 and also a decrease between ages 10 and 14. Although the differences in correlation between the complete data set and ignorable analyses are carried across the simplified correlation structures, they are, in fact, very modest. For example, the complete data exchangeable correlation of 0.6178 changes to 0.6054 here.

It is interesting to consider the covariance structure of random-effects Model 6 in a bit more detail. The matrix D is estimated to be

$$\widehat{D} = \begin{pmatrix} 6.7853 & -0.3498 \\ -0.3498 & 0.0337 \end{pmatrix}$$

and $\hat{\sigma}^2 = 1.7700$. Thus, all entries in the random-effects covariance matrix as well as the measurement error σ^2 seem to have increased slightly in absolute value in comparison to the complete data analysis version (17.12). The resulting covariance matrix is now

$$\widehat{V}_i = Z_i \widehat{D} Z_i' + \hat{\sigma}^2 I_4 = \begin{pmatrix} 5.1140 & 3.1833 & 3.0226 & 2.8620 \\ 3.1833 & 4.9274 & 3.1315 & 3.1055 \\ 3.0226 & 3.1315 & 5.0103 & 3.3491 \\ 2.8620 & 3.1055 & 3.3491 & 5.3626 \end{pmatrix}.$$

In conclusion, a likelihood ignorable analysis is preferable since it uses all available information, without the need neither to delete nor to impute measurements or entire subjects. It is theoretically justified whenever the missing data mechanism is MAR, which is a more relaxed assumption than MCAR, necessary for simple analyses (complete case, frequentist available case, and single-imputation based analyses, with the exception of Buck's method of conditional mean imputation). There is no statistical information distortion, since observations are neither removed (such as in complete case analysis) nor added (such as in single imputation). There is no additional programming involved to perform an ignorable analysis in PROC MIXED, provided the order of the measurements is correctly specified. This can be done either by supplying records with missing data in the input data set or by properly indicating the order of the measurement in the REPEATED and/or RANDOM statements in PROC MIXED.

When the scientific interest is directed to the missing data mechanism as well, then a simple ignorable analysis is generally not sufficient and has to be supplemented with a model for missingness. Also, when the missingness mechanism is nonrandom, then an ignorable analysis is not valid and more complex modeling (e.g., Diggle and Kenward 1994) is required. This topic is of vital importance and will be discussed further in Sections 17.5 and 17.6.

There are still a few issues with the estimation of precision and with hypothesis testing, related to this type of analysis. These will be discussed in Chapter 21. We will first study the missingness mechanism for the growth data.

17.4.4 Missingness Process for the Growth Data

The only prior information we have about the nonresponse mechanism is that Little and Rubin (1987) conceived it to depend on the measurement at age 8 in such a way that lower values led to a higher nonresponse. Let us assume that the missingness probability follows a logistic model. For example, if dropout would depend solely on the measurement at age 8, a candidate model would be

$$\ln\left(\frac{P(R_i = 0|\boldsymbol{y}_i)}{1 - P(R_i = 0|\boldsymbol{y}_i)}\right) = \psi_0 + \psi_1 y_{i1},$$

where $R_i = 1$ for complete observations and 0 otherwise, and Y_{i1} is the measurement at age 8. Of course, this model is easily adapted to include a different subset of measurements into the linear predictor. Table 17.10 shows the model fit for a few choices. In each of the four models, missingness depends on a single outcome. When this dependence is on Y_{i2}, the process is nonrandom; it is MAR otherwise. We used the complete data set to estimate the parameters of the nonrandom model (with linear predictor including Y_{i2}). This is generally not possible. Ways to overcome this problem are discussed in Sections 17.2.2 and 17.5. The only important covariates are the measurements at ages 8 and 12 (i.e., the ones adjacent to the possibly missing measurement). A backward logistic regression model retains only Y_{i1}. Its coefficients (standard errors) are estimated as

$$\ln\left(\frac{P(R_i = 0|y_{i1})}{1 - P(R_i = 0|y_{i1})}\right) = 41.22(18.17) - 1.94(0.85)y_{i1}.$$

TABLE 17.10. *Growth Data. Model fit for logistic nonresponse models.*

Type	Effects	Deviance	p
MAR	Y_{i1}	19.51	<0.0001
MAR	Y_{i3}	7.43	0.0064
MAR	Y_{i4}	2.51	0.1131
MNAR	Y_{i2}	2.55	0.1105

This model implies that the missingness probability decreases with increasing Y_{i1}.

It is important to note that omitting a relevant predictor from the nonresponse model might lead to the wrong conclusions, even about the nature of the nonresponse mechanism itself. For example, a MCAR mechanism might be classified as nonrandom if a crucial covariate is omitted from the model. Therefore, it is wise to examine all available information, including covariates, with the greatest care. In the growth example, the only covariate is sex (x_i). A model including the age 8 measurement Y_{i1} together with sex x_i as an analysis by sex group leads to a complete separation in the covariate space, resulting in parameter estimates at infinity. Examining the fit carefully, we deduce the following mechanism:

$$\text{boys}: \quad \ln\left(\frac{P(R_i = 0|y_{i1}, x_i = 0)}{1 - P(R_i = 0|y_{i1}, x_i = 0)}\right) = \infty(22 - y_{i1}),$$

$$\text{girls}: \quad \ln\left(\frac{P(R_i = 0|y_{i1}, x_i = 1)}{1 - P(R_i = 0|y_{i1}, x_i = 1)}\right) = \infty(20.75 - y_{i1}).$$

The model for boys is interpreted as follows:

$$P(R_i = 0|y_{i1}, x_i = 0) = \begin{cases} 1 & \text{if } y_{i1} < 22 \\ 0.5 & \text{if } y_{i1} = 22 \\ 0 & \text{if } y_{i1} > 22. \end{cases}$$

This is exactly what is seen in Table 2.5. The same is true for girls, with the sole difference that the cut point lies halfway between two observable outcome values (20.75):

$$P(R_i = 0|y_{i1}, x_i = 1) = \begin{cases} 1 & \text{if } y_{i1} < 20.75 \\ 0 & \text{if } y_{i1} > 20.75. \end{cases}$$

The models are displayed in Figure 17.5. Thus, the missingness mechanism used by Little and Rubin (1987) is, in fact, *deterministic* (given the outcomes at age 8). This should not be confused with the fact that nonresponse depends (very clearly) on the observed outcomes and the observed outcomes only, whence it is missing at *random* ! A similar mechanism is employed in Section 21.3.

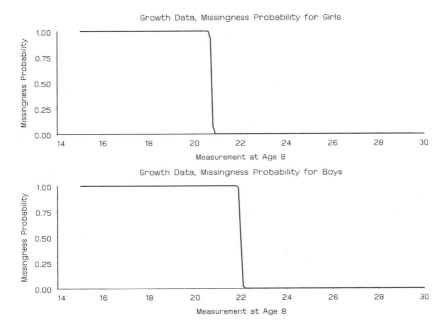

FIGURE 17.5. *Growth Data. Logistic nonresponse models for growth data.*

17.5   A Selection Model for Nonrandom Dropout

In Section 17.2, the toenail data were analyzed assuming both MAR (Section 17.2.1) and MNAR (Section 17.2.2). The responses were modeled using linear mixed models and the dropout process was described by means of logistic regression. This is in agreement with the model proposed by Diggle and Kenward (1994). Using the notation laid out in Section 15.9, we will now slightly generalize this model.

We assume the measurement model to be of the linear mixed model (3.8) form. Assuming that the first measurement Y_{i1} is obtained for every subject in the study, the model for the dropout process is based on a logistic regression for the probability of dropout at occasion j, given the subject was still in the study up to occasion j. We denote this probability by $g(\boldsymbol{h}_{ij}, y_{ij})$, in which $\boldsymbol{h}_{ij}$ is a vector containing all responses observed up to but not including occasion j, as well as relevant covariates. We then assume that $g(\boldsymbol{h}_{ij}, y_{ij})$ satisfies

$$\text{logit}[g(\boldsymbol{h}_{ij}, y_{ij})] = \text{logit}\left[\text{pr}(D_i = j | D_i \geq j, \boldsymbol{y}_i)\right] = \boldsymbol{h}_{ij}\boldsymbol{\psi} + \omega y_{ij} \quad (17.17)$$

$(i = 1, \ldots, N)$. When ω equals zero, the dropout model is random, and all parameters can be estimated using standard software since the measurement model for which we use a linear mixed model and the dropout model, assumed to follow a logistic regression, can then be fitted separately. If $\omega \neq 0$, the dropout process is assumed to be nonrandom. A special case is given by (17.3). Then, (17.4) can be used to calculate the dropout probability at a given occasion.

In line with our discussion in Sections 17.1 and 17.2, Rubin (1994) points out that such analyses heavily depend on the assumed dropout process, whereas it is impossible to find evidence for or against the model, unless supplemental information on the dropouts is available. In practice, a dropout model may be found to be nonrandom solely because one or a few influential subjects have driven the analysis. This and related issues will be studied in Chapter 19, which is devoted to sensitivity analysis for the selection model.

17.6   A Selection Model for the Vorozole Study

We will assume a model of the form described in the previous section. Since we are modeling change versus baseline, all models are forced to pass through the origin. The following covariates were considered for the measurement model: baseline value, treatment, dominant site, stage, and time in months. Second-order interactions were considered as well. For design reasons, treatment was kept in the model in spite of its nonsignificance. An F-test for treatment effect produces a p-value of 0.5822. Apart from baseline, no other time-stationary covariates were kept. A quadratic time effect provided an adequate description of the time trend. Based on the variogram (Figure 10.3), we confined the random-effects structure to random intercepts and supplemented this with a Gaussian serial correlation component and measurement error. The final model is presented in Table 17.11.

Fitted profiles are displayed in Figure 17.6 and Figure 17.7. In Figure 17.7, empirical Bayes estimates of the random effects are included, whereas in Figure 17.6, the purely marginal mean is used. For each treatment group, we obtain three sets of profiles. The fitted complete profile is the average curve that would be obtained had all individuals been completely observed. If we use only those predicted values that correspond to occasions at which an observation was made, then the fitted incomplete profiles are obtained. The latter are somewhat above the former when the random effects are included, and somewhat below when they are not, suggesting that individuals with lower measurements are more likely to disappear from the study. In addition, although the fitted complete curves are very close (the treatment effect was not significant), the fitted incomplete curves are not, suggesting

TABLE 17.11. *Vorozole Study. Selection model parameter estimates and standard errors.*

Effect	Parameter	Estimate (s.e.)
Fixed-Effect Parameters:		
Time	β_0	7.78 (1.05)
Time*baseline	β_1	−0.065 (0.009)
Time*treatment	β_2	0.086 (0.157)
Time^2	β_3	−0.30 (0.06)
Time^2*baseline	β_4	0.0024 (0.0005)
Variance Parameters:		
Random intercept	d	105.42
Serial variance	τ^2	77.96
Serial association	λ	7.22
Measurement error	σ^2	77.83

that there is more dropout in the standard arm than in the treatment arm. This is in agreement with the dropout rate, displayed in Figure 14.1, and should not be seen as evidence of a bad fit. Finally, the observed curves, based on the measurements available at each time point, are displayed. These are higher than the fitted ones, but this should be viewed with the standard errors of the observed means in mind (see Figure 4.2).

The fitted variance structure is represented by means of the fitted variogram in Figure 10.3. The total correlation between two measurements, 1 month apart, equals 0.696. The residual correlation, which remains after accounting for the random effects, is still equal to 0.491. The serial correlation, obtained by further ignoring the measurement error, equals $\rho = \exp(-1/7.22^2) = 0.981$.

Next, we will study factors which influence dropout. A logistic regression model, described by (17.17) and (17.4) is used. To start, we restrict attention to MAR processes, whence $\omega = 0$. The first model includes treatment, dominant side, stage group, baseline, and the previous measurement, but only the last two are significant, producing

$$\begin{aligned} \text{logit}[g(\boldsymbol{h}_{ij})] &= 0.080(0.341) - 0.014(0.003)\text{base}_i \\ &\quad -0.033(0.004)y_{i,j-1}. \end{aligned} \tag{17.18}$$

With larger data sets such as this one, convergence of nonrandom models can be painstakingly difficult in, for example, OSWALD, and one has to

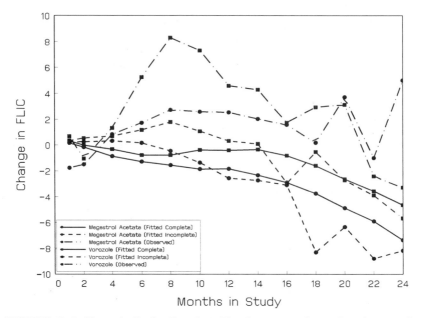

FIGURE 17.6. *Vorozole Study. Fitted profiles (averaging the predicted means for the incomplete and complete measurement sequences, without the random effects).*

worry about apparent convergence. Therefore, we first proceed in an alternative way. Both Diggle and Kenward (1994) and Molenberghs, Kenward, and Lesaffre (1997) observed that in nonrandom models, dropout tends to depend on the increment (i.e., the difference between the current and previous measurements $y_{ij} - y_{i,j-1}$). Clearly, a very similar quantity is obtained as $y_{i,j-1} - y_{i,j-2}$, but a major advantage of such a model is that it fits within the MAR framework. In our case, we obtain

$$
\begin{aligned}
\mathrm{logit}[g(\boldsymbol{h}_{ij})] \;=\; & 0.033(0.401) - 0.013(0.003)\mathrm{base}_i \\
& + 0.012(0.006)y_{i,j-2} - 0.035(0.005)y_{i,j-1} \\[6pt]
\;=\; & 0.033(0.401) - 0.013(0.003)\mathrm{base}_i \\
& - 0.023(0.005)\frac{y_{i,j-2} + y_{i,j-1}}{2} \\
& - 0.047(0.010)\frac{y_{i,j-1} - y_{i,j-2}}{2}, \qquad (17.19)
\end{aligned}
$$

indicating that both size and increment are significant predictors for dropout. We conclude that dropout increases with a decrease in baseline, in overall level of the outcome variable, as well as with a decreasing evolution in the outcome. Recall that fitting the dropout model could be done using a logistic regression of the type (17.19), given (17.4) and the discussion following this equation.

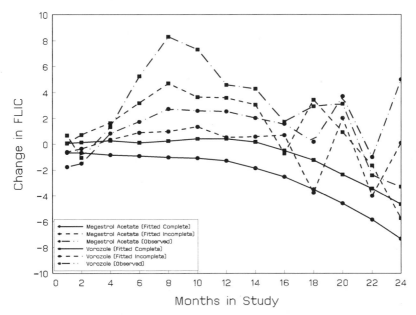

FIGURE 17.7. *Vorozole Study. Fitted profiles (averaging the predicted means for the incomplete and complete measurement sequences, including the random effects).*

Using OSWALD, both dropout models (17.18) and (17.19) can be compared with their nonrandom counterparts, where y_{ij} is added to the linear predictor. The first one becomes

$$\text{logit}[g(\boldsymbol{h}_{ij}, y_{ij})] = 0.53 - 0.015\text{base}_i - 0.076y_{i,j-1}$$
$$+0.057y_{ij}, \tag{17.20}$$

and the second one becomes

$$\text{logit}[g(\boldsymbol{h}_{ij}, y_{ij})] = 1.38 - 0.021\text{base}_i - 0.0027y_{i,j-2}$$
$$-0.064y_{i,j-1} + 0.035y_{ij}. \tag{17.21}$$

It turns out that model (17.21) is not significantly better than (17.19) and, hence, we retain (17.19) as the most plausible description of the dropout process we have so far obtained.

18
Pattern-Mixture Models

18.1   Introduction

The high sensitivity of selection modeling results to the correct specification of the measurement model as well as the dropout model, about which little is often known, has been extensively documented. See also Sections 15.3, 15.4, 17.1, 17.2.2, and 17.5. This has lead to growing interest in pattern-mixture modeling, based on the factorization (15.2) (Little 1993, Glynn, Laird and Rubin 1986, Hogan and Laird 1997). After initial mention of pattern-mixture models (Glynn, Laird, and Rubin 1986, Little and Rubin 1987), they are receiving more attention lately (Little 1993, 1994a, 1995, Hogan and Laird 1997, Ekholm and Skinner 1998, Molenberghs, Michiels, Kenward, and Diggle 1998, Molenberghs, Michiels, and Kenward 1998).

18.1.1   A Simple Illustration

We will first illustrate the idea of pattern-mixture modeling using a simple setting. Let us adopt pattern-mixture decomposition (15.2) and suppress dependence on covariates:

$$f(\boldsymbol{y}_i, \boldsymbol{r}_i | \boldsymbol{\theta}, \boldsymbol{\psi}) \quad = \quad f(\boldsymbol{y}_i | \boldsymbol{r}_i, \boldsymbol{\theta}) f(\boldsymbol{r}_i, \boldsymbol{\psi}),$$

with notation as laid out in Chapter 15. Restricting attention to dropout (Section 15.9), we obtain, using (15.7),

$$f(\boldsymbol{y}_i, d_i | \boldsymbol{\theta}, \boldsymbol{\psi}) \quad = \quad f(\boldsymbol{y}_i | d_i, \boldsymbol{\theta}) f(d_i | \boldsymbol{\psi}). \tag{18.1}$$

Equivalently, using (15.8),

$$f(\boldsymbol{y}_i, t_i | \boldsymbol{\theta}, \boldsymbol{\psi}) \quad = \quad f(\boldsymbol{y}_i | t_i, \boldsymbol{\theta}) f(t_i | \boldsymbol{\psi}). \tag{18.2}$$

Consider a continuous response at three times of measurement which will be modeled using a trivariate Gaussian distribution. Assume that there may be dropout at time 2 or 3, and let the dropout indicator T_i take the values 1 and 2 to indicate that the last observation occurred at these times and 3 to indicate no dropout. Then, in the first instance, the model implies a different distribution for each time of dropout. We can write

$$\boldsymbol{y}_i \mid t_i \quad \sim \quad N(\boldsymbol{\mu}(t_i), \Sigma(t_i)), \tag{18.3}$$

where

$$\boldsymbol{\mu}(t) \quad = \quad \begin{pmatrix} \mu_1(t) \\ \mu_2(t) \\ \mu_3(t) \end{pmatrix} \quad \text{and} \quad \Sigma(t) = \begin{pmatrix} \sigma_{11}(t) & \sigma_{21}(t) & \sigma_{31}(t) \\ \sigma_{21}(t) & \sigma_{22}(t) & \sigma_{32}(t) \\ \sigma_{31}(t) & \sigma_{32}(t) & \sigma_{33}(t) \end{pmatrix},$$

for $t = 1, 2, 3$. Recall that t indicates length of sequences, as defined in Section 15.9, rather than time points of measurements actually taken. Let $P(t) = \pi_t = f(t_i | \boldsymbol{\psi})$, then the marginal distribution of the response is a mixture of normals with, for example, mean

$$\boldsymbol{\mu} \quad = \quad \sum_{t=1}^{3} \pi_t \boldsymbol{\mu}(t).$$

Its variance can be derived by application of the delta method (see Sections 18.3, 18.4, 20.6.2, and 24.4.2).

However, although the π_t can be simply estimated from the observed proportions in each dropout group, only 16 of the 27 response parameters can be identified from the data without making further assumptions. These 16 comprise all the parameters from the completers plus those from the following two submodels. For $t = 2$

$$N \left(\begin{pmatrix} \mu_1(2) \\ \mu_2(2) \end{pmatrix}; \begin{pmatrix} \sigma_{11}(2) & \sigma_{21}(2) \\ \sigma_{31}(2) & \sigma_{32}(2) \end{pmatrix} \right),$$

and for $t = 1$

$$N \left(\mu_1(1); \sigma_{11}(1) \right).$$

This is a *saturated* pattern-mixture model and the representation makes it very clear what information each dropout group provides and, consequently, the assumptions that need to be made if we are to predict the behavior of the unobserved responses, and so obtain marginal models for the response. If the three sets of parameters $\boldsymbol{\mu}(t)$ are simply equated, with the same holding for the corresponding variance components, then this implies that dropout is completely random. Progress can be made with less stringent restrictions however. Little (1993) introduces so-called *complete case missing value* (CCMV) restrictions. These can be defined in terms of conditional distributions. Let $\boldsymbol{y} = (y_1, y_2, \ldots, y_n)'$. Then the CCMV restrictions imply that for any $T = t < j$

$$f(y_j \mid y_1, \ldots, y_{j-1}, T = t) \quad = \quad f(y_j \mid y_1, \ldots, y_{j-1}, T = n).$$

Little (1993) shows how these constraints can be used to identify all the parameters in the model and so obtain estimates for these and the marginal probabilities. The CCMV restrictions essentially equate conditional distributions beyond time t (i.e., those unidentifiable from this dropout group), with the same conditional distributions from the completers. A stronger restriction is to identify the former conditional distributions and all conditional distributions from those who drop out after t. This has been called the *available case missing value* (ACMV) restrictions and it has been shown (Molenberghs, Michiels, Kenward, and Diggle 1998; see also Section 20.2) that for dropout, these conditions are equivalent to MAR in the selection model framework. Again, such constraints can be used to develop methods of estimation or to set up schemes for sensitivity analysis. A detailed account is given in Chapter 20.

In practice, choice of restrictions will need to be guided by the context. In addition, the form of the data will typically be more complex, requiring, for example, a more structured model for the response with the incorporation of covariates. Hence, models for $f(t_i|\boldsymbol{\psi})$ can be constructed in many ways. Most authors assume the dropout process is fully observed and that T_i satisfies a parametric model (Wu and Bailey 1988, 1989, Little 1993, DeGruttola and Tu 1994). Hogan and Laird (1997) extend this to cases where the dropout time is allowed to be right censored and no parametric restrictions are put on the dropout times. Their conditional model for Y_i^o given T_i is a linear mixed model with dropout time as one of the covariates in the mean structure. Due to the right censoring, the estimation method must handle incomplete covariates. Hogan and Laird (1997) use the EM algorithm (Dempster, Laird, and Rubin 1977) for ML estimation.

At this point, a distinction between so-called *outcome-based* and *random-coefficient-based* models is useful. In the context of the former, Little (1995) and Little and Wang (1996) consider the restrictions implied by a selection dropout model in the pattern-mixture framework. For example, with two time points and a Gaussian response, Little proposes a general form of dropout model:

$$P(\text{dropout} \mid \boldsymbol{y}) \;=\; g(y_1 + \lambda y_2), \qquad (18.4)$$

with the function $g(\cdot)$ left unspecified. In a selection modeling context, (18.4) is often assumed to have a logistic form [Chapter 17, e.g., (17.17)]. This relationship implies that the conditional distribution of Y_1 given $Y_1 + \lambda Y_2$ is the same for those who drop out and those who do not. With this restriction and given λ, the parameters of the full distribution of the dropouts is identified. The "weight" λ can then be used as a sensitivity parameter, its size determining dependence of dropout on the past and present, as in the selection models. Such a procedure can be extended to more general problems (Little 1995, Little and Wang 1996). It is instructive in this very simple setting to compare the sources of identifiability in the pattern-mixture and selection models. In the former, the information comes from the assumption that the dropout probability is some function of a linear combination of the two observations with known coefficients. In the latter, it comes from the shape of the assumed conditional distribution of Y_2 given Y_1 (typically Gaussian), together with the functional form of the dropout probability. The difference is highlighted if we consider a sensitivity analysis for the selection model that varies λ in the same way as with the pattern-mixture model. Such sensitivity analysis is much less convincing because the data can, through the likelihood, distinguish between the fit associated with different values of λ.

Therefore, identifiability problems in the selection context tend to be masked. Indeed, there are always unidentified parameters, although a related "problem" seems absent in the selection model. This apparent paradox has been observed by Glynn, Laird, and Rubin (1986). Let us discuss this paradox in some detail.

18.1.2   A Paradox

Assume we have two measurements where Y_1 is always observed and Y_2 is either observed ($t = 2$) or missing ($t = 1$). Let us further simplify the notation by suppressing dependence on parameters and additionally adopting the following definitions:

$$
\begin{aligned}
g(t|y_1, y_2) &:= f(t|y_1, y_2), \\
p(t) &:= f(t), \\
f_t(y_1, y_2) &:= f(y_1, y_2|t).
\end{aligned}
$$

Equating the selection model and pattern-mixture model factorizations
yields

$$f(y_1, y_2)g(d = 2|y_1, y_2) = f_2(y_1, y_2)p(t = 2),$$
$$f(y_1, y_2)g(d = 1|y_1, y_2) = f_1(y_1, y_2)p(t = 1).$$

Since we have only two patterns, this obviously simplifies further to

$$f(y_1, y_2)g(y_1, y_2) = f_2(y_1, y_2)p,$$
$$f(y_1, y_2)[1 - g(y_1, y_2)] = f_1(y_1, y_2)[1 - p],$$

of which the ratio yields

$$f_1(y_1, y_2) = \frac{1 - g(y_1, y_2)}{g(y_1, y_2)} \frac{p}{1 - p} f_2(y_1, y_2).$$

All selection model factors are identified, as are the pattern-mixture quantities on the right-hand side. However, the left-hand side is not entirely identifiable. We can further separate the identifiable from the nonidentifiable quantities:

$$f_1(y_2|y_1) = f_2(y_2|y_1)\frac{1 - g(y_1, y_2)}{g(y_1, y_2)} \frac{p}{1 - p}\frac{f_2(y_1)}{f_1(y_1)}. \qquad (18.5)$$

In other words, the conditional distribution of the second measurement given the first one, *in the incomplete first pattern*, about which there is no information in the data, is identified by equating it to its counterpart from the complete pattern, modulated via the ratio of the "prior" and "posterior" odds for dropout [$p/(1-p)$ and $g(y_1, y_2)/(1 - g(y_1, y_2))$, respectively] and via the ratio of the densities for the first measurement.

Thus, although an identified selection model is seemingly less arbitrary than a pattern-mixture model, it incorporates *implicit* restrictions. Indeed, precisely these are used in (18.5) to identify the component for which there is no information.

This clearly illustrates the need for sensitivity analysis. Due to the different nature of the selection and pattern-mixture models, specific forms for each of the two contexts will be presented in Chapters 19 and 20, respectively.

In Section 18.2, we will describe a general strategy for fitting pattern-mixture models. The remainder of this chapter is devoted to pattern-mixture models for the toenail data (Section 18.3) and for the Vorozole data (Section 18.4). Chapter 20 is devoted to a formal juxtaposition of several strategies for pattern-mixture modeling.

18.2   Pattern-Mixture Models

As indicated in Section 18.1, this family is based on factorization (15.2), which, for dropout, can be rewritten as (18.1) or (18.2). The conditional density of the measurements given the dropout pattern is combined with the marginal density describing the dropout mechanism. Note that the second factor can depend on covariates, but not on outcomes. It is, of course, possible to have different covariate dependencies in both components of the factorization. For example, dropout can vary with treatment arm and age of the respondent, whereas the measurement model can depend on treatment arm, sex, and measurement time.

Thus, the dropout process $f(t_i|X_i, \psi)$ is just a, possibly covariate-dependent, model for the probability to belong to a particular pattern. If it is expressed in analogy with (17.17), then $g(h_{ij})$ will describe the dropout rate at each occasion.

The measurement model has to reflect dependence on dropout. In its most general form, this implies that (3.8) is replaced by

$$
\begin{cases}
Y_i = X_i\beta(t_i) + Z_i b_i + \varepsilon_i, \\[2mm]
b_i \sim N(0, D(t_i)), \\[2mm]
\varepsilon_i \sim N(0, \Sigma_i(t_i)).
\end{cases}
\tag{18.6}
$$

Thus, the fixed effects as well as the covariance parameters are allowed to change with dropout pattern and a priori no restrictions are placed on the structure of this change.

It immediately follows from (15.2) that the likelihood contribution of the ith subject, based on the observed data (y_i^o, t_i), is proportional to

$$
f(y_i^o, t_i) = f(t_i)f(y_i^o|t_i),
\tag{18.7}
$$

which only requires specifying a marginal model for the dropout process and a conditional model for the observed outcomes, given the dropout pattern as in (18.6). Further, as for ignorable selection models, both models can be fitted separately, provided separability of their parameters.

Model family (18.6) contains underidentified members since it describes the full set of measurements in pattern t_i, even though there are not measurements after occasion t_i, as was pointed out in Section 18.1.2 for the simple case of two measurements. Several routes can be taken to solve this problem. They are described in detail in Section 20.4. Let us briefly sketch them.

Little (1993, 1994a) advocated the use of identifying restrictions which works well in relatively simple settings. Molenberghs, Michiels, Kenward, and Diggle (1998) proposed a particular set of restrictions for the monotone case which correspond to MAR. Alternatively, several types of simplified (identified) models can be considered. The advantage is that the number of parameters decreases, which is generally an issue with pattern-mixture models. This route will be followed in Sections 18.3 and 18.4.

Hogan and Laird (1997) noted that in order to estimate the large number of parameters in general pattern-mixture models, one has to make the awkward requirement that each dropout pattern is sufficiently "filled"; in other words, one has to require large numbers of dropouts. This problem is less prominent in simplified models. Note however that simplified models, qualified as "assumption rich" by Sheiner, Beal, and Dunne (1997), are also making untestable assumptions and therefore illustrate that even pattern-mixture models do not provide a free lunch. A main advantage however is that the need of assumptions and their implications are more obvious. For example, it is not possible to assume an unstructured time trend in incomplete patterns, except if one restricts attention to the time range from onset until dropout. In contrast, assuming a linear time trend allows estimation in all patterns containing at least two measurements.

In general, we distinguish between two types of simplification to identify pattern-mixture models. First, functional model forms can be restricted to those which are supported by the information available within a pattern. For example, a linear time trend with a fixed treatment effect, together with a compound symmetry covariance structure, is identifiable as soon as there are two time points. This will be illustrated in Section 18.3. Second, one can let the parameters vary across patterns in a parametric way. Thus, rather than estimating a separate time trend in each pattern, one could assume that the time evolution is unstructured within a pattern, but parallel across patterns. The available data can be used to assess whether such simplifications are supported *within the range of the observed data*. Using the so-obtained profiles past the time of dropout still requires extrapolation or, in other words, a leap of faith. This is the route chosen in Section 18.4.

18.3   Pattern-Mixture Model for the Toenail Data

For the TDO data, we will assume the dropout patterns T_i to be sampled from a multinomial distribution with support $\{1, 2, 3, 4, 5, 6, 7\}$, where the class $T_i = 7$ contains all completers. The associated multinomial probability vector is denoted by $\boldsymbol{\pi} = (\pi_1, \pi_2, \ldots, \pi_7)'$. Further, our model for $\boldsymbol{Y}_i^o$, conditional on $T_i = t_i$, is assumed to be of the same form as model (17.1),

TABLE 18.1. *Toenail Data. Fitted dropout probabilities under the multinomial dropout model.*

Dropout occasion d:	1	2	3	4	5	6	7
Fitted prob. $\hat{\pi}_t = P(T_i = t)$:	0.02	0.02	0.04	0.05	0.10	0.01	0.76

but with different parameters for each possible value of T_i. Obviously, no quadratic average evolution can be fitted whenever $t_i = 1$ or $t_i = 2$. We then only fit a constant term or a linear average evolution, respectively. More specifically, our model for Y_i^o, conditional on $T_i = t_i$, equals

$$
Y_{ij}^o = \begin{cases}
(\beta_{A0}(t_i) + b_i(t_i)) + \beta_{A1}(t_i)t_{ij} \\
\quad + \beta_{A2}(t_i)t_{ij}^2 + \varepsilon_{(2)ij}(t_i) & \text{group A} \\
(\beta_{B0}(t_i) + b_i(t_i)) + \beta_{B1}(t_i)t_{ij} \\
\quad + \beta_{B2}(t_i)t_{ij}^2 + \varepsilon_{(2)ij}(t_i) & \text{group B}.
\end{cases} \tag{18.8}
$$

Recall that t_i indicates pattern and t_{ij} indicates time of measurement. For the patterns $t_i = 1$ and $t_i = 2$, we only have information in the data to fit constant average trends or linear average trends, respectively. Therefore, we need to restrict the parameters in model (18.8). A possible restriction is $\beta_{A1}(1) = \beta_{B1}(1) = \beta_{A2}(1) = \beta_{B2}(1) = \beta_{A2}(2) = \beta_{B2}(2) = 0$. For $t_i = 1$, there is only one measurement per subject such that no random intercepts can be included into the model. As before, it is assumed that the $b_i(t_i)$ and the $\varepsilon_{(2)ij}(t_i)$ are normally distributed with means zero, but we allow their variance to depend on the dropout pattern t_i. More parsimonious models can be obtained by putting additional restrictions on the parameters, such as assuming that all variance components are independent of t_i. The most extreme case assumes that none of the parameters in model (18.8) depends on t_i. We then have that the measurement model is statistically independent of the dropout model, implying completely random dropout.

The separability of the parameters in the measurement model and the dropout model, together with the separability of the parameters in the measurement models across dropout occasions, implies that fitting the above model reduces to fitting a linear mixed model to each dropout pattern separately and to the calculation of the observed dropout probabilities at the various occasions. Table 18.1 contains the fitted dropout probabilities, whereas Figure 18.1 shows the fitted average profiles for each dropout pattern, obtained from fitting the linear mixed models (18.8). Each panel of the figure corresponds to a specific dropout pattern, and the number of subjects in the pattern is denoted by n_A and n_B for group A and group B, respectively. Note that there is very little information to fit some of the

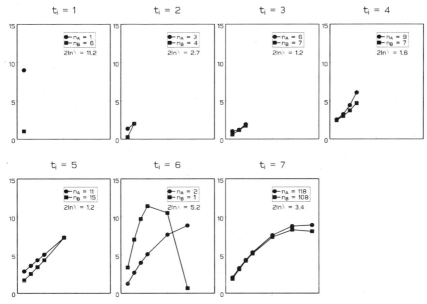

FIGURE 18.1. *Toenail Data. Fitted average profiles for each dropout pattern, obtained from fitting the mixed models (18.8). For each pattern, the number of subjects in group A and group B is denoted by n_A and n_B, respectively.*

models in (18.8). This explains the unexpected behavior observed for $t_i = 1$ and $t_i = 6$.

For each pattern, the likelihood ratio statistic $2 \ln \lambda$ measures the difference between the two treatment groups with respect to the fitted average profile. The sum of all of these statistics could be used to test whether there is any treatment difference at all for any of the dropout patterns. In our example, this sum equals 26.5, on 18 degrees of freedom. When compared to a χ^2-distribution, there is no evidence for any treatment effect ($p = 0.089$). However, this should be interpreted with care since the χ^2-approximation may not be accurate due to the small numbers of subjects observed in some of the dropout patterns.

One of the main advantages of pattern-mixture models is that a conditional model for the observed responses Y_i^o only is needed, rather than for the complete vector Y_i. However, this changes when interest is in inferences for the marginal distribution of Y_i instead of the conditional distribution. Suppose we want to compare the marginal average time trends between both treatment groups, as was done under the selection models of Section 17.2. We then need to evaluate the marginal expectation

$$E[Y_{ij}] \quad = \quad E[Y_{ij} \mid T_i = 1] \; P(T_i = 1) \; + \; E[Y_{ij} \mid T_i = 2] \; P(T_i = 2)$$

$$+ \cdots + E[Y_{ij} \mid T_i = 7] \, P(T_i = 7), \tag{18.9}$$

which requires specification of $E(\mathbf{Y}_i^m \mid T_i)$ (i.e., of the expected evolution of the subjects after they dropped out). It should be emphasized that the data do not contain any information on these average profiles beyond the time of dropout. Hence, the estimation of (18.9) entirely relies on the extrapolation of the fitted average profiles shown in Figure 18.1 to time points where no data were observed.

For the dropout patterns where at least three measurements are available per subject $(t_i \geq 3)$, we extrapolate the quadratic trend over time, fitted from the observed data. Borrowing the linear and quadratic time effects, or just the quadratic time effect, from pattern $t_i = 3$, we can extrapolate the patterns $t_i = 2$ and $t_i = 3$ as well. More precisely, our extrapolation assumes that

$$\begin{cases} \beta_{A1}(1) = \beta_{A1}(3), \\ \beta_{A2}(1) = \beta_{A2}(2) = \beta_{A2}(3), \\ \beta_{B1}(1) = \beta_{B1}(3), \\ \beta_{B2}(1) = \beta_{B2}(2) = \beta_{B2}(3). \end{cases} \tag{18.10}$$

This expresses our belief that the average behavior in the first two patterns is likely to be similar to the third pattern. The obtained extrapolations are indicated by dashed lines in Figure 18.2. Note that the strongly positive estimated average quadratic time effect for the patterns $t_i = 3$ and $t_i = 4$ implies extremely steep extrapolated curves for all patterns with $t_i \leq 4$.

The marginal expectation (18.9) for treatment group A now becomes

$$\begin{aligned} E[Y_{ij}] = \ & \left(\beta_{A0}(1) + \beta_{A1}(3)t_{ij} + \beta_{A2}(3)t_{ij}^2 \right) \pi_1 \\ & + \left(\beta_{A0}(2) + \beta_{A1}(2)t_{ij} + \beta_{A2}(3)t_{ij}2 \right) \pi_2 \\ & + \left(\beta_{A0}(3) + \beta_{A1}(3)t_{ij} + \beta_{A2}(3)t_{ij}^2 \right) \pi_3 + \cdots \\ & + \left(\beta_{A0}(7) + \beta_{A1}(7)t_{ij} + \beta_{A2}(7)t_{ij}^2 \right) \pi_7, \end{aligned} \tag{18.11}$$

which is estimated by replacing all parameters by their estimates obtained by our original pattern-mixture model. The marginal expectation for treatment group B is obtained from replacing the subscript A in (18.11) by B. The so-obtained estimates of the marginal average profiles are shown in Figure 18.3. Note the completely different behavior of treatment group A when compared to the fitted average trends obtained from selection modeling, shown in Figure 16.1. Obviously, this is a consequence of the extremely steep extrapolations for the dropout patterns $t_i \leq 4$, which was more pronounced for group A than for group B. One might argue that these extrapolations are unrealistic. However, as discussed previously, such conclusions cannot be supported by the collected data, since they do not contain any

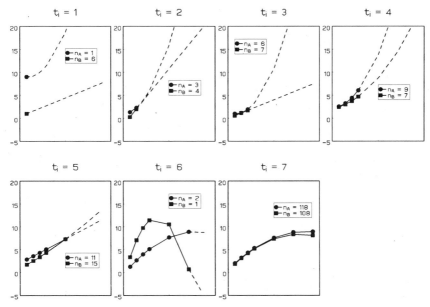

FIGURE 18.2. *Toenail Data. Extrapolated fitted average profiles for each dropout pattern, obtained from fitting the mixed models (18.8), imposing the restrictions (18.10). For each pattern, the number of subjects in group A and group B is denoted by n_A and n_B, respectively.*

information on the trends beyond dropout. Moreover, our analyses in Section 17.2.2 have shown that subjects with large increments in unaffected nail length are more susceptible to dropout than others, suggesting that early dropouts are subjects for which the response increases quickly over time. In this respect, the extrapolations used in Figure 18.2 become less unrealistic. Chapter 20 will discuss various strategies which can then be considered simultaneously.

When interest is in testing for marginal average differences between both treatment groups, the null hypothesis of interest is

$$H_0 : \begin{cases} \displaystyle\sum_{t=1}^{7} \pi_t\, \beta_{A0}(t) \;-\; \sum_{t=1}^{7} \pi_t\, \beta_{B0}(t) \;=\; 0, \\[2mm] \displaystyle\sum_{t=1}^{7} \pi_t\, \beta_{A1}(t) \;-\; \sum_{t=1}^{7} \pi_t\, \beta_{B1}(t) \;=\; 0, \\[2mm] \displaystyle\sum_{t=1}^{7} \pi_t\, \beta_{A2}(t) \;-\; \sum_{t=1}^{7} \pi_t\, \beta_{B2}(t) \;=\; 0, \end{cases} \qquad (18.12)$$

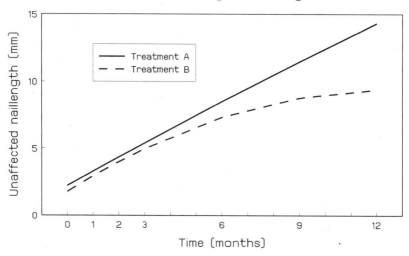

FIGURE 18.3. *Toenail Data. Fitted marginal average profiles (18.11) for both treatment groups, obtained from fitting the pattern-mixture model (18.8), under the restrictions (18.10).*

versus the alternative hypothesis that H_0 does not hold. Following Little (1993), we tested the above hypothesis using a Wald-type test, where the asymptotic covariance matrix of the estimators of the three functions in (18.12) is estimated via the delta method. Other methods for precision estimation in pattern-mixture models are described by Michiels, Molenberghs and Lipsitz (1999). These authors use multiple imputation in the context of categorical data.

Let $\beta^{(t)}$ denote the vector of all six fixed effects in the linear mixed model corresponding to pattern $t_i = t$. The asymptotic covariance matrix

$$\text{var}(\widehat{\pi}, \widehat{\beta}(1), \dots, \widehat{\beta}(7))$$

of all parameters involved in (18.12) is block-diagonal with blocks $\text{var}(\widehat{\pi})$, $\text{var}(\widehat{\beta}(1))$, ..., $\text{var}(\widehat{\beta}(7))$. All $\text{var}(\widehat{\beta}(t))$, $t = 1, \dots, 7$, are readily available from the statistical package (e.g., the SAS procedure MIXED) used for fitting the models (18.8) to each pattern separately. Further,

$$\text{var}(\widehat{\pi}) \quad = \quad \text{diag}(\pi) - \pi\pi'$$

(Bickel and Doksum 1977, Section A.13).

In our example, the average difference in intercepts, linear time effects, and quadratic time effects [the three functions in (18.12)] are estimated by

0.446, -0.131, and 0.042, respectively. The asymptotic covariance matrix for these estimators is estimated by

$$\begin{pmatrix} 0.174 & -0.021 & 0.001 \\ -0.021 & 0.018 & -0.003 \\ 0.001 & -0.003 & 0.001 \end{pmatrix}.$$

The resulting observed test statistic equals 2.464, which is not significant ($p = 0.482$) when compared to a χ^2-distribution with 3 degrees of freedom. This may seem counterintuitive in view of the large difference between both treatment groups, seen in Figure 18.3. Again, it should be emphasized that many parameters were estimated with very little precision. Further, the observed difference is, to a large extent, a function of extrapolation rather than observation.

18.4 A Pattern-Mixture Model for the Vorozole Study

In Chapter 14, the individual and average profiles for the Vorozole study were plotted in a pattern-specific way (Figures 14.5 and 14.6).

Figure 14.6 suggests that pattern-specific profiles are of a quadratic nature; with a sharp decline prior to dropout in most cases. Note that this is in line with the fitted dropout mechanism (17.19). Therefore, it seems reasonable to expect reflection of this feature in the pattern-mixture model. In analogy with our selection model, the profiles are forced to pass through the origin. This is done by allowing only time main effects and interactions of other covariables with time in the model.

The most complex pattern-mixture model we consider includes a different parameter vector for each of the observed patterns. This is done by having all effects in the model interact with *pattern*, a factor variable. We then proceed by backward selection in order to simplify the model. First, we found that the covariance structure is common to all patterns, encompassing random intercept, a serial exponential process, and measurement error.

For the fixed effects, we proceeded as follows. A backward selection procedure was conducted, starting from a model that includes a main effect of time and time^2, as well as interactions of time with baseline value, treatment effect, dominant site and pattern, and the interaction of pattern with time^2. This procedure revealed main effects of time and time^2, as well as interactions of time with baseline value, treatment effect, and pattern, and

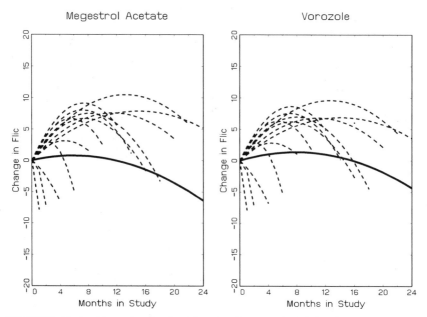

FIGURE 18.4. *Vorozole Study. Fitted selection (solid line) and first pattern-mixture models (dashed lines).*

the interaction of pattern with time^2. This reduced model can be found in Table 18.2. As was the case with the selection model in Table 17.11, the treatment effect is nonsignificant. Indeed, a single degree of freedom F-test yields a p-value of 0.687. Note that such a test is possible since treatment effect does not interact with pattern, in contrast to the model which we will describe next. The fitted profiles are displayed in Figure 18.4. We observe that the profiles for both arms are very similar. This is due to the fact that treatment effect is not significant but perhaps also because we did not allow a more complex treatment effect. For example, we might consider an interaction of treatment with the square of time and, more importantly, a treatment effect which is pattern-specific. Some evidence for such an interaction is seen in Figure 14.6.

Our second, expanded model allowed for up to cubic time effects, the interaction of time with dropout pattern, dominant site, baseline value and treatment, as well as their two- and three-way interactions. After a backward selection procedure, the effects included are time and time^2, the two-way interaction of time and dropout pattern, as well as three-factor interactions of time and dropout pattern with (1) baseline, (2) group, and (3) dominant site. Finally, time^2 interacts with dropout pattern and with the interaction of baseline and dropout pattern. No cubic time effects were necessary, which is in agreement with the observed profiles in Figure 14.6.

TABLE 18.2. *Vorozole Study. Parameter estimates and standard errors for the first pattern-mixture model.*

Effect	Estimate (s.e.)	Effect	Estimate (s.e.)
Fixed-effect Parameters:			
Time	4.671 (0.844)	$Time^2$	−0.034 (0.029)
Time∗Pattern 1	−8.856 (2.739)	$Time^2$∗Pattern 1	
Time∗Pattern 2	−0.796 (2.958)	$Time^2$∗Pattern 2	−1.918 (1.269)
Time∗Pattern 3	−1.959 (1.794)	$Time^2$∗Pattern 3	−0.145 (0.365)
Time∗Pattern 4	1.600 (1.441)	$Time^2$∗Pattern 4	−0.541 (0.197)
Time∗Pattern 5	0.292 (1.295)	$Time^2$∗Pattern 5	−0.107 (0.133)
Time∗Pattern 6	1.366 (1.035)	$Time^2$∗Pattern 6	−0.181 (0.080)
Time∗Pattern 7	1.430 (1.045)	$Time^2$∗Pattern 7	−0.132 (0.071)
Time∗Pattern 8	1.176 (1.025)	$Time^2$∗Pattern 8	−0.118 (0.061)
Time∗Pattern 9	0.735 (0.934)	$Time^2$∗Pattern 9	−0.083 (0.049)
Time∗Pattern 10	0.797 (1.078)	$Time^2$∗Pattern 10	−0.078 (0.055)
Time∗Pattern 11	0.274 (0.989)	$Time^2$∗Pattern 11	−0.023 (0.046)
Time∗Pattern 12	0.544 (1.087)	$Time^2$∗Pattern 12	−0.026 (0.049)
Time∗Baseline	−0.031 (0.004)	Time∗Treatment	−0.067 (0.166)
Variance Parameters:			
Random intercept	78.45		
Serial variance	95.38		
Serial association	8.85		
Measurement error	73.77		

The parameter estimates of this model are displayed in Tables 18.3 and 18.4. The model is graphically represented in Figure 18.5.

Because a pattern-specific parameter has been included, we have several options for the assessment of treatment. Since there are 13 patterns (remember we cut off the patterns at 2 years), one can test the global hypothesis, based on 13 degrees of freedom, of no treatment effect. We obtain $F = 1.25$, producing $p = 0.240$, indicating that there is no overall treatment effect. Each of the treatment effects separately is at a nonsignificant level. Alternatively, the *marginal* effect of treatment can be calculated, which is the weighted average of the pattern-specific treatment effects, with weights given by the probability of occurrence of the various patterns. Its standard error is calculated using a straightforward application of the delta

TABLE 18.3. *Vorozole Study. Parameter estimates and standard errors for the second pattern-mixture model (part I). Each column represents an effect, for which a main effect is given, as well as interactions with the dropout patterns.*

Fixed-effect parameters [estimate (s.e.)]			
Effect	Time	Time*Baseline	Time^2
Main	5.468 (5.089)	−0.034 (0.040)	−0.271 (0.206)
Pattern 1	7.616 (21.908)	−0.119 (0.175)	
Pattern 2	44.097 (17.489)	−0.440 (0.148)	−18.632 (7.491)
Pattern 3	22.471 (10.907)	−0.218 (0.089)	−5.871 (2.143)
Pattern 4	10.578 (9.833)	−0.055 (0.079)	−1.429 (1.276)
Pattern 5	14.691 (8.424)	−0.123 (0.069)	−1.571 (0.814)
Pattern 6	7.527 (6.401)	−0.061 (0.052)	−0.827 (0.431)
Pattern 7	−12.631 (7.367)	0.086 (0.058)	0.653 (0.454)
Pattern 8	14.827 (6.467)	−0.126 (0.053)	−0.697 (0.343)
Pattern 9	5.667 (6.050)	−0.049 (0.049)	−0.315 (0.288)
Pattern 10	12.418 (6.473)	−0.093 (0.051)	−0.273 (0.296)
Pattern 11	1.934 (6.551)	−0.022 (0.053)	−0.049 (0.289)
Pattern 12	6.303 (6.426)	−0.052 (0.050)	−0.182 (0.259)
Effect	Time^2*Baseline	Time*Treatment	
Main	0.002 (0.002)		
Pattern 1		0.445 (5.095)	
Pattern 2	0.1458 (0.0644)	0.867 (1.552)	
Pattern 3	0.0484 (0.0178)	−1.312 (0.808)	
Pattern 4	0.0080 (0.0107)	−0.249 (0.686)	
Pattern 5	0.0127 (0.0069)	−0.184 (0.678)	
Pattern 6	0.0058 (0.0036)	0.527 (0.448)	
Pattern 7	−0.0065 (0.0038)	0.782 (0.502)	
Pattern 8	0.0052 (0.0029)	−0.809 (0.464)	
Pattern 9	0.0021 (0.0023)	−0.080 (0.443)	
Pattern 10	0.0016 (0.0024)	0.331 (0.579)	
Pattern 11	0.0003 (00024)	−0.679 (0.492)	
Pattern 12	0.0015 (0.0021)	0.433 (0.688)	
Pattern 13		−1.323 (0.706)	

method, as in Section 18.3. See also Section 20.6.2. This effect is equal to −0.286(0.288), producing a p-value of 0.321, which is still nonsignificant.

TABLE 18.4. *Vorozole Study. Parameter estimates and standard errors for the second pattern-mixture model (part II). Each column represents an effect, for which a main effect is given, as well as interactions with the dropout patterns.*

	Fixed-effect parameters [estimate (s.e.)]		
Effect	Time*Domsite (1)	Time*Domsite (2)	Time*Domsite (3)
Main	−0.873 (1.073)	0.941 (0.845)	0.023 (0.576)
Pattern 1	−5.822 (17.401)	−9.320 (9.429)	1.431 (9.878)
Pattern 2	2.024 (3.847)	4.393 (2.690)	5.681 (2.642)
Pattern 3	2.937 (2.596)	0.940 (1.697)	1.414 (1.633)
Pattern 4	−1.378 (2.699)	−4.366 (2.367)	−3.237 (2.289)
Pattern 5	−0.547 (1.917)	−1.099 (1.456)	−1.015 (1.344)
Pattern 6	1.302 (1.130)	−0.914 (0.811)	
Pattern 7	3.881 (1.485)	1.733 (1.226)	4.548 (1.218)
Pattern 8	2.359 (1.241)	−0.436 (0.843)	
Pattern 9	1.138 (1.128)	−0.326 (0.753)	
Pattern 10		−3.595 (0.996)	
Pattern 11	0.317 (1.152)	0.182 (0.825)	
Pattern 12		−1.694 (0.972)	
Variance parameters			
Random intercept	98.93		
Serial variance	38.86		
Serial association	6.10		
Measurement error	73.65		

The various assessment of treatment effect, based on the results obtained in this section and in Section 17.6, are summarized in Table 18.5.

Thus, we obtain a nonsignificant treatment effect from all our different models, which gives more weight to this conclusion.

18.5   Some Reflections

Pattern-mixture modeling does strictly speaking not require modeling of the unobserved outcomes. Indeed, in its simplest form, a chosen measurement model (e.g., a linear mixed-effects model) can be fitted to the *observed* data in each of the dropout patterns separately. Together with estimating

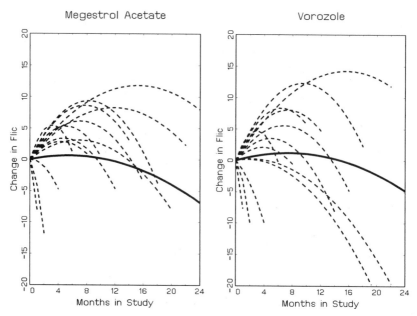

FIGURE 18.5. *Vorozole Study. Fitted selection (solid line) and second pattern-mixture models (dashed lines).*

the (possibly covariate-dependent) probabilities of membership to each of the dropout patterns, the model is completed.

However, there are several reasons why more complex manipulations may be needed. First, there will often be interest in the marginal distribution of the responses, for which a mixture of an effect over the different dropout patterns is needed, such as in (18.9).

Second, interest can be placed on the prediction of pattern-specific quantities, such as average profiles, *beyond* the time of dropout. This is where the underidentification of pattern-mixture models manifests itself. Several solutions have been proposed. Little (1993) suggested identifying restrictions (see Section 18.1). Alternatively, relatively simple models can be constructed such as linear or quadratic time evolutions, which allow easy extrapolation (Section 18.3). Finally, incorporating dropout time as a covariate into the model, information can be borrowed across patterns (Section 18.4). Thus, an advantage is that the assumptions are always very explicit, in contrast to selection modeling. This simplifies performing sensitivity analyses by investigating the effect of various assumptions on the final results. This advantage will be exploited in Chapter 20.

TABLE 18.5. *Vorozole Study. Summary of treatment effect assessment.*

Method	d.f.	p-value
Selection model	1	0.582
First pattern-mixture model	1	0.687
Second pattern-mixture model	13	0.240
Second pattern-mixture model	1	0.321

Finally, as illustrated in our analyses, pattern-mixture models often comprise large numbers of parameters, some of which may be estimated very inefficiently, thereby possibly distorting asymptotics.

19

Sensitivity Analysis for Selection Models

19.1 Introduction

In the previous chapters, it was indicated on various occasions (see Sections 15.4, 16.1, 16.5, 17.1, and 17.2) that incomplete longitudinal data pose specific challenges related to sensitivity to modeling assumptions. Even when the linear mixed model would beyond any doubt be the choice of preference to describe the measurement process *should the data be complete*, then the analysis of the actually observed, incomplete version is subject still to further untestable modeling assumptions. The terminology which is useful to this end has been reviewed in Chapter 15.

The methodologically simplest case is discussed in Chapter 16, where it is assumed that the missing data are MCAR. Simple techniques such as a complete case analysis, simple forms of imputation, and so forth may be advised in some cases. However, the MCAR assumption is a strong one and made too often in practice. Thus, simple forms of analysis are certainly too common in applied statistical practice.

When more flexible assumptions, such as MAR or even MNAR, are considered several choices have to be made. For example, one has to choose between selection and pattern-mixture models, or an alternative framework such as shared-parameter models (Wu and Bailey 1988, 1989, Wu and Carroll 1988, DeGruttola and Tu 1994). For a review, see Little (1995). A more complete literature review can be found in Section 15.4. Selection models have been studied in Chapter 17, and pattern-mixture models are the subject of Chapters 18 and 20.

Particularly within the selection modeling framework, there has been an increasing literature on nonrandom missing data. At the same time, concern has been growing precisely about the fact that models often rest on strong assumptions and relatively little evidence from the data themselves. This point was already raised by Glynn, Laird and Rubin (1986), who indicate that this is typical for so-called selection models, whereas it is much less so for a pattern-mixture model (Section 18.1.2). In Section 17.1 attention was drawn to the fact that much of the debate on selection models is rooted in the econometrics literature, in particular Heckman's selection model (Heckman 1976). Draper (1995) and Copas and Li (1997) provide useful insight in model uncertainty and nonrandomly selected samples. Vach and Blettner (1995) study the case of incompletely observed covariates in logistic regression.

Because the model of Diggle and Kenward (1994) fits within the class of selection models, it is fair to say that it raised, at first, too high expectations. This was made clear by many discussants of the paper. It implies that, for example, formal tests for the null hypothesis of random missingness, although technically possible, should be approached with caution. See also Section 18.1.1. In Section 17.2.2, it was shown, using the toenail data, that excluding a small amount of measurement error, can have a serious impact on the rest of the model parameters. In particular, the likelihood ratio test statistics for the random dropout null hypothesis changes drastically (Table 17.1).

In response to these concerns, there is growing awareness of the need for methods that investigate the sensitivity of the results with respect to the model assumptions. See, for example, Nordheim (1984), Little (1994a), Rubin (1994), Laird (1994), Fitzmaurice, Molenberghs, and Lipsitz (1995), Molenberghs, Goetghebeur, Lipsitz, and Kenward (1999), and Kenward and Molenberghs (1999). Still, only few actual proposals have been made. Moreover, many of these are to be considered useful but ad hoc approaches. Whereas such informal sensitivity analyses are an indispensable step in the analysis of incomplete longitudinal data, it is desirable to conduct more formal sensitivity analyses.

In any case, fitting a nonrandom dropout model should be subject to careful scrutiny. The modeler needs to pay attention, not only to the assumed distributional form of her model (Little 1994b, Kenward 1998; see also Section 19.5.1), but also to the impact one or a few influential subjects may have on the dropout and/or measurement model parameters (Section 19.5). Because fitting a nonrandom dropout model is feasible by virtue of strong assumptions, such models are likely to pick up a wide variety of influences in the parameters describing the nonrandom part of the dropout mechanism. Hence, a good level of caution is in place.

We could define a sensitivity analysis as one in which several statistical models are considered simultaneously and/or where a statistical model is further scrutinized using specialized tools (such as diagnostic measures). This rather loose and very general definition encompasses a wide variety of useful approaches. The simplest procedure is to fit a selected number of (nonrandom) models which are all deemed plausible or one in which a preferred (primary) analysis is supplemented with a number of variations. The extent to which conclusions (inferences) are stable across such ranges provides an indication about the belief that can be put into them. Variations to a basic model can be constructed in different ways. The most obvious strategy is to consider various dependencies of the missing data process on the outcomes and/or on covariates. Alternatively, the distributional assumptions of the models can be changed.

Section 19.2 adapts the model of Diggle and Kenward (1994) to a form useful for sensitivity analysis. Such a sensitivity analysis method, based on local influence (Cook 1986; Thijs, Molenberghs, and Verbeke 2000; see also Section 11.2) is introduced in Section 19.3 and applied to the rats data in Section 19.4. Note that in Section 24.4, a comparison is made with a more conventional global influence analysis (Chatterjee and Hadi 1988). Both informal and formal methods of sensitivity are applied to the mastitis data in Section 19.5. Note that a sensitivity analysis of the milk protein contents data is given in Section 24.4. An outlook on alternative approaches is given in Section 19.6. Random-coefficient-based models are discussed in Section 19.7. We will conclude that caution is needed with selection models and that a mechanical use, perhaps stimulated by the availability of software such as the SPlus suite of functions termed OSWALD (Smith, Robertson, and Diggle 1996), should be avoided.

19.2   A Modified Selection Model for Nonrandom Dropout

In Section 17.5, the selection model of Diggle and Kenward (1994) was presented, specifically with a linear mixed model for the measurement process and a logistic regression, based dropout model (see also Sections 17.2 and 17.6).

In this chapter, we investigate the sensitivity of the estimation of quantities of interest, such as treatment effect, growth parameters, or the dropout parameters, with respect to assumptions about the dropout model. To this end, we consider the following perturbed version of (17.17):

$$\text{logit}(g(\boldsymbol{h}_{ij}, y_{ij})) = \text{logit}\left[\text{pr}(D_i = j | D_i \geq j, \boldsymbol{y}_i)\right] = \boldsymbol{h}_{ij}\boldsymbol{\psi} + \omega_i y_{ij} \quad (19.1)$$

$(i = 1, \ldots, N)$, in which different subjects give different weights to the response at occasion j to predict dropout at occasion j. If all ω_i equal zero, the model reduces to a MAR model; hence, (19.1) can be seen as an extension of the MAR model, which allows some individuals to drop out in a "less random" way ($|\omega_i|$ large) than others ($|\omega_i|$ small). It is important to note that we will not consider ω_i to be (subject-specific) parameters in the usual sense. Rather, they have to be seen as perturbations around a null model, which in this case will be MAR ($\omega_i = 0$). Then, studying the effect of extending an MAR model to the nonrandom case on the parameters of interest can be achieved by investigating the effect of perturbing the ω_i's around zero. This will be done using the local influence approach of Cook (1986) which has been described in Section 11.2. In the next section, the theory is summarized and then adapted to the current setting.

19.3   Local Influence

George Box has a famous quote saying that all statistical models are wrong, but some are useful. Cook (1986) uses this idea to motivate his assessment of local influence. He suggests that more confidence can be put in a model which is relatively stable under small modifications. The best known perturbation schemes are based on case deletion (Cook and Weisberg 1982), in which the effect is studied of completely removing cases from the analysis. A quite different paradigm is the local influence approach where one investigates how the results of an analysis are changed under small perturbations of the model. In the framework of the linear mixed model, Beckman, Nachtsheim, and Cook (1987) used local influence to assess the effect of perturbing the error variances, the random-effects variances, and the response vector. In the same context, Lesaffre and Verbeke (1998) have shown that the local influence approach is also useful for the detection of influential subjects in a longitudinal data analysis. Moreover, because the resulting influence diagnostics can be expressed analytically, they often can be decomposed in interpretable components, which yield additional insights in the reasons why some subjects are more influential than others.

In our case, we are interested in the influence the nonrandomness of dropout exerts on the parameters of interest, which will most often be the fixed-effects parameters, possibly supplemented with the variance components. This can be done in a meaningful way by considering (19.1) as the dropout model. Indeed, $\omega_i = 0$, for all i, corresponds to an MAR process and such a process cannot influence the measurement model parameters. When small perturbations in a specific ω_i lead to relatively large differences in the model parameters, then this suggests that these subjects may have a large impact on the final analysis. Therefore, even though we may be tempted to conclude that such subjects drop out nonrandomly, this conclusion is misguided because we are not aiming to detect (groups of) subjects that drop out nonrandomly but rather subjects that have a considerable impact on the dropout and measurement model parameters.

In Section 19.3.1, a general introduction is given about the local influence methodology as introduced by Cook (1986). In Section 19.3.2, we will apply it to the dropout model presented in Section 17.5, and a special but important case will be discussed in Section 19.3.3.

19.3.1 Review of the Theory

We denote the log-likelihood function corresponding to model (19.1) by $\ell(\boldsymbol{\gamma}|\boldsymbol{\omega}) = \sum_{i=1}^{N} \ell_i(\boldsymbol{\gamma}|\omega_i)$, in which $\ell_i(\boldsymbol{\gamma}|\omega_i)$ is the contribution of the ith individual to the log-likelihood and where $\boldsymbol{\gamma} = (\boldsymbol{\theta}, \boldsymbol{\psi})$ is the s-dimensional vector, grouping the parameters of the measurement model and the dropout model, not including the $N \times 1$ vector $\boldsymbol{\omega} = (\omega_1, \omega_2, \ldots, \omega_N)'$ of weights defining the perturbation of the MAR model. This expression arises from taking the logarithm of (15.10), the model components of which are described in Section 17.5. It is assumed that $\boldsymbol{\omega}$ belongs to an open subset Ω of $\mathbb{R}^N$. For $\boldsymbol{\omega}$ equal to $\boldsymbol{\omega_0} = (0, 0, \ldots, 0)'$, $\ell(\boldsymbol{\gamma}|\boldsymbol{\omega_0})$ is the log-likelihood function which corresponds to a MAR dropout model.

Let $\widehat{\boldsymbol{\gamma}}$ be the maximum likelihood estimator for $\boldsymbol{\gamma}$, obtained by maximizing $\ell(\boldsymbol{\gamma}|\boldsymbol{\omega_0})$, and let $\widehat{\boldsymbol{\gamma}}_\omega$ denote the maximum likelihood estimator for $\boldsymbol{\gamma}$ under $\ell(\boldsymbol{\gamma}|\boldsymbol{\omega})$. The local influence approach now compares $\widehat{\boldsymbol{\gamma}}_\omega$ with $\widehat{\boldsymbol{\gamma}}$. Similar estimates indicate that the parameter estimates are robust with respect to perturbations of the MAR model in the direction of nonrandom dropout. Strongly different estimates suggest that the model is highly sensitive to such perturbations, which suggests that the choice between an MAR model and a nonrandom dropout model highly affects the results of the analysis. Recall that Cook (1986) proposed to measure the distance between $\widehat{\boldsymbol{\gamma}}_\omega$ and $\widehat{\boldsymbol{\gamma}}$ by the likelihood displacement, defined by $\mathrm{LD}(\boldsymbol{\omega}) = 2[\ell(\widehat{\boldsymbol{\gamma}}|\boldsymbol{\omega_0}) - \ell(\widehat{\boldsymbol{\gamma}}_\omega|\boldsymbol{\omega_0})]$. This takes into account the variability of $\widehat{\boldsymbol{\gamma}}$. Indeed, $\mathrm{LD}(\boldsymbol{\omega})$ will be large if $\ell(\boldsymbol{\gamma}|\boldsymbol{\omega_0})$ is strongly curved at $\widehat{\boldsymbol{\gamma}}$, which means that $\boldsymbol{\gamma}$ is estimated with high precision, and small otherwise. Therefore, a graph of $\mathrm{LD}(\boldsymbol{\omega})$ versus $\boldsymbol{\omega}$ contains essential information on the influence of perturbations. It is useful to view this graph as the geometric surface formed by the values of the $N+1$ dimensional vector $\boldsymbol{\xi}(\boldsymbol{\omega}) = (\boldsymbol{\omega}', \text{LD}(\boldsymbol{\omega}))'$ as $\boldsymbol{\omega}$ varies throughout Ω. See Figure 11.1.

Since this so-called influence graph can only be depicted when $N = 2$, Cook (1986) proposed looking at local influence [i.e., at the normal curvatures C_h of $\boldsymbol{\xi}(\boldsymbol{\omega})$ in $\boldsymbol{\omega}_0$], in the direction of some N dimensional vector $\boldsymbol{h}$ of unit length. Let $\boldsymbol{\Delta}_i$ be the s-dimensional vector defined by

$$\boldsymbol{\Delta}_i = \left. \frac{\partial^2 \ell_i(\boldsymbol{\gamma}|\omega_i)}{\partial \omega_i \partial \boldsymbol{\gamma}} \right|_{\boldsymbol{\gamma}=\widehat{\boldsymbol{\gamma}}, \omega_i=0}$$

and define $\boldsymbol{\Delta}$ as the $(s \times N)$ matrix with $\boldsymbol{\Delta}_i$ as its ith column. Further, let $\ddot{L}$ denote the $(s \times s)$ matrix of second-order derivatives of $\ell(\boldsymbol{\gamma}|\boldsymbol{\omega}_0)$ with respect to $\boldsymbol{\gamma}$, also evaluated at $\boldsymbol{\gamma} = \widehat{\boldsymbol{\gamma}}$. Cook (1986) has then shown that C_h can be easily calculated by $C_h = 2|\boldsymbol{h}' \boldsymbol{\Delta}' \ddot{L}^{-1} \boldsymbol{\Delta} \boldsymbol{h}|$.

Obviously, C_h can be calculated for any direction $\boldsymbol{h}$. One evident choice is the vector $\boldsymbol{h}_i$ containing 1 in the ith position and 0 elsewhere, corresponding to the perturbation of the ith weight only. This reflects the influence of allowing the ith subject to drop out nonrandomly, whereas the others can only drop out at random. The corresponding local influence measure, denoted by C_i, then becomes $C_i = 2|\boldsymbol{\Delta}'_i \ddot{L}^{-1} \boldsymbol{\Delta}_i|$. Another important direction is the direction $\boldsymbol{h}_{\max}$ of maximal normal curvature $C_{\max}$. It shows how to perturb the MAR model to obtain the largest local changes in the likelihood displacement. It is readily seen that $C_{\max}$ is the largest eigenvalue of $-2\boldsymbol{\Delta}' \ddot{L}^{-1} \boldsymbol{\Delta}$, and that $\boldsymbol{h}_{\max}$ is the corresponding eigenvector.

When a subset $\boldsymbol{\gamma}_1$ of $\boldsymbol{\gamma} = (\boldsymbol{\gamma}'_1, \boldsymbol{\gamma}'_2)'$ is of special interest, a similar approach can be used, replacing the log-likelihood by the profile log-likelihood for $\boldsymbol{\gamma}_1$, and the methods discussed above for the full parameter vector directly carry over. There are many possible choices for the vector $\boldsymbol{h}$. For example, $C_i(\boldsymbol{\gamma}_1)$, corresponding to $\boldsymbol{h} = \boldsymbol{h}_i$, defined above, expresses the local influence of allowing the ith subject to drop out nonrandomly on the estimation of $\boldsymbol{\gamma}_1$.

19.3.2   Applied to the Model of Diggle and Kenward

As discussed in the previous section, calculation of local influence measures merely reduces to the evaluation of $\boldsymbol{\Delta}$ and $\ddot{L}$. The components of $\ddot{L}$ follow from a standard likelihood optimization routine. For the linear mixed model with $\Sigma_i = \sigma^2 I_{n_i}$, expressions are derived by Lesaffre and Verbeke (1998) and can easily be extended to the more general case considered here. Let us use $\ell_{i\omega}$ as shorthand for $\ell_i(\boldsymbol{\gamma}|\omega_i)$. It is shown in the Appendix (Section B.1), which can be skipped without problem for the less technically interested reader, that the components of the columns Δ_i of Δ are given by

$$\frac{\partial^2 \ell_{i\omega}}{\partial\boldsymbol{\theta}\partial\omega_i}\bigg|_{\omega_i=0} = 0, \tag{19.2}$$

$$\frac{\partial^2 \ell_{i\omega}}{\partial\boldsymbol{\psi}\partial\omega_i}\bigg|_{\omega_i=0} = -\sum_{j=2}^{n_i} \boldsymbol{h}_{ij} y_{ij} g(\boldsymbol{h}_{ij})[1 - g(\boldsymbol{h}_{ij})], \tag{19.3}$$

for complete sequences (no dropout) and by

$$\frac{\partial^2 \ell_{i\omega}}{\partial\boldsymbol{\theta}\partial\omega_i}\bigg|_{\omega_i=0} = [1 - g(\boldsymbol{h}_{id})]\frac{\partial\lambda(y_{id}|\boldsymbol{h}_{id})}{\partial\boldsymbol{\theta}}, \tag{19.4}$$

$$\frac{\partial^2 \ell_{i\omega}}{\partial\boldsymbol{\psi}\partial\omega_i}\bigg|_{\omega_i=0} = -\sum_{j=2}^{d-1} \boldsymbol{h}_{ij} y_{ij} g(\boldsymbol{h}_{ij})[1 - g(\boldsymbol{h}_{ij})]$$
$$-\boldsymbol{h}_{id}\lambda(y_{id}|\boldsymbol{h}_{id})g(\boldsymbol{h}_{id})[1 - g(\boldsymbol{h}_{id})]. \tag{19.5}$$

for incomplete sequences, where all of the above expressions are evaluated at $\widehat{\boldsymbol{\gamma}}$ and where $g(\boldsymbol{h}_{ij}) = g(\boldsymbol{h}_{ij}, y_{ij})|_{\omega_i=0}$, is the MAR version of the dropout model. By $\lambda(y_{id}|\boldsymbol{h}_{id})$, we denote the expected value of y_{id}, given the history and the fitted MAR model parameters. It is understood that $\boldsymbol{h}_{ij}$ is restricted to a relevant subset of the history. For example, one can restrict attention to the previous measurement, in which case $\boldsymbol{h}_{ij}$ is taken to be $y_{i,j-1}$.

Let $V_{i,11}$ be the predicted covariance matrix for the observed vector given by $(y_{i1}, \ldots, y_{i,d-1})'$, $V_{i,22}$ is the predicted variance for the missing observation y_{id}, and $V_{i,12}$ is the vector of predicted covariances between the elements of the observed vector and the missing observation. It then follows from the linear mixed model (3.8) that the conditional expectation for the observation at dropout, given the history, equals

$$\lambda(y_{id}|\boldsymbol{h}_{id}) = \lambda(y_{id}) + V_{i,21}V_{i,11}^{-1}[\boldsymbol{h}_{id} - \lambda(\boldsymbol{h}_{id})]. \tag{19.6}$$

The derivatives of (19.6) w.r.t. the fixed effects and variance components in the measurement model are

$$\frac{\partial\lambda(y_{id}|\boldsymbol{h}_{id})}{\partial\boldsymbol{\beta}} = \boldsymbol{x}_{id} - V_{i,21}V_{i,11}^{-1}X_{i,(d-1)},$$

$$\frac{\partial\lambda(y_{id}|\boldsymbol{h}_{id})}{\partial\boldsymbol{\alpha}} = \left[\frac{\partial V_{i,21}}{\partial\boldsymbol{\alpha}} - V_{i,21}V_{i,11}^{-1}\frac{\partial V_{i,11}}{\partial\boldsymbol{\alpha}}\right]V_{i,11}^{-1}[\boldsymbol{h}_{id} - \lambda(\boldsymbol{h}_{id})],$$

respectively, where $\boldsymbol{x}'_{id}$ is the dth row of X_i and where $X_{i,(d-1)}$ indicates the first $(d-1)$ rows of X_i.

In practice, the parameter $\boldsymbol{\theta}$ in the measurement model is often of primary interest. Since $\ddot{L}$ is block-diagonal with blocks $\ddot{L}(\boldsymbol{\theta})$ and $\ddot{L}(\boldsymbol{\psi})$, we have that for any unit vector h, C_h equals $C_h(\theta) + C_h(\psi)$, with

$$C_h(\theta) = -2h'\Delta'\ddot{L}^{-1}(\theta)\Delta h,$$

$$C_h(\psi) = -2h'\Delta'\ddot{L}^{-1}(\psi)\Delta h,$$

evaluated at $\gamma = \hat{\gamma}$. It now immediately follows from (19.2) and (19.4) that influence on θ only arises from those measurement occasions at which dropout occurs. This implies that complete sequences cannot be influential $(C_i(\theta) = 0)$ and that incomplete sequences only contribute at the actual dropout time. This is intuitively clear from the following consideration. Suppose the model is fitted using the EM algorithm (Dempster, Laird, and Rubin 1977), then the E step determines the expected values of the missing measurements and the M step fits an ignorable model to the so completed set of data. The only way in which the dropout process can influence the measurement model parameters is by predicting a value which deviates from prediction under ignorability, which is simply the conditional mean of the missing measurement, given the history. From expression (19.4), it is clear that the corresponding contribution is large only if (1) the dropout probability was small but the subject disappeared nevertheless and (2) the conditional mean "strongly depends" on the parameter of interest.

Additional insight can be gained from comparing two incomplete sequences, with equal history, which drop out at the same time point. They then have the same contribution $1 - g(h_{id})$ to (19.4). Hence, different influences on θ can be ascribed to differences for the second factor of (19.4). For example, for the fixed effects, we have that

$$\frac{\partial\lambda(y_{id}|h_{id})}{\partial\beta} - \frac{\partial\lambda(y_{jd}|h_{jd})}{\partial\beta} = x_{id} - x_{jd} - (V_{i,21} - V_{j,21})V_{i,11}^{-1}X_{i,(d-1)}.$$

Hence, if the estimated covariance matrix for the complete data is the same for both sequences, the above expression reduces to $x_{id} - x_{jd}$, indicating that differences with respect to $C_i(\theta)$ can be entirely ascribed to differences in time-varying covariates for the mean structure. A similar interpretation can be obtained for the variance components α.

19.3.3   Special Case: Compound Symmetry

A special but enlightening case is the compound symmetry covariance structure. It arises from assuming that $Z_i = 1_{n_i}$, a vector of ones. The matrix D then reduces to ν^2. Assuming further that $\Sigma_i = \sigma^2 I_{n_i}$ the covariance matrix becomes $V_i = \sigma^2 I_{n_i} + \nu^2 J_{n_i}$, where J_{n_i} is an $(n_i \times n_i)$ matrix of ones. We will now study $C_i(\theta)$ and $C_i(\psi)$ in turn.

As discussed earlier, $\ddot{L}$ is block-diagonal, from which it follows that

$$C_i(\boldsymbol{\theta}) \quad = \quad 2[1 - g(\boldsymbol{h}_{id})]^2 \frac{\partial \lambda(y_{id}|\boldsymbol{h}_{id})'}{\partial \boldsymbol{\theta}} \ddot{L}^{-1}(\boldsymbol{\theta}) \frac{\partial \lambda(y_{id}|\boldsymbol{h}_{id})}{\partial \boldsymbol{\theta}}, \quad (19.7)$$

in which the first factor is large for a small dropout probability at the time of dropout,—in other words, for an unlikely event. This is intuitively appealing, since g_{id} then has the potential of being improved by including dependence on y_{id}. For such a subject, apparent "nonrandomness" would help.

The second factor of (19.7) involves $\ddot{L}(\boldsymbol{\theta})$ and is therefore harder to study. However, we can still make progress if we are prepared to make some approximations. The off-diagonal block of the observed information matrix $\ddot{L}(\boldsymbol{\theta})$ pertaining to the mixed derivatives w.r.t. $\boldsymbol{\beta}$ and $\boldsymbol{\alpha}$ is not equal to zero. The corresponding block of the expected information matrix is zero for a complete data problem, but is not for an incomplete data set, unless the missing data are MCAR (Kenward and Molenberghs 1998). However, these authors also argue that in many practical settings, the difference might be small (see also Chapter 21). Therefore, we will assume that $\ddot{L}(\boldsymbol{\theta})$ is block-diagonal and that $C_i(\boldsymbol{\theta}) \simeq C_i^{\mathrm{ap}}(\boldsymbol{\beta}) + C_i^{\mathrm{ap}}(\sigma^2, \nu^2)$.

Let us consider $C_i^{\mathrm{ap}}(\boldsymbol{\beta})$ first. With some algebra, we arrive at

$$\frac{\partial \lambda(y_{id}|\boldsymbol{h}_{id})}{\partial \boldsymbol{\beta}} \quad = \quad \xi_{id}\boldsymbol{x}_{id} + (1 - \xi_{id})\boldsymbol{\rho}_{id} \quad (19.8)$$

with

$$\xi_{id} \quad = \quad \frac{\sigma^2}{\sigma^2 + (d - 1)\nu^2},$$

$$\boldsymbol{\rho}_{id} \quad = \quad \boldsymbol{x}_{id} - \frac{1}{d - 1} X'_{i(d-1)} \mathbf{1}_{n_i d - 1}.$$

Note that (19.8) is a weighted average of the covariate $\boldsymbol{x}_{id}$ and the within-series residual covariate $\boldsymbol{\rho}_{id}$, at time d. Further, the matrix of second order derivatives $\ddot{L}^{-1}(\boldsymbol{\beta})$ equals

$$\ddot{L}^{-1}(\boldsymbol{\beta}) \quad = \quad \sum_{i=1}^{N} X'_{i(d-1)}\left(I_{d-1} + \frac{\nu^2}{\sigma^2 + (d-1)\nu^2} J_{d-1}\right) X_{i(d-1)},$$

from which it follows that

$$\begin{aligned}
C_i^{\mathrm{ap}}&(\boldsymbol{\beta}) \\
&= \quad 2[1 - g(\boldsymbol{h}_{id})]^2 (\xi_{id}\boldsymbol{x}_{id} + (1 - \xi_{id})\boldsymbol{\rho}_{id})' \\
&\quad \times \sigma^2 \left[\sum_{i=1}^{N}\left(\xi_{id}X'_{i(d-1)}X_{i(d-1)} + (1 - \xi_{id})R'_{i(d-1)}R_{i(d-1)}\right)\right]^{-1} \\
&\quad \times (\xi_{id}\boldsymbol{x}_{id} + (1 - \xi_{id})\boldsymbol{\rho}_{id}), \quad (19.9)
\end{aligned}$$

where

$$R_{i,d-1} = X_{i(d-1)} - \mathbf{1}_{n_i d-1}\overline{X_{i(d-1)}}.$$

Here,

$$\overline{X_{i(d-1)}} = \frac{1}{d-1}\mathbf{1}'_{n_i d-1}X_{i(d-1)}.$$

Expression (19.9) is the product of the factor which purely depends on the dropout probability and a factor which has the structure of a leverage. When $\xi_{id} = 1$ for all individuals, we have a classical leverage where each measurement is an independent contribution. When $\xi_{id} = 0$, each subject presents a single independent contribution. The general case is a weighted combination of the between- and within-individual contributions. These arguments motivate to call the second factor of $C_i^{\mathrm{ap}}(\beta)$ a generalized leverage, not only for compound symmetry but also for general covariance structures.

Similar calculations can be performed for the variance components (σ^2, ν^2), yielding

$$C_i^{\mathrm{ap}}(\sigma^2, \nu^2) = 2[1 - g(\mathbf{h}_{id})]^2 \xi_{id}^2 (1 - \xi_{id})^2 \overline{[\mathbf{h}_{id} - \lambda(\mathbf{h}_{id})]}^2$$
$$\times \left(-1, \frac{1}{\nu^2}\right) \ddot{L}^{-1}(\sigma^2, \nu^2) \begin{pmatrix} -1 \\ \frac{1}{\nu^2} \end{pmatrix}, \qquad (19.10)$$

where

$$\ddot{L}(\sigma^2, \nu^2)$$
$$= \sum_{i=1}^{N} \frac{d-1}{2(\sigma^2 + (d-1)\nu^2)^2}$$
$$\times \begin{pmatrix} [\sigma^2 + (d-1)\nu^2]^2 - \nu^2[2\sigma^2 + (d-1)\nu^2] & 1 \\ 1 & (d-1) \end{pmatrix},$$

and $\overline{[\mathbf{h}_{id} - \lambda(\mathbf{h}_{id})]}$ represents the average difference of the $\mathbf{h}_{id}$ column and its predicted value.

It is important to note that, even though $\ddot{L}^{-1}(\sigma^2, \nu^2)$ has a somewhat complicated form, it occurs in (19.10) only through a scalar. Thus, $C_i^{\mathrm{ap}}(\sigma^2, \nu^2)$ can in practice be decomposed into three interpretable components. The first factor is shared with $C_i^{\mathrm{ap}}(\beta)$ and has the same interpretation. The second factor disappears when either the measurement error variance or the variance of the random intercept is reduced to zero. It is maximal when there is "balance" between both components of variability ($\xi_{id} = 0.5$). The third factor is large when the squared average residual of the history at the time of dropout is large.

For the dropout model parameters, there are no approximations involved, and we have that

$$
C_i(\psi) = 2 \left(\sum_{j=2}^{d} h_{ij} y_{ij} v_{ij} \right)' \left(\sum_{i=1}^{N} \sum_{j=2}^{d} v_{ij} h_{ij} h'_{ij} \right)^{-1}
$$
$$
\times \left(\sum_{j=2}^{d} h_{ij} y_{ij} v_{ij} \right), \tag{19.11}
$$

in which $d = n_i$ for a complete case and where y_{id} needs to be replaced with

$$
\lambda(y_{id} | \boldsymbol{h}_{id}) = \lambda(y_{id}) + (1 - \xi_{id}) \overline{[\boldsymbol{h}_{id} - \lambda(\boldsymbol{h}_{id})]}
$$

for incomplete sequences. Further, v_{ij} equals $g(h_{ij})[1 - g(h_{ij})]$, which is the variance of the estimated dropout probability under MAR. Expression (19.11) bears some resemblance with the hat-matrix diagonal, used for diagnostic purposes in logistic regression (Hosmer and Lemeshow 1989). One of the differences is that the contributions from a single individual are summed in the first and third factors of (19.11), even though they contribute independent pieces of information to the logistic regression. This is because each individual is given a single weight ω_i for an entire sequence of measurements.

To get a good feel for when $C_i(\psi)$ is large, simplify $h_{ij} y_{ij} v_{ij}$ to

$$
F(y) = y^2 g(1 - g), \tag{19.12}
$$

which is based on the assumption that previous and current measurements are approximately equal. Given estimates for ψ, it is easy to determine at which value this function is maximal.

An even greater resemblance would be obtained by using an alternative weighting scheme which replaces ω_i in (19.1) by ω_{ij}, hereby giving different weights to the different observations within subjects. This alternative perturbation scheme would not imply any differences in the influence contributions for the measurement model, but for the dropout parameters, we would obtain

$$
C_{ij}(\psi) = 2(v_{ij} y_{ij}^2) \left\{ v_{ij} h'_{ij} \left(\sum_{i=1}^{N} \sum_{j=2}^{d} v_{ij} h_{ij} h'_{ij} \right)^{-1} h_{ij} \right\}, \tag{19.13}
$$

where the factor in curly braces equals the hat-matrix diagonal. In the case of dropout, the same replacement as before for y_{id} has to be made. When the length of a measurement sequence is restricted to 2, then (19.13) and (19.11) coincide. Note that this alternative perturbation scheme assigns weights to observations rather than subjects, changing the interpretation of the results of the analysis. Also, the graphical representation of the results is more involved since series of influence measures C_{ij} now have to be studied and interpreted.

19.3.4 Serial Correlation

Thus far, the development has focused on the standard linear mixed model, with random effects and measurement error. If, in addition, a serially correlated part of the variance structure is thought to be present, the variance parameter α needs to be extended with the serial correlation parameters, as was done in Section 3.3.4, and thus encompasses d_{jk}, the components of the variance-covariance matrix of the random effects, τ^2, the variance of the serial process, φ, the serial correlation parameter, and σ^2, the variance of the measurement error process. The general model is spelled out in (3.11).

Most of the development in Section 19.3.2 remains the same. We will briefly outline the changes that have to be made. For the various variance parameters, expression (19.7) can be written as

$$\frac{\partial\lambda(y_{id}|h_{id})}{\partial d_{jk}} = \left[[Z_iD^{jk}Z_i']_{21} - V_{i,21}V_{i,11}^{-1}[Z_iD^{jk}Z_i']_{11}\right]V_{i,11}^{-1}[h_{id} - \lambda(h_{id})],$$

$$\frac{\partial\lambda(y_{id}|h_{id})}{\partial\tau^2} = [H_{i,21} - V_{i,21}V_{i,11}^{-1}H_{i,11}]V_{i,11}^{-1}[h_{id} - \lambda(h_{id})],$$

$$\frac{\partial\lambda(y_{id}|h_{id})}{\partial\varphi} = [\tau^2K_{i,21} - V_{i,21}V_{i,11}^{-1}\tau^2K_{i,11}]V_{i,11}^{-1}[h_{id} - \lambda(h_{id})],$$

$$\frac{\partial\lambda(y_{id}|h_{id})}{\partial\sigma^2} = -V_{i,21}V_{i,11}^{-2}[h_{id} - \lambda(h_{id})].$$

Here,

$$[D^{jk}]_{\ell m} = \delta_{j\ell}\delta_{km} + \delta_{jm}\delta_{k\ell} - \delta_{jk}.$$
$$K_{i,\ell m} = |t_{i\ell} - t_{im}|e^{\phi|t_{i\ell}-t_{im}|^u},$$

with obvious subscript and superscript use. The exponent $u = 1$ in the exponential case and $u = 2$ in the Gaussian case.

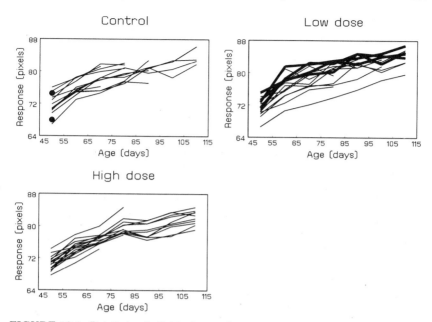

FIGURE 19.1. *Rat Data. Individual growth curves for the three treatment groups separately. Influential subjects are highlighted by bold lines or dots.*

19.4 Analysis of the Rat Data

In order to illustrate the above methodology, we will apply the local influence approach to data from a randomized experiment, designed to study the effect of the inhibition of the testosterone production in rats. The data were introduced in Section 2.1. The profiles were explored in Section 4.3.3. A linear mixed model with random intercepts was fitted in Section 6.3.3 (equation (6.12)).

The individual profiles are shown in Figure 19.1. They can be linearized by using the logarithmic transformation $t = \ln(1 + (\text{age} - 45)/10)$ for the time scale. This is also the scale we will use from now on in all statistical analyses. Note that the transformation was chosen such that $t = 0$ corresponds to the start of the treatment. We assume a linear mixed model for the response with common average intercept β_0 for all three groups, with average slopes β_1, β_2, and β_3 for the three treatment groups, respectively, and assuming compound symmetry covariance structure, with common variance $\sigma^2 + \nu^2$ and common covariance ν^2. These models are estimated under MCAR, MAR, and MNAR processes, using the PCMID function in the Splus suite of functions called OSWALD (Smith, Robertson, and Diggle 1996). The estimates are displayed in Table 19.1 (original data). Following these models, and if we are prepared to believe the assumptions on which

TABLE 19.1. *Rat Data. Maximum likelihood estimates (standard errors) of completely random, random and nonrandom dropout models, fitted to the rat data set, with and without modification.*

		Original Data		
Effect	Parameter	MCAR	MAR	MNAR
Measurement model:				
Intercept	β_0	68.61	68.61	68.61
Slope control	β_1	7.51	7.51	7.50
Slope low dose	β_2	6.87	6.87	6.86
Slope high dose	β_3	7.31	7.31	7.30
Random intercept	ν^2	3.44	3.44	3.44
Measurement error	σ^2	1.43	1.43	1.43
Dropout model:				
Intercept	ψ_0	−1.98	−8.48	−8.05
Prev. measurement	ψ_1		0.084	0.096
Curr. measurement	$\omega = \psi_2$			−0.017
−2 log-likelihood		1777.3	1774.5	1774.5
		Modified Data		
Effect	Parameter	MCAR	MAR	MNAR
Measurement model:				
Intercept	β_0	70.20	70.20	70.26
Slope control	β_1	7.52	7.52	7.39
Slope low dose	β_2	6.97	6.97	6.88
Slope high dose	β_3	7.21	7.21	6.98
Random intercept	ν^2	40.38	40.38	40.83
Measurement error	σ^2	1.42	1.42	1.46
Dropout model:				
Intercept	ψ_0	−2.20	−0.79	3.23
Prev. measurement	ψ_1		−0.015	0.32
Curr. measurement	$\omega = \psi_2$			−0.38
−2 log-likelihood		1906.6	1894.6	1890.2

they rest, there is little evidence of MAR and no evidence for MNAR. The estimates in Table 19.1 differ from those in Table 6.4, since the latter were obtained with the REML method.

Figure 19.2 displays overall C_i, as well as influences for subvectors θ, β, α, and ψ. In addition, the direction $h_{\max}$ corresponding to maximal local influence is given. We observe large absolute scale differences for different influence graphs. As is clear from such expressions as (19.9) and (19.11), the

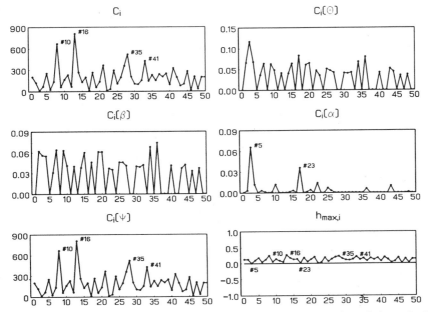

FIGURE 19.2. *Rat Data. Index plots of* C_i, $C_i(\theta)$, $C_i(\beta)$, $C_i(\alpha)$, $C_i(\psi)$, *and of the components of the direction* $h_{\max}$ *of maximal curvature.*

absolute magnitude of $C_i(\cdot)$ depends upon the scale on which the measurements and/or covariates are expressed, and hence influence graphs should be interpreted in a relative fashion.

The largest C_i are observed for rats #10, #16, #35, and #41, and virtually the same picture holds for $C_i(\psi)$. They are highlighted in Figure 19.1. All four belong to the low-dose group. Arguably, their relatively large influence is caused by an interplay of three facts. First, the profiles are relatively high, and hence y_{ij} and h_{ij} in (19.11) are large. Second, since all four profiles are complete, the first factor in (19.11) contains a maximal number of large terms. Third, the computed v_{ij} are relatively large, which is implied by the MAR dropout model parameter estimates in Table 19.1. Indeed, for these measurements, the logit of the dropout probability is closest to 0 and hence v_{ij} is fairly close to its maximal value of 0.25.

Turning attention to $C_i(\alpha)$ reveals peaks for rats #5 and #23. Both belong to the control group and drop out after a single measurement occasion. They are highlighted (by means of a bullet) in the first panel of Figure 19.1. To explain this, observe that the relative magnitude of $C_i(\alpha)$, approximately given by (19.10), is determined by $1 - g(h_{id})$ and $h_{id} - \lambda(h_{id})$. The first term is large when the probability of dropout is small. Now, when dropout occurs early in the sequence, the measurements are still relatively low, implying that the dropout probability is rather small (cf.

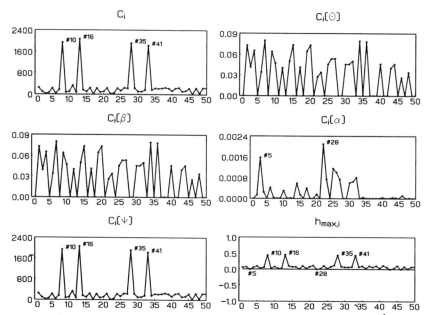

FIGURE 19.3. *Rat Data. Index plots of* C_i, $C_i(\theta)$, $C_i(\beta)$, $C_i(\alpha)$, $C_i(\psi)$, *and of the components of the direction* h_{max} *of maximal curvature where four profiles have been shifted upward.*

Table 19.1). This feature is built into the model by writing the dropout probability in terms of the raw measurements with time-independent coefficients rather than, for example, in terms of residuals. Alternatively, the dropout model parameters could be made time dependent. Further, the residual $h_{id} - \lambda(h_{id})$ is large since these two rats are somewhat distant from their group-by-time mean.

All deviations discussed are fairly moderate. This conclusion is supported by the observation that the components of the normalized vector h_{max} do not deviate much from $1/\sqrt{N}$ *and* it is consistent with the observation that the likelihood ratio statistics for MNAR versus MAR did not reject the null hypothesis.

To further explore the properties of the influence diagnostics, we consider a second analysis where all responses for rats #10, #16, #35, and #41 have been increased by 20 units. A graphical display of the local influence measures is given in Figure 19.3. The parameter estimates for all three models are also shown in Table 19.1. The peaks in C_i and $C_i(\psi)$ observed earlier have become much clearer. Thus, the fact that the test statistics for MAR versus MCAR and for MNAR versus MAR have become significant is correctly explained by the influence analysis to have been driven by the four extreme profiles.

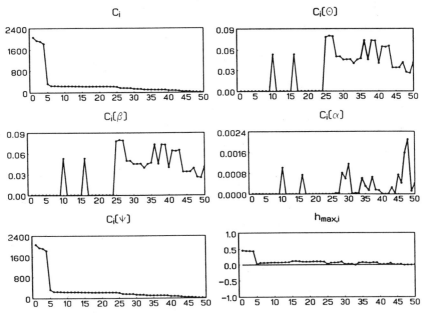

FIGURE 19.4. *Rat Data. Index plots of* C_i, $C_i(\boldsymbol{\theta})$, $C_i(\beta)$, $C_i(\alpha)$, $C_i(\psi)$, *and of the components of the direction* $h_{\max}$ *of maximal curvature, where 4 profiles have been shifted upward and the components have been ordered in decreasing order of* C_i.

Graphical representations such as Figure 19.3 are sometimes judged misleading since the apparent magnitude of a subject is influenced by its neighbors. On the other hand, it preserves the order across all six index plots. One way to overcome this problem is by ordering one plot (e.g., according to C_i) and keeping this order across all six panels. This is done in Figure 19.4. Alternatively, scatter plots of (1) the measurement versus dropout components and (2) fixed-effects versus variance component elements can be used. An example of the latter is presented in Figure 19.5. In this figure, the axes are extended slightly below zero for ease of display, even though these values are always non-negative.

The analysis of the rat data set supports the claim that the influence measures are easy to interpret. In addition to the advantages quoted earlier, we claim that a careful study of the conditions under which the diagnostics become large can shed some light on the adequacy of the model formulation. For example, the Diggle and Kenward (1994) model usually writes the logit of the dropout probability as a function of the raw measurements, with time-independent coefficients. This implies that an expression such as (19.11) depends directly on the magnitude of the responses. An alternative parameterization of the dropout probability in terms of residuals $(Y_{ij} - \mu_{ij})/\sigma_{ij}$ would obviously yield a different picture. However, this pa-

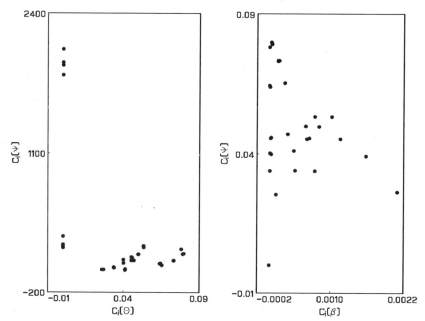

FIGURE 19.5. *Rat Data. Scatter plots of (1) $C_i(\boldsymbol{\theta})$ versus $C_i(\boldsymbol{\psi})$ and (2) $C_i(\boldsymbol{\beta})$ versus $C_i(\boldsymbol{\alpha})$, where four profiles have been shifted upward.*

rameterization has one important drawback in the sense that parameters are shared between the measurement and dropout models, thus destroying the separability (see Section 15.8). As a consequence, such a parameterization would require an entire new and much more complicated theoretical development.

19.5   Mastitis in Dairy Cattle

19.5.1   Informal Sensitivity Analysis

The data have been introduced in Section 2.7. Diggle and Kenward (1994) and Kenward (1998) performed several analyses of these data. In Diggle and Kenward (1994), a separate mean for each group defined by the year of first lactation and a common time effect was considered, together with an unstructured 2×2 covariance matrix. The dropout model included both Y_{i1} and Y_{i2} and was reparameterized in terms of the size variable $(Y_{i1} + Y_{i2})/2$ and the increment $Y_{i2} - Y_{i1}$. It turned out that the increment was important, in contrast to a relatively small contribution of the size. If this model were

TABLE 19.2. *Mastitis in Dairy Cattle. Maximum likelihood estimates (standard errors) of random and nonrandom dropout models, under several deletion schemes.*

Parameter	All	(53,54,66,69)	(4,5)	(66)	(4,5,66)
			Random dropout		
Measurement model:					
β_0	5.77(0.09)	5.69(0.09)	5.81(0.08)	5.75(0.09)	5.80(0.09)
β_d	0.72(0.11)	0.70(0.11)	0.64(0.09)	0.68(0.10)	0.60(0.08)
σ_1^2	0.87(0.12)	0.76(0.11)	0.77(0.11)	0.86(0.12)	0.76(0.11)
σ_2^2	1.30(0.20)	1.08(0.17)	1.30(0.20)	1.10(0.17)	1.09(0.17)
ρ	0.58(0.07)	0.45(0.08)	0.72(0.05)	0.57(0.07)	0.73(0.05)
Dropout model:					
ψ_0	$-2.65(1.45)$	$-3.69(1.63)$	$-2.34(1.51)$	$-2.77(1.47)$	$-2.48(1.54)$
ψ_1	0.27(0.25)	0.46(0.28)	0.22(0.25)	0.29(0.24)	0.24(0.26)
$\omega = \psi_2$	0	0	0	0	0
-2 log-likelihood	280.02	246.64	237.94	264.73	220.23
			Nonrandom dropout		
Parameter	All	(53,54,66,69)	(4,5)	(66)	(4,5,66)
Measurement model:					
β_0	5.77(0.09)	5.69(0.09)	5.81(0.08)	5.75(0.09)	5.80(0.09)
β_d	0.33(0.14)	0.35(0.14)	0.40(0.18)	0.34(0.14)	0.63(0.29)
σ_1^2	0.87(0.12)	0.76(0.11)	0.77(0.11)	0.86(0.12)	0.76(0.11)
σ_2^2	1.61(0.29)	1.29(0.25)	1.39(0.25)	1.34(0.25)	1.10(0.20)
ρ	0.48(0.09)	0.42(0.10)	0.67(0.06)	0.48(0.09)	0.73(0.05)
Dropout model:					
ψ_0	0.37(2.33)	$-0.37(2.65)$	$-0.77(2.04)$	0.45(2.35)	$-2.77(3.52)$
ψ_1	2.25(0.77)	2.11(0.76)	1.61(1.13)	2.06(0.76)	0.07(1.82)
$\omega = \psi_2$	$-2.54(0.83)$	$-2.22(0.86)$	$-1.66(1.29)$	$-2.33(0.86)$	0.20(2.09)
-2 log-likelihood	274.91	243.21	237.86	261.15	220.23
G^2 for MNAR	5.11	3.43	0.08	3.57	0.005

assumed plausible, MAR would be rejected on the basis of a likelihood ratio test statistic of $G^2 = 5.11$ on 1 degree of freedom.

Kenward (1998) carried out what we could term a data-driven sensitivity analysis. He started from the original model in Diggle and Kenward (1994), albeit with a common intercept, since there was no evidence for a dependence on first lactation year. The right-hand panel of Figure 2.6. reveals that there appear to be two cows, #4 and #5, with unusually large increments. He conjectures that this might mean that these animals were ill during the first lactation year, producing an unusually low yield, whereas a normal yield was obtained during the second year. He then fitted t-distributions to Y_{i2} given $Y_{i1} = y_{i1}$. Not surprisingly, his finding was that the heavier the tails of the t-distribution, the better the outliers were accommodated. As a result, the difference between the MAR and nonrandom models vanished ($G^2 = 1.08$ for a t_2-distribution). Alternatively, removing these two cows and refitting the normal model shows complete lack of evidence for nonrandom dropout ($G^2 = 0.08$). This latter procedure is similar to a global influence analysis by means of deleting two observations. Parameter estimates and standard errors for random and nonrandom dropout, under several deletion schemes, are reproduced in Table 19.2. It is clear that the influence on the measurement model parameters is small in the random dropout case, although the gap on the time effect β_d between the random and nonrandom dropout models is reduced when #4 and #5 are removed.

Next, these informal but insightful forms of sensitivity analysis will be presented. A sensitivity analysis based on local influence, as introduced in Section 19.3, is performed in Section 19.5.2.

Kenward's Sensitivity Analysis

A simple multivariate Gaussian linear model is used to represent the marginal milk yield in the 2 years (i.e., the yield that would be, or was, observed in the absence of mastitis):

$$\begin{pmatrix} Y_1 \\ Y_2 \end{pmatrix} = N\left(\begin{pmatrix} \mu \\ \mu + \Delta \end{pmatrix}, \begin{pmatrix} \sigma_1^2 & \rho\sigma_1\sigma_2 \\ \rho\sigma_1\sigma_2 & \sigma_2^2 \end{pmatrix} \right).$$

Note that the parameter Δ represents the change in average yield between the 2 years. The probability of mastitis is assumed to follow the logistic regression model:

$$P(\text{dropout}) = \frac{e^{\psi_0 + \psi_1 y_1 + \psi_2 y_2}}{1 + e^{\psi_0 + \psi_1 y_1 + \psi_2 y_2}}. \tag{19.14}$$

The combined response/dropout model was fitted to the milk yields by maximum likelihood using a generic function maximization routine. In addition, the MAR model ($\psi_2 = 0$) was fitted. This latter is equivalent to fitting separately the Gaussian linear model for the milk yields and logistic regression model for the occurrence of mastitis. These fits produced the parameter estimates as displayed in the "all" column of Table 19.2, standard errors and minimized value of twice the negative log-likelihood.

Using the likelihoods to compare the fit of the two models, we get a difference $G^2 = 5.11$. The corresponding tail probability from the χ_1^2 is 0.02.

This test essentially examines the contribution of ψ_2 to the fit of the model. Using the Wald statistic for the same purpose gives a statistic of $(-2.53)^2/0.83 = 9.35$, with corresponding χ_1^2 probability of 0.002. The discrepancy between the results of the two tests suggests that the asymptotic approximations on which these are based are not very accurate in this setting and the standard error probably underestimates the true variability of the estimate of ψ_2. Nevertheless, there is a suggestion from the change in likelihood that ψ_2 is making a real contribution to the fit of the model. The dropout model estimated from the MNAR setting is as follows:

$$\text{logit}[P(\text{mastitis})] \;=\; 0.37 + 2.25 y_1 - 2.54 y_2. \tag{19.15}$$

Some insight into this fitted model can be obtained by rewriting it in terms of the milk yield totals $(Y_1 + Y_2)$ and increments $(Y_2 - Y_1)$:

$$\text{logit}[P(\text{mastitis})] \;=\; 0.37 - 0.145(y_1 + y_2) - 2.395(y_2 - y_1) \tag{19.16}$$

The probability of mastitis increases with larger negative increments; that is, those animals who showed (or would have shown) a greater decrease in yield over the 2 years have a higher probability of getting mastitis. The other differences in parameter estimates between the two models are consistent with this: The MNAR dropout model predicts a smaller average increment in yield (Δ), with larger second year variance and smaller correlation caused by greater negative imputed differences between yields.

To gain some additional insight into these two fitted models, we now take a closer look at the raw data and the predictive behavior of the Gaussian MNAR model. Under an MNAR model, the predicted, or imputed, value of a missing observation is given by the ratio of expectations:

$$\hat{\boldsymbol{y}}_m \;=\; \frac{E_{Y_m|Y_o}[\boldsymbol{y}_m P(\boldsymbol{r} \mid \boldsymbol{y}_o, \boldsymbol{y}_m)]}{E_{Y_m|Y_o}[P(\boldsymbol{r} \mid \boldsymbol{y}_o, \boldsymbol{y}_m)]}. \tag{19.17}$$

Recall that the fitted dropout model (19.15) implies that the probability of mastitis increases with decreasing values of the increment $Y_2 - Y_1$. We therefore plot the 27 imputed values of this quantity together with the 80 observed increments against the first year yield Y_1. This is presented in Figure 19.6, in which the imputed values are indicated with triangles and the observed values with crosses. Note how the imputed values are almost linear in Y_1: This is a well-known property of the ratio (19.17) within this range of observations. The imputed values are all negative, in contrast to the observed increments, which are nearly all positive. With animals of this age, one would normally expect an increase in yield between the 2 years. The dropout model is imposing very atypical behavior on these animals and this corresponds to the statistical significance of the MNAR component of the model (ψ_2) but, of course, necessitates further scrutiny.

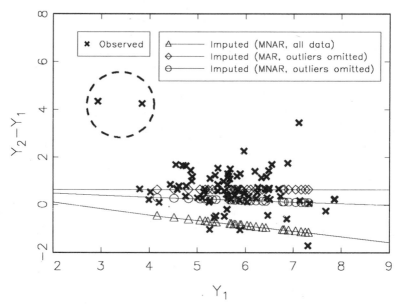

FIGURE 19.6. *Mastitis in Dairy Cattle. Plot of observed and imputed year 2 −
year 1 yield differences against year 1 yield. Two outlying points are circled.*

Another feature of this plot is the pair of outlying observed points circled in
the top left-hand corner. These two animals have the lowest and third lowest
yields in the first year, but moderately large yields in the second, leading
to the largest positive increments. In a well-husbanded dairy herd, one
would expect approximately Gaussian joint milk yields, and these two then
represent outliers. It is likely that there is some anomaly, possibly illness,
leading to their relatively low yields in the first year. One can conjecture
that these two animals are the cause of the structure identified by the
Gaussian MNAR model. Under the joint Gaussian assumption, the MNAR
model essentially "fills in" the missing data to produce a complete Gaussian
distribution. To counterbalance the effect of these two extreme positive
increments, the dropout model predicts negative increments for the mastitic
cows, leading to the results observed. As a check on this conjecture, we
omit these two animals from the data set and refit the MAR and MNAR
Gaussian models. The resulting estimates are presented in the $(4, 5)$ column
of Table 19.2.

The deviance is minimal and the MNAR model now shows no improvement
in fit over MAR. The estimates of the dropout parameters, although still
moderately large in an absolute sense, are of the same size as their standard errors which, as mentioned earlier, are probably underestimates. In
the absence of the two anomalous animals, the structure identified earlier
in terms of the MNAR dropout model no longer exists. The increments

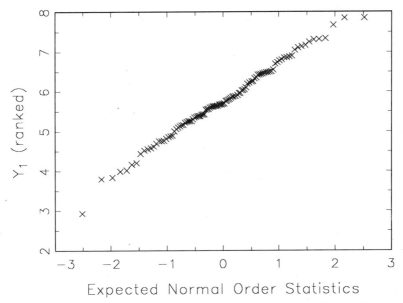

FIGURE 19.7. *Mastitis in Dairy Cattle. Normal probability plot of the year 1 milk yields.*

imputed by the fitted model are also plotted in Figure 19.6, indicated by circles. Although still lying among the lower region of the observed increments, these are now all positive and lie close to the increments imputed by the MAR model (diamonds). Thus, we have a plausible representation of the data in terms of joint Gaussian milk yields, two pairs of outlying yields and no requirement for an MNAR dropout process.

The two key assumptions underlying the outcome-based MNAR model are, first, the form chosen for the relationship between dropout probability and response and, second, the distribution of the response or, more precisely, the conditional distribution of the possibly unobserved response given the observed response. In the current setting for the first assumption, if there is dependence of mastitis occurrence on yield, experience with logistic regression tells us that the exact form of the link function in this relationship is unlikely to be critical. In terms of sensitivity, we therefore consider the second assumption, the distribution of the response.

All the data from the first year are available, and a normal probability plot of these, Figure 19.7, does not show great departures from the Gaussian assumption. Leaving this distribution unchanged, we therefore examine the effect of changing the conditional distribution of Y_2 given Y_1. One simple and obvious choice is to consider a heavy-tailed distribution, and for this, we use the translated and scaled t_m-distribution with density:

$$f(y_2 \mid y_1) = \{\sigma\sqrt{m}B(1/2, m/2)\}^{-1}\left\{1 + \frac{1}{m}\left(\frac{y_2 - \mu_{2|1}}{\sigma}\right)^2\right\}^{-(m+1)/2},$$

where

$$\mu_{2|1} = \mu + \Delta + \frac{\rho\sigma_2(y_1 - \mu)}{\sigma_1}$$

is the conditional mean of $Y_2 \mid y_1$. The corresponding conditional variance is

$$\frac{m}{m-2}\sigma^2.$$

Relevant parameter estimates from the fits of both MAR and MNAR models are presented in Table 19.3 for three values of m: 2, 10, and 25. Smaller values of m correspond to greater kurtosis and, as m becomes large, the model approaches the Gaussian one used in the previous section. It can be seen from the results for the MNAR model in Table 19.3, that as the kurtosis increases the estimate of ψ_2 decreases. Also, the maximized likelihoods of the MAR and MNAR models converge. With 10 and 2 degrees of freedom, there is no evidence at all to support the inclusion of ψ_2 in the model; that is, the MAR model provides as good a description of the observed data as the MNAR, in contrast to the Gaussian-based conclusions. Further, as m decreases, the estimated yearly increment in milk yield Δ from the MNAR model increases to the value estimated under the MAR model. In most applications of outcome-based selection models (see Section 19.7), it will be quantities of this type that will be of prime interest, and it is clearly seen in this example how the dropout model can have a crucial influence on the estimate of this. Comparing the values of the deviance from the t-based model with those from the original Gaussian model, we also see that the former with $m = 10$ or 2 produces a slightly better fit, although no meaning can be attached to the statistical significance of the difference in these likelihood values.

The results observed here are consistent with those from the deletion analysis. The two outlying pairs of measurements identified earlier are not inconsistent with the heavy-tailed t-distribution; so it would require no "filling in" and hence no evidence for nonrandomness in the dropout process under the second model. In conclusion, if we consider the data with outliers included, we have two models that effectively fit equally well to the observed data. The first assumes a joint Gaussian distribution for the responses and a MNAR dropout model. The second assumes a Gaussian distribution for the first observation and a conditional t_m-distribution (with small m) for the second given the first, with no requirement for a MNAR dropout component. Each provides a different explanation for what has been observed,

TABLE 19.3. *Mastitis in Dairy Cattle. Details of the fit of MAR and MNAR dropout models, assuming a t_m-distribution for the conditional distribution of Y_2 given Y_1. Maximum likelihood estimates (standard errors) are shown.*

t DF	Par.	MAR	MNAR
25	Δ	0.69(0.10)	0.35(0.13)
	ψ_1	0.27(0.24)	2.11(0.78)
	ψ_2		$-2.33(0.88)$
-2 log-likelihood		275.54	271.77
10	Δ	0.67(0.09)	0.38(0.14)
	ψ_1	0.27(0.24)	1.84(0.82)
	ψ_2		$-1.96(0.95)$
-2 log-likelihood		271.22	269.12
2	Δ	0.61(0.08)	0.54(0.11)
	ψ_1	0.27(0.24)	0.80(0.66)
	ψ_2		$-0.65(0.73)$
-2 log-likelihood		267.87	266.79

with quite a different biological interpretation. In likelihood terms, the second model fits a little better than the first, but a key feature of such dropout models is that the distinction between them should not be based on the observed data likelihood alone. It is always possible to specify models with identical maximized observed data likelihoods that differ with respect to the unobserved data and dropout mechanism and such models can have very different implications for the underlying mechanism generating the data. Finally, the most plausible explanation for the observed data is that the pairs of milk yields have joint Gaussian distributions, with no need for an MNAR dropout component, and that two animals are associated with anomalous pairs of yields.

19.5.2 Local Influence Approach

In the previous section, the sensitivity to distributional assumptions of conclusions concerning the randomness of the dropout process has been established in the context of the mastitis data. Such sensitivity has led some to conclude that such modeling should be avoided. We argue that this conclusion is too strong. First, repeated measures tend to be incomplete and therefore the consideration of the dropout process is simply unavoidable.

Second, if a nonrandom dropout component is added to a model and the maximized likelihood changes appreciably, then some real structure in the data has been identified that is not encompassed by the original model. The MNAR analysis may tell us about inadequacies of the original model rather than the adequacy of the MNAR model. It is the interpretation of the identified structure that cannot be made unequivocally from the data under analysis. The mastitis data clearly illustrated that, using external information on the distribution of the response, a plausible explanation of the structure so identified *might* be made in terms of the outlying responses from two animals.

However, it should also be noted that absence of structure in the data associated with an MNAR process does not imply that an MNAR process is not operating: Different models with similar maximized likelihoods (i.e., with similar plausibility with respect to the observed data), may have completely different implications for the dropout process and the unobserved data. These points together suggest that the appropriate role of such modeling is as a component of a sensitivity analysis.

The analysis of the previous section is characterized by its basis within substantive knowledge about the data. The local influence approach, presented in Section 19.3 and applied to the rats data in Section 19.4, may appear to be "blindly" applicable (i.e., without departing from specific information about the data). In this section, we will apply the technique to the mastitis data, confront the results with those found in Section 19.5.1, and suggest that here a combination of methodology and substantive insight will be the most fruitful approach.

Applying the method to the mastitis data produces Figure 19.8, which suggests that there are four influential subjects: #53, #54, #66, and #69. The most striking feature of this analysis is that #4 and #5 are *not* recovered. See also Figure 2.6. It is interesting to consider an analysis with these four cows removed. Details are given in Table 19.2. Unlike removing #4 and #5, the influence on the likelihood ratio test is rather small: $G^2 = 3.43$ instead of the original 5.11. The influence on the measurement model parameters under both random and nonrandom dropout is small.

It is very important to realize that one should not expect agreement between deletion and our local influence analysis. The latter focuses on the sensitivity of the results with respect to the assumed dropout model; more specifically, how the results change when the MAR model is extended into the direction of nonrandom dropout. In particular, all subjects singled out so far are complete and hence $C_i(\boldsymbol{\theta}) \equiv 0$, placing all influence on $C_i(\boldsymbol{\psi})$ and $\boldsymbol{h}_{\max,i}$. A comparison between local influence and deletion is given in Section 24.4.

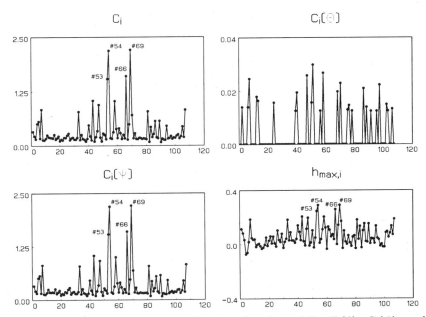

FIGURE 19.8. *Mastitis in Dairy Cattle. Index plots of C_i, $C_i(\boldsymbol{\theta})$, $C_i(\boldsymbol{\psi})$, and of the components of the direction $\mathbf{h}_{\max}$ of maximal curvature, when the dropout model is parameterized in function of Y_{i1} and Y_{i2}.*

More insight can also be obtained by studying (19.13). The contribution for subject i is made up of three factors. The first factor, V_i, is small for extreme dropout probabilities. The subjects with a very high probability to either remain in the study or disappear will be less influential. Cows #4 and #5 have dropout probabilities equal to 0.13 and 0.17, respectively. The 107 cows in the study span the dropout probability interval $[0.13, 0.37]$. Thus, this component rather deflates the influence of subjects #4 and #5. Second, (19.13) contains a leverage factor in curly braces. Third, a subject is relatively more influential when both milk yields are high. We now need to question whether this is plausible or relevant. Since both measurements are positively correlated, measurements with both milk yields high or low will not be unusual. In Section 19.5.1, we observed that cows #4 and #5 are unusual on the basis of their *increment*. This is in line with several other applications of similar dropout models (Diggle and Kenward 1994, Molenberghs, Kenward, and Lesaffre 1997) where it was found that a strong incremental component pointed to genuine nonrandomness. In contrast, the size variable can often be replaced by just the history, and hence the corresponding model is very close to random dropout.

Even though a dropout model in the outcomes themselves, termed direct variables model, is equivalent to a model in the first variable Y_{i1} and the increment $Y_{i2} - Y_{i1}$, termed incremental variables representation, we will

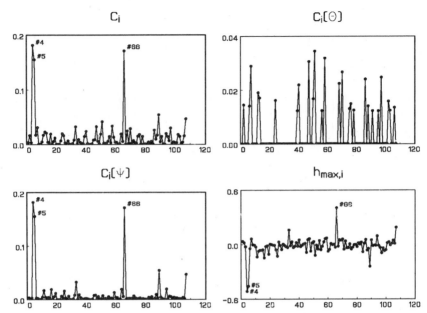

FIGURE 19.9. *Mastitis in Dairy Cattle. Index plots of C_i, $C_i(\boldsymbol{\theta})$, $C_i(\boldsymbol{\psi})$, and of the components of the direction $\boldsymbol{h}_{\max}$ of maximal curvature, when the dropout model is parameterized in function of Y_{i1} and $Y_{i2} - Y_{i1}$.*

show that they lead to different perturbation schemes of the form (19.1). At first, this feature can be seen as both an advantage and a disadvantage. The fact that reparameterizations of the linear predictor of the dropout model leads to different perturbation schemes requires careful reflection based on substantive knowledge in order to guide the analysis, such as the considerations on the incremental variable made earlier.

We will present the results of the incremental analysis and then offer further comments on the rationale behind this particular transformation. From the diagnostic plots in Figure 19.9, it is obvious that we recover three influential subjects: #4, #5, and #66. Although Kenward (1998) did not consider #66 to be influential, it does appear to be somewhat distant from the bulk of the data (Figure 2.6). The main difference between both types is that the first two were likely sick during year 1, and this is not necessarily so for #66. An additional feature is that in all cases, both $C_i(\boldsymbol{\psi})$ and $\boldsymbol{h}_{\max}$ show the same influential animals. In addition, $\boldsymbol{h}_{\max}$ suggests that the influence for #66 is different than for the others. It could be conjectured that the latter one pulls the coefficient ω in a different direction than the other two. The other values are all relatively small. This could indicate that for the remaining 104 subjects, MAR is plausible, whereas a deviation in the direction of the incremental variable, *with differing signs*, appears to be necessary for the other three subjects. At this point, a comparison between

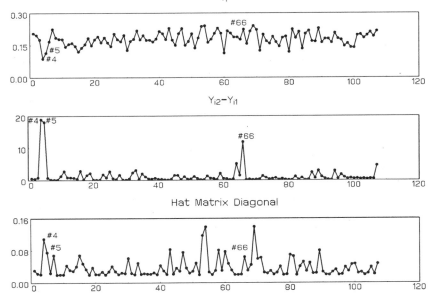

FIGURE 19.10. *Mastitis in Dairy Cattle. Index plots of the three components of* $C_i(\psi)$ *when the dropout model is parameterized in function of* Y_{i1} *and* $Y_{i2} - Y_{i1}$.

$h_{\max}$ for the direct variable and incremental analyses is useful. Since the contributions h_i sum to 1, these two plots are directly comparable. There is no pronounced influence indication in the direct variables case and perhaps only random noise is seen. A more formal way to distinguish between signal and noise needs to be developed.

In Figure 19.10, we have decomposed (19.13) in its three components: the variance of the dropout probability V_i, the incremental variable $Y_{i2} - Y_{i1}$, which is replaced by its predicted value for a dropout, and the hat-matrix diagonal. In agreement with the preceding discussion, the influence clearly stems from an unusually large increment, which survives the fact that V_i actually downplays the influence because Y_{41} and Y_{51} are comparatively small and dropout increases with the milk yield in the first year. Further, the sign difference of $h_{\max,4}$ and $h_{\max,5}$ versus $h_{\max,66}$ can be interpreted better.

We noted already that cows #4 and #5 have relatively small dropout probabilities. In contrast, the dropout probability of #66 is large within the observed range $[0.13; 0.37]$. Since for those subjects the increment is large, changing its perturbation ω_i can have a large impact on the other dropout parameters ψ_0 and ψ_1. In order to avoid that the effects of the change for #4 and #5 will cancel with the effect for #66, the corresponding signs need to be opposite. Such a change implies either that all three dropout probabilities move toward the center of the range or are pulled away from it. (Note that $-\boldsymbol{h}_{\max}$ is another normalized eigenvector corresponding to the largest eigenvalue.)

In the informal approach, extra analyses where considered with #4 and #5 removed. The resulting likelihood ratio statistic reduces to $G^2 = 0.08$. When only #66 is removed, the likelihood ratio for nonrandom dropout is $G^2 = 3.57$, very similar to the one when #53, #54, #66, and #69 were removed. Removing all three (#4, #5, and #66) results in $G^2 = 0.005$ (i.e., complete disappearance of all evidence for nonrandom dropout). Details are given in Table 19.2.

We now provide insight into why the transformation of direct outcomes to increments is useful. We noted already that the associated perturbation schemes (19.1) are different. An important device in this respect is the equality

$$\psi_0 + \psi_1 y_{i1} + \psi_2 y_{i2} \;\; = \;\; \psi_0 + (\psi_1 + \psi_2) y_{i1} + \psi_2 (y_{i2} - y_{i1}). \tag{19.18}$$

Equation (19.18) shows that the direct variables model checks the influence on the random dropout parameter ψ_1, whereas the random dropout parameter in the incremental model is $\psi_1 + \psi_2$. Not only is this a different parameter, it is also estimated with higher precision. One often observes that $\hat{\psi}_1$ and $\hat{\psi}_2$ exhibit a similar variance and negative correlation, in which case, the linear combination with smallest variance is approximately in the direction of the sum $\psi_1 + \psi_2$. When the correlation is negative, the difference direction $\psi_1 - \psi_2$ is obtained instead. Let us assess this in case all 107 observations are included. The estimated covariance matrix is

$$\begin{pmatrix} 0.59 & -0.54 \\ & 0.70 \end{pmatrix},$$

with correlation -0.84. The variance of $\widehat{\psi}_1 + \widehat{\psi}_2$, on the other hand, is estimated to be 0.21. In this case, the direction of minimal variance is along $(0.74; 0.67)$, which is indeed close to the sum direction. When all three influential subjects are removed, the estimated covariance matrix becomes

$$\begin{pmatrix} 3.31 & -3.77 \\ & 4.37 \end{pmatrix},$$

with correlation -0.9897. Removing only #4 and #5 yields an intermediate situation of which the results are not shown. The variance of the sum is 0.15, which is a further reduction and still close to the direction of minimal variance. These considerations reinforce the claim that an incremental analysis is highly recommended. It might therefore be interesting to routinely construct a plot such as in Figure 2.6 or Figure 19.6, even with longer measurement sequences. On the other hand, transforming the dropout model to a size variable $(Y_{i1} + Y_{i2})/2$ will worsen the problem since an insensitive parameter for Y_{i1} will result.

Finally, observe that a transformation of the dropout model to a size and incremental variable at the same time for the model with all three influential subjects removed gives a variance of the size and increment variables of 0.15 and 15.22, respectively. In other words, there is no evidence for an incremental effect, confirming that random dropout is plausible.

Although local and global influence are, strictly speaking, not equivalent, it is insightful to see how the global influence on $\boldsymbol{\theta}$ can be linked to the behavior of $C_i(\boldsymbol{\psi})$. We observed earlier that all locally influential subjects are completers and hence $C_i(\boldsymbol{\theta}) \equiv 0$. Yet, removing #4, #5, and #66 shows some effect on the discrepancy between the random dropout (MAR) and nonrandom dropout (MNAR) estimates of the time effect β_d. In particular, MAR and MNAR estimates with all three subjects removed are virtually identical (0.60 and 0.63, respectively). Let us do a small thought experiment. Since these subjects are influential in $C_i(\boldsymbol{\psi})$, the MAR model could be improved by including incremental terms for these three subjects. Such a model would still imply random dropout. In contrast, allowing a dependence on the increment in *all* subjects will influence $E(Y_{i2}|y_{i1}, \text{dropout})$ for all incomplete observations; hence, the measurement model parameters under MNAR will change. In conclusion, this provides a way to assess the *indirect* influence of the dropout mechanism on the measurement model parameters through local influence methods. In the milk data set, this influence is likely due to the fact that an exceptional increment which is caused by a different mechanism, perhaps a diseased animal during the first year, is nevertheless treated on equal footing with the other observations within the dropout model. Such an investigation cannot be done with the case-deletion method because it is not possible to disentangle the various sources of influence.

In conclusion, it is found that an incremental variable representation of the dropout mechanism is beneficial over a direct variable representation. Contrasting our local influence approach with a case-deletion scheme as applied in Kenward (1998), it is argued that the former approach is advantageous since it allows one to assess direct and indirect influences on the dropout and measurement model parameters, stemming from perturbing the random dropout model in the direction of nonrandom dropout. In contrast, a case-deletion scheme does not allow one to disentangle the various sources of influence.

19.6   Alternative Local Influence Approaches

The perturbation scheme used throughout this chapter has several elegant properties. The perturbation is around the MAR mechanism, which is often deemed a sensible starting point. Extra calculations are limited and free of numerical integration. Influence decomposes in a measurement and dropout part, the first of which is zero in the case of a complete observation. Finally, if the special case of compound symmetry is assumed, the measurement part can approximately be written in interpretable components for the fixed effect and variance component parts.

However, other schemes are worthwhile considering as well. Most of the developments presented here can be adapted to such alternatives, although not all schemes will preserve the remarkable computational convenience. Also, interpretation of the influence expressions in an alternative scheme will require additional work.

Apart from MAR, often MCAR also is considered a useful model. It is then natural to consider departures from the MCAR model, rather than from the MAR model. This would change (19.1) to

$$
\begin{aligned}
\text{logit}(g(\boldsymbol{h}_{ij}, y_{ij})) &= \text{logit}\left[\text{pr}(D_i = j | D_i \geq j, \boldsymbol{y}_i)\right] \\
&= \boldsymbol{h}_{ij}\boldsymbol{\psi} + \omega_{i1}y_{i,j-1} + \omega_{i2}y_{ij}, \qquad (19.19)
\end{aligned}
$$

with obvious change in the definition of $\boldsymbol{h}_{ij}$. This way, the perturbation parameter becomes a two-component vector $\boldsymbol{\omega}_i = (\omega_{i1}, \omega_{i2})$. As a result, the ith subject produces a pair (h_{i1}, h_{i2}), which is a normalized vector and hence main interest lies in its direction. Also, $C_{\boldsymbol{h}} = C_i$ is the local influence on $\widehat{\boldsymbol{\gamma}}$ of allowing the ith subject to drop out randomly or nonrandomly. Figure 19.11 shows the result of this procedure, applied to the mastitis data. Pairs (h_{i1}, h_{i2}) are plotted. The main diagonal corresponds to the size direction, whereas the diagonal represents the purely incremental direction. The circles are used to indicate the minimal and maximal distances to the origin. Finally, squares rather than bullets are used for cows #4, #5, and #66. Most cows lie in the *size direction*, but it is noticeable that #4, #5, and #66 tend toward the nonrandom direction. Further, no extremely large C_i are seen in this case.

Another extension would result from the observation that the choice of the incremental analysis in Section 19.5.2 may, although motivated by substantive insight, seem rather arbitrary. Hence, it would be desirable to have a more automatic, data-driven selection of a direction. One way of doing this is by considering

$$
\begin{aligned}
\text{logit}(g(\boldsymbol{h}_{ij}, y_{ij})) &= \text{logit}\left[\text{pr}(D_i = j | D_i \geq j, \boldsymbol{y}_i)\right] \\
&= \boldsymbol{h}_{ij}\boldsymbol{\psi} + \omega_i(\sin\theta\, y_{i,j-1} + \cos\theta\, y_{ij}). \qquad (19.20)
\end{aligned}
$$

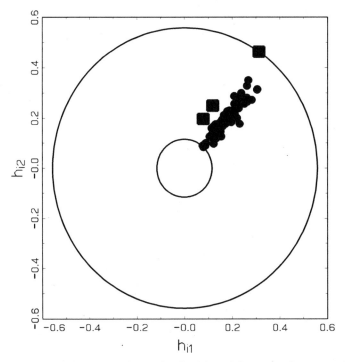

FIGURE 19.11. *Mastitis in Dairy Cattle. Plot of C_i in the direction of h_i.*

Now, it is possible to apply (19.20) for a selected number of angles θ, to range through a fine grid covering the entire circle, or to consider θ as another influence parameter. In the latter case, θ becomes subject-specific and the pair (ω_i, θ_i) is essentially a reparameterization of the pair $\omega_i = (\omega_{i1}, \omega_{i2})$ in (19.20).

A completely different local influence approach would modify the general form (15.5) as follows:

$$
\begin{aligned}
& f(\boldsymbol{y}_i^o, \boldsymbol{r}_i | \boldsymbol{\theta}, \boldsymbol{\psi}, \omega_i) \\
& = \int f(\boldsymbol{y}_i^o, \boldsymbol{y}_i^m | X_i, Z_i, \boldsymbol{\theta}) f(\boldsymbol{r}_i | \boldsymbol{y}_i^o, \boldsymbol{y}_i^m, X_i, \boldsymbol{\psi})^{\omega_i} d\boldsymbol{y}_i^m.
\end{aligned}
\tag{19.21}
$$

Now, if $\omega_i = 0$, then the missing data process is considered ignorable and only the measurement process is considered. If $\omega_i = 1$, the posited, potentially nonrandom, model is considered. Other values of ω_i correspond to partial case weighting.

19.7   Random-coefficient-based Models

It has been seen in the mastitis data, and observed elsewhere, such as in Diggle and Kenward (1994) and Molenberghs, Kenward, and Lesaffre (1997), that apparent nonrandom dropout in a selection model often manifests itself in terms of a dependence of dropout on the increment or change in response, $C = Y_{ij} - Y_{i,j-1}$, say. If a subject is exhibiting a clear trend in response, then we might regard C as an estimate, with error, of an underlying trend. This is supported particularly in examples in which a MNAR model with dependence on C fits a little better than an MAR model with dependence on a change calculated wholly from past values. A better way of approaching such a situation may be to attempt to model the underlying trend directly using a latent, or random-coefficient, model and allow both response and dropout model to depend on this. Little (1995) uses the term *random-coefficient based* to distinguish such models from those used in this chapter in which the probability of dropout depends directly on the response $\boldsymbol{Y}_i$, which he terms *outcome based*. We will now discuss these types of model in some detail.

Suppose that the latent variable $\boldsymbol{b}_i$ (possibly vector valued) describes some aspect of an individual's response. This leads naturally to the linear mixed model (3.8). Let us consider a model with random intercept and random slope:

$$Y_{ij} \mid b_{0i}, b_{1i} \sim N(\beta_0 + b_{0i} + (\beta_1 + b_{1i})t_i; \sigma^2), \qquad j = 1, \ldots, n_i. \quad (19.22)$$

The subject's random coefficients $\boldsymbol{b}_i = (b_{0i}, b_{1i})'$ are assumed to be normally distributed with zero mean and covariance matrix D. Also, the dropout model is assumed to depend on the random effects: $P(\boldsymbol{r} \mid \boldsymbol{b})$.

In general, such a selection model can be written, conditional on $\boldsymbol{b}$,

$$f(\boldsymbol{y}, \boldsymbol{r} \mid \boldsymbol{b}) \quad = \quad f(\boldsymbol{y} \mid \boldsymbol{b}) P(\boldsymbol{r} \mid \boldsymbol{b}). \quad (19.23)$$

Integration is still required to obtain the marginal distribution of $(\boldsymbol{y}_o, \boldsymbol{r})$, now with respect to both $\boldsymbol{y}_m$ and $\boldsymbol{b}$:

$$f(\boldsymbol{y}_o, \boldsymbol{r}) \quad = \quad \int \int f(\boldsymbol{y} \mid \boldsymbol{b}) P(\boldsymbol{r} \mid \boldsymbol{b}) f(\boldsymbol{b}) d\boldsymbol{b} d\boldsymbol{y}_m,$$

but the dependence of dropout on $\boldsymbol{y}_m$ through $\boldsymbol{b}$ allows some simplification, given appropriate regularity conditions:

$$f(\boldsymbol{y}_o, \boldsymbol{r}) \quad = \quad \int \left\{ \int f(\boldsymbol{y} \mid \boldsymbol{b}) d\boldsymbol{y}_m \right\} P(\boldsymbol{r} \mid \boldsymbol{b}) f(\boldsymbol{b}) d\boldsymbol{b}$$

$$= \quad \int f(\boldsymbol{y}_o \mid \boldsymbol{b}) P(\boldsymbol{r} \mid \boldsymbol{b}) f(\boldsymbol{b}) d\boldsymbol{b}. \quad (19.24)$$

The response y now enters the joint distribution only through the observed data and the latent variable, and the nonrandom dropout model will typically be identified.

Wu and Carroll (1988) proposed such a model for what they termed informative right censoring. The situation they cover extends the earlier setting to accommodate right censoring of the dropout times. Although this complicates the likelihood to some degree, it does not fundamentally change the structure of the model and could equally well be used for outcome-based models. For a continuous response, Wu and Carroll suggested using a conventional Gaussian random-coefficient model (Laird and Ware 1982, Longford 1993) combined with an appropriate model for time to dropout, such as proportional hazards, logistic or probit regression. The combination of probit and Gaussian response allows explicit solution of the integral and was used in their application.

In a slightly different approach to modeling dropout time as a continuous variable in the latent variable setting, Schluchter (1992) and DeGruttola and Tu (1994) proposed joint multivariate Gaussian distributions for the latent variable(s) of the response process and a variable representing time to dropout. The correlation between these variables induces dependence between dropout and response. To permit more realistic distributions for dropout time, Schluchter proposed that dropout time itself should be some monotone transformation of the corresponding Gaussian variable. The use of a joint Gaussian representation does simplify computational problems associated with the likelihood. There are clear links here with the Tobit model and this is made explicit by Cowles, Carlin, and Connett (1996), who use a number of correlated latent variables to represent various aspects of an individual's behavior, such as compliance and attendance at scheduled visits.

These random-coefficient-based models do have the advantage over the earlier outcome-based models in providing a simpler framework for non-dropout patterns of missing values and allowing very general patterns of observation time among individuals. There are many ways in which such models can be extended and generalized. Follman and Wu (1995) introduce the idea of shared-parameter models in which generalized linear models are defined for both response and dropout that share latent variable(s). With some exceptions, the random-coefficient-based models share the main drawbacks of the outcome-based models: (1) Computational algorithms can be complex and problem specific, although the EM has been applied to some problems (Schluchter 1992, DeGruttola and Tu 1994, Molenberghs, Kenward, and Lesaffre 1997), and (2) inferences are necessarily highly dependent on parametric modeling assumptions that cannot be assessed from the data under analysis. In answer to these concerns, several authors have considered to replace parametric assumptions, at least partially, by a non-parametric approach (Robins, Rotnitzky, and Zhao 1995, Robins and Rotnitzky 1995, Rotnitzky and Robins 1995, 1997, Robins 1997, Robins and Gill 1997, Robins, Rotnitzky, and Scharfstein 1998).

19.8   Concluding Remarks

Since all models for incomplete longitudinal data rest on unverifiable assumptions, this chapter argues that a careful investigation of the model output is in place. Using two examples, both informal and formal sensitivity analyses have been conducted. The former are based on insight into the modeling process and the distributional assumptions made, as well as on background knowledge of the data problem.

Needless to say that a variety of different approaches to formal sensitivity analysis are potentially possible. We focused on global and local influence measures, being but one way of assessing sensitivity.

The sensitivity problem at large is receiving a lot of attention and we believe that a number of methodologies will emerge over the coming years.

20

Sensitivity Analysis for Pattern-Mixture Models

20.1 Introduction

Chapter 18 is devoted to the study of pattern-mixture models, thus providing an alternative formulation for the common selection model factorization (see also Section 15.4). In Section 18.1, we observed that pattern-mixture models are chronically underidentified, which is clearly seen by means of the Glynn, Laird, and Rubin (1986) "paradox" (Section 18.1.2). Consequently, Little (1993, 1994a, 1995) suggested the use of so-called identifying restrictions to overcome this underidentification. Choosing a set of different restriction schemes, rather than a single one, is an obvious way to pass from a standard approach to sensitivity analysis.

The need to use identifying restrictions is often quoted as an advantage for pattern-mixture models since it forces careful reflection on the nature of the assumptions made. On the other hand, neither of the two case studies in Chapter 18 (the toenail data in Section 18.3 and the Vorozole study in Section 18.4) made use of identifying restrictions. The reason is different for both studies. The pattern-specific models for the toenail data were simple enough (quadratic curves) to allow extrapolation beyond the last measurement obtained in a particular pattern. Only the first two patterns, with a single or only two measurements, posed problems and an ad hoc solution was employed. In the Vorozole study, pattern was included as a covariate in both the fixed-effects and variance portions of the models. Subsequent simplification lead to a model which was easy to extrapolate.

Thus, in line with the reflections made in Section 18.5, we have three strategies to build a full data model in the pattern-mixture context: identifying restrictions, simple within-pattern models, and the inclusion of pattern as a covariate, the latter of which allows for the combination of information across patterns. A few observations are in place. First, although identifying restrictions impose a careful reflection on the unidentified part of the distribution, the other strategies are more implicit about the assumptions made to identify the full distribution. In this respect, they are open to some of the criticisms toward selection models. Second, the identifying-restrictions strategy is harder to implement, unless in fairly simple settings, such as a single normal sample or contingency tables (Little 1993, 1994a). This chapter provides tools to conduct such a strategy in realistic longitudinal settings. Third, in the selection modeling framework, the MAR assumption plays a crucial role. It can be seen as a compromise between the very rigid and unrealistic MCAR assumption and the complex and fundamentally problematic MNAR assumptions. A counterpart to the MAR assumption is provided in Section 20.2, which can be exploited as the basis for specific identifying-restrictions strategies.

The identifying-restrictions strategy requires some theoretical justification, which is provided in this chapter. In spite of some long and tedious derivations, the resulting procedure is relatively simple to implement and a set of SAS macros has been provided which can be downloaded from the website. GAUSS functions are available at the same location.

The first couple of sections provide background material. Section 20.2 describes the relationship between MAR and the pattern-mixture framework. Multiple imputation, a tool used in the identifying restrictions strategy, is reviewed in Section 20.3. The three strategies to fit pattern-mixture models, mentioned earlier, are described in Section 20.4. The identifying-restrictions strategy is described in detail in Section 20.5. Application to the Vorozole study is discussed in Sections 20.6. Reflections and suggestions for alternative routes of sensitivity are offered in Section 20.7.

20.2 Pattern-Mixture Models and MAR

The missing data taxonomy of Rubin (1976) and Little and Rubin (1987), which distinguishes between missing completely at random, missing at random, and nonrandom missingness, is widely used (Section 15.5). It is usually presented in the selection modeling framework rather than in the pattern-mixture context.

Although selection and pattern-mixture models are interchangeable from a probabilistic point of view, in the sense that they represent different factorizations of the *same* joint distribution, in practice they encourage different kinds of simplifying assumptions. For this reason, it is important to consider their relative merits as scientific models, especially when the probability of missingness depends on the unobserved outcomes. One attraction of selection models is that they fit naturally into Little and Rubin's taxonomy, whereas pattern-mixture models appear not to do so. Here, we show that pattern-mixture models can be classified similarly, and further that the intermediate MAR category is connected to particular kinds of restrictions on the parameters of a pattern-mixture model in the case of monotone missingness. This suggests to us that a purely philosophical debate about the relative merits of the selection and pattern-mixture paradigms is not helpful. Instead, the focus of debate should shift to two other issues.

First, a consideration of the statistical and scientific merits of proposed missing value models on their own terms is needed. For example, if the question of scientific interest regards the treatment effect, averaged over all dropout patterns, then choosing a selection model seems to be obvious. On the other hand, if one is interested in the treatment effect for various dropout patterns separately, then a pattern-mixture model is a natural choice. Second, selection models and pattern-mixture models can be combined into a sensitivity analysis (Section 20.4). For example, one can select a model family of primary focus and fit a model in the other one as well. In addition, insight gained from both model families can be combined into a richer data-analytic picture.

20.2.1 *MAR and ACMV*

Assume a complete measurement sequence is of length n. Recall (Section 15.7) that in a pattern-mixture model, the joint density of $f(\boldsymbol{y}, d)$ is factorized as

$$f(\boldsymbol{y}, d) \quad = \quad f(d)f(\boldsymbol{y}|d).$$

We will now show how pattern-mixture models can be classified using exactly the same taxonomy as is used for selection models (MCAR, MAR, MNAR). Furthermore, we enable a link between this classification and the identifying restrictions proposed in Little (1993).

Clearly, selection models and pattern-mixture models coincide under the MCAR assumption, since, in either case, the joint density simplifies to

$f(\boldsymbol{y})f(d)$. Next, we show that MAR can be expressed in a pattern-mixture framework through restrictions, related to the *complete case missing value* (CCMV) restrictions (Little 1993), which we call *available case missing value* (ACMV) restrictions. Little's CCMV restrictions set a conditional density of unobserved components given a particular set of observed components equal to the corresponding conditional density in the subgroup of completers. Our ACMV restrictions equate this conditional density to the one calculated from the subgroup of all patterns for which all required components have been observed.

In our setting of longitudinal data with dropouts, CCMV can be defined formally as the condition that for each $t \geq 2$ and for $j < t$,

$$f(y_t|y_1,\ldots,y_{t-1},d=j+1) \quad = \quad f(y_t|y_1,\ldots,y_{t-1},d=n+1),$$

whereas ACMV is the condition that for all $t \geq 2$ and $j < t$,

$$f(y_t|y_1,\ldots,y_{t-1},d=j+1) \quad = \quad f(y_t|y_1,\ldots,y_{t-1},d>t). \quad (20.1)$$

If there are only two time points ($n=2$), then ACMV and CCMV coincide. With these definitions, we obtain:

Theorem 20.1 *For longitudinal data with dropouts, MAR $\Longleftrightarrow$ ACMV.*

The proof of Theorem 20.1 is given in the Appendix (Section B.2).

An interesting aside of this theorem is that, since MAR corresponds to a set of (untestable) restrictions (ACMV) in the pattern-mixture framework, MAR itself is also untestable. Precisely, *given* MAR, standard (observed data) methods can be used, but the assumption of MAR itself cannot be tested. This fact is often overlooked in the selection framework.

Little (1993) suggested the possibility of using more than the completers to construct identifying restrictions for two practical reasons: (1) The set of completers may be small and (2) there may be a closer similarity between the conditional distributions given $d = t + 1$ and some other incomplete pattern $d = s + 1$, or set of patterns, than between those for $d = t+1$ and the completers, $d = n + 1$.

One way to proceed is as follows. First, restrict the data set to the first two components only. Then, missing data patterns $d = 3,\ldots,n+1$ collapse into a single pattern $d > 2$. Applying ACMV restrictions to $d = 2$ and $d > 3$ leads to the construction of the density $f(y_2|y_1,d=2) = f(y_2|y_1,d>2)$, as in (20.1). Multiplying by $f(y_1|d=2)$ leads to $f(y_1,y_2|d=2)$, thus determining the joint densities of $f(y_1,y_2|d)$ for all $d = 2,\ldots,n+1$. Next, $f(y_3|y_1,y_2,d)$ ($d=2,3$) can be calculated from $f(y_3|y_1,y_2,d>3)$. We then proceed by induction to construct all joint densities.

A precise formulation of this and related strategies is discussed in Section 20.5. The next section is devoted to an insightful counterexample in nonmonotone patterns.

20.2.2 Nonmonotone Patterns: A Counterexample

Note that the result of Theorem 20.1 does not hold for general missing data patterns. Consider a bivariate outcome (Y_1, Y_2) where missingness can occur in both components. Let (R_1, R_2) be the corresponding bivariate missingness indicator, where $R_j = 0$ if Y_j is missing and 1 otherwise ($j = 1, 2$). Consider the following MAR mechanism:

$$f(r|y) = \Pr(r_1, r_2|y_1, y_2) = \begin{cases} p & \text{if } (r_1, r_2) = (0, 0) \\ q_{y_1} & \text{if } (r_1, r_2) = (1, 0) \\ s_{y_2} & \text{if } (r_1, r_2) = (0, 1) \\ 1 - p - q_{y_1} & \\ \quad - s_{y_2} & \text{if } (r_1, r_2) = (1, 1). \end{cases} \quad (20.2)$$

We need to indicate how the concept of ACMV would be translated to this setting. Several proposals can be considered. A trivial extension of the ACMV restrictions in the monotone case implies for the patterns $r = (1, 0)$ and $r = (0, 1)$:

$$r = (1, 0): \quad f(y_1, y_2|r = (1, 0)) \\ = f(y_1|r = (1, 0)).f(y_2|y_1, r = (1, 1)), \quad (20.3)$$

$$r = (0, 1): \quad f(y_1, y_2|r = (0, 1)) \\ = f(y_2|r = (0, 1)).f(y_1|y_2, r = (1, 1)). \quad (20.4)$$

The idea is that the density of missing components, given observed components, is replaced by the corresponding density of patterns for which both are available. Restrictions for the pattern $r = (0, 0)$ will be discussed further.

From condition (20.3) we derive

$$\frac{f(r = (1, 0)|y_1, y_2)f(y_1, y_2)}{f(r = (1, 0))} = \frac{f(r = (1, 0)|y_1)f(y_1)}{f(r = (1, 0))}$$

$$\times \frac{f(r = (1, 1)|y_1, y_2)f(y_1, y_2)}{f(r = (1, 1)|y_1)f(y_1)}$$

$$\Updownarrow$$

$$f(r = (1, 1)|y_1, y_2) = f(r = (1, 1)|y_1),$$

since $f(r = (1,0)|y_1, y_2) = f(r = (1,0)|y_1) = q_{y_1}$, implying that s_{y_2} is constant. Similarly, condition (20.4) implies that q_{y_1} is constant.

Clearly, since both q_{y_1} and s_{y_2} have to be constant, the mechanism needs to be MCAR. In other words, ACMV≡MCAR, independent of the restrictions for $f(y_1, y_2|r = (0,0))$, and hence ACMV and MAR differ.

20.3 Multiple Imputation

In Section 20.5, multiple imputation will be used as a tool in developing identifying-restrictions strategies. For this reason, and for its central place in the incomplete-data literature, this section reviews the principles. Multiple imputation was formally introduced by Rubin (1978). Rubin (1987) provides a comprehensive treatment. Several other sources, such as Rubin and Schenker (1986), Little and Rubin (1987), Tanner and Wong (1987), and Schafer's (1997) book give excellent and easy-to-read descriptions of the technique. Efron (1994) discusses connections between multiple imputation and the bootstrap. An important review, containing an extensive list of references and a large bibliography, is given in Rubin (1996).

The concept of multiple imputation refers to replacing each missing value with more than one imputed value. The goal is to combine the simplicity of imputation strategies, with unbiasedness in both point estimates and measures of precision. In Section 16.3, we have seen that some simple imputation procedures may yield inconsistent point estimates as soon as the missingness mechanism surpasses MCAR. This could be overcome to a large extent with conditional mean imputation, but the problem of under-estimating the variability of the estimators is common to all methods since they all treat imputed values as observed values. By imputing several values for a single missing component, this uncertainty is explicitly acknowledged.

Rubin (1987) points to another very useful application of multiple imputation. Rather than merely accounting for *sampling uncertainty*, the method can be used to incorporate *model uncertainty*. Indeed, when a measurement is missing but the researcher has a good idea about the probabilistic measurement and missingness mechanisms, constructing the appropriate distribution to draw imputations from is, at least in principle, relatively straightforward. In practice, there may be considerable uncertainty about some parts of the joint model. In that case, several mechanisms for drawing imputations might seem equally plausible. They can be combined in a single multiple imputation analysis. As such, multiple imputation can be used as a tool for sensitivity analysis.

Suppose we have a sample of N, i.i.d. $n \times 1$ random vectors $\boldsymbol{Y}_i$. Our interest lies in estimating some parameter vector $\boldsymbol{\theta}$ of the distribution of $\boldsymbol{Y}_i$. Assume the notation is as in Section 15.5. Multiple imputation fills in $\boldsymbol{Y}^m$ using the observed data $\boldsymbol{Y}^o$, several times, and then the completed data are used to estimate $\boldsymbol{\theta}$.

As discussed by Rubin and Schenker (1986), the theoretical justification for multiple imputation is most easily understood using Bayesian concepts, but a likelihood-based treatment of the subject is equally possible. If we knew the joint distribution of $\boldsymbol{Y}_i = (\boldsymbol{Y}_i^o, \boldsymbol{Y}_i^m)$ with parameter vector $\boldsymbol{\gamma}$ say, then we could impute $\boldsymbol{Y}_i^m$ by drawing a value of $\boldsymbol{Y}_i^m$ from the conditional distribution

$$f(\boldsymbol{y}_i^m | \boldsymbol{y}_i^o, \boldsymbol{\gamma}). \tag{20.5}$$

Note that we explicitly distinguish the parameter of scientific interest $\boldsymbol{\theta}$ from the parameter $\boldsymbol{\gamma}$ in (20.5). Since $\boldsymbol{\gamma}$ is unknown, we must estimate it from the data, say $\hat{\boldsymbol{\gamma}}$, and use

$$f(\boldsymbol{y}_i^m | \boldsymbol{y}_i^o, \hat{\boldsymbol{\gamma}}) \tag{20.6}$$

to impute the missing data. In Bayesian terms, $\boldsymbol{\gamma}$ in (20.5) is a random variable, of which the distribution is a function of the data. In particular, we first obtain the distribution of $\boldsymbol{\gamma}$ from the data, depending on $\hat{\boldsymbol{\gamma}}$. The construction of model (20.5) is referred to by Rubin (1987) as the *Modeling Task*.

After formulating the distribution of $\boldsymbol{\gamma}$, the imputation algorithm is as follows:

1. Draw $\boldsymbol{\gamma}^*$ from the distribution of $\boldsymbol{\gamma}$.

2. Draw $\boldsymbol{Y}_i^{m*}$ from $f(\boldsymbol{y}_i^m | \boldsymbol{y}_i^o, \boldsymbol{\gamma}^*)$.

3. Using the completed data, $(\boldsymbol{Y}^o, \boldsymbol{Y}^{m*})$, and the method of choice (i.e., maximum likelihood, restricted maximum likelihood, method of moments, partial likelihood), estimate the parameter of interest $\hat{\boldsymbol{\theta}} = \hat{\boldsymbol{\theta}}(\boldsymbol{Y}) = \hat{\boldsymbol{\theta}}(\boldsymbol{Y}^o, \boldsymbol{Y}^{m*})$ and its variance (called *within*-imputation variance) $U = \widehat{\text{var}}(\hat{\boldsymbol{\theta}})$.

4. Independently repeat steps 1–3, M times. The M data sets give rise to $\hat{\boldsymbol{\theta}}^{(m)}$ and $U^{(m)}$, for $m = 1, ..., M$.

Steps 1 and 2 are referred to as the *Imputation Task*. Step 3 is the *Estimation Task*. Of course, one wants to combine the M inferences into a single one. Both parameter and precision estimation, on the one hand, and hypothesis testing, on the other hand, will be discussed next.

20.3.1 Parameter and Precision Estimation

The M within-imputation estimates for $\boldsymbol{\theta}$ are pooled to give the multiple imputation estimate:

$$\hat{\boldsymbol{\theta}}^* \;=\; \frac{1}{M}\sum_{m=1}^{M}\hat{\boldsymbol{\theta}}^{(m)}.$$

Suppose that complete data inference about $\boldsymbol{\theta}$ would be made by $(\boldsymbol{\theta}-\hat{\boldsymbol{\theta}}) \sim N(\mathbf{0},U)$. Then, one can make normal-based inferences for $\boldsymbol{\theta}$ based upon

$$(\boldsymbol{\theta}-\hat{\boldsymbol{\theta}}^*) \;\sim\; N(\mathbf{0},V), \qquad (20.7)$$

where

$$V \;=\; \hat{W} + \left(\frac{M+1}{M}\right)\hat{B}, \qquad (20.8)$$

$$\hat{W} \;=\; \frac{\sum_{m=1}^{M} U^{(m)}}{M} \qquad (20.9)$$

is the average within-imputation variance, and

$$\hat{B} \;=\; \frac{\sum_{m=1}^{M}(\hat{\boldsymbol{\theta}}^{(m)}-\hat{\boldsymbol{\theta}}^*)(\hat{\boldsymbol{\theta}}^{(m)}-\hat{\boldsymbol{\theta}}^*)'}{M-1} \qquad (20.10)$$

is the between-imputation variance (Rubin 1987). Rubin and Schenker (1986) report that a small number of imputations ($M = 2, 3$) already yields a major improvement over single imputation. Upon noting that the factor $(M+1)/M$ approaches 1 for large M, (20.8) is approximately the sum of the within- and the between-imputations variability.

Multiple imputation is most useful in situations where γ is an easily estimated set of parameters characterizing the distribution of $\boldsymbol{Y}_i$, whereas $\boldsymbol{\theta}$ is complicated to estimate in the presence of missing data and/or when obtaining a correct estimate for the variance is nontrivial with incomplete data.

20.3.2 Hypothesis Testing

Testing hypotheses could be based on the asymptotic normality results (20.7) and (20.8). However, the rationale for using asymptotic results and hence χ^2 reference distributions is not just a function of the sample size, N, but also of the number of imputations, M. Therefore, Li, Raghunathan, and Rubin (1991) propose the use of an F reference distribution. Precisely, to test the hypothesis $H_0 : \boldsymbol{\theta} = \boldsymbol{\theta}_0$, they advocate the following method to calculate p-values:

$$ p \;=\; P(F_{k,w} > F), \tag{20.11} $$

where k is the length of the parameter vector $\boldsymbol{\theta}$, $F_{k,w}$ is an F random variable with k numerator and w denominator degrees of freedom, and

$$ F \;=\; \frac{(\boldsymbol{\theta}^* - \boldsymbol{\theta}_0)'W^{-1}(\boldsymbol{\theta}^* - \boldsymbol{\theta}_0)}{k(1+r)}, \tag{20.12} $$

$$ w \;=\; 4 + (\tau - 4)\left[1 + \frac{(1 - 2\tau^{-1})}{r}\right]^2, $$

$$ r \;=\; \frac{1}{k}\left(1 + \frac{1}{M}\right) \operatorname{tr}(BW^{-1}), \tag{20.13} $$

$$ \tau \;=\; k(M - 1). $$

It is interesting to note that when $M \to \infty$, the reference distribution of F approaches an $F_{k,\infty} = \chi^2/k$-distribution, in line with intuition. Good operational characteristics of this procedure are reported in Li, Raghunathan, and Rubin (1991), which combines nicely with computational ease.

Clearly, procedure (20.11) can be used as well when not the full vector $\boldsymbol{\theta}$, but one component, a subvector, or a set of linear contrasts, is the subject of hypothesis testing. When a subvector is of interest (a single component being a special case), the corresponding submatrices of B and W need to be used in (20.12) and (20.13). For a set of linear contrasts $L\boldsymbol{\theta}$, one should use the appropriately transformed covariance matrices: $\tilde{W} = LWL'$, $\tilde{B} = LBL'$, and $\tilde{V} = LVL'$.

20.4   Pattern-Mixture Models and Sensitivity Analysis

Sensitivity analysis for pattern-mixture models can be conceived in many different ways. Crucial guiding questions are whether a pattern-mixture model is the central focus of interest or should rather be viewed as complementary to another model, such as a selection model.

Following the initial mention of pattern-mixture models (Glynn, Laird, and Rubin 1986, Little and Rubin 1987) and the renewed interest in recent years (Little 1993, 1994a, Hogan and Laird 1997), several authors have contrasted selection models and pattern-mixture models. This is done to either (1) answer the same scientific question, such as marginal treatment effect or time evolution, based on these two rather different modeling strategies, or (2) to gain additional insight by supplementing the selection model results with those from a pattern-mixture approach. Examples can be found in Verbeke, Lesaffre, and Spiessens (2000), Curran, Pignatti, and Molenberghs (1997), and Michiels *et al.* (1999) for continuous outcomes. The categorical outcome case has been treated in Molenberghs, Michiels, and Lipsitz (1999), and Michiels, Molenberghs, and Lipsitz (1999). Further references include Cohen and Cohen (1983), Muthén, Kaplan, and Hollis (1987), Allison (1987), McArdle and Hamagani (1992) Little and Wang (1996), Hedeker and Gibbons (1997), Siddiqui and Ali (1998), Ekholm and Skinner (1998), Molenberghs, Michiels, and Kenward (1998), and Park and Lee (1999).

We want to stress the usefulness of these two modeling strategies and also refer to the toenail (Sections 17.2 and 18.3) and Vorozole studies (Sections 17.6 and 18.4). In the Vorozole case, the treatment effect assessment is virtually identical under both strategies, but useful additional insight is obtained from the pattern-specific average profiles. Of course, such bilateral comparisons are not confined to the selection and pattern-mixture model families. For example, shared-parameter models can be included as well (see Section 15.4 and the very complete review by Little 1995).

On the other hand, the sensitivity analysis we propose here can also be conducted *within* the pattern-mixture family, in analogy with the selection model case (Chapter 18). The key area on which sensitivity analysis should focus is the unidentified components of the model and the way(s) in which this is handled. Indeed, recall that model family (18.6) contains underidentified members since it describes the full set of measurements in pattern t_i, even though there are no measurements after occasion $t_i - 1$. In the introduction, we mentioned three strategies to deal with this underidentification. Let us describe these in turn.

- **Strategy 1.** Little (1993, 1994a) advocated the use of identifying restrictions, which work well in relatively simple settings, and presented several examples. Perhaps the best known ones are so-called complete case missing value restrictions (CCMV), where, for a given pattern, the conditional distribution of the missing data, given the observed data, is equated to its counterpart in the completers.

  Based in part on the pattern-mixture formulation of MAR as described in Section 20.2, we will outline a general framework for identifying restrictions. An important case is available case missing value (ACMV), where the conditional distribution of the unobserved outcomes given the observed outcomes in a specific pattern is equated to combined information from all patterns on which the outcomes are observed, in such a way that it corresponds to MAR. This is useful if one wants to assign special value to the MAR case, just as in the selection model context. It also provides a way to compare ignorable selection models with their counterpart in the pattern-mixture setting. Molenberghs, Michiels, and Lipsitz (1999) and Michiels, Molenberghs, Lipsitz (1999) took up this idea in the context of binary outcomes, with a marginal global odds ratio model to describe the measurement process (Molenberghs and Lesaffre 1994).

It will be clear that ACMV is but one particular way of combining information from patterns on which a given set of outcomes is observed. Among the family of such approaches, special emphasis can be put on the so-called *neighboring* pattern which, in a monotone dropout setting, is the pattern with one additional measurement obtained. They are referred to as neighboring case missing value (NCMV) restrictions. In a sense, they are opposite to CCMV.

A full account of identifying restrictions is provided in Section 20.5.

- We will now introduce the other two strategies. As opposed to identifying restrictions, model simplification can be done in order to identify the parameters. The advantage is that the number of parameters decreases, which is desirable since the length of the parameter vector is a general issue with pattern-mixture models. Indeed, Hogan and Laird (1997) noted that in order to estimate the large number of parameters in general pattern-mixture models, one has to make the awkward requirement that each dropout pattern occurs sufficiently often. In other words, one has to require large amounts of dropout.

Broadly, we distinguish between two types of simplifications to identify pattern-mixture models.

  - **Strategy 2.** First, trends can be restricted to functional forms supported by the information available within a pattern. For example, a linear time trend is easily extrapolated beyond the last obtained measurement. As discussed in the introduction, this was the strategy followed in the toenail study (Section 18.3). It is relatively simple to apply when, for example, a quadratic curve is assumed for each of the patterns. One only needs to provide an ad hoc solution for the first or the first few patterns. However, as was seen in the toenail study, some of the extrapolations, especially when based on traditional polynomials and/or in sparse patterns, can appear to be unrealistic (Figure 18.2) and may then require further scrutiny.

  In order to fit such models, one simply has to carry out a model-building exercise in each of the patterns separately. Each of these comes down to fitting a standard linear mixed model and therefore entails no additional complexity.

- **Strategy 3.** Second, one can let the parameters vary across patterns in a parametric way. Thus, rather than estimating a separate time trend in each pattern, one could assume that the time evolution within a pattern is unstructured, but parallel across patterns. This is effectuated by treating pattern as a covariate in the model. The available data can be used to assess whether such simplifications are supported within the time ranges for which there is information. Using the so-obtained profiles past the time of dropout still requires extrapolation.

  From a model-building perspective, this modeling strategy can be viewed as a standard linear mixed model with the pattern indicator as an additional covariate.

  This approach was recently adopted by Park and Lee (1999) in the context of count data for which generalized estimating equations (Liang and Zeger 1986) are used.

While the second and third strategies are computationally simple, it is important to note that there is a price to pay. Indeed, simplified models, qualified as "assumption rich" by Sheiner, Beal and Dunne (1997), are also making untestable assumptions, just as in the selection model case. Still, the need of assumptions and their implications are more obvious. It is, for example, not possible to assume an unstructured time trend in incomplete patterns, except if one restricts attention to the time range from onset until dropout. In contrast, assuming a linear time trend is possible in all patterns containing at least two measurements. Whereas such an approach is very simple from a modeler's point of view, it is then less obvious what the precise nature of the dropout mechanism is. In any case, the dropout model is not explicitated in this way. This is in contrast with the identifying-restrictions route, where the assumptions have to be clear from the start and, importantly, the MAR (ACMV) case is available as a special case.

A final observation, applying to all strategies, is that pattern-mixture models do not necessarily provide estimates and standard errors of marginal quantities of interest, such as overall treatment effect or overall time trend. Hogan and Laird (1997) provided a way to derive selection model quantities from the pattern-mixture model. Several authors have followed this idea to formally compare the conclusions from a selection model with the selection model parameters in a pattern-mixture model. Verbeke, Lesaffre, and Spiessens (2000), Curran, Pignatti, and Molenberghs (1997), and Michiels, Molenberghs, Bijnens, and Vangeneugden (1999) applied this approach in the context of linear mixed models for continuous data. We refer to Sections 18.3, 18.4, and 24.4.2 for illustrations in the toenail study, the Vorozole study, and the milk protein trial, respectively.

In Section 20.5, we describe identifying restriction strategies, with MAR (ACMV), CCMV, and NCMV as special cases. They are applied in Section 20.6 to the Vorozole study.

20.5   Identifying Restrictions Strategies

In line with the results obtained in Section 20.2, we restrict attention to monotone patterns. In general, indicate dropout patterns by $t = 1, \ldots, n$, where, as in Section 15.9, the dropout indicator is $d = t + 1$. For pattern t, the complete data density is given by

$$f_t(y_1, \ldots, y_n) \;=\; f_t(y_1, \ldots, y_t) f_t(y_{t+1}, \ldots, y_n | y_1, \ldots, y_t). \quad (20.14)$$

The first factor on the right-hand side of (20.14) is clearly identifiable from the observed data, whereas the second factor is not. Therefore, identifying restrictions are applied in order to identify the second component.

20.5.1   Strategy Outline

The above observations suggest the following strategy:

1. Fit a (linear mixed) model to the pattern-specific identifiable densities:

$$f_t(y_1, \ldots, y_t). \quad (20.15)$$

   This results in a parameter estimate $\hat{\gamma}_t$ which, for example, consists of fixed-effects parameters and variance components.

2. Select an identification method of choice. This will be discussed in full detail in Section 20.5.2.

3. Using this identification method, determine the conditional distributions of the unobserved outcomes, given the observed ones:

$$f_t(y_{t+1}, \ldots, y_n | y_1, \ldots, y_t). \quad (20.16)$$

In the case that the observed densities (20.15) are assumed to be normal, the conditional densities are normal or finite mixtures of normal densities. This feature will be taken up in Section 20.5.2 as well.

4. Using the methodology outlined in Section 20.3, draw multiple imputations for the unobserved components, given the observed outcomes
   and the correct pattern-specific density (20.16). We will study this
   further in Section 20.5.4.

5. Analyze the multiply-imputed sets of data using the method of choice.
   It would be most natural to consider a pattern-mixture model, but it
   is also possible to use selection models. One has to ensure, however,
   that the multiple imputation is *proper*, in the sense described by
   Rubin (1987, 1996), Rubin and Schenker (1986), and Schafer (1997).
   Informally, an important concern is that relations between variables
   of scientific interest should not be excluded in the imputation stage.
   This implies that the original observed-data models, fitted in the first
   step, should be rich enough to carry the relations included in the final
   completed-data models.

6. Inferences can be conducted in the way described in Sections 20.3.1
   and 20.3.2.

In the next sections, a more technical treatment is given to items 2 and 3 of
the above strategy. Section 20.5.2 describes the identifying restrictions in
detail. Section 20.5.4 is dedicated to handling the conditional distributions
of the unobserved components given the observed ones, as a preparation
for the multiple imputation draws.

20.5.2 Identifying Restrictions

Although, in principle, completely arbitrary restrictions can be used by
means of any valid density function over the appropriate support, strategies
which relate back to the observed data deserve privileged interest. Little
(1993) proposes CCMV, which uses the following identification:

$$f_t(y_{t+1}, \ldots, y_n | y_1, \ldots, y_t) \;=\; f_n(y_{t+1}, \ldots, y_n | y_1, \ldots, y_t). \quad (20.17)$$

In other words, information which is unavailable is always borrowed from
the completers. This strategy can be defended in cases where the bulk of
the subjects are complete and only small proportions are assigned to the
various dropout patterns. Also, extension of this approach to nonmonotone
patterns is particularly easy. On the other hand, the completers may be
rather "distant" in some sense from incomplete patterns, especially in the
case of early dropouts.

In such cases, it is useful to borrow information from other or even all
available patterns. To this end, we first rewrite the unidentified component

as

$$f_t(y_{t+1}, \ldots, y_n | y_1, \ldots, y_t) = \prod_{s=t+1}^{n} f_t(y_s | y_1, \ldots, y_{s-1}). \quad (20.18)$$

Now, a very general formulation is obtained by allowing each factor on the right-hand side of (20.18) to be identified from a selected number of patterns. For example, CCMV follows from using f_n for each unidentified factor. Alternatively, the nearest identified pattern can be used:

$$f_t(y_s | y_1, \ldots, y_{s-1}) = f_s(y_s | y_1, \ldots, y_{s-1}), \qquad s = t+1, \ldots, n. \quad (20.19)$$

We will refer to these restrictions as *neighboring case missing values* or NCMV. In what follows, we will provide some motivation for this terminology. However, using information from only one pattern can be seen as wasteful and therefore one can base identification on all patterns for which the sth component is identified:

$$f_t(y_s | y_1, \ldots, y_{s-1}) = \sum_{j=s}^{n} \omega_{sj} f_j(y_s | y_1, \ldots, y_{s-1}), \qquad s = t+1, \ldots, n. \quad (20.20)$$

For simplicity, we will use $\boldsymbol{\omega}_s = (\omega_{ss}, \ldots, \omega_{sn})'$. Every $\boldsymbol{\omega}_s$ which sums to 1 and consists of nonnegative components provides a valid identification scheme. Obviously, (20.20) generalizes the earlier schemes. CCMV restrictions are obtained by setting $\omega_{sn} = 1$ and all others equal to zero. The NCMV system follows from setting $\omega_{ss} = 1$ and the others equal to zero. Further, we will show that there always is a unique choice for $\boldsymbol{\omega}_s$ which corresponds to ACMV and hence to MAR.

MOTIVATION FOR NCMV

Neighboring case missing value restrictions, as outlined in (20.19), can be introduced in two slightly different but perhaps more intuitive ways. A *downward* approach identifies (20.16) as

$$f_t(y_{t+1}, \ldots, y_n | y_1, \ldots, y_t) = f_{t+1}(y_{t+1}, \ldots, y_n | y_1, \ldots, y_t), \quad (20.21)$$

for $t = 1, \ldots, n-1$. The idea is to start with $t = n-1$ and then gradually step down until $t = 1$. Thus, the distribution in a given pattern is identified from the pattern with one more component observed. Similarly, an *upward* strategy identifies

$$f_s(y_{t+1} | y_1, \ldots, y_t) = f_{t+1}(y_{t+1} | y_1, \ldots, y_t) \quad (20.22)$$

for $t = 1, \ldots, n$ and $s = 1, \ldots, t$.

We now state the connection between these strategies.

Result 20.1 *Strategies (20.21) and (20.22) are equivalent to neighboring case missing value restrictions (20.19).*

The result is easily shown. First, the equivalence between (20.22) and (20.19) is immediate. Second, for (20.21) we proceed by inductive reasoning. Clearly, the equivalence holds for $t = n - 1$. Assume now that it holds for $t + 1$. Then,

$$f_t(y_{t+1}, \ldots, y_n | y_1, \ldots, y_t)$$
$$= f_{t+1}(y_{t+1}, \ldots, y_n | y_1, \ldots, y_t)$$
$$= f_{t+1}(y_{t+1} | y_1, \ldots, y_t) \prod_{s=2}^{n-t} f_{t+1}(y_{t+s} | y_1, \ldots, y_{t+s-1})$$
$$= f_{t+1}(y_{t+1} | y_1, \ldots, y_t) \prod_{s=2}^{n-t} f_{t+s}(y_{t+s} | y_1, \ldots, y_{t+s-1}),$$

where the last equality holds by virtue of the induction hypothesis. This establishes the result.

USING ALL PATTERNS

Similarly to the NCMV case, we can formulate (20.20) in two alternative ways. The downward strategy is formulated as

$$f_t(y_{t+1}, \ldots, y_n | y_1, \ldots, y_t)$$
$$= \sum_{j=t+1}^{n} \omega_{t+1,j} f_j(y_{t+1}, \ldots, y_n | y_1, \ldots, y_t). \qquad (20.23)$$

Similarly, we write the upward strategy as

$$f_s(y_{t+1} | y_1, \ldots, y_t) = \sum_{j=t+1}^{n} \omega_{t+1,j} f_j(y_{t+1} | y_1, \ldots, y_t) \qquad (20.24)$$

for $t = 1, \ldots, n$ and $s = 1, \ldots, t$. Now, by setting

$$g_{t+1}(\cdot | y_1, \ldots, y_t) \equiv \sum_{j=t+1}^{n} \omega_{t+1,j} f_j(\cdot | y_1, \ldots, y_t)$$

and reproducing the proof of Result 20.1 for the $g(\cdot)$ functions, it is clear that the following result holds:

Result 20.2 *Strategies (20.23) and (20.24) are equivalent to restrictions (20.20).*

Note that Result 20.2 includes Result 20.1 as a special case.

20.5.3   ACMV Restrictions

A general class of restrictions, based on the information available from other patterns, is provided by (20.20). Equivalent formulations are (20.23) and (20.24), and (20.19), (20.21) and (20.22) provide special cases, and so do (20.17). In this section, we will derive expressions for the $\boldsymbol{\omega}_s$ which correspond to ACMV, as defined in Section 20.2. This will then constitute a third special case.

The definition of ACMV is presented in (20.1). We will now show that it provides an easy way to derive an expression for $\boldsymbol{\omega}_s$ in (20.20). Indeed, (20.20) can be restated, using notation of this section, as

$$f_t(y_s|y_1,\ldots,y_{s-1}) \quad = \quad f_{(\geq s)}(y_s|y_1,\ldots,y_{s-1}), \tag{20.25}$$

for $s = t+1,\ldots,n$. Here,

$$f_{(\geq s)}(\cdot|\cdot) \equiv f(\cdot|\cdot, d > s) \equiv f(\cdot|\cdot, t \geq s),$$

with $d = t+1$ indicating time of dropout. Now, we can transform (20.25) as follows:

$$f_t(y_s|y_1,\ldots,y_{s-1})$$

$$= \quad f_{(\geq s)}(y_s|y_1,\ldots,y_{s-1})$$

$$= \quad \frac{\sum_{j=s}^{n} \alpha_j f_j(y_1,\ldots,y_s)}{\sum_{j=s}^{n} \alpha_j f_j(y_1,\ldots,y_{s-1})} \tag{20.26}$$

$$= \quad \sum_{j=s}^{n} \left(\frac{\alpha_j f_j(y_1,\ldots,y_{s-1})}{\sum_{j=s}^{n} \alpha_j f_j(y_1,\ldots,y_{s-1})} \right) f_j(y_s|y_1,\ldots,y_{s-1}). \tag{20.27}$$

Now, comparing (20.27) to (20.20) yields

$$\omega_{sj} \quad = \quad \frac{\alpha_j f_j(y_1,\ldots,y_{s-1})}{\sum_{\ell=s}^{n} \alpha_\ell f_\ell(y_1,\ldots,y_{s-1})}. \tag{20.28}$$

We have derived two equivalent explicit expressions of (20.1). Expression (20.26) is the conditional density of a mixture, whereas (20.20) with (20.28) is a mixture of conditional densities.

Clearly, $\boldsymbol{\omega}_s$ defined by (20.28) consists of components which are nonnegative and sum to 1, thus defining a valid density function as soon as its components are genuine densities.

Let us incorporate (20.20) into (20.14):

$$f_t(y_1,\ldots,y_n)$$

$$
\begin{aligned}
&= f_t(y_1, \ldots, y_t) \\
&\quad \times \prod_{s=0}^{n-t-1} \left[\sum_{j=n-s}^{n} \omega_{n-s,j} f_j(y_{n-s}|y_1, \ldots, y_{n-s-1}) \right].
\end{aligned}
\tag{20.29}
$$

Expression (20.29) clearly shows which information is used to complement the observed data density in pattern t in order to establish the complete data density.

The practical use of (20.20) in multiple imputation, with the CCMV, NCMV, and ACMV strategies as special cases, will be studied in Section 20.5.4. First, we will study the simple case of sequences of length 3 and 4.

SPECIAL CASE: THREE MEASUREMENTS

In this case, there are only three patterns, and identification (20.29) takes the following form:

$$
\begin{aligned}
f_3(y_1, y_2, y_3) &= f_3(y_1, y_2, y_3), & (20.30) \\
f_2(y_1, y_2, y_3) &= f_2(y_1, y_2) f_3(y_3|y_1, y_2), & (20.31) \\
f_1(y_1, y_2, y_3) &= f_1(y_1) \left[\omega f_2(y_2|y_1) + (1 - \omega) f_3(y_2|y_1) \right] \\
&\quad \times f_3(y_3|y_1, y_2). & (20.32)
\end{aligned}
$$

Since $f_3(y_1, y_2, y_3)$ is completely identifiable from the data, and for density $f_2(y_1, y_2, y_3)$ there is only one possible identification, given (20.20), the only room for choice is in pattern 1. Setting $\omega = 1$ corresponds to NCMV, whereas $\omega = 0$ implies CCMV. Using (20.28), ACMV boils down to

$$
\omega = \frac{\alpha_2 f_2(y_1)}{\alpha_2 f_2(y_1) + \alpha_3 f_3(y_1)}.
\tag{20.33}
$$

The conditional density $f_1(y_2|y_1)$ in (20.32) can be rewritten as

$$
f_1(y_2|y_1) = \frac{\alpha_2 f_2(y_1, y_2) + \alpha_3 f_3(y_1, y_2)}{\alpha_2 f_2(y_1) + \alpha_3 f_3(y_1)},
$$

which is, of course, a special case of (20.26). Upon division by $\alpha_2 + \alpha_3$, both numerator and denominator are mixture densities, hence producing a legitimate conditional density.

SPECIAL CASE: FOUR MEASUREMENTS

The counterparts of (20.30)–(20.32) are

$$
f_4(y_1, y_2, y_3, y_4) = f_4(y_1, y_2, y_3, y_4),
\tag{20.34}
$$

$$f_3(y_1, y_2, y_3, y_4) = f_3(y_1, y_2, y_3) f_4(y_4|y_1, y_2, y_3), \tag{20.35}$$

$$f_2(y_1, y_2, y_3, y_4) = f_2(y_1, y_2) \left[\omega_{33} f_3(y_3|y_1, y_2) + \omega_{34} f_4(y_3|y_1, y_2)\right]$$
$$\times f_4(y_4|y_1, y_2, y_3), \tag{20.36}$$

$$f_1(y_1, y_2, y_3, y_4) = f_1(y_1)$$
$$\times \left[\omega_{22} f_2(y_2|y_1) + \omega_{23} f_3(y_2|y_1) + \omega_{24} f_4(y_2|y_1)\right]$$
$$\times \left[\omega_{33} f_3(y_3|y_1, y_2) + \omega_{34} f_4(y_3|y_1, y_2)\right]$$
$$\times f_4(y_4|y_1, y_2, y_3). \tag{20.37}$$

Now, setting $\omega_{33} = \omega_{22} = 1$ corresponds to NCMV, and $\omega_{34} = \omega_{24} = 1$ implies CCMV. ACMV corresponds to the system

$$\omega_{33} = \frac{\alpha_3 f_3(y_1, y_2)}{\alpha_3 f_3(y_1, y_2) + \alpha_4 f_4(y_1, y_2)},$$

$$\omega_{34} = \frac{\alpha_4 f_4(y_1, y_2)}{\alpha_3 f_3(y_1, y_2) + \alpha_4 f_4(y_1, y_2)},$$

$$\omega_{22} = \frac{\alpha_2 f_2(y_1)}{\alpha_2 f_2(y_1) + \alpha_3 f_3(y_1) + \alpha_4 f_4(y_1)},$$

$$\omega_{23} = \frac{\alpha_3 f_3(y_1)}{\alpha_2 f_2(y_1) + \alpha_3 f_3(y_1) + \alpha_4 f_4(y_1)},$$

$$\omega_{24} = \frac{\alpha_4 f_4(y_1)}{\alpha_2 f_2(y_1) + \alpha_3 f_3(y_1) + \alpha_4 f_4(y_1)}.$$

Explicit ACMV expressions for those conditional densities which involve mixtures are

$$f_1(y_3|y_1, y_2) = f_2(y_3|y_1, y_2)$$
$$= \omega_{33} f_3(y_3|y_1, y_2) + \omega_{34} f_4(y_3|y_1, y_2) \tag{20.38}$$
$$= \left[\frac{\alpha_3 f_3(y_1, y_2, y_3) + \alpha_4 f_4(y_1, y_2, y_3)}{\alpha_3 f_3(y_1, y_2) + \alpha_4 f_4(y_1, y_2)}\right], \tag{20.39}$$

$$f_1(y_2|y_1)$$
$$= \omega_{22} f_2(y_2|y_1) + \omega_{23} f_3(y_2|y_1) + \omega_{24} f_4(y_2|y_1)$$
$$= \left[\frac{\alpha_2 f_2(y_1, y_2) + \alpha_3 f_3(y_1, y_2) + \alpha_3 f_3(y_2, y_1)}{\alpha_2 f_2(y_1) + \alpha_3 f_3(y_1) + \alpha_4 f_4(y_1)}\right]. \tag{20.40}$$

In general, (20.34)–(20.37) provide three free parameters which can be exploited as a natural basis for sensitivity analysis.

20.5.4   Drawing from the Conditional Densities

In the previous section, we have seen how general identifying restrictions (20.20), with CCMV, NCMV, and ACMV as special cases, lead to the conditional densities for the unobserved components, given the observed ones. This came down to deriving expressions for ω_s, such as in (20.28) for ACMV. This endeavor corresponds to items 2 and 3 of the strategy outline (Section 20.5.1). In order to carry out item 4 (drawing multiple imputations), we need to draw imputations from these conditional densities.

Let us focus on the special case of three measurements. To this end, we consider identification scheme (20.30)–(20.32). At first, we leave the parametric form of these densities unspecified. The following steps are required:

1. Estimate the parameters of the identifiable densities: $f_3(y_1, y_2, y_3)$, $f_2(y_1, y_2)$, and $f_1(y_1)$. Then, for each of the M imputations, we have to execute the following steps.

2. Draw from the parameter vectors as in the first step on p. 337. It will be assumed that in all densities from which we draw next, this parameter vector is used.

3. **For pattern 2.** Given an observation (y_1, y_2) in this pattern, calculate the conditional density $f_3(y_3|y_1, y_2)$ and draw from it.

4. **For pattern 1.** We now have to distinguish three substeps.

   (a) Given y_1 and the proportions α_2 and α_3 of observations in the second and third patterns, respectively, determine ω. Every ω in the unit interval is valid. Special cases are as follows:
   - For NCMV, $\omega = 1$.
   - For CCMV, $\omega = 0$.
   - For ACMV, ω is calculated from (20.33). Note that, given y_1, this is a constant.

   If $0 < \omega < 1$, generate a random uniform variate, U say. Note that, strictly speaking, this draw is unnecessary for the boundary NCMV ($\omega = 1$) and CCMV ($\omega = 0$) cases.

   (b) If $U \le \omega$, calculate $f_2(y_2|y_1)$ and draw from it. Otherwise, do the same based on $f_3(y_2|y_1)$.

   (c) Given the observed y_1 and given y_2, which has just been drawn, calculate the conditional density $f_3(y_3|y_1, y_2)$ and draw from it.

All steps but the first one have to be repeated M times to obtain the same number of imputed data sets. Inference then proceeds as outlined in Sections 20.3.1 and 20.3.2.

When the observed densities are estimated using linear mixed models, $f_3(y_1, y_2, y_3)$, $f_2(y_1, y_2)$, and $f_1(y_1)$ produce fixed effects and variance components. Let us group all of them in γ and assume a draw is made from their distribution, γ^* say. To this end, their precision estimates need to be computed. In SAS, they are provided from the 'covb' option in the MODEL statement and the 'asycov' option in the PROC MIXED statement.

Let us illustrate this procedure for (20.31). Assume that the ith subject has only two measurements, and hence belongs to the second pattern. Let its design matrices be X_i and Z_i for the fixed effects and random effects, respectively. Its mean and variance for the *third* pattern are

$$\begin{aligned} \mu_i(3) &= X_i\beta^*(3), & (20.41) \\ V_i(3) &= Z_iD^*(3)Z_i' + \Sigma_i^*(3), & (20.42) \end{aligned}$$

where (3) indicates that the parameters are specific to the third pattern, as in (18.6). The asterisk is reminiscent of these parameters being part of the draw γ^*.

Now, based on (20.41)–(20.42) and the observed values y_{i1} and y_{i2}, the parameters for the conditional density follow immediately:

$$\begin{aligned} \mu_{i,2|1}(3) &= \mu_{i,2}(3) + V_{i,21}(3)[V_{i,11}(3)]^{-1}(y_i - \mu_{i,2}(3)), \\ V_{i,2|1}(3) &= V_{i,22}(3) - V_{i,21}(3)[V_{i,11}(3)]^{-1}V_{i,12}(3), \end{aligned}$$

where a subscript 1 indicates the first two components and a subscript 2 refers to the third component.

Based on (20.20), it is clear that the above procedure readily generalizes. This holds in particular for NCMV and CCMV, in which case no random uniform variates are required. Using (20.28), it is also clear that such a generalization is straightforward in the ACMV case. Formally, drawing from (20.20) consists of two steps:

- Draw a random uniform variate U to determine which of the $n - s + 1$ components from which one is going to draw. Specifically, the kth component is chosen if

$$\sum_{j=s}^{k-1} \omega_{sj} \leq U < \sum_{j=s}^{k} \omega_{sj},$$

where $k = s, \ldots, n$. Note that if $k = 1$, the left-hand sum is set equal to zero.

- Draw from the kth component. Since every component of the mixture is normal, only draws from uniform and normal random variables are required.

All of these steps have been combined in a SAS macro, which is available from the web site.

20.6   Analysis of the Vorozole Study

In order to concisely illustrate the methodology described in this chapter, we will apply it to the Vorozole study, restricted to those subjects with one, two, and three follow-up measurements, respectively. Thus, 190 subjects are included into the analysis, with subsample sizes 35, 86, and 69, respectively. The corresponding pattern probabilities are

$$\widehat{\pi} = (0.184, 0.453, 0.363)', \tag{20.43}$$

with asymptotic covariance matrix

$$\widehat{\text{var}}(\widehat{\pi}) = \begin{pmatrix} 0.000791 & -0.000439 & -0.000352 \\ -0.000439 & 0.001304 & -0.000865 \\ -0.000352 & -0.000865 & 0.001217 \end{pmatrix}. \tag{20.44}$$

These figures, apart from giving a feel for the relative importance of the various patterns, will be needed to calculate marginal effects (such as marginal treatment effect) from pattern-mixture model parameters, as was done in Sections 18.3 and 18.4.

We will apply each of the three strategies, presented in Section 20.5.1, to these data. First, within each of the strategies, starting models will be fitted (Section 20.6.1). Second, it will be illustrated how hypothesis testing can be performed, given the pattern-mixture parameter estimates and their estimated covariance matrix (Section 20.6.2). Third, model simplification will be discussed and applied (Section 20.6.3).

20.6.1   Fitting a Model

STRATEGIES 2 AND 1

In order to apply the identifying restriction Strategy 1, one needs to fit a model to the observed data first. We will opt for a simple model, with parameters specific to each pattern. Using extrapolation, such a model can be used for the second strategy as well.

We include time and time^2 effects, as well as their interactions with treatment. Further, time by baseline value interaction is included. Recall from Section 18.4 that all effects interact with time, in order to force profiles to pass through the origin, since we are studying change versus baseline. An unstructured 3×3 covariance matrix is assumed for each pattern.

Parameter estimates are presented in Table 20.1, in the "initial" column. Of course, not all parameters are estimable. This holds for the variance components, where in pattern 1 and 2 only the upper 1×1 block and the upper 2×2 block are identified, respectively. In the first pattern, the effects in time^2 are unidentified. The linear effects are identified by virtue of the absence of an intercept term.

Let us present this model graphically. Since there is one binary (treatment arm) and one continuous covariate (baseline level of FLIC score), and there are three patterns, we can represent the models using 3×2 surface plots, as in Figure 20.1. Similar plots for the other strategies are displayed in Figures 20.3, 20.5, 20.7, and 20.9. More insight on the effect of extrapolation can be obtained by presenting time plots for selected values of baseline value, the only continuous covariate. Precisely, we chose the minimum, average, and maximum values (Figure 20.2). Bold type is used for the range over which data are obtained within a particular pattern, and extrapolation is indicated using thinner type. Note that the extrapolation can have surprising effects, even with these relatively simple models. Thus, although this form of extrapolation is simple, its plausibility can be called into question. Note that this is in line with the experience gained from the toenail data analysis in Section 18.3 (Figure 18.2).

This initial model provides a basis, and its graphical representation extra motivation, to consider identifying-restrictions models. Using the methodology detailed in Section 20.5, a GAUSS macro and a SAS macro (available from the web) were written to conduct the multiple imputation, fitting of imputed data sets, and combination of the results into a single inference. Results are presented in Table 20.1, for each of the three types of restrictions (CCMV, NCMV, ACMV). For patterns 1 and 2, there is some variability in the parameter estimates across the three strategies, although this is often consistent with random variation (see the standard errors). Since the data in pattern 3 are complete, there is, of course, no difference between the initial model parameters and those obtained with each of the identifying-restrictions techniques. Again, a better impression can be obtained from a graphical representation. In all of the two-dimensional plots, the same mean response scale as in Figure 20.2 was retained, illustrating that the identifying-restrictions strategies extrapolate much closer to

TABLE 20.1. *Vorozole Study. Multiple imputation estimates and standard errors for CCMV, NCMV, and ACMV restrictions.*

Effect	Initial	CCMV	NCMV	ACMV
Pattern 1:				
Time	3.40(13.94)	13.21(15.91)	7.56(16.45)	4.43(18.78)
Time*base	−0.11(0.13)	−0.16(0.16)	−0.14(0.16)	−0.11(0.17)
Time*treat	0.33(3.91)	−2.09(2.19)	−1.20(1.93)	−0.41(2.52)
Time^2		−0.84(4.21)	−2.12(4.24)	−0.70(4.22)
Time^2*base		0.01(0.04)	0.03(0.04)	0.02(0.04)
σ_{11}	131.09(31.34)	151.91(42.34)	134.54(32.85)	137.33(34.18)
σ_{12}		59.84(40.46)	119.76(40.38)	97.86(38.65)
σ_{22}		201.54(65.38)	257.07(86.05)	201.87(80.02)
σ_{13}		55.12(58.03)	49.88(44.16)	61.87(43.22)
σ_{23}		84.99(48.54)	99.97(57.47)	110.42(87.95)
σ_{33}		245.06(75.56)	241.99(79.79)	286.16(117.90)
Pattern 2:				
Time	53.85(14.12)	29.78(10.43)	33.74(11.11)	28.69(11.37)
Time*base	−0.46(0.12)	−0.29(0.09)	−0.33(0.10)	−0.29(0.10)
Time*treat	−0.95(1.86)	−1.68(1.21)	−1.56(2.47)	−2.12(1.36)
Time^2	−18.91(6.36)	−4.45(2.87)	−7.00(3.80)	−4.22(4.20)
Time^2*base	0.15(0.05)	0.04(0.02)	0.07(0.03)	0.05(0.04)
σ_{11}	170.77(26.14)	175.59(27.53)	176.49(27.65)	177.86(28.19)
σ_{12}	151.84(29.19)	147.14(29.39)	149.05(29.77)	146.98(29.63)
σ_{22}	292.32(44.61)	297.38(46.04)	299.40(47.22)	297.39(46.04)
σ_{13}		57.22(37.96)	89.10(34.07)	99.18(35.07)
σ_{23}		71.58(36.73)	107.62(47.59)	166.64(66.45)
σ_{33}		212.68(101.31)	264.57(76.73)	300.78(77.97)
Pattern 3:				
Time	29.91(9.08)	29.91(9.08)	29.91(9.08)	29.91(9.08)
Time*base	−0.26(0.08)	−0.26(0.08)	−0.26(0.08)	−0.26(0.08)
Time*treat	0.82(0.95)	0.82(0.95)	0.82(0.95)	0.82(0.95)
Time^2	−6.42(2.23)	−6.42(2.23)	−6.42(2.23)	−6.42(2.23)
Time^2*base	0.05(0.02)	0.05(0.02)	0.05(0.02)	0.05(0.02)
σ_{11}	206.73(35.86)	206.73(35.86)	206.73(35.86)	206.73(35.86)
σ_{12}	96.97(26.57)	96.97(26.57)	96.97(26.57)	96.97(26.57)
σ_{22}	174.12(31.10)	174.12(31.10)	174.12(31.10)	174.12(31.10)
σ_{13}	87.38(30.66)	87.38(30.66)	87.38(30.66)	87.38(30.66)
σ_{23}	91.66(28.86)	91.66(28.86)	91.66(28.86)	91.66(28.86)
σ_{33}	262.16(44.70)	262.16(44.70)	262.16(44.70)	262.16(44.70)

the observed data mean responses. There are some differences among the identifying-restrictions methods. Roughly speaking, CCMV extrapolates rather toward a rise whereas NCMV seems to predict more of a decline, at least for baseline value 53. Further, ACMV predominantly indicates a steady state. For the other baseline levels, a status quo or a mild increase

Pattern 1, Megestrol Acetate, Extrapolated

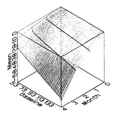

Pattern 1, Vorozole, Extrapolated

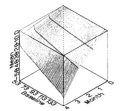

Pattern 2, Megestrol Acetate, Extrapolated

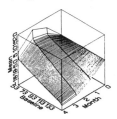

Pattern 2, Vorozole, Extrapolated

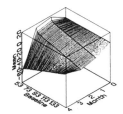

Pattern 3, Megestrol Acetate, Extrapolated

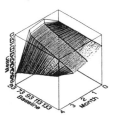

Pattern 3, Vorozole, Extrapolated

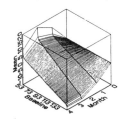

FIGURE 20.1. *Vorozole Study. Extrapolation based on initial model fitted to observed data (Strategy 2). Per pattern and per treatment group, the mean response surface is shown as a function of time (month) and baseline value.*

is predicted. This conclusion needs to be considered carefully. Since these patients drop out mainly because they relapse or die, it seems unlikely to expect a rise in quality of life. Hence, it is very possible that the dropout mechanism is not CCMV, since this strategy always refers to the "best" group, in the sense that it groups patients who stay longer in the study and hence have, on average, a better prognosis. ACMV, which compromises between all strategies, may be more realistic, but NCMV may be even better since information is borrowed from the nearest pattern, which is then based on the nearest patients in terms of dropout time and perhaps prognosis and quality of life evolution. However, recall that the identification is done sequentially, and hence even under NCMV, the first pattern is identified borrowing from both the second and the third patterns.

Nevertheless, the NCMV prediction looks more plausible since the worst baseline value shows declining profiles, whereas the best one leaves room for

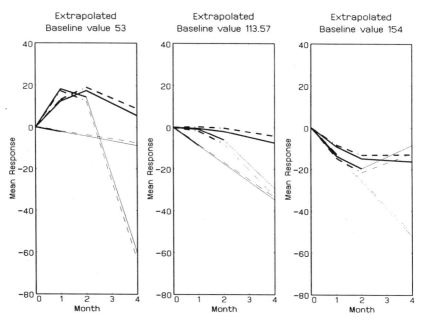

FIGURE 20.2. *Vorozole Study. Extrapolation based on initial model fitted to observed data (Strategy 2). For three levels of baseline value (minimum, average, maximum), plots of mean profiles over time are presented. The bold portion of the curves runs from baseline until the last obtained measurement, and the extrapolated piece is shown in thin type. The dashed line refers to megestrol acetate; the solid line is the Vorozole arm.*

improvement. Should one want to explore the effect of assumptions beyond the range of (20.20), one can allow ω_s to include components outside of the unit interval. In that situation, one has to ensure that the resulting density is still non-negative over its entire support. Finally, completely different restrictions can be envisaged as well.

SAS CODE FOR STRATEGIES 1 AND 2

We first present an example of a Strategy 2 model, which is also the starting point for the identifying restriction Strategy 1.

```
proc mixed data = vor01 method = ml
            noclprint asycov info covtest;
title 'Strategy 2 / Initial Model Strategy 1';
class treat timedisc id;
by pattern;
```

Pattern 1, Megestrol Acetate, CCMV

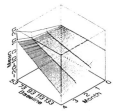

Pattern 1, Vorozole, CCMV

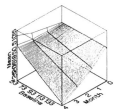

Pattern 2, Megestrol Acetate, CCMV

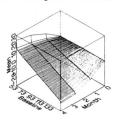

Pattern 2, Vorozole, CCMV

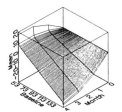

Pattern 3, Megestrol Acetate, CCMV

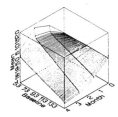

Pattern 3, Vorozole, CCMV

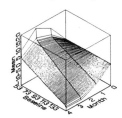

FIGURE 20.3. *Vorozole Study. Complete case missing value restrictions analysis. Per pattern and per treatment group, the mean response surface is shown as a function of time (month) and baseline value.*

```
model y = time base*time treat*time
          time*time base*time*time
          / noint solution covb;
repeated timedisc / subject = id type = un;
run;
```

Clearly, the essential bit is to include the BY statement in order to achieve a pattern-specific analysis. To apply the delta method (Agresti 1990) for the estimation and testing of marginal effects, it is important to generate the asymptotic covariance matrix of the parameters. This is done by including the 'covb' option in the MODEL statement for the fixed-effects parameters and the 'asycov' option in the PROC MIXED statement for the covariance parameters.

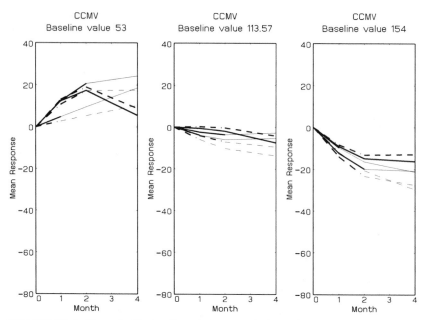

FIGURE 20.4. *Vorozole Study. Complete case missing value restrictions analysis. For three levels of baseline value (minimum, average, maximum), plots of mean profiles over time are presented. The bold portion of the curves runs from baseline until the last obtained measurement, whereas the extrapolated piece is shown in thin line type. The dashed line refers to megestrol acetate, the solid line is the Vorozole arm.*

It is useful to note that an alternative parameterization is possible as well:

```
proc mixed data = vor01 method = ml
             noclprint asycov info covtest;
title 'Strategy 2 / Initial Model Strategy 1';
title2 'Alternative parameterization';
class treat timedisc id pattern;
model y = time*pattern base*time*pattern
          treat*time*pattern time*time*pattern
          base*time*time*pattern
          / noint solution covb;
repeated timedisc / subject = id type = un
                    group = pattern;
run;
```

Exactly the same results are obtained as before, by ensuring that every effect interacts with the class variable pattern, and 'group=pattern' is in-

Pattern 1, Megestrol Acetate, NCMV

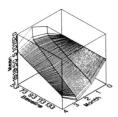

Pattern 1, Vorozole, NCMV

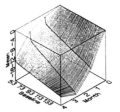

Pattern 2, Megestrol Acetate, NCMV

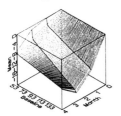

Pattern 2, Vorozole, NCMV

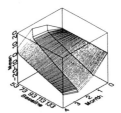

Pattern 3, Megestrol Acetate, NCMV

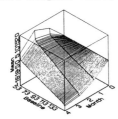

Pattern 3, Vorozole, NCMV

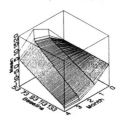

FIGURE 20.5. *Vorozole Study. Neighboring case missing value restrictions analysis. Per pattern and per treatment group, the mean response surface is shown as a function of time (month) and baseline value.*

cluded in the REPEATED statement. The first parameterization is useful to clearly separate the model elements for each of the patterns, whereas the second one is advantageous when further calculations are needed across patterns, such as the assessment of all treatment effect parameters simultaneously. See also Section 20.6.2.

Based on the model output, multiple imputation can be conducted. Details on the macro used to do this are suppressed. After multiple imputation has been conducted, a single meta-data set is obtained, which contains M completed data sets. Although it is possible to create M different copies, it is more convenient to paste them together, so that a single analysis program suffices to analyze all of them, which can be done as follows:

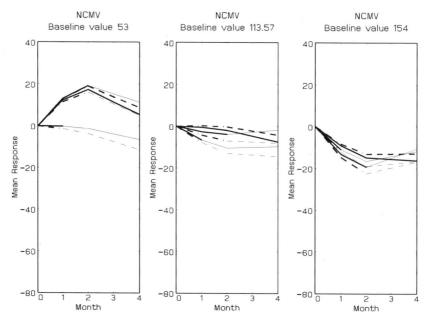

FIGURE 20.6. *Vorozole Study. Neighboring case missing value restrictions analysis. For three levels of baseline value (minimum, average, maximum), plots of mean profiles over time are presented. The bold portion of the curves runs from baseline until the last obtained measurement, and the extrapolated piece is shown in thin type. The dashed line refers to megestrol acetate; the solid line is the Vorozole arm.*

```
proc sort data = m.vor02nc;
by imput pattern;
run;

proc mixed data = m.vor02nc method = ml covtest
                  noclprint asycov info;
title 'NCMV';
class id imput pattern;
by imput pattern;
model y = time base*time treat*time time*time
        base*time*time / noint solution covb;
repeated timedisc / subject = id type = un;
make 'solutionf' out = m.fixednc;
make 'CovParms'  out = m.covparnc;
make 'COVB'      out = m.covbnc;
make 'asycov'    out = m.asycovnc;
run;
```

Pattern 1, Megestrol Acetate, ACMV

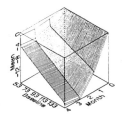

Pattern 1, Vorozole, ACMV

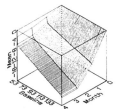

Pattern 2, Megestrol Acetate, ACMV

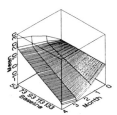

Pattern 2, Vorozole, ACMV

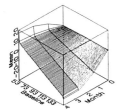

Pattern 3, Megestrol Acetate, ACMV

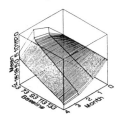

Pattern 3, Vorozole, ACMV

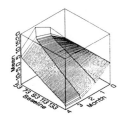

FIGURE 20.7. *Vorozole Study. Available case missing value restrictions analysis. Per pattern and per treatment group, the mean response surface is shown as a function of time (month) and baseline value.*

In order to conduct parameter estimation (Section 20.3.1) and hypothesis testing (Section 20.3.2), parameter estimates and estimated covariance matrices for the fixed effects and variance components need to be conserved. This is done by means of four MAKE statements. The actual combination of these into a single inference is done in a separate macro.

STRATEGY 3

In this strategy, *pattern* is included as a covariate, as was done in Section 18.4. An initial model is considered with the following effects: time, the interaction between time and treatment, baseline value, pattern, treatment∗baseline, treatment∗pattern, and baseline∗pattern. Further, time^2 is included, as well as its interaction with baseline, treatment, and pattern.

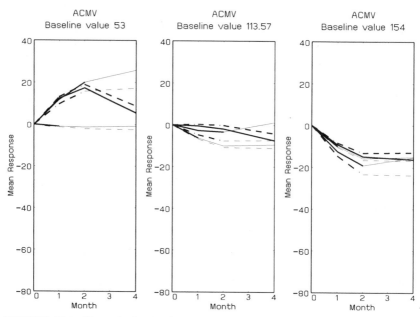

FIGURE 20.8. *Vorozole Study. Available case missing value restrictions analysis. For three levels of baseline value (minimum, average, maximum), plots of mean profiles over time are presented. The bold portion of the curves runs from baseline until the last obtained measurement, and the extrapolated piece is shown in thin type. The dashed line refers to megestrol acetate; the solid line is the Vorozole arm.*

No higher order interactions are included, and the unstructured covariance structure is common to all three patterns. This implies that the current model is *not* equivalent to a Strategy 1 model, where all parameters are pattern-specific. In order to achieve this goal, every effect would have to be made pattern-dependent.

The estimated model parameters are presented in Table 20.2. Graphical representations are given in Figures 20.9 and 20.10. From the latter, we can make two distinct observations. First, the behavior of the model over the range of the observed data is very similar to the behavior seen in the analysis of these data in Section 18.4. Precisely, early dropouts decline immediately, whereas those who stay longer in the study first show a rise and then decline thereafter. However, this is less pronounced for higher baseline values. Second, the extrapolation based on the fitted model is very unrealistic, in the sense that for the early dropout sharp rises are predicted, which is extremely implausible.

TABLE 20.2. *Vorozole Study. Parameter estimates and standard errors for Strategy 3.*

Effect	Pattern	Estimate (s.e.)
Time	1	7.29(15.69)
Time	2	37.05(7.67)
Time	3	39.40(9.97)
Time*treat	1	5.25(6.41)
Time*treat	2	3.48(5.46)
Time*treat	3	3.44(6.04)
Time*base	1	−0.21(0.15)
Time*base	2	−0.34(0.06)
Time*base	3	−0.36(0.08)
Time*treat*base		−0.06(0.04)
$Time^2$	1	
$Time^2$	2	−9.18(2.47)
$Time^2$	3	−7.70(2.29)
$Time^2$*treat		1.10(0.74)
$Time^2$*base		0.07(0.02)
σ_{11}		173.63(18.01)
σ_{12}		117.88(17.80)
σ_{22}		233.86(26.61)
σ_{13}		89.59(24.56)
σ_{23}		116.12(34.27)
σ_{33}		273.98(48.15)

These findings suggest, again, that a more careful reflection on the extrapolation method is required. This is very well possible in a pattern-mixture context, but then the first strategy, rather than the second or third strategy, has to be used, either as model of choice or to supplement insight gained from another model.

SAS CODE FOR STRATEGY 3

The following code can be used:

```
proc mixed data = m.vor01 method = ml noclprint asycov
                covtest info;
title 'Strategy 3';
class treat timedisc id pattern;
model y = time
        time*base
```

Pattern 1, Megestrol Acetate, Strategy 3

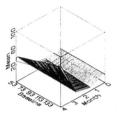

Pattern 1, Vorozole, Strategy 3

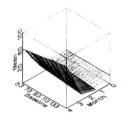

Pattern 2, Megestrol Acetate, Strategy 3

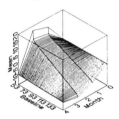

Pattern 2, Vorozole, Strategy 3

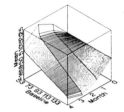

Pattern 3, Megestrol Acetate, Strategy 3

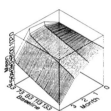

Pattern 3, Vorozole, Strategy 3

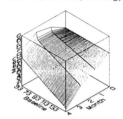

FIGURE 20.9. *Vorozole Study. Models with pattern used as a covariate (Strategy 3). Per pattern and per treatment group, the mean response surface is shown as a function of time (month) and baseline value.*

```
                    time*treat
                    time*treat*base
                    time*pattern
                    time*pattern*base
                    time*pattern*treat
                    time*time
                    time*time*base
                    time*time*treat
                    time*time*pattern
                    / noint solution covb;
          repeated timedisc / subject = id type = un;
          run;
```

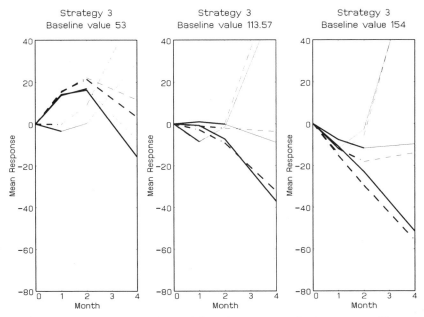

FIGURE 20.10. *Vorozole Study. Models with* pattern *used as a covariate (Strategy 3). For three levels of baseline value (minimum, average, maximum), plots of mean profiles over time are presented. The bold portion of the curves runs from baseline until the last obtained measurement, and the extrapolated piece is shown in thin type. The dashed line refers to megestrol acetate; the solid line is the Vorozole arm.*

The above program uses a hierarchical parameterization, implying that all interactions represent contrasts versus the main effects. For example, treatment effect is represented as time*treat and in addition time*treat*pattern. An equivalent but more parsimonious program, which includes exactly one treatment effect parameter for each dropout group, is

```
proc mixed data = m.vor01 method = ml noclprint asycov
                   covtest info;
title 'Strategy 3';
class   timedisc id pattern;
model y = time(pattern)
         time*treat(pattern)
         time*base(pattern)
         time*treat*base
         time*time(pattern)
         time*time*base
         time*time*treat
```

```
        / noint solution covb;
repeated timedisc / subject = id type = un;
run;
```

This program is also more parsimonious in the sense that it treats treatment as a continuous variable which, when there are only two modalities, comes down to exactly the same model as when it is treated as a class variable, but a number of structural zeros are removed from the parameter vector.

20.6.2   Hypothesis Testing

For ease of exposition, let us assume we are interested in a single effect, treatment effect, say. In the particular case of the Vorozole data, this translates into the time∗treatment interaction parameter. For simplicity, we will generically refer to the parameter of interest as *treatment effect*. For ease of exposition, we ignore the time^2∗treatment interaction. It is a simple extension to include both into the test. In the simplest case of a single parameter for the effect of interest, the corresponding selection model would contain exactly this single treatment effect parameter, turning the hypothesis testing task into a very straightforward one. If there were several treatment effect parameters, such as in a three-armed trial or in an analysis where interactions between treatment and other effects are included, standard hypothesis testing theory, such as in Section 6.2, could be applied.

Some pattern-mixture models will have a treatment effect parameter specific to each pattern. This is the case for all five models in Tables 20.1 and 20.2. Let us note in passing that this does not need to be the case. For example, in the final Strategy 3 analysis in Section 20.6.3, treatment effect is reduced to a single parameter. In such cases, the assessment of treatment effect is no more difficult than in a corresponding selection model. Therefore, this section will focus on the situation where there are pattern-dependent treatment effects.

It is useful to point out a strong analogy with post hoc stratification, where pattern plays the role of a stratifying variable. A selection model corresponds to a pooled analysis, where data from all patterns (strata) are pooled, without correction for the "confounding effect" stemming from heterogeneity across dropout patterns. A pattern-mixture model on the other hand does correct for pattern and hence, in a sense, for the confounding effect arising from pattern. If treatment effect does not interact with pattern, such as in the Strategy 3 analysis in Section 20.6.3, then a simple, so-called *corrected*, treatment effect estimate is obtained. Finally, if treatment effect interacts with pattern, such as in all five models above (although it is not significant in this case), there is heterogeneity of treatment effect across patterns (cf. heterogeneity of the relative risks in epidemiological studies).

In the latter case, two distinct routes are possible. The more "epidemiologic" viewpoint is to direct inferences toward the vector of treatment effects. For example, if treatment effects are heterogeneous across patterns, then it may be deemed better to avoid combining such effects into a single measure. In our case, this implies, for example, testing for the treatment by time interaction to be zero in all three patterns simultaneously. Alternatively, one can calculate the same quantity as would be obtained in the corresponding selection model. Then, the *marginal* treatment effect is calculated, based on the pattern-specific treatment effects and the weighting probabilities, perhaps irrespective of whether the treatment effects are homogeneous across patterns or not. This was done in (18.9) and (18.11) for the toenail data, and in Section 18.4. See also Section 24.4.2 (Eq. (20.45)).

Precisely, let $\beta_{\ell t}$ represent the treatment effect parameter estimates $\ell = 1, \ldots, g$ (assuming there are $g + 1$ groups) in pattern $t = 1, \ldots, n$ and let π_t be the proportion of patients in pattern t. Then, the estimates of the marginal treatment effects β_ℓ are

$$\beta_\ell = \sum_{t=1}^{n} \beta_{\ell t} \pi_t, \qquad \ell = 1, \ldots, g. \tag{20.45}$$

The variance is obtained using the delta method. Precisely, it assumes the form

$$\text{var}(\beta_1, \ldots, \beta_g) = AVA', \tag{20.46}$$

where

$$V = \left(\begin{array}{c|c} \text{var}(\beta_{\ell t}) & 0 \\ \hline 0 & \text{var}(\pi_t) \end{array} \right) \tag{20.47}$$

and

$$A = \frac{\partial(\beta_1, \ldots, \beta_g)}{\partial(\beta_{11}, \ldots, \beta_{ng}, \pi_1, \ldots, \pi_n)}. \tag{20.48}$$

The estimate of the variance-covariance matrix of the $\widehat{\beta}_{\ell t}$ is obtained from statistical software (e.g., the 'covb' option in the MODEL statement of the SAS procedure MIXED). The multinomial quantities are easy to obtain from the pattern-specific sample sizes. In the case of the Vorozole data, these quantities are presented in (20.43) and (20.44). A Wald test statistic for the null hypothesis $H_0 : \beta_1 = \ldots = \beta_g = 0$ is then given by

$$\beta_0'(AVA')^{-1}\beta_0, \tag{20.49}$$

where $\boldsymbol{\beta}_0 = (\beta_1, \ldots, \beta_g)'$.

We will now apply both testing approaches to the models presented in Tables 20.1 and 20.2. All three pattern-mixture strategies will be considered. Since the identifying restrictions strategies are slightly more complicated than the others, we will consider the other strategies first.

STRATEGY 2

Recall that the parameters are presented in Table 20.1 as the initial model. The treatment effect vector is $\boldsymbol{\beta} = (0.33, -0.95, 0.82)'$ with, since the patterns are analyzed separately, diagonal covariance matrix:

$$V = \begin{pmatrix} 15.28 & & \\ & 3.44 & \\ & & 0.90 \end{pmatrix}.$$

These quantities are obtained as either the square of the standard errors reported in the 'Solution for Fixed Effects' panel in the output of the SAS procedure MIXED or, directly, as the appropriate diagonal elements of the 'Covariance Matrix for Fixed Effects' panel, produced by means of the 'covb' option in the MODEL statement. This leads to the test statistic $\boldsymbol{\beta}'V^{-1}\boldsymbol{\beta} = 1.02$ on 3 degrees of freedom, producing $p = 0.796$.

In order to calculate the marginal treatment effect, we apply (20.46)–(20.49). The single (since there are only two groups) marginal effect is estimated as $\widehat{\beta_0} = -0.07$ (s.e. 1.16). The corresponding observed asymptotic p-value is $p = 0.95$.

Both approaches agree on the nonsignificance of the treatment effect.

STRATEGY 3

The parameters are presented in Table 20.2. The treatment effect vector is $\boldsymbol{\beta} = (5.25, 3.48, 3.44)'$ with nondiagonal covariance matrix

$$V = \begin{pmatrix} 41.12 & 23.59 & 25.48 \\ 23.59 & 29.49 & 30.17 \\ 25.48 & 30.17 & 36.43 \end{pmatrix}.$$

These quantities are obtained as the appropriate block of the 'Covariance Matrix for Fixed Effects' panel. The information provided by the square of the standard errors is insufficient since they fail to report the covariances between the parameter estimates. In this case, the correlation between them is quite substantial. The reason is that some parameters, in particular, the other treatment effects (three-way interaction with baseline and time, interaction with time^2), are common to all three patterns, inducing dependence across patterns.

This leads to the test statistic $\beta'V^{-1}\beta = 0.70$ on 3 degrees of freedom, producing $p = 0.874$. The same p-value up to three digits is reported in the 'Tests of Fixed Effects' panel, where the same test is conducted using an F-statistic.

Calculating the marginalized treatment effect, we obtain $\widehat{\beta_0} = 3.79$ (s.e. 5.44). The corresponding asymptotic p-value is $p = 0.49$. The different numerical value of the treatment effects, as compared to those obtained with the other strategies, is entirely due to the presence of a quadratic treatment effect, which, for ease of exposition, is left out of the picture in testing here. If deemed necessary, and often it will, it is straightforward to add this parameter to the contrast(s) being considered.

STRATEGY 1

The CCMV case will be discussed in detail. The two other restriction types are entirely similar.

There are three treatment effects, one for each pattern. Hence, multiple imputation produces five vectors of three treatment effects which are averaged to produce a single treatment effect vector. In addition, the within, between, and total covariance matrices are calculated:

$$\beta_{CC} = (-2.09, -1.68, 0.82)', \tag{20.50}$$

$$W_{CC} = \begin{pmatrix} 1.67 & 0.00 & 0.00 \\ 0.00 & 0.59 & 0.00 \\ 0.00 & 0.00 & 0.90 \end{pmatrix}, \tag{20.51}$$

$$B_{CC} = \begin{pmatrix} 2.62 & 0.85 & 0.00 \\ 0.85 & 0.72 & 0.00 \\ 0.00 & 0.00 & 0.00 \end{pmatrix}, \tag{20.52}$$

and

$$T_{CC} = \begin{pmatrix} 4.80 & 1.02 & 0.00 \\ 1.02 & 1.46 & 0.00 \\ 0.00 & 0.00 & 0.90 \end{pmatrix}. \tag{20.53}$$

In the stratified case, we want to test the hypothesis $H_0 : \beta = 0$. Using (20.50)–(20.52), we can apply the multiple imputation results described in Section 20.3.2.

TABLE 20.3. *Vorozole Study. Tests of treatment effect for CCMV, NCMV, and ACMV restrictions.*

Parameter	CCMV	NCMV	ACMV
Stratified analysis:			
k	3	3	3
τ	12	12	12
denominator d.f. w	28.41	17.28	28.06
r	1.12	2.89	1.14
F-statistic	1.284	0.427	0.946
p-value	0.299	0.736	0.432
Marginal Analysis:			
Marginal effect (s.e.)	$-0.85(0.77)$	$-0.63(1.22)$	$-0.74(0.85)$
k	1	1	1
τ	4	4	4
denominator d.f. w	4	4	4
r	1.49	4.57	1.53
F-statistic	0.948	0.216	0.579
p-value	0.385	0.667	0.489

Note that, even though the analysis is done per pattern, the between and total matrices have nonzero off-diagonal elements. This is because imputation is done based on information from *other* patterns, hence introducing interpattern dependence. Results are presented in Table 20.3. All results are nonsignificant, in line with earlier evidence from Strategies 2 and 3, although the p-values for CCMV and ACMV are somewhat smaller.

For the marginal parameter, the situation is more complicated here than with Strategies 2 and 3. Indeed, the theory of Section 20.3.2 assumes inference is geared toward the original vector, or linear contrasts thereof. Formula (20.45) displays a nonlinear transformation of the parameter vector and therefore needs further development. First, consider π to be part of the parameter vector. Since there is no missingness involved in this part, it contributes to the within matrix, but not to the between matrix. Then, using (20.46), the approximate within matrix for the marginal treatment effect is

$$W_0 \;=\; \pi'W\pi + \beta'\mathrm{var}(\pi)\beta,$$

with, for the between matrix, simply

$$B_0 \;=\; \pi'B\pi.$$

The latter formula consists of only one term, since there is no between variance for π.

The results are presented in the second panel of Table 20.3. All three p-values are, again, nonsignificant, in agreement with Strategies 2 and 3. Of course, all five agree on the nonsignificance of the treatment effect. The reason for the differences is to be found in the way the treatment effect is extrapolated beyond the period of observation. Indeed, the highest p-value is obtained for Strategy 2, and from Figure 20.2, we learn that virtually no separation between both treatment arms is projected. On the other hand, wider separations are seen in Figure 20.10.

20.6.3   Model Reduction

Model building guidelines for the standard linear mixed-effects model can be found in Chapter 9. These guidelines can be used without any problem in a selection model context, but the pattern-mixture case is more complicated. Of course, the same general principles can be applied, taking into account the intertwining between the mean or fixed-effects structure, on the one hand, and the components of variability on the other hand, as graphically represented in Figure 9.1.

In addition to these principles, one has to reflect on the special status of *pattern* in a pattern-mixture model. Broadly, we can distinguish between two cases, reflecting Strategy 2 (a per-pattern analysis) and Strategy 3 (use pattern as a covariate). In fact, the identifying-restrictions strategy leaves the method of analysis to be used after multiple imputation unspecified, as mentioned in item 6 of the strategy outline (Section 20.5.1). In our analysis, we have chosen to conduct a per-pattern analysis (Table 20.1) as in Strategy 1, but it is possible to conduct a global analysis, using pattern as a covariate, or even to use selection modeling. The only requirement is that the *proper* nature of the imputation be preserved (Rubin 1987). It is therefore sufficient to discuss and illustrate model reduction using the second and third strategies only.

STRATEGY 3

Model reduction in a context where pattern is used as a covariate is clearly of the same level of complexity as with complete data or as for a selection model. Let us reduce the model presented in Table 20.2. It is convenient to use a hierarchical representation of the model as with the second SAS program (p. 365). The following effects are removed using a hierarchical sequence of models, and using F-test statistics: the time by pattern by treatment interaction ($p = 0.934$), the time by pattern interaction ($p = 0.776$), the time by pattern by baseline value interaction ($p = 0.707$), the time by baseline by treatment interaction ($p = 0.165$), and the time^2 by treatment

TABLE 20.4. *Vorozole Study. Strategy 3. Parameter estimates and standard errors of a reduced model.*

Effect	Pattern	Estimate (s.e.)
Time		33.06(6.67)
Time*treat		0.40(0.84)
Time*base		−0.29(0.06)
$Time^2$	1	−16.71(3.46)
$Time^2$	2	−8.56(1.90)
$Time^2$	3	−7.09(1.78)
$Time^2$*base		0.06(0.01)
σ_{11}		178.02(18.46)
σ_{12}		121.75(18.30)
σ_{22}		238.31(26.98)
σ_{13}		88.75(24.94)
σ_{23}		121.10(34.70)
σ_{33}		274.58(48.32)

interaction ($p = 0.093$). Note that one cannot necessarily conclude that these parameters automatically are jointly nonsignificant (see Section 5.5). The reduced model is displayed in Table 20.4.

STRATEGY 2

For Strategy 2, where a per-pattern analysis is conducted, there are several model building decisions to be made:

- In the process of simplifying, one can allow that effects be shared between two or more patterns. For example, a baseline effect, common to all patterns, can be included. By doing so, this strategy effectively reduces to Strategy 3 and there is no need to discuss this any further here.

- When simplifying the model, effects are either absent or common to all patterns. Again, this approach is close to Strategy 3 and can be conducted within that framework without any problem if one starts with a model where all effects, including the covariance parameters, depend on pattern. For this reason, we will not pursue it further.

- Finally, model reduction is done entirely separately in each of the patterns. This may yield different levels of simplification for each pattern and certainly a pattern-specific set of covariates, which is found to influence the response profile. This strategy will be illustrated.

In order to enable treatment effect assessment, the interaction between time and treatment will not be removed from the models. In pattern 1, there is one simplification possible in the sense that the interaction between time and baseline is not significant ($p = 0.415$). Thus, the only effects that remain in the model are time and the time by treatment interaction. For patterns 2 and 3, there are no non-significant effects to be removed. In conclusion, baseline FLIC score influences the follow-up scores in patterns 2 and 3, but not in pattern 1.

20.7   Thoughts

In this chapter, we have illustrated three distinct strategies to fit pattern-mixture models. In this way, we have brought together several existing practices. Little (1993, 1994a) has proposed identifying restrictions, which we formalized here using the connection with MAR (Section 20.2) and multiple imputation (Section 20.3). Strategies 2 and 3 refer to fitting a model per pattern, as in the toenail data (Section 18.3), and using pattern as a covariate, as in the Vorozole study (Section 18.4).

By contrasting these strategies on a single set of data, one obtains a range of conclusions rather than a single one, which provides insight into the sensitivity to the assumptions made. Especially with the identifying restrictions, one has to be very explicit about the assumptions and, moreover, this approach offers the possibility to consider several forms of restrictions. Special attention should go to the ACMV restrictions, since they are the MAR counterpart within the pattern-mixture context.

In addition, a comparison between the selection and pattern-mixture modeling approaches is useful to obtain additional insight into the data and/or to assess sensitivity. This has been done, informally, in Chapters 17 and 18, using the toenail data and the Vorozole study.

Section 24.4 offers a case study on the milk protein contents trial (Diggle and Kenward 1994). Several sensitivity analysis tools, both informal and formal, are employed to gain insight into the data and into the missingness mechanism.

The identifying-restrictions strategy provides further opportunity for sensitivity analysis, beyond what has been presented here. Indeed, since CCMV and NCMV are extremes for the ω_s vector in (20.20), it is very natural to consider the idea of *ranges* in the allowable space of ω_s. Clearly, any ω_s which consists of non-negative elements that sum to 1 is allowable, but also the idea of extrapolation could be useful, where negative components are allowed, given they provide valid conditional densities.

As in the previous chapter, we underscore that the strategies presented here are but one approach to sensitivity analysis in the pattern-mixture context. Surely, more will be developed and more work is needed in this area.

The SAS and GAUSS macros which have been used to carry out the multiple imputation related tasks are available from the authors' web pages.

21

How Ignorable Is Missing At Random ?

21.1   Introduction

For over two decades, following the pioneering work of Rubin (1976) and Little (1976), there has been a growing literature on incomplete data, with a lot of emphasis on longitudinal data. Following the original work of Rubin and Little, there has evolved a general view that "likelihood methods" that ignore the missing value mechanism are valid under an MAR process, where likelihood is interpreted in a frequentist sense. The availability of flexible standard software for incomplete data, such as PROC MIXED, and the advantages quoted in Section 17.3 contribute to this point of view. This statement needs careful qualification however. Kenward and Molenberghs (1998) provided an exposition of the precise sense in which frequentist methods of inference are justified under MAR processes.

As discussed in Section 15.8, Rubin (1976) has shown that MAR (and parameter distinctness) is necessary and sufficient to ensure validity of *direct-likelihood* inference when ignoring the process that causes missing data. Here, direct-likelihood inference is defined as an "inference that results solely from ratios of the likelihood function for various values of the parameter," in agreement with the definition in Edwards (1972). In the concluding section of the same paper, Rubin remarks:

> One might argue, however, that this apparent simplicity of likelihood and Bayesian inference really buries the important issues. (...) likelihood inferences are at times surrounded with references to the sampling distributions of likelihood statistics. Thus, practically, when there is the possibility of missing data, some interpretations of Bayesian and likelihood inference face the same restrictions as sampling distribution inference. The inescapable conclusion seems to be that when dealing with real data, the practicing statistician should explicitly consider the process that causes missing data far more often than he does.

In essence, the problem from a frequentist point of view is that of identifying and using the appropriate sampling distribution. This is obviously relevant for determining distributions of test statistics, expected values of the information matrix, and measures of precision.

Little and Rubin (1987) discuss several aspects of this problem and propose, using the observed information matrix, to circumvent problems associated with the determination of the correct expected information matrix. Laird (1988) makes a similar point in the context of incomplete longitudinal data analysis.

In a variety of settings, several authors have reexpressed this preference for the observed information matrix and derived methods to compute it: Meng and Rubin (1991), the supplemented EM algorithm; Baker (1992), composite link models; Fitzmaurice, Laird, and Lipsitz (1994), incomplete longitudinal binary data; and Jennrich and Schluchter (1986). A group of authors has used the observed information matrix, without reference to the problems associated with the expected information: Louis (1982), Meilijson (1989), and Kenward, Molenberghs, and Lesaffre (1994).

However, others, while claiming validity of analysis under MAR mechanisms, have used expected information matrices and other measures of precision that do not account for the missingness mechanism (Murray and Findlay 1988, Patel 1991). A number of references is given in Baker (1992). It is clear that the problem as identified in the initial work of Rubin (1976) is not fully appreciated in the more recent literature. An exception to this is Heitjan's (1994) clear restatement of the problem.

A recent exchange of correspondence (Diggle 1992, Heitjan 1993, Diggle 1993) indicates a genuine interest in these issues and suggests a need for clarification. We will build on the framework of likelihood inference under an MAR process, sketched in Section 15.5. The difference between the expected information matrix with and without taking the missing data mechanism into account is elucidated and the relevance of this for Wald and score statistics is elaborated upon. Analytic and numerical illustrations of this difference are provided using a bivariate Gaussian setting. A longitudinal example is used for practical illustration.

21.2 Information and Sampling Distributions

In this section, we will drop the subject subscript i from the notation. We assume that the joint distribution of the full data $(\boldsymbol{Y}, \boldsymbol{R})$ is regular in the sense of Cox and Hinkley (1974, p. 281). We are concerned here with the sampling distributions of certain statistics under MCAR and MAR mechanisms. Under an MAR process, the joint distribution of $\boldsymbol{Y}^o$ (the observed components) and $\boldsymbol{R}$ factorizes as in (15.6). In terms of the log-likelihood function, we have

$$\ell(\boldsymbol{\theta}, \boldsymbol{\psi}; \boldsymbol{y}^o, \boldsymbol{r}) \quad = \quad \ell_1(\boldsymbol{\theta}; \boldsymbol{y}^o) + \ell_2(\boldsymbol{\psi}; \boldsymbol{r}, \boldsymbol{y}^o). \tag{21.1}$$

It is assumed that $\boldsymbol{\theta}$ and $\boldsymbol{\psi}$ satisfy the separability condition. This partition of the likelihood has, with important exceptions, been taken for granted to mean that, under an MAR mechanism, likelihood methods based on ℓ_1 alone are valid for inferences about $\boldsymbol{\theta}$ *even when interpreted in the broad frequentist sense*. We now consider more precisely the sense in which the different elements of the frequentist likelihood methodology can be regarded as valid in general under the MAR mechanism. It is now well known that such inferences are valid under an MCAR mechanism (Rubin 1976, Section 6).

First, we note that under the MAR mechanism, $\boldsymbol{r}$ is *not* an ancillary statistic for $\boldsymbol{\theta}$ in the extended sense of Cox and Hinkley (1974, p. 35). (A statistic $S(\boldsymbol{Y}, \boldsymbol{R})$ is ancillary for $\boldsymbol{\theta}$ if its distribution does not depend upon $\boldsymbol{\theta}$.) Hence, we are not justified in restricting the sample space from that associated with the pair $(\boldsymbol{Y}, \boldsymbol{R})$. In considering the properties of frequentist procedures below, we therefore define the appropriate sampling distributions to be that determined by this pair. We call this the *unconditional* sampling framework. By working within this framework, we do need to consider the missing value mechanism. We shall be comparing this with the sampling distribution that would apply if $\boldsymbol{r}$ were fixed by design [i.e., if we repeatedly sampled using the distribution $f(\boldsymbol{y}^o; \boldsymbol{\theta})$]. If this sampling distribution were appropriate, this would lead directly to the use of ℓ_1 as a basis for inference. We call this the *naive* sampling framework.

Little (1976), in a comment on the paper by Rubin (1976), mentions explicitly the role played by the nonresponse pattern. He argues:

> For sampling based inferences, a first crucial question concerns when it is justified to condition on the observed pattern,

that is on the event $R = r$ (...). A natural condition is that R should be ancillary (...). Otherwise the pattern on its own carries at least some information about $\boldsymbol{\theta}$, which should in principle be used.

Certain elements of the frequentist methodology can be justified immediately from (21.1). The maximum likelihood estimator obtained from maximizing $l_1(\boldsymbol{\theta}; \boldsymbol{y}^o)$ alone is identical to that obtained from maximizing the complete log-likelihood function. Similarly, the maximum likelihood estimator of ψ is functionally independent of $\boldsymbol{\theta}$ and so any maximum likelihood ratio concerning $\boldsymbol{\theta}$, with common ψ, will involve ℓ_1 only. Because these statistics are identical whether derived from ℓ_1 or the complete log-likelihood, it follows that they have the required properties under the naive sampling framework. See, for example, Rubin (1976), Little (1976), and Little and Rubin (1987, Section 5.2).

An important element of likelihood-based frequentist inference is the derivation of measures of precision of the maximum likelihood estimators from the information. For this, either the observed information, i_o, can be used where

$$i_O(\theta_j, \theta_k) = -\frac{\partial^2 \ell(\cdot)}{\partial \theta_j \partial \theta_k}$$

or the expected information, i_E, where

$$i_E(\theta_j, \theta_k) = E\{i_O(\theta_j, \theta_k)\}. \tag{21.2}$$

The above argument justifying the use of the maximum likelihood estimators from $\ell_1(\boldsymbol{\theta}; \boldsymbol{y}^o)$ applies equally well to the use of the inverse of the *observed* information derived from ℓ_1 as an estimate of the asymptotic variance-covariance matrix of these estimators. This has been pointed out by Little and Rubin (1987, Section 8.2.2) and Laird (1988, p. 307). In addition, there are other reasons for preferring the observed information matrix (Efron and Hinkley 1978).

The use of the expected information matrix is more problematical. The expectation in (21.2) needs to be taken over the *unconditional* sampling distribution (the *unconditional information i_U*) and, consequently, the use of the naive sampling framework (producing the *naive information i_N*) can lead to inconsistent estimates of precision. In the next section, we give an example of the bias resulting from the use of the naive framework. It is possible however, as we show below, to calculate the unconditional information by taking expectations over the appropriate distribution and so correct this bias. Although this added complication is generally unnecessary in practice, given the availability of the observed information, it does allow a direct examination of the effect of ignoring the missing value mechanism on the expected information.

As part of the process of frequentist inference, we also need to consider the sampling distribution of the test statistics. Provided that use is made of the likelihood ratio, or Wald and score statistics based on the observed information, then reference to a null asymptotic χ^2-distribution will be appropriate because this is derived from the *implicit* use of the unconditional sampling framework. Only in those situations in which the sampling distribution is explicitly constructed must care be taken to ensure that the unconditional framework is used; that is, account must be taken of the missing data mechanism.

21.3   Illustration

For an incomplete multivariate normal sample, Little and Rubin (1987) state:

> If the data are MCAR, the expected information matrix of $\theta = (\mu, \Sigma)$ represented as a vector is block diagonal. (...) The observed information matrix, which is calculated and inverted at each iteration of the Newton-Raphson algorithm, is not block diagonal with respect to μ and Σ, so this simplification does not occur if standard errors are based on this matrix. On the other hand, the standard errors based on the observed information matrix are more conditional and thus valid when the data are MAR but not MCAR, and hence should be preferable to those based on [the expected information] in applications.

Suppose now that we have N independent pairs of observations (Y_{i1}, Y_{i2}), each with a bivariate Gaussian distribution with mean vector $\mu = (\mu_1, \mu_2)'$ and variance-covariance matrix

$$\Sigma = \begin{pmatrix} \sigma_{11} & \sigma_{12} \\ \sigma_{12} & \sigma_{22} \end{pmatrix}.$$

It is assumed that m complete pairs, and only the first member (Y_{i1}) of the remaining pairs are observed. This implies that the dropout process can be represented by a scalar indicator R_i which is 1 if the second component is observed and 0 otherwise. The log-likelihood can be expressed as the sum of the log-likelihoods for the complete and incomplete pairs:

$$\ell = \sum_{i=1}^{m} \ln f(y_{i1}, y_{i2} \mid \mu_1, \mu_2, \sigma_{11}, \sigma_{12}, \sigma_{22}) + \sum_{i=m+1}^{N} \ln f(y_{i1} \mid \mu_1, \sigma_{11}),$$

which, in the Gaussian setting, has kernel,

$$
\ell = -\frac{N-m}{2}\ln\sigma_{11} - \frac{m}{2}\ln|\Sigma| - \frac{1}{2\sigma_{11}}\sum_{i=m+1}^{N}(y_{i1}-\mu_1)^2
$$

$$
-\frac{1}{2}\sum_{i=1}^{m}\left(\begin{array}{c} y_{i1}-\mu_1 \\ y_{i2}-\mu_2 \end{array}\right)'\left(\begin{array}{cc} \sigma_{11} & \sigma_{12} \\ \sigma_{12} & \sigma_{22} \end{array}\right)^{-1}\left(\begin{array}{c} y_{i1}-\mu_1 \\ y_{i2}-\mu_2 \end{array}\right).
$$

Straightforward differentiation produces the elements of the observed information matrix that relate to $\boldsymbol{\mu}$:

$$
i_o(\boldsymbol{\mu},\boldsymbol{\mu}) = (N-m)\left(\begin{array}{cc} \sigma_{11}^{-1} & 0 \\ 0 & 0 \end{array}\right) + m\Sigma^{-1},
$$

and

$$
i_o(\mu_1,\sigma_{11}) = \sum_{i=m+1}^{N}\frac{y_{i1}-\mu_1}{\sigma_{11}^2}
$$

$$
+\sum_{i=1}^{m}e_1'\Sigma^{-1}E_{11}\Sigma^{-1}\left(\begin{array}{c} y_{i1}-\mu_1 \\ y_{i2}-\mu_2 \end{array}\right) \qquad (21.3)
$$

and, when at least one of the indices j, k, or ℓ is different from 1,

$$
i_o(\mu_j,\sigma_{k\ell}) = \sum_{i=1}^{m}e_j'\Sigma^{-1}E_{k\ell}\Sigma^{-1}\left(\begin{array}{c} y_{i1}-\mu_1 \\ y_{i2}-\mu_2 \end{array}\right) \qquad (21.4)
$$

for

$$
e_1 = \left(\begin{array}{c} 1 \\ 0 \end{array}\right), \quad e_2 = \left(\begin{array}{c} 0 \\ 1 \end{array}\right)
$$

and

$$
E_{11} = \left(\begin{array}{cc} 1 & 0 \\ 0 & 0 \end{array}\right), \quad E_{12} = \left(\begin{array}{cc} 0 & 1 \\ 1 & 0 \end{array}\right), \quad E_{22} = \left(\begin{array}{cc} 0 & 0 \\ 0 & 1 \end{array}\right).
$$

For the naive information, we just take expectations of these quantities over $(Y_{i1},Y_{i2})' \sim N(\boldsymbol{\mu},\Sigma)$ for $i = 1,\ldots m$ and $Y_{i1} \sim N(\mu_1,\sigma_{11})$ for $i = m+1,\ldots,N$. It follows at once that the cross-terms linking the mean and variance-covariance parameters vanish, establishing the familiar orthogonality property of these sets of parameters in the Gaussian setting. We now examine the behavior of the expected information under the actual sampling process implied by the MAR mechanism.

We need to consider first the conditional expectation of these quantities given the occurrence of R, the dropout pattern. Because (Y,R) enters the expression for $i_U(\boldsymbol{\mu}, \boldsymbol{\mu})$ only through m, the naive and unconditional information matrices for $\boldsymbol{\mu}$ are effectively equivalent. However, we show now that this is not true for the cross-term elements of the information matrices. Define $\alpha_j = E(Y_{i1} \mid r_i = j) - \mu_1$, $j = 0, 1$. For the conditional expectation of Y_{i2} in the complete subgroup, we have

$$
E(Y_{i2} \mid r_i = 1) = \int \left\{ y_{i2} \int f(y_{i2} \mid y_{i1}) dy_{i2} \right\} f(y_{i1} \mid r_i = 1) dy_{i1}
$$

$$
= \mu_2 - \sigma_{12}\sigma_{11}^{-1}\mu_1 + \frac{\sigma_{12}}{\sigma_{11}P(r_i = 1)} \int y_{i1} f(y_{i1}, r_i = 1) dy_{i1}
$$

$$
= \mu_2 + \sigma_{12}\sigma_{11}^{-1}\{E(Y_{i1} \mid r_i = 1) - \mu_1\}
$$

or

$$
E_{Y|R}(Y_{i2} - \mu_2) = \beta\alpha_1
$$

for $\beta = \sigma_{12}\sigma_{11}^{-1}$. Hence,

$$
E_{Y|R}\left\{ \begin{pmatrix} Y_{i1} - \mu_1 \\ Y_{i2} - \mu_2 \end{pmatrix} \right\} = \alpha_1 \begin{pmatrix} 1 \\ \beta \end{pmatrix}.
$$

Noting that

$$
\Sigma^{-1} \begin{pmatrix} 1 \\ \beta \end{pmatrix} = \begin{pmatrix} \sigma_{11}^{-1} \\ 0 \end{pmatrix} = \sigma_{11}^{-1}\boldsymbol{e_1},
$$

we then have from (21.3) and (21.4)

$$
E_{Y|R}\{i_o(\mu_j, \sigma_{kl})\} = \begin{cases} (N - m)\dfrac{\alpha_0}{\sigma_{11}^2} \\[2mm] +m\dfrac{\alpha_1}{\sigma_{11}}\boldsymbol{e_1}'\Sigma^{-1}\boldsymbol{E_{11}e_1}, \quad j = k = \ell = 1 \\[3mm] m\dfrac{\alpha_1}{\sigma_{11}}\boldsymbol{e_j}'\Sigma^{-1}\boldsymbol{E_{k\ell}e_1} \qquad \text{otherwise.} \end{cases}
$$

Finally, taking expectations over R, we get for the cross-terms of the unconditional information matrix

$$
i_U(\boldsymbol{\mu}, \sigma_{11}) = \frac{N}{\sigma_{11}} \left\{ \frac{(1 - \pi)\alpha_0}{\sigma_{11}} \begin{pmatrix} 1 \\ 0 \end{pmatrix} + \frac{\pi\alpha_1}{\sigma_{11}\sigma_{22} - \sigma_{12}^2} \begin{pmatrix} \sigma_{22} \\ -\sigma_{12} \end{pmatrix} \right\}, \quad (21.5)
$$

$$
i_U(\boldsymbol{\mu}, \sigma_{12}) = \frac{N\pi\alpha_1}{\sigma_{11}\sigma_{22} - \sigma_{12}^2} \begin{pmatrix} -\beta \\ 1 \end{pmatrix}, \quad (21.6)
$$

$$
i_U(\boldsymbol{\mu}, \sigma_{22}) = \begin{pmatrix} 0 \\ 0 \end{pmatrix}, \quad (21.7)
$$

for $\pi = P(r_i = 1)$. In contrast to the naive information, these cross-terms do not all vanish, and the orthogonality of mean and variance-covariance parameters is lost under the MAR mechanism. One implication of this is that although the information relating to the linear model parameters alone is not affected by the move from an MCAR to an MAR mechanism, the asymptotic variance-covariance matrix *is* affected due to the induced non-orthogonality and, therefore, the dropout mechanism cannot be regarded as ignorable as far as the estimation of precision of the estimators of the linear model parameters is concerned. It can also be shown that the expected information for the variance-covariance parameters is not equivalent under the MCAR and MAR dropout mechanisms, but the expressions are more involved. Assuming that π is nonzero, it can be seen that the necessary and sufficient condition for the terms in (21.5) and (21.6) to be equal to zero is that $\alpha_0 = \alpha_1 = 0$, the condition defining, as expected, an MCAR mechanism.

We now illustrate these findings with a few numerical results. The off-diagonal unconditional information elements (21.5)–(21.7) are computed for sample size $N = 1000$, mean vector $(0,0)'$, and two covariance matrices: (1) $\sigma_{11} = \sigma_{22} = 1$ and correlation $\rho = \sigma_{12} = 0.5$ and (2) $\sigma_{11} = 2$, $\sigma_{33} = 3$, and $\rho = 0.5$ leading to $\sigma_{12} = \sqrt{6}/2$. Further, two MAR dropout mechanisms are considered. They are both of the logistic form

$$P(R_1 = 0|y_{i1}) = \frac{\exp(\gamma_0 + \gamma_1 y_{i1})}{1 + \exp(\gamma_0 + \gamma_1 y_{i1})}.$$

We choose $\gamma_0 = 0$ and (a) $\gamma_1 = 1$ or (b) $\gamma_1 = -\infty$. The latter mechanism implies $r_i = 0$ if $y_{i1} \geq 0$ and $r_i = 1$ otherwise. Both dropout mechanisms yield $\pi = 0.5$. In all cases, $\alpha_1 = -\alpha_0$, with α_1 in the four possible combinations of covariance and dropout parameters: (1a) 0.4132, (1b) 0.7263, (2a) $\sqrt{2/\pi}$, and (2b) $2/\sqrt{\pi}$. Numerical values for (21.5)–(21.7) are presented in Table 21.1, as well as the average from the observed information matrices in a simulation with 500 replicates.

Obviously, these elements are far from zero, as would be found with the naive estimator. They are of the same order of magnitude as the upper left block of the information matrix (pertaining to the mean parameters), which are

$$\begin{pmatrix} 1166.67 & -333.33 \\ -333.33 & 666.67 \end{pmatrix}.$$

We performed a limited simulation study to verify the coverage probability for the Wald tests under the unconditional and a selection of conditional frameworks. The hypotheses considered are $H_{01} : \mu_1 = 0$, $H_{02} : \mu_2 = 0$, and $H_{03} : \mu_1 = \mu_2 = 0$. The simulations have been restricted to the first covariance matrix used in Table 21.1 and to the second dropout mechanism ($\gamma_1 = -\infty$). Results are reported in Table 21.2. The coverages for the

TABLE 21.1. *Bivariate Normal Data. Computed and simulated values for the off-diagonal block of the unconditional information matrix. Sample size is* $N =$ *1000 (500 replications). (The true model has zero mean vector. Two true covariances* Σ *and two dropout parameters* γ_1 *are considered.)*

Parameters			Uncond. $i_U(\boldsymbol{\mu}, \cdot)$			Simulated $\widehat{i_O}(\boldsymbol{\mu}, \cdot)$		
Σ		γ_1	σ_{11}	σ_{12}	σ_{22}	σ_{11}	σ_{12}	σ_{22}
1	0.5	1	−68.87	137.75	0.00	−69.36	137.95	−0.04
0.5	1		137.75	−275.49	0.00	137.88	−276.83	−0.04
2	$\sqrt{6}/2$	1	−30.26	49.42	0.00	−30.21	49.54	0.04
$\sqrt{6}/2$	3		49.42	−80.70	0.00	49.52	−81.31	0.06
1	0.5	$-\infty$	132.98	−265.96	0.00	135.67	−267.66	0.16
0.5	1		−265.96	531.92	0.00	−267.73	537.58	−0.02
2	$\sqrt{6}/2$	$-\infty$	47.02	−76.78	0.00	49.52	−78.73	−0.02
$\sqrt{6}/2$	3		−76.78	125.38	0.00	−78.58	126.91	0.02

unconditional framework are in good agreement with a χ^2 reference distribution; the first naive framework (500 complete cases) leads to a conservative procedure, whereas the second and the third lead to extreme liberal behavior, that is most marked for hypotheses H_{01} and H_{03}. This is to be expected because by fixing $m = 500$, the proportion of positive first outcomes is constrained to be equal to its predicted value. This has the effect of reducing the variability of $\hat{\mu}_1$. The second and the third frameworks also suppress the variability, but introduce bias at the same time. The comparative insensitivity of the behavior of the test for H_{02} to the sampling framework is because μ_1 has only an indirect influence through the correlation between the outcomes on both occasions. It should be noted that due to numerical problems, not all simulations led to 500 successful estimations. On average, 489 convergencies were observed, the lowest value being 460 for H_{02} in the first naive sampling frame.

21.4 Example

We will now consider a relatively small example with a continuous response, analyzed in Crépeau *et al.* (1985). Fifty-four rats were divided into five treatment groups corresponding to exposure to increasing doses of

TABLE 21.2. *Bivariate Normal Data. True values are as in the third model of Table 21.1. Coverage probabilities ($\times$ 1000) for Wald test statistics. Sample size is $N = 1000$ (500 replications). The null hypotheses are $H_{01} : \mu_1 = 0$, $H_{02} : \mu_2 = 0$, $H_{03} : \mu_1 = \mu_2 = 0$. For the naive sampling frameworks, m denotes the fixed number of complete cases.*

Hypothesis	Uncond.	$m = 500$	$m = 450$	$m = 400$
H_{01}	933	996	187	0
H_{02}	953	952	913	830
H_{03}	952	992	338	0

halothane (0%, 0.25%, 0.5%, 1%, and 2%). The groups were of sizes 11, 10, 11, 11, and 11 rats, respectively. Following an induced heart attack in each rat, the blood pressure was recorded on nine unequally spaced occasions. A number of rats died during the course of the experiment, including all rats from group 5 (2% halothane). Following the original authors we omit this group from the analysis since they contribute no information at all, leaving 43 rats, of which 23 survived the experiment.

Examination of the data from these four groups does not provide any evidence of an MAR dropout process, although this observation must be considered in the light of the small sample size. A Gaussian multivariate linear model with an unconstrained covariance matrix was fitted to the data. There was very little evidence of a treatment by time interaction and the following results are based on the use of a model with additive effects for treatment and time. The Wald statistics for the treatment main effect on 3 degrees of freedom are equal to 46.95 and 30.82 respectively using the expected and observed information matrices. Although leading to the same qualitative conclusions, the figures are notably discrepant. A first reaction may be to attribute this difference to the incompleteness of the data. However, the lack of evidence for an MAR process together with the relatively small sample size points to another cause. The equivalent analysis of the 24 completers produces Wald statistics of 45.34 and 26.35, respectively; that is, the effect can be attributed to a combination of small-sample variation and possible model misspecification. A theoretical reason for this difference might be that the expected value of the off-diagonal block of the information matrix of the maximum likelihood estimates (describing covariance between mean and covariance parameters) has expectation zero but is likely to depart from this in small samples. As a consequence, the variances of the estimated treatment effects will be higher when derived from the observed information, thereby reducing the Wald statistic.

To summarize, this example provides an illustration of an alternative source of discrepancy between the expected and observed information matrices, which is likely to be associated with the use, in smaller samples, of covariance matrices with many parameters.

21.5   Implications for PROC MIXED

The literature indicates an early awareness of problems with conventional likelihood-based frequentist inference in the MAR setting. Specifically, several authors point to the use of the observed information matrix as a way to circumvent issues with the expected information matrix. In spite of this, it seems that a broad awareness of this problem has diminished while the number of methods formulated to deal with the MAR situation has risen dramatically in recent years. We therefore feel that a restatement and exposition of this important problem is timely, especially since PROC MIXED allows routine fitting of ignorable models with likelihood-based methods.

The MIXED procedure allows both Newton-Raphson and Fisher scoring algorithms. Specifying the 'scoring' option in the PROC MIXED statement requests the Fisher scoring algorithm in conjunction with the method of estimation for a specified number of iterations (1 by default). If convergence is reached before scoring is stopped, then the expected Hessian is used to compute approximate standard errors rather than the observed Hessian. In both cases, the standard errors for the fixed effects are based on inverting a single block of the Hessian matrix. Since we have shown in Section 21.3 that the off-diagonal block, pertaining to the covariance between the fixed effects and covariance parameters, does not have expectation zero, this procedure is, strictly speaking, incorrect. Correction factors to overcome this problem have been proposed (e.g., Prasad and Rao 1990) but they tend to be small for fairly well-balanced data sets. It has to be noted that a substantial amount of (randomly) missing data will destroy this balance. The extent to which all this is problematic is illustrated in Table 21.3. Model 7 for the growth data is reconsidered for both the complete data set, as well as for the incomplete data on the basis of an ignorable analysis. The fixed-effects parameters are as in (17.10), whereas the covariance structure consists of the residual variance σ^2 and the variance of the random intercept d. Apart from the parameter estimates, two sets of standard errors are shown: (1) taken from inverting the fixed-effects block from the observed Hessian and (2) taken from inverting the entire observed Hessian. The first set is found from the MIXED output, whereas the second one was constructed using the numerical optimizer OPTMUM of GAUSS (Edlefsen).

TABLE 21.3. *Maximum likelihood estimates and standard errors (in parentheses) for the parameters in Model 7, fitted to the growth data (complete data set and ignorable analysis).*

Parameter	Complete Data Estimate (s.e.)[1] (s.e.)[2]	Ignorable Estimate (s.e.)[1] (s.e.)[2]
β_0	17.3727 (1.1615,1.1645)	17.2218 (1.2220,1.2207)
β_{01}	-1.0321 (1.5089,1.5156)	-0.9188 (1.5857,1.5814)
β_{10}	0.4795 (0.0923,0.0925)	0.4890 (0.0969,0.0968)
β_{11}	0.7844 (0.0765,0.0767)	0.7867 (0.0802,0.0801)
σ^2	1.8746 (0.2946,0.2946)	2.0173 (0.3365,0.3365)
d	3.0306 (0.9552,0.9550)	3.0953 (1.0011,1.0011)

[1] Standard error based on the Newton-Raphson algorithm of PROC MIXED
[2] Standard error obtained from inverting the entire observed information matrix.

Clearly, there are only minor differences between the two sets of standard errors and the analysis on an incomplete set of data does not seem to widen the gap.

We can conclude from this that, with the exception of the expected information matrix, conventional likelihood-based frequentist inference, including standard hypothesis testing, is applicable in the MAR setting. Standard errors based on inverting the *entire* Hessian are to be preferred, and in this sense, it is a pity that this option is presently not available in PROC MIXED.

22

The Expectation-Maximization Algorithm

Although the models in Table 17.5 are fitted using direct observed data likelihood maximization in PROC MIXED, Little and Rubin (1987) obtained these same results using the *Expectation-Maximization* algorithm. Special forms of the algorithm, designed for specific applications, had been proposed for about half a century (e.g., Yates 1933), but the first unifying and formal account was given by Dempster, Laird, and Rubin (1977). McLachlan and Krishnan (1997) devoted a whole volume to the EM algorithm and its extensions.

Even though the SAS procedure MIXED uses direct likelihood maximization, the EM algorithm is generally useful to maximize certain complicated likelihood functions. For example, it has been used to maximize mixture likelihoods in Section 12.3. Liu and Rubin (1995) used it to estimate the t-distribution, based on EM, its extension ECM (expectation conditional maximization), and ECME (expectation conditional maximization, either), which are described in Meng and Rubin (1993), Liu and Rubin (1994), and van Dyk, Meng, and Rubin (1995). EM methods specifically for mixed-effects models are discussed in Meng and van Dyk (1998). A nice review is given in Meng (1997), where the focus is on EM applications in medical studies.

We will first give a brief description of the algorithm, with emphasis on incomplete data problems. Suppose we are interested in maximizing the ignorable observed-data log-likelihood $\ell(\boldsymbol{\theta}; \boldsymbol{y}^o)$. Let $\boldsymbol{\theta}^{(0)}$ be an initial guess, which can be found from, for example, a complete case analysis, an available case analysis, single imputation, or any other convenient method.

The EM algorithm consists of an *expectation step* (E step) and a *maximization step* (M step).

The E Step. Given the current value $\boldsymbol{\theta}^{(t)}$ of the parameter vector, the E step computes the expected value of the complete data log-likelihood, given the observed data and the current parameters, which is called the *objective function*:

$$
\begin{aligned}
Q(\boldsymbol{\theta}|\boldsymbol{\theta}^{(t)}) &= \int \ell(\boldsymbol{\theta}, \boldsymbol{y}) f(\boldsymbol{y}^m|\boldsymbol{y}^o, \boldsymbol{\theta}^{(t)}) d\boldsymbol{y}^m \\
&= E\left\{\ell(\boldsymbol{\theta}|\boldsymbol{y})|\boldsymbol{y}^o, \boldsymbol{\theta}^{(t)}\right\}.
\end{aligned}
\tag{22.1}
$$

In the case that the log-likelihood is linear in sufficient statistics, this procedure comes down to substituting the expected value of the sufficient statistics, given $\boldsymbol{Y}^o$ and $\boldsymbol{\theta}^{(t)}$. In particular, for exponential family models, the E step reduces to the computation of complete-data sufficient statistics.

The M Step. Next, the M step determines $\boldsymbol{\theta}^{(t+1)}$, the parameter vector maximizing the objective function (22.1). Formally, $\boldsymbol{\theta}^{(t+1)}$ satisfies

$$
Q(\boldsymbol{\theta}^{(t+1)}|\boldsymbol{\theta}^{(t)}) \geq Q(\boldsymbol{\theta}|\boldsymbol{\theta}^{(t)}), \qquad \text{for all } \boldsymbol{\theta}.
$$

One then iterates between the E and M steps until convergence.

Consider, for example, a multivariate normal sample. Then, $\boldsymbol{\theta} = (\boldsymbol{\mu}, \boldsymbol{\Sigma})$. The E step computes the sufficient statistics

$$
E\left(\sum_{i=1}^N Y_{ij}|\boldsymbol{y}^o, \boldsymbol{\theta}^{(t)}\right) \quad \text{and} \quad E\left(\sum_{i=1}^N Y_{ij}Y_{ik}|\boldsymbol{y}^o, \boldsymbol{\theta}^{(t)}\right).
$$

From these, computation of $\boldsymbol{\mu}^{(t+1)}$ and $\boldsymbol{\Sigma}^{(t+1)}$ is straightforward.

When the covariance matrix is structured or patterned, the E step remains the same, but the M step is slightly modified. This situation arose when Little and Rubin (1987) fitted the incomplete growth data Models 5–7 using the EM algorithm. Let us sketch their procedure. Emphasis is placed on the patterns in the covariance matrix. The outcomes $\boldsymbol{Y}_i$ are assumed to follow a normal model $\boldsymbol{Y}_i \sim N(\boldsymbol{\mu}_i, \boldsymbol{\Sigma})$ where $\boldsymbol{\mu}_i = X_i\boldsymbol{\beta}$ and $\boldsymbol{\Sigma} = \boldsymbol{\Sigma}(\boldsymbol{\alpha})$, a known function of $\boldsymbol{\alpha}$, such as a banded, AR(1), or exchangeable model, or a model induced by random effects. The complete-data log-likelihood is linear in $\boldsymbol{y}_i$ and $\boldsymbol{y}_i'\boldsymbol{y}_i$. The E step is restricted to computing

$$
E\left(\boldsymbol{Y}_i|\boldsymbol{y}_i^o, X_i, Z_i, \boldsymbol{\theta}\right) \quad \text{and} \quad E\left(\boldsymbol{Y}_i'\boldsymbol{Y}_i|\boldsymbol{y}_i^o, X_i, Z_i, \boldsymbol{\theta}\right).
$$

These computations can easily be done using the *sweep* operator (Little and Rubin 1987).

The M step consists of a standard estimation procedure for complete data. Whereas for simple and unstructured covariance models, the M step may be available in closed form, it is usually iterative for patterned covariance matrices, turning the EM algorithm into a doubly iterative scheme. To make the M step noniterative, a GEM (generalized EM) algorithm can be used. A GEM algorithm merely increases the likelihood in the M step, rather than maximizing it. For example, a single scoring step can be used rather than full convergence. Under general conditions, the convergence of the GEM is the same as for the EM (Dempster, Laird, and Rubin 1977).

Let us write Σ as a function of α for the covariance matrices in the growth example.

Unstructured: $\Sigma = \alpha$.

Banded: $\sigma_{jk} = \alpha_r$ with $r = |j - k| + 1$.

Autoregressive: $\sigma_{jk} = \alpha_1 \alpha_2^{|j-k|}$.

Compound symmetry: $\Sigma = \alpha_1 J_4 + \alpha_2 I_4$ with J_4 a matrix of ones.

Random effects: $\Sigma = Z\alpha Z' + \sigma^2 I$ with α the covariance matrix of the random effects.

Independence: $\Sigma = \alpha I_4$.

The mean structure design matrices X_i are as discussed in Section 17.4.1.

The main drawbacks of the EM algorithm are its typically slow rate of convergence. The double iterative structure of many implementations adds to the problem. Further, the algorithm does not automatically provide precision estimators. Proposals for overcoming these limitations have been made by, for example, Louis (1982), Meilijson (1989), Meng and Rubin (1991), Baker (1992), Meng and van Dyk (1997), and Liu, Rubin, and Wu (1998).

In the light of these observations, one might argue that the existence of PROC MIXED, enabling the use of Newton-Raphson or Fisher scoring algorithms to maximize the observed data likelihood, is fortunate. Although this statement is certainly warranted for a wide range of applications, there may be situations where the EM algorithm is beneficial. Baker (1994) mentions advantages of starting with an EM algorithm and then switching to Newton-Raphson, if necessary, including less sensitivity to poor starting values and more reliable convergence to a boundary when the maximum likelihood estimators is indeed a boundary value. In the latter situation,

Newton-Raphson and Fisher scoring algorithms exhibit a tendency to converge to values outside the allowable parameter space. Further, the EM algorithm can be easily extended for use with nonignorable problems, such as discussed by Diggle and Kenward (1994). This route was explicitly chosen by Molenberghs, Kenward, and Lesaffre (1997) for a comparable categorical data setting.

Many of the issues briefly touched upon in this section are discussed at length in McLachlan and Krishnan (1997). This includes the definition and basic principles and theory of EM, various ways of obtaining standard errors and improving the speed of convergence, as well as extensions, including those mentioned earlier, but also stochastic EM and Gibbs sampler versions.

23

Design Considerations

23.1 Introduction

In the first part of this book (Chapters 3 to 13), emphasis was on the formulation and the fitting of, as well as on inference and diagnostics for linear mixed models in general. Later (Chapters 14 to 22), the problem of missing data was discussed in full detail, with emphasis on how to obtain valid inferences from observed longitudinal data and how to perform sensitivity analyses with respect to assumptions made about the dropout process.

In this chapter, we will reflect on the design of longitudinal studies. In Section 23.2, we will briefly discuss how power calculations can be performed based on linear mixed models. We refer to Mentré, Mallet and Baccar (1997) for a note on D-optimal designs in random-effects regression models, and to Liu and Liang (1997), where sample-size calculations are treated in the context of generalized estimating equations (Liang and Zeger 1986).

In practice (see, e.g., the rat experiment and the Vorozole study introduced in Chapter 2), longitudinal experiments often do not yield the amount of information hoped for at the design stage, due to dropout. This results in realized experiments with (possibly much) less power than originally planned. In Section 23.4, it will be shown how expected dropout can be taken into account in sample-size calculations. The basic idea behind this is that two designs with equal power under the absence of dropout are not necessarily equally likely to yield realized experiments with high power. The main question then is how to design experiments with minimal risk of huge losses in efficiency due to dropout. In Section 23.5, this will be extensively illustrated in the context of the rat experiment.

23.2   Power Calculations Under Linear Mixed Models

Chapter 6 was devoted to inference in the marginal linear mixed model (5.1). Several testing procedures were discussed, including approximate Wald tests, approximate t-tests, approximate F-tests, and likelihood ratio tests (based on ML as well as REML estimation), for the fixed effects as well as for the variance components in the model. Obviously, any of these testing procedures can be used in power calculations. Unfortunately, the distribution of many of the corresponding test statistics is only known under the null hypothesis. In practice, this means that if such tests are to be used in sample-size calculations, extensive simulations would be required. One then would have to sample data sets under the alternative hypothesis of interest, analyze each of them using the selected testing procedure, and estimate the probability of correctly rejecting the null hypothesis. Finally, this whole procedure would have to be repeated for every new design under consideration.

When interest is in testing a general linear hypothesis of the form

$$H_0 : \boldsymbol{\xi} \equiv L\boldsymbol{\beta} - \boldsymbol{\xi_0} = 0 \quad \text{versus} \quad H_A : \boldsymbol{\xi} \neq 0 \qquad (23.1)$$

for some known matrix L and known vector $\boldsymbol{\xi_0}$, a simplified procedure can be followed. As explained in Section 6.2.2, (23.1) can be tested based on the fact that, under the null hypothesis, the test statistic

$$F = \widehat{\boldsymbol{\xi}}' \left[L \left(\sum_i X_i' \widehat{V}_i^{-1} X_i \right)^{-1} L' \right]^{-1} \widehat{\boldsymbol{\xi}} \, / \, \text{rank}(L)$$

follows approximately an F-distribution. The numerator degrees of freedom equals $\text{rank}(L)$, and several methods can be used to estimate the denominator degrees of freedom from the data.

Helms (1992) reports simulation results which show that, under the alternative hypothesis H_A, the distribution of F can also be approximated by an F-distribution, now with $\text{rank}(L)$ and $\sum_i n_i - \text{rank}[X|Z]$ degrees of freedom, and with noncentrality parameter

$$\delta = \xi' \left[L \left(\sum_i X_i' V_i^{-1} X_i \right)^{-1} L' \right]^{-1} \xi.$$

The matrices X and Z are as previously defined in Section 5.3.3 [i.e., $X = (X_1'|\ldots|X_N')'$ and $Z = \text{Diag}(Z_1, \ldots, Z_N)$]. Hence, under H_A, we get a noncentral F-distribution, from which power calculations immediately follow.

An example in which the above results are used for power calculations can be found in Helms (1992), where it has been shown empirically that intentionally incomplete designs, where some subjects are intentionally not measured at all time points, can have more power while being less expensive to conduct. Another example will be given in the next section. Finally, the noncentral F-approximation will also be used in Section 23.4 to perform power calculations, taking into account that dropout is to be expected.

23.3   Example: The Rat Data

We reconsider here the rat experiment, introduced in Section 2.1. Recall that our final model, derived in Section 6.3.3, is given by

$$Y_{ij} = \begin{cases} \beta_0 + b_i + \beta_1 t_{ij} + \varepsilon_{ij}, & \text{if low dose} \\ \beta_0 + b_i + \beta_2 t_{ij} + \varepsilon_{ij}, & \text{if high dose} \\ \beta_0 + b_i + \beta_3 t_{ij} + \varepsilon_{ij}, & \text{if control dose}, \end{cases} \quad (23.2)$$

where t_{ij} represents the logarithmically transformed ages, $t_{ij} = \ln(1 + (\text{Age}_{ij} - 45)/10))$, at which the repeated measurements have been taken. As before, the residual components ε_{ij} only contain measurement error (i.e., $\varepsilon_i = \varepsilon_{(1)i}$, see Section 3.3.4). Estimates of all parameters in the corresponding marginal model have been provided in Table 6.4. The estimated average profiles are shown in Figure 6.2.

The hypothesis of primary interest is $H_0 : \beta_1 = \beta_2 = \beta_3$, which has already been tested in Section 6.3.3, yielding a nonsignificant approximate Wald statistic ($p = 0.0987$). A similar result ($p = 0.1010$) is obtained using an approximate F-test, with Satterthwaite approximation for the denominator degrees of freedom. We conclude from this that there is little evidence for any treatment effects. However, the power for detecting the observed differences (as described in Table 6.4) at the 5% level of significance and calculated using the F-approximation described in the previous section is as low as 56%.

Note that, as already mentioned in Section 2.1 and shown in Table 2.1, this rat experiment suffers from a severe degree of dropout, since many rats do not survive anesthesia needed to measure the outcome. Indeed, although 50 rats have been randomized at the start of the experiment, only 22 of them survived the 6 first measurements, so measurements on only 22 rats are available in the way anticipated at the design stage. For example, at the second occasion (age = 60 days), only 46 rats were available, implying that for 4 rats, only 1 measurement has been recorded. As can be expected, this high dropout rate inevitably leads to severe losses in efficiency of the statistical inferential procedures. Indeed, if no dropout had occurred (i.e., if all 50 rats would have withstood the 7 measurements), the power for detecting the observed differences at the 5% level of significance would have been 74%, rather than the 56% previously reported for the realized experiment.

In the rat example, dropout was not entirely unexpected since it is inherently related to the way the response of interest is actually measured (anesthesia cannot be avoided) and should therefore have been taken into account at the design stage. In the next section, we will discuss a general, computationally simple method, proposed by Verbeke and Lesaffre (1999) for the design of longitudinal experiments, when dropout is to be expected. Afterward, in Section 23.5, the proposed approach is applied to the rat data.

23.4   Power Calculations When Dropout Is to Be Expected

In order to fully understand how the dropout process can be taken into account at the design stage, we first investigate how it affects the power of a realized experiment. Note that the power of the F-test described in Section 23.2 not only depends on the true parameter values β, D, and σ^2 (or, more generally, Σ_i) but also on the covariates X_i and Z_i. Usually, in designed experiments, many subjects will have the same covariates, such that there are only a small number of different sets (X_i, Z_i). For the rat data, for example, all 15 rats in the control group have X_i and Z_i equal to

$$
X_i = \begin{pmatrix} 1 & 0 & 0 & \ln[1 + (50 - 45)/10] \\ 1 & 0 & 0 & \ln[1 + (60 - 45)/10] \\ \vdots & \vdots & \vdots & \vdots \\ 1 & 0 & 0 & \ln[1 + (110 - 45)/10] \end{pmatrix}, \quad Z_i = \begin{pmatrix} 1 \\ 1 \\ \vdots \\ 1 \end{pmatrix}.
$$

However, due to the dropout mechanism, the above matrices have been realized for only four of them. Indeed, for a rat that drops out early, say at the kth occasion, the realized design matrices equal the first k rows of the above planned matrices; that is,

$$
X_i = \begin{pmatrix} 1 & 0 & 0 & \ln[1 + (50 - 45)/10] \\ \vdots & \vdots & \vdots & \vdots \\ 1 & 0 & 0 & \ln[1 + (40 + k \times 10 - 45)/10] \end{pmatrix}, \quad Z_i = \begin{pmatrix} 1 \\ \vdots \\ 1 \end{pmatrix}.
$$

Note that the number of rats that drop out at each occasion is a realization of the stochastic dropout process, from which it follows that the power of the realized experiment is also a realization of a random variable, the distribution of which depends on the planned design and on the dropout process. From now on, we will denote this random power function by $\mathcal{P}$.

In general, a planned design is characterized by a small number of triplets $(\mathcal{X}_j, \mathcal{Z}_j, M_j)$, $j = 1, \ldots, M$, in which it is indicated that the $(n_j \times p)$ design matrix $\mathcal{X}_j$ for the fixed effects and the $(n_j \times q)$ design matrix $\mathcal{Z}_j$ for the random effects is repeated for M_j subjects, $\sum_j M_j = N$, the total number of subjects in our sample. Due to the dropout process, the realized covariate matrices are $(\mathcal{X}_j^{[k]}, \mathcal{Z}_j^{[k]})$ with multiplicity $M_{j,k}$, $k = 1, \ldots, n_j$, $\sum_k M_{j,k} = M_j$, $j = 1, \ldots, M$, where $\mathcal{X}_j^{[k]}$ and $\mathcal{Z}_j^{[k]}$ denote the first k rows of $\mathcal{X}_j$ and $\mathcal{Z}_j$, respectively. Note that once all realized values for $M_{j,k}$ are known, the corresponding realization of the power $\mathcal{P}$ can be calculated.

Since, in the presence of dropout, the power $\mathcal{P}$ becomes a stochastic variable, it is not obvious how two different designs with two different associated power functions $\mathcal{P}_1$ and $\mathcal{P}_2$ should be compared in practice. Several criteria can be used, such as the average power, $E(\mathcal{P})$, the median power, median($\mathcal{P}$), the risk of having a final analysis with power less than for example 70%, $P(\mathcal{P} \leq 70\%)$, and so forth.

Note that all of the above criteria are based on only one specific aspect of the distribution of $\mathcal{P}$. A criterion which takes into account the full distribution selects the second design over the first one if $\mathcal{P}_1$ is stochastically smaller than $\mathcal{P}_2$, $\mathcal{P}_1 \prec \mathcal{P}_2$, which is defined as (Lehmann and D'Abrera 1975, p. 66)

$$
\mathcal{P}_1 \prec \mathcal{P}_2 \iff P(\mathcal{P}_1 \leq p) \geq P(\mathcal{P}_2 \leq p), \quad \forall p.
$$

This means that, for any power p, the risk of ending up with a final analysis with power less than p is smaller for the second design than for the first design. Obviously, if this criterion is to be used, one needs to assess the complete power distribution function for all designs which are to be compared. We propose doing this via sampling methods in which, for each design under consideration, a large number of realized values p_s, $s = 1, \ldots, S$, are sampled from $\mathcal{P}$ and used to construct the empirical distribution function

$$\widehat{P}(\mathcal{P} \leq p) = \frac{1}{S} \sum_{s=1}^{S} I[p_s \leq p]$$

in which $I[A]$ equals one if A is true and zero otherwise.

As indicated above, sampling from $\mathcal{P}$ actually comes down to sampling realized values for all $M_{j,k}$, $k = 1, \ldots, n_j$, $j = 1, \ldots, M$, and constructing all necessary realized matrices $\mathcal{X}_j^{[k]}$ and $\mathcal{Z}_j^{[k]}$. One then can easily calculate the implied noncentrality parameter

$$\delta = \boldsymbol{\xi}' \left\{ L \left[\sum_{j=1}^{M} \sum_{k=1}^{n_j} M_{j,k} \, \mathcal{X}_j^{[k]'} \left(\mathcal{Z}_j^{[k]} D \mathcal{Z}_j^{[k]'} + \sigma^2 I_k \right)^{-1} \mathcal{X}_j^{[k]} \right]^{-1} L' \right\}^{-1} \boldsymbol{\xi}.$$

and the appropriate numbers of degrees of freedom for the F-statistic, from which a realized power follows. Note that the dropout process associates with each triplet $(\mathcal{X}_j, \mathcal{Z}_j, M_j)$ in the design a vector $\boldsymbol{p_j} = (p_{j,1}, \ldots, p_{j,n_j})'$ in which $p_{j,k}$ equals the marginal probability that exactly k measurements are taken on a subject with planned design matrices $\mathcal{X}_j$ and $\mathcal{Z}_j$. Once all vectors $\boldsymbol{p_j}$ have been specified, we have that all sets $(M_{j,1}, \ldots, M_{j,n_j})$ follow a multinomial distribution with M_j trials and probabilities given by the elements of the vectors $\boldsymbol{p_j}$. This implies that the sampling procedure basically reduces to sampling from multinomial distributions, from which it follows that the implementation is straightforward. This allows one to explore many different combinations of models for the dropout process with models for the actual responses and to investigate the effect of possible model misspecifications. Further, it can easily be seen that the computing time does not increase with the planned sample size. It only depends on the number of triplets $(\mathcal{X}_j, \mathcal{Z}_j, M_j)$ in the design rather than on the total sample size $\sum_j M_j$.

It should be emphasized that the above approach is not restricted to any particular statistical test. The idea of sampling designs under specific dropout patterns is applicable for any testing procedure, as long as it remains possible to evaluate the power associated to each realized design. Note also that the only additional information needed, in comparison to classical power analyses, are the vectors $\boldsymbol{p_j}$ of marginal dropout probabilities $p_{j,k}$. This does not require full knowledge of the underlying dropout

TABLE 23.1. *Rat Data. Observed conditional dropout rates* $\widehat{p}_{j,k|\geq k}$ *at each occasion, for all treatment groups simultaneously.*

Age (days):	50	60	70	80	90	100
Observed rate:	0.08	0.07	0.12	0.24	0.17	0.08

process. We only need to make assumptions about the dropout rate at each occasion where observations are designed to be taken. For example, we do not need to know whether the dropout mechanism is "completely at random" or "at random" (see Section 15.7). Still, we have to assume that dropout is "not informative" in the sense that it does not depend on the response values which would have been recorded if no dropout had occurred, since otherwise our final analysis based on the linear mixed model would not yield valid results (see Section 15.8 and Chapter 21).

Finally, the proposed method can be used in combination with techniques, such as those proposed by Helms (1992), which would allow the costs of performing the designs under consideration to be taken into account. This could yield less costly experiments with minimal risk of large efficiency losses due to dropout. This will not be explored any further here.

23.5 Example: The Rat Data

In this section, we will use the sampling procedure described in the previous section to compare the design that was used in the rat experiment with alternative designs which could be used in future similar experiments. The assumption of random dropout is supported by our analyses in Section 19.4. In order to be able to specify realistic marginal dropout probabilities $p_{j,k}$, we first study the dropout rates observed in the data set at hand. According to the clinicians who collected the data, there is a strong belief that there is constant probability for a rat to drop out, given that the rat survived up to that moment. This suggests that the conditional probability $p_{j,k|\geq k}$ that a rat with planned covariates $(\mathcal{X}_j, \mathcal{Z}_j)$ does not survive the kth occasion, given that it survived all previous measurements, does not depend on k. Further, it is believed that the dropout process does not depend on treatment (i.e., that $p_{j,k}$ and $p_{j,k|\geq k}$ do not depend on j), as long as measurements are planned at the same occasions for the three treatment groups. Table 23.1 shows the conditional observed dropout rates $\widehat{p}_{j,k|\geq k}$ at each occasion, for the total sample. For example, 3 rats, out of the 46 rats who survived the first measurement, died at the second occasion, leading

TABLE 23.2. *Rat Data. Marginal probabilities $p_{j,k}$ for several conditional dropout models and designs. Empty entries correspond to occasions at which no observations are planned in the design.*

Logits of conditional probabilities $p_{j,k|\geq k}$	Occasions						
	50	60	70	80	90	100	110
logit(0.12)	0.12	0.11	0.09	0.08	0.07	0.06	0.46
logit(0.12)	0.12		0.11		0.09		0.68
logit(0.12)	0.12			0.11			0.77
logit(0.12)	0.12						0.88
logit(0.12)	0.12	0.11					0.77
logit(0.12)	0.12					0.11	0.77
$-3 + 0.06(\text{Age} - 45)$	0.06	0.10					0.83
$-3 + 0.06(\text{Age} - 45)$	0.06			0.27			0.67
$-3 + 0.06(\text{Age} - 45)$	0.06					0.54	0.40
$-3 + 0.02(\text{Age} - 45)$	0.05	0.06					0.89
$-3 + 0.02(\text{Age} - 45)$	0.05			0.09			0.86
$-3 + 0.02(\text{Age} - 45)$	0.05					0.13	0.82

to an observed dropout rate of $\hat{p}_{j,2|\geq 2} = 3/46 = 0.07$. Using a likelihood ratio test, we tested whether it is reasonable to assume the dropout rates in Table 23.1 to be observed values of one common dropout probability, as has been hypothesized by the clinicians. The likelihood ratio statistic equals $2\ln\lambda = 7.37$, on 5 degrees of freedom, which is clearly not significant at the 5% level. Further, the maximum likelihood estimate for the common probability equals $\hat{p}_{j,k|\geq k} = 0.122$, suggesting that each time a rat is anesthetized, there is about 12% chance that the rat will not survive anesthesia, independent of the treatment.

In Sections 23.5.1 and 23.5.2, we will first compare several designs under the assumption that $p_{j,k|\geq k} = 0.12$. Afterward, in Section 23.5.3, the results will be compared with those obtained under two alternative dropout models which assume that $p_{j,k|\geq k}$ increases over time. In all designs, the three treatment groups are measured on the same occasions, and all designs plan their first and last observation at the age of 50 and 110 days, respectively. The marginal probabilities are shown in Table 23.2, and the actual designs are summarized in Table 23.3, together with their power if no dropout

TABLE 23.3. *Rat Data. Summary of the designs compared in the simulation study.*

Design	Occasions Age (days)	Number of subjects (M_1, M_2, M_3)	Power if no dropout
A	50-60-70-80-90-100-110	(15, 18, 17)	0.74
B	50-70-90-110	(15, 18, 17)	0.63
C	50-80-110	(15, 18, 17)	0.59
D	50-110	(15, 18, 17)	0.53
E	50-70-90-110	(22, 22, 22)	0.74
F	50-80-110	(24, 24, 24)	0.74
G	50-110	(27, 27, 27)	0.75
H	50-60-110	(26, 26, 26)	0.74
I	50-100-110	(20, 20, 20)	0.73

would occur. All calculations are done under the assumption that the true parameter values are given by the estimates in Table 6.4 and all simulated power distributions are based on 1000 draws from the correct distribution.

23.5.1 Constant $p_{j,k|\geq k}$, Varying n_j

Since, at each occasion, rats may die, it seems natural to reduce the number of occasions at which measurements are taken. We have therefore simulated the power distribution of four designs in which the number of rats assigned to each treatment group is the same as in the original experiment, but the planned number of measurements per subject is seven, four, three, and two, respectively. These are the designs A to D in Table 23.3. Note that design A is the design used in the original rat experiment. The simulated power distributions are shown in Figure 23.1.

First, note that the solid line is an estimate for the power function of the originally designed rat experiment under the assumption of constant dropout probability $p_{j,k|\geq k}$ equal to 12%. It shows that there was more than 80% chance for the final analysis to have realized power less than the 56% which was observed in the actual experiment. Comparing the four designs under consideration, we observe that the risk of high power losses increases as the planned number of measurements per subject decreases. On the other hand, it should be emphasized that the four designs are, strictly speaking, not comparable in the sense that, in the absence of dropout, they have very different powers ranging from 74% for design A to only 53% for

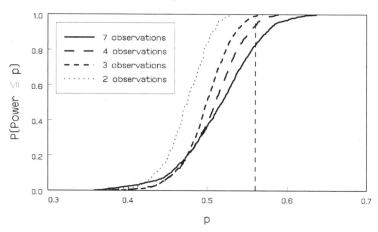

FIGURE 23.1. *Rat Data. Comparison of the simulated power distributions for designs with seven, four, three, or two measurements per rat, with equal number of rats in each design (designs A, B, C, and D, respectively), under the assumption of constant $p_{j,k|\geq k}$ equal to 12%. The vertical dashed line corresponds to the power which was realized in the original rat experiment (56%).*

design D (Table 23.3). In order to make fair comparisons, we will from now on only consider designs with comparable powers if no dropout would occur. As shown in Table 23.3, this can be achieved by considering designs with different sample sizes.

Designs E, F, and G are the same as designs B, C, and D, but with sample sizes such that their power is approximately the same as the power of design A, in the absence of dropout. The simulated power distributions are shown in Figure 23.2. The figure suggests that $\mathcal{P}_A \prec \mathcal{P}_E \prec \mathcal{P}_F \prec \mathcal{P}_G$, from which it follows that, in practice, the design in which subjects are measured only at the beginning and at the end of the study is to be preferred, under the assumed dropout process. This can be explained by the fact that the probability for surviving up to the age of 110 days is almost twice as high for design G (88%) as for the original design (46%) (Table 23.2). Note also that the parameters of interest [β_1, β_2, and β_3 in model (23.2)] are slopes in a linear model such that two measurements are sufficient for the parameters to be estimable. On the other hand, design G does not allow testing for possible nonlinearities in the average evolutions. It also follows from Figure 23.2 that if design E, F, or G had been taken, it would have been very unlikely to have a final analysis with such small power as in the original experiment.

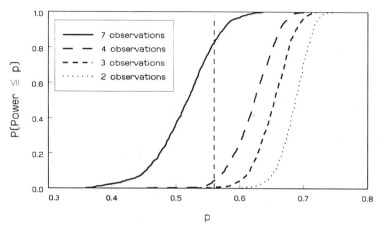

Simulated power distributions

FIGURE 23.2. *Rat Data. Comparison of the simulated power distributions for designs with seven, four, three, or two measurements per rat, with equal power if no dropout would occur (designs A, E, F, and G, respectively), under the assumption of constant $p_{j,k|\geq k}$ equal to 12%. The vertical dashed line corresponds to the power which was realized in the original rat experiment (56%).*

23.5.2   Constant $p_{j,k|\geq k}$, Constant n_j

The results in Section 23.5.1 suggest that similar future experiments should plan less measurements for each rat. From now on, we will, therefore, only consider designs with only three measurements per rat. As before, the first and last measurement are planned to be taken at the beginning and at the end of the study, respectively (at the age of 50 and 110 days). Hence, only the second observation needs to be specified. Designs H, F, and I have their second observation planned early in the study (at the age of 60 days), halfway through the study (at the age of 80 days) and late in the study (at the age of 100 days), respectively. Note that design H needs 18 more subjects than design I in order to get comparable power in the absence of dropout (Table 23.3). This is due to the fact that our linear mixed model is linear as a function of $t = \ln(1 + (\text{Age} - 45)/10))$ instead of the original Age scale and because maximal spread in the values t_{ij} is obtained by taking the second measurement at the end rather than at the beginning of the experiment.

Figure 23.3 shows the simulated power distributions for designs F, H, and I. As in Section 23.5.1, these are obtained under the assumption that $p_{j,k|\geq k}$ is constant and equal to 12%. This implies that, under each of these designs, there is 12% chance for a subject to have only one measurement, 11% chance for two measurements, and 77% chance that all three observations will be available at the time of the analysis (Table 23.2).

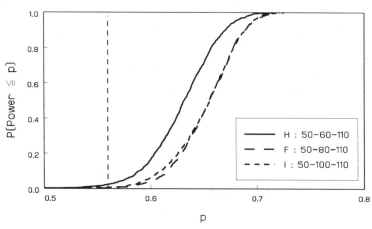

FIGURE 23.3. *Rat Data. Comparison of the simulated power distributions for designs with three measurements per rat, with equal power if no dropout would occur (designs H, F, and I), under the assumption of constant $p_{j,k|\geq k}$ equal to 12%. The vertical dashed line corresponds to the power which was realized in the original rat experiment (56%).*

First, under all three designs, it is very unlikely to have a final analysis with such small power as observed in the rat experiment. Further, we have that, from the perspective of efficiency, designs F and I are almost identical, but clearly superior to design H, that is, we have that $\mathcal{P}_H \prec \mathcal{P}_F \approx \mathcal{P}_I$. The relatively poor behavior of design H can be explained as follows. For subjects which drop out at the first occasion (and therefore have only one observation available), there is no difference between the three designs. This is in contrast with subjects which drop out at the second occasion, since subjects from design H then contain less information on the parameters of interest than subjects from design I. Note that although the designs F and I are equivalent with respect to efficiency of the final analysis, design I is to be preferred since it requires the randomization of less subjects.

23.5.3 Increasing $p_{j,k|\geq k}$ over Time, Constant n_j

In order to investigate the effect of the assumed dropout model on the simulated power distributions, we reconsider designs F, H, and I which have been investigated in Section 23.5.2 under the assumption of constant $p_{j,k|\geq k}$. However, we will now assume that $p_{j,k|\geq k}$ increases as a function of the time (days) elapsed since the start of the treatment. More specifically, it will be assumed that the conditional dropout probabilities satisfy the

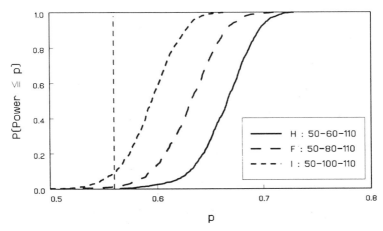

FIGURE 23.4. *Rat Data. Comparison of the simulated power distributions for designs with three measurements per rat, with equal power if no dropout would occur (designs H, F, and I), under the assumption that $p_{j,k|\geq k}$ satisfies $logit(p_{j,k|\geq k}) = -3 + 0.06(Age - 45)$. The vertical dashed line corresponds to the power which was realized in the original rat experiment (56%).*

logistic regression model

$$p_{j,k|\geq k} \quad = \quad \frac{\exp(\psi_0 + \psi_1(\text{Age} - 45))}{1 + \exp(\psi_0 + \psi_1(\text{Age} - 45))},$$

with $\psi_0 = -3$ and $\psi_1 = 0.06$.

The simulated power distributions are shown in Figure 23.4. We now clearly get that $\mathcal{P}_I \prec \mathcal{P}_F \prec \mathcal{P}_H$, which is opposite to our results under the assumption of constant $p_{j,k|\geq k}$. The most efficient design is now the one where the second occasion is planned immediately after the first measurement. This can be explained by observing that there is 83% chance under design H that a subject will have all measurements available as planned at the design stage. For design I, this probability drops to only 40%. Hence, the efficiency gained by having more spread in the covariate values t_{ij} for our linear model is lost by the severely increased risk of dropping out.

The fact that our conclusions are opposite to those in Section 23.5.2 suggests that there exists a dropout model under which designs F, H, and I are equivalent with respect to efficiency of the final analysis. One such model is obtained by setting $\psi_0 = -3$ and $\psi_1 = 0.02$ in the above logistic regression model. The corresponding simulated power distributions are shown in Figure 23.5. We now have that the gain in efficiency due to more spread in the covariate values t_{ij} is in balance with the loss in efficiency due to an

Simulated power distributions

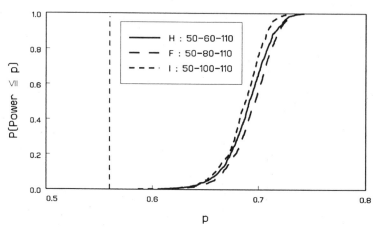

FIGURE 23.5. *Rat Data. Comparison of the simulated power distributions for designs with three measurements per rat, with equal power if no dropout would occur (designs H, F, and I), under the assumption that $p_{j,k|\geq k}$ satisfies $logit(p_{j,k|\geq k}) = -3 + 0.02(Age - 45)$. The vertical dashed line corresponds to the power which was realized in the original rat experiment (56%).*

increased risk of dropping out. In this case, one would prefer design I since it requires fewer subjects to conduct the study.

Note that the results presented here fully rely on the assumed linear mixed model (23.2). For example, the simulation results reported in Section 23.5.1 show that design G, with only two observations per subject, is to be preferred over designs A, E, and F, with more than two observations scheduled for each subject. Obviously, the assumption of linearity is crucial here, and design G will not allow testing for nonlinearities. Hence, when interest would be in providing support for model (23.2), more simulations would be needed comparing the behavior of different designs under different models for the outcome under consideration, and design G should no longer be taken into account. As for any sample-size calculation, it would be advisable to perform some informal sensitivity analysis to investigate the impact of model assumptions and imputed parameter values on the final results.

24

Case Studies

Building on the methodology developed in this text, the current chapter presents five case studies. In Section 24.1, we study the extension of univariate longitudinal data technology to the multivariate setting, where several measurements (i.e., systolic and diastolic blood pressure) are obtained at each measurement occasion. Section 24.2 is devoted to a developmental toxicology experiments where, due to litter effects, fetuses are clustered within dams. The flexibility of the linear mixed model to combine cluster effects with individual-specific covariates is illustrated. Even though time is not a factor in these data, we are able to establish a close connection with longitudinal modeling. In Section 24.3, we describe how bivariate outcomes from multicenter trials can be used to study the validity of one outcome as a surrogate endpoint for the other. A sensitivity analysis on incomplete longitudinal data on milk protein content is conducted in Section 24.4. The chapter is concluded with the analysis of hepatitis B vaccination data. It is shown how rather irregular sequences can be handled within the linear mixed-models context.

24.1 Blood Pressures

As an illustration of the use of linear mixed models for the analysis of repeated measures of a bivariate outcome, we analyze data reported by

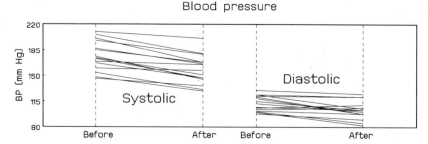

FIGURE 24.1. *Blood Pressure Data. Systolic and diastolic blood pressure in patients with moderate essential hypertension, immediately before and 2 hours after taking captopril.*

Hand *et al.* (1994), data set #72. For 15 patients with moderate essential (unknown cause) hypertension, the supine (measured while patient is lying down) systolic and diastolic blood pressure was measured immediately before and 2 hours after taking the drug captopril. The individual profiles are shown in Figure 24.1. The objective of the analysis is to investigate the effect of treatment on both responses.

Note that since we only have two measurements available for each response, there is no need for modeling the variance or the mean as continuous functions of time. Also, saturated mean structures and covariance structures can easily be fitted because of the balanced nature of the data. No transformation to normality is needed since none of the four responses has a distribution which shows clear deviations from normality.

In order to explore the covariance structure of our data, we fitted several linear mixed models with the saturated mean structure. Table 24.1 shows minus twice the maximized REML log-likelihood value, the Akaike and Schwarz information criteria, and the number of degrees of freedom in the corresponding covariance model. Our first model has a general unstructured covariance matrix. The REML-estimated covariance matrix and corresponding standard errors equal

$$\begin{pmatrix} 423\ (160) & 371\ (148) & 143\ (69) & 105\ (75) \\ 371\ (148) & 400\ (151) & 153\ (69) & 166\ (81) \\ 143\ (69) & 153\ (69) & 110\ (41) & 97\ (44) \\ 105\ (75) & 166\ (81) & 97\ (44) & 157\ (60) \end{pmatrix}, \quad (24.1)$$

for measurements ordered as indicated in Figure 24.1. Note that if one would model the diastolic and systolic blood pressures separately, a random-intercepts model would probably fit the data well. This would implicitly assume that for both responses, the variance before and after the treatment with captopril is the same, which is not unreasonable in view of the large standard errors shown in (24.1). We therefore reparameterize the unstruc-

TABLE 24.1. *Blood Pressure Data. Summary of the results of fitting several covariance models. All models include a saturated mean structure. Notations RI_{dia}, RS_{dia}, RI_{sys}, RS_{sys}, RI, and RS are used for random intercepts and slopes for the diastolic blood pressures, random intercepts and slopes for the systolic blood pressures, and random intercepts and slopes which are the same for both blood pressures, respectively.*

Model	$-2\ell_{\text{REML}}$	AIC	SBC	df
Unstructured	420.095	−220.047	−230.174	10
$RI_{dia} + RI_{sys} + RS_{dia} + RS_{sys}$	420.095	−221.047	−232.187	10
$RI_{dia} + RI_{sys} + RS$	420.656	−217.328	−224.417	7
$RI_{dia} + RI_{sys}$	433.398	−220.699	−224.749	4
$RI + RI_{sys} + RS$	420.656	−217.328	−224.417	7
$RI + RI_{sys} + RS$, uncorrelated	424.444	−216.222	−220.273	4

tured covariance model as a random-effects model with random intercepts and random slopes for the two responses separately. The random slopes are random coefficients for the dummy covariate defined as 0 before the treatment and as 1 after the treatment, and therefore represent subject-specific deviations from the average effect of the treatment, for systolic and diastolic blood pressure, respectively. Since this model includes four linearly independent covariates in the Z_i matrix, it is equivalent to the first model in Table 24.1, yielding the same maximized log-likelihood value. The advantage is that we now have expressed the model in terms of interpretable components. This will prove to be convenient for reducing the number of variance components. As discussed previously, SAS always adds a residual component $\varepsilon_{(1)i}$ of measurement error to any random-effects model. In our model this component is not identified. Still, the reported AIC and SBC values are based on 11 variance components and are therefore not the same as the ones found for the first model in Table 24.1.

In a first attempt to reduce the covariance structure, we fitted a linear mixed model assuming equal random slopes for diastolic and systolic blood pressure. This model, which is the third model in Table 24.1, assumes that for each subject, the treatment effect additional to the average effect is the same for both responses. Note that there is only a difference of 3 degrees of freedom when compared to the unstructured model, which is due to the estimation of the residual variance which was not included in the first model. Therefore, we can, strictly speaking, not apply the theory of Section 6.3.4 for testing random effects. However, the very small difference in twice the maximized REML log-likelihood clearly suggests that the random slopes may be assumed equal for both responses.

As a second step, we refit our model, not including the random slopes. This is the fourth model in Table 24.1. The p-value calculated using the theory of Section 6.3.4 on testing the significance of the random slopes equals

$$
\begin{aligned}
P(\chi^2_{2:3} \geq 12.742) &= 0.5\, P(\chi^2_2 \geq 12.742) + 0.5\, P(\chi^2_3 \geq 12.742) \\
&= 0.5 \times 0.0017 + 0.5 \times 0.0052 \\
&= 0.0035,
\end{aligned}
$$

indicating that the treatment effect is not the same for all subjects.

We therefore investigate our third model further. The REML-estimated random-effects covariance matrix D and the corresponding estimated standard errors are

$$
\begin{array}{l}
\mathrm{RI}_{sys} \\
\mathrm{RI}_{dia} \\
\mathrm{RS}
\end{array}
\begin{array}{l}
\longrightarrow \\
\longrightarrow \\
\longrightarrow
\end{array}
\left(
\begin{array}{ccc}
409\ (158) & 146\ (68) & -38\ (46) \\
146\ (68) & 92\ (39) & 4\ (22) \\
-38\ (46) & 4\ (22) & 51\ (25)
\end{array}
\right),
$$

where the random effects are ordered as indicated in front of the matrix. Clearly, there is no significant correlation between either one of the random intercepts on one side and the random slopes on the other side ($p = 0.8501$ and $p = 0.4101$ for the diastolic and systolic blood pressure, respectively), meaning that the treatment effect does not depend on the initial value. On the other hand, there is a significant positive correlation ($p = 0.0321$) between the random intercepts for the diastolic blood pressure and the random intercepts for the systolic blood pressure, meaning that a patient with an initial diastolic blood pressure higher than average is likely to have an initial systolic blood pressure which is also higher than average.

This suggests that an overall subject effect may be present in the data. We can easily reparameterize our fourth model such that an overall random intercept is included, but a correction term for either systolic or diastolic blood pressure is then needed. In view of the larger variability in systolic blood pressures than in diastolic blood pressures, we decided to reparameterize our model as a random-effects model, with overall random intercepts, random intercepts for systolic blood pressure, and random slopes. The overall random intercepts can then be interpreted as the random intercepts for the diastolic blood pressures. The random intercepts for systolic blood pressure are corrections to the overall intercepts, indicating, for each patient, what its deviation from the average initial systolic blood pressure is, additional to its deviation from the average initial diastolic blood pressure. These correction terms then explain the additional variability for systolic blood pressure, in comparison to diastolic blood pressure. Information on the model fit for this fifth model is also shown in Table 24.1. Since this model is merely a reparameterization of our third model, we obtain the same results for both models. The REML-estimated random-effects covariance matrix D and the corresponding estimated standard errors are now

$$
\begin{array}{rl}
\text{RI} & \longrightarrow \\
\text{RI}_{dia} & \longrightarrow \\
\text{RS} & \longrightarrow
\end{array}
\left(
\begin{array}{rrr}
92\ (39) & 54\ (42) & 4\ (22) \\
54\ (42) & 209\ (84) & -42\ (34) \\
4\ (22) & -42\ (34) & 51\ (25)
\end{array}
\right),
$$

suggesting that there are no pairwise correlations between the three random effects (all p-values larger than 0.19). We therefore fit a sixth model assuming independent random effects. The results are also shown in Table 24.1. Minus twice the difference in maximized REML log-likelihood between the sixth and fifth model equals 3.788, which is not significant when compared to a chi-squared distribution with 3 degrees of freedom ($p = 0.2853$). We will preserve this covariance structure (four independent components of stochastic variability: a random subject effect, a random effect for the overall difference between the systolic and diastolic blood pressures, a random effect for the overall treatment effect, and a component of measurement error) in the models considered next. Note that this covariance structure is also the one selected by the AIC as well as the SBC criterion (see Table 24.1).

Using the above covariance structure, we can now try to reduce our saturated mean structure. The average treatment effect was found to be significantly different for the two blood pressure measurements, and we found significant treatment effects for the systolic as well as diastolic blood pressures (all p-values smaller than 0.0001). Our final model is now given by

$$
Y_{ij} =
\begin{cases}
\beta_1 + b_{1i} + \varepsilon_{(1)ij} & \text{diastolic, before,} \\[2mm]
\beta_2 + b_{1i} + b_{2i} + \varepsilon_{(1)ij} & \text{systolic, before,} \\[2mm]
\beta_3 + b_{1i} + b_{3i} + \varepsilon_{(1)ij} & \text{diastolic, after,} \\[2mm]
\beta_4 + b_{1i} + b_{2i} + b_{3i} + \varepsilon_{(1)ij} & \text{systolic, after.}
\end{cases}
\tag{24.2}
$$

As previously, we assume the random effects to be uncorrelated, and the $\varepsilon_{(1)ij}$ represent independent components of measurement error with equal variance σ^2. The program needed to fit this model in SAS is

```
data blood;
set blood;
slope = (time = 'after');
intsys = (meas = 'systolic');
run;

proc mixed data = blood covtest;
class time meas id;
```

TABLE 24.2. *Blood Pressure Data. Results from fitting the final model (24.2), using restricted maximum likelihood estimation.*

Effect	Parameter	Estimate (s.e.)
Intercepts:		
Diastolic, before	β_1	112.333 (2.687)
Systolic, before	β_2	176.933 (4.648)
Diastolic, after	β_3	103.067 (3.269)
Systolic, after	β_4	158.000 (5.007)
Treatment effects:		
Diastolic	$\beta_3 - \beta_1$	9.267 (2.277)
Systolic	$\beta_4 - \beta_2$	18.933 (2.277)
Covariance of b_i:		
$\mathrm{var}(b_{1i})$	d_{11}	95.405 (39.454)
$\mathrm{var}(b_{2i})$	d_{22}	215.728 (86.315)
$\mathrm{var}(b_{3i})$	d_{33}	52.051 (24.771)
Measurement error variance:		
$\mathrm{var}(\varepsilon_{(1)ij})$	σ^2	12.862 (4.693)
Observations		60
REML log-likelihood		-212.222
-2 REML log-likelihood		424.444
Akaike's Information Criterion		-216.222
Schwarz's Bayesian Criterion		-220.273

```
model bp = meas*time / noint s;
random intercept intsys slope / type = un(1) subject = id;
estimate 'trt_sys' meas*time 0 -1 0 1 / cl alpha = 0.05;
estimate 'trt_dia' meas*time -1 0 1 0 / cl alpha = 0.05;
contrast 'trt_sys = 2xtrt_dia' meas*time 2 -1 -2 1;
run;
```

The variables *meas* and *time* are factors with levels "systolic" and "diastolic" and with levels "before" and "after," respectively. The ESTIMATE statements are included to estimate the average treatment effect for the systolic and diastolic blood pressures separately. The CONTRAST statement is used to compare these two effects.

The REML estimates for all parameters in the marginal model are shown in Table 24.2. The 95% confidence intervals for the average treatment effect on diastolic and systolic blood pressure are [4.383; 14.151] and [14.050; 23.817], respectively. Further, the parameter estimates in Table 24.2 suggest that the average treatment effect on systolic blood pressure is twice the average treatment effect on diastolic blood pressure. This hypothesis, tested with the CONTRAST statement in the above program, was indeed not rejected at the 5% level of significance ($p = 0.9099$).

24.2   The Heat Shock Study

24.2.1   Introduction

In the last several decades, teratology and developmental toxicity studies conducted in laboratory animals have served as an important strategy for evaluating the potential risk of chemical compounds and other environmental agents on fertility, reproduction, and fetal development. Standard protocols for conducting developmental toxicity studies have been in use since shortly after the thalidomide tragedy of the early 1960s, yet methods for quantitative risk assessment are still being developed and refined (Schwetz and Harris, 1993). One issue that requires further consideration is the fact that these types of animal study often result in multiple outcomes of interest. For example, developmental toxicity studies may record several different types of malformation on each embryo, such as skeletal, visceral, and gross malformations. Teratology studies also consider multiple outcomes, such as the extent of implantations, resorptions, and several different clinical signs of maternal toxicity. These responses are usually examined individually, and risk assessment is based on the most sensitive outcome. However, statistical methods which incorporate the important multiple outcomes have several advantages: They can increase the power of detecting effects of exposure under a common dose effect model, they allow investigation of the association among the multiple outcomes, and they provide a more biologically informative and realistic description of the range of exposure effects on various responses.

A typical study design for evaluating adverse effects of exposure on the developing fetus, referred to as a "Segment II design," involves 20 to 30 pregnant rodents randomly assigned to each of several exposure groups and a control group. The pregnant dams are usually exposed for 1 or more days during the early part of the gestational cycle and are then sacrificed just prior to delivery to examine the fetuses for abnormalities. The standard approach for conducting risk assessment for developmental endpoints has been to use the results of such experimental animal studies to estimate the "No observed adverse effect level" (NOAEL), by determining the highest dose level that shows no significant difference from the control group in the rate of malformed embryos or fetal deaths (Gaylor 1989). Several limitations of this approach have been widely recognized, and new regulatory guidelines emphasize the use of quantitative methods similar to those developed for cancer risk assessment to estimate reference concentrations and benchmark doses (U.S. EPA 1991). Thus, more recent techniques for risk assessment in this area are based on fitting dose-response models and estimating the dose that leads to a certain increase in risk of some type of adverse developmental effect over that of the control group.

Although a wide variety of statistical methods have been developed for cancer risk assessment, the issue of multiple endpoints does not present quite the degree of complexity in this area as it does for developmental toxicity studies. The endpoint of interest in an animal cancer bioassay is typically the occurrence of a particular type of tumor, whereas in developmental toxicity studies, there is no clear choice for a single type of adverse outcome. In fact, an entire array of outcomes are needed to define certain birth defect syndromes (Khoury *et al.* 1987, Holmes 1988). Ryan (1992a) describes the data resulting from a developmental toxicity study as consisting of a series of hierarchical outcomes and has proposed a modeling framework which incorporates the hierarchical structure and allows for the combination of data on fetal death and resorption. Catalano *et al.* (1993) extend this approach to account for low birth weight, by modeling this continuous outcome conditionally on other outcomes. In the most general situation, the multiple outcomes recorded in the course of a developmental toxicity study may represent a combination of binary, ordinal, and continuous measurements (Ryan 1992a, 1992b).

Risk assessment methods for developmental toxicity endpoints have traditionally been based on daily exposure levels and have not taken into account either the duration or timing of exposure. However, organogenesis is a very sensitive process and it is acknowledged that exposure during early periods of organ development will lead to different types of adverse effect than later exposures. In addition, it is usually anticipated that short acute exposures at higher concentrations will result in more severe damage to the developing fetus than longer chronic exposures. One test system developed to explore the joint effects of exposure levels and durations is a "heat shock" study, as described by Brown and Fabro (1981) and Kimmel *et al.* (1993). In this type of developmental toxicity study, mice are exposed *in vitro* to various combinations of heat stress levels and durations. The outcomes of interest are a series of morphological responses which are measured on an ordinal scale. Data from such a heat shock study are studied here. Focusing on continuous responses, we will show that linear mixed models are advantageous in combining clustering effects with covariates of interest, which are both duration and level of heat stress exposure.

TABLE 24.3. *Heat Shock Study. Study design: number of (viable) embryo's exposed to each combination of duration and temperature.*

	Duration of Exposure							
Temperature	5	10	15	20	30	45	60	Total
37.0	11	11	12	13	12	18	11	88
40.0	11	9	9	8	11	10	11	69
40.5	9	8	10	9	11	10	7	64
41.0	10	9	10	11	9	6	0	55
41.5	9	8	9	10	10	7	0	53
42.0	10	8	10	5	7	6	0	46
Total	60	53	60	56	60	57	29	375

In these heat shock experiments, the embryos are explanted from the uterus of the maternal dam during the gestation period and cultured *in vitro*. Each individual embryo is subjected to a short period of heat stress by placing the culture vial into a water bath, usually involving an increase over body temperature of 4°C to 5°C for a duration of 5 to 60 minutes. The embryos are examined 24 hours later for signs of impaired or accelerated development.

This type of developmental toxicity test system has several advantages over the standard Segment II design. First of all, the exposure is administered directly to the embryo, so controversial issues regarding the unknown (and often nonlinear) relationship between the level of exposure to the maternal dam and that received by the developing embryo need not be addressed. Although genetic factors are still expected to exert an influence on the vulnerability to injury of embryos from a common dam, direct exposure to individual embryos reduces the need to account for such litter effects, but does not remove it. Second, the exposure pattern can be much more easily controlled than in most developmental toxicity studies, since it is possible to achieve target temperature levels in the water bath within 1 to 2 minutes. Whereas the typical Segment II study requires waiting 8 to 12 days after exposure to assess its impact, information regarding the effects of exposure are quickly obtained in heat shock studies. Finally, this animal test system provides a convenient mechanism for examining the joint effects of both duration of exposure and exposure levels, which, until recently, have received little attention. The actual study design for the set of experiments conducted by Kimmel *et al.* (1994) is shown in Table 24.3. For the experiment, 71 dams (clusters) were available, yielding a total of 375 embryos. The distribution of cluster sizes is given in Table 24.4.

TABLE 24.4. *Heat Shock Study. Distribution of cluster sizes.*

Cluster size n_i	1	2	3	4	5	6	7	8	9	10	11
Number of clusters of size n_i	6	3	6	12	13	11	8	5	2	3	2

Historically, the strategy for comparing responses among exposures of different durations to a variety of environmental agents (e.g., radiation, inhalation, chemical compounds) has relied on a conjecture called Haber's Law, which states that adverse response levels should be the same for any equivalent level of dose times duration (Haber 1924). In other words, a 15-minute exposure to an increase of 3 degrees should produce the same response as a 45-minute exposure to an increase of 1 degree. Clearly, the appropriateness of applying Haber's Law depends on the pharmacokinetics of the particular agent, the route of administration, the target organ, and the dose/duration patterns under consideration. Although much attention has been focused on documenting exceptions to this rule, it is often used as a simplifying assumption in view of limited testing resources and the multitude of exposure scenarios. However, given the current desire to develop regulatory standards for a range of exposure durations, models flexible enough to describe the response patterns over varying levels of both exposure concentration and duration are greatly needed.

For the heat shock studies, the vector of exposure covariates must incorporate both exposure level (also referred to as temperature or dose), d_{ij}, and duration (time), t_{ij}, for the jth embryo within the ith cluster. Furthermore, models must be formulated in such a way that departures from Haber's premise of the same adverse response levels for any equivalent multiple of dose times duration can easily be assessed. The exposure metrics in these models are the cumulative heat exposure, $(dt)_{ij} = d_{ij}t_{ij}$, referred to as *durtemp* and the effect of duration of exposure at positive increases in temperature (the increase in temperature over the normal body temperature of 37°C):

$$(pd)_{ij} \quad = \quad t_{ij}I(d_{ij} > 37).$$

We refer to the latter as *posdur*. There are measurements on 13 morphological variables. Some are binary; others are measured on a continuous scale. Even though we will focus on continuous outcomes, as in Geys, Molenberghs, and Williams (1999), it is worth mentioning that a lot of work has been done in the area of clustered binary outcomes and combined binary and continuous outcomes.

Indeed, as a result of the research activity over the past 10 to 15 years, there are presently several different schools of thought regarding the best approach to the analysis of correlated binary data. Unlike in the normal setting, marginal, conditional, and random-effects approaches tend to give dissimilar results, as do likelihood, quasi-likelihood, and GEE-based inferential methods. There are many excellent reviews (Prentice 1988, Fitzmaurice, Laird and Rotnitzky 1993, Diggle, Liang, Zeger 1994, Pendergast *et al.* 1996).

Several likelihood-based methods have been proposed. Fitzmaurice and Laird (1993) incorporate marginal parameters for the main effects in this model and quantify the degree of association by means of conditional odds ratios. Fully marginal models are presented by Bahadur (1961) and Cox (1972), using marginal correlations, and by Ashford and Sowden (1970), using a dichotomized version of a multivariate normal to analyze multivariate binary data. Alternatively, marginal odds ratios can be used, as shown by Dale (1986) and Molenberghs and Lesaffre (1994). Cox (1972) also describes a model whose parameters have interpretations in terms of conditional probabilities. Similar models were proposed by Rosner (1984) and Liang and Zeger (1989). A full exponential family model was proposed by Molenberghs and Ryan (1999) and Ryan and Molenberghs (1999). Pseudo-likelihood methods were developed by Geys, Molenberghs, and Ryan (1997, 1999). Random-effects approaches have been studied by Stiratelli, Laird, and Ware (1984), Zeger, Liang, and Albert (1988), Breslow and Clayton (1993), and Wolfinger and O'Connell (1993). Generalized estimating equations were developed in Liang and Zeger (1986). A thorough account is given in Fahrmeir and Tutz (1994). Williams, Molenberghs, and Lipsitz (1996) consider ordinal outcomes in the context of the heat shock study.

Early work on the combination of continuous and discrete outcomes can be found in Olkin and Tate (1961). We also refer to Cox and Wermuth (1992, 1994) and to Schafer (1997). There is a substantial amount of work which focuses on clustered data: Catalano and Ryan (1992), Catalano *et al.* (1993), Chen (1993), Fitzmaurice and Laird (1995, 1997), Regan and Catalano (1999a, 1999b), and Geys, Regan, Catalano, and Molenberghs (1999). Molenberghs, Geys, and Buyse (1998) studied such models for use in surrogate marker evaluation.

24.2.2   Analysis of Heat Shock Data

There are several continuous outcomes recorded in the heat shock study, such as size measures on crown rump, yolk sac, and head. We will focus on crown rump length (CRL).

It will be shown that the three components of variability customarily incorporated in a linear mixed-effects model of the form (3.11) can usefully be applied here as well, even in the absence of a repeated-measures structure.

Although there will be no doubt that random effects are used to model interdam variability and also the role of the measurement error time is unambiguous, it is less obvious what the role of the serial association would be. Generally, serial association results from the fact that within a cluster, residuals of individuals closer to each other are often more similar than residuals for individuals further apart. Although this distance concept is clear in longitudinal and spatial applications, it is less so in this context. However, covariates like duration and temperature, or relevant transformations thereof, can play a similar role. This distinction is very useful since random effects capture the correlation structure which is attributable to the dam and hence includes genetic components. The serial correlation, on the other hand, is entirely design driven. If one conjectures that the latter component is irrelevant, then translation into a statistical hypothesis and, consequently, testing for it are relatively straightforward. Note that such a model is not possible in conventional developmental toxicity studies, where exposure applies at the dam level, not at the individual fetus level.

The model we consider is based on Haber's Law and controlled deviations thereof, in the sense that the fixed-effects structure includes *durtemp* and *posdur*. For computational convenience, the ranges of these covariates are transformed to the unit interval. The maximal values correspond to 225 min°C for *durtemp* and 60 minutes for *posdur*. The random-effects structure includes a random intercept and a random slope for dt. The residual covariance structure is decomposed into a Gaussian serial process in dt and measurement error. Formally,

$$Y_{ij} = (\beta_1 + b_{i1}) + (\beta_2 + b_{i2})(dt)_{ij} + \beta_3(pd)_{ij} + \varepsilon_{(1)ij} + \varepsilon_{(2)ij}, \qquad (24.3)$$

where the $\varepsilon_{(1)ij}$ are uncorrelated and follow a normal distribution with zero mean and variance σ^2. The $\varepsilon_{(2)ij}$ have zero mean, variance τ^2, and serial correlation

$$h_{ijk} \quad = \quad \exp\left\{-\phi[(dt)_{ij} - (dt)_{ik}]^2\right\}.$$

The random-effects vector (b_{i1}, b_{i2}) is assumed to be a zero-mean normal variable with covariance matrix D. SAS code to fit this model is

```
proc mixed data = heatshock method = ml covtest;
class idnr;
model crl = durtemp posdur / solution;
random intercept durtemp / subject = idnr g v type = un;
repeated / subject = idnr local type = sp(gau)(durtemp) r;
parms (0.01) (0.07) (-0.03) (0.04) (4.26) (0.09)
      / nobound;
run;
```

TABLE 24.5. *Heat Shock Study. Parameter estimates (standard errors) for initial and final model.*

Effect	Parameter	Initial	Final
Fixed effects:			
Intercept	β_1	3.622 (0.034)	3.627 (0.042)
Durtemp $(dt)_{ij}$	β_2	−1.558 (0.376)	−1.331 (0.353)
Posdur $(pd)_{ij}$	β_3	0.019 (0.006)	0.015 (0.006)
Random-effects parameters:			
var(b_{1i})	d_{11}	0.010 (0.014)	0.046 (0.014)
var(b_{12})	d_{12}	−0.038 (0.065)	
cov(b_{1i}, b_{2i})	d_{22}	0.071 (0.032)	
Residual variance parameters:			
var$(\varepsilon_{(1)ij})$	σ^2	0.097 (0.014)	0.097 (0.014)
var$(\varepsilon_{(2)ij})$	τ^2	0.044 (0.017)	0.042 (0.017)
Spatial corr. parameter	ρ	4.268 (5.052)	4.143 (3.772)

Since the 'nobound' option is added to the PARMS statement, variance components are allowed to assume values on the whole real line. This implies that conventional χ^2 tests can be used, rather than the mixtures described in Section 6.3.4. The initial model is reproduced in Table 24.5.

First, the covariance model is simplified. The covariance between both random effects is not significant and can be removed ($G^2 = 3.35$ on 1 degree of freedom, $p = 0.067$). Next, the random *durtemp* effect is removed ($G^2 = 3.63$, 2 df, $p = 0.057$). The serial process cannot be removed ($G^2 = 6.19$, 2 df, $p = 0.045$). Finally, both fixed effects are highly significant and cannot be removed. The final model is given in Table 24.5. SAS code for this model is

```
proc mixed data = heatshock method = ml covtest;
class idnr;
model crl = durtemp posdur / solution;
random intercept / subject = idnr g v vcorr;
repeated / subject = idnr local
```

Final Model   Posdur Removed

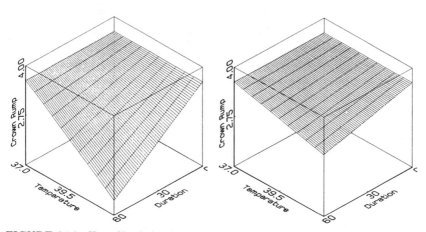

FIGURE 24.2. *Heat Shock Study. Fixed-effects structure for (a) the final model and (b) the model with* posdur *removed.*

Fitted Variogram

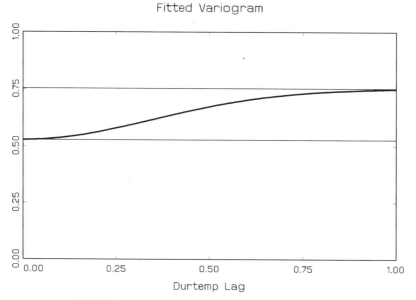

FIGURE 24.3. *Heat Shock Study. Fitted variogram.*

```
        type = sp(gau)(durtemp) r rcorr;
parms (0.032) (0.043) (2.015) (0.094) / nobound;
run;
```

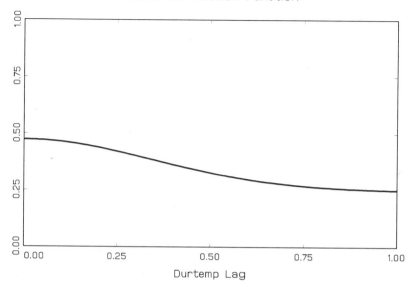

Fitted Correlation Function

FIGURE 24.4. *Heat Shock Study. Fitted correlation function.*

The fixed-effects structure is presented in Figure 24.2. The left-hand panel shows the fixed-effects structure of the final model, as listed in Table 24.5. The coefficient of *durtemp* is negative, indicating a decreasing crown rump length with increasing exposure. This effect is reduced by a positive coefficient for *posdur*. Fitting this model with *posdur* removed shows qualitatively the same trend, but the effect of exposure is much less pronounced, underscoring that there is a significant deviation from Haber's Law.

The fitted variogram is presented in Figure 24.3. Roughly half of the variability is attributed to measurement error, and the remaining half is divided equally over the random intercept and the serial process. The corresponding fitted correlation function is presented in Figure 24.4. The correlation is about 0.50 for two fetuses that are at the exact same level of exposure. It then decreases to 0.25 when the distance between exposures is maximal. This reexpresses that half of the correlation is due to the random effect, and the other half is attributed to the serial process in *durtemp*.

24.3 The Validation of Surrogate Endpoints from Multiple Trials

24.3.1 Introduction

Surrogate endpoints are referred to as endpoints that can be used in lieu of other endpoints in the evaluation of experimental treatments or other interventions. Surrogate endpoints are useful when they can be measured earlier, more conveniently, or more frequently than the endpoints of interest, which are referred to as the "true" or "final" endpoints (Ellenberg and Hamilton 1989). Biological markers of the disease process are often proposed as surrogate endpoints for clinically meaningful endpoints, the hope being that if a treatment showed benefit on the markers, it would ultimately also show benefit upon the clinical endpoints of interest. Before a surrogate endpoint can replace a final endpoint in the evaluation of an experimental treatment, it must be formally "validated," a process that has caused a number of controversies and has not been fully elucidated so far.

In a landmark paper, Prentice (1989) proposed a formal definition of surrogate endpoints, outlined how they could be validated, and, at the same time, discussed intrinsic limitations in the surrogate marker validation quest. Much debate ensued, since some authors perceived a formal criteria-based approach as too stringent and not straightforward to verify (Fleming *et al.* 1994). Freedman, Graubard, and Schatzkin (1992) took Prentice's approach one step further by introducing the *proportion explained*, which is the proportion of the treatment effect mediated by the surrogate. Buyse and Molenberghs (1998) and Molenberghs, Buyse *et al.* (1999) discussed some issues with the proportion explained and proposed to enhance insight by means of two new measures. The first, defined at the population level and termed *relative effect*, is the ratio of the overall treatment effect on the true endpoint over that on the surrogate endpoint. The second is the individual-level association between both endpoints, after accounting for the effect of treatment, and referred to as *adjusted association*.

Buyse *et al.* (2000) extended these concepts to situations in which data are available from several randomized experiments. The individual-level association between the surrogate and final endpoints carries over naturally, the only change required being an additional stratification to account for the presence of multiple experiments. The experimental unit can be the center in a multicenter trial, or the trial in a meta-analysis context. We emphasize the latter situation, because a sufficiently informative validation of a surrogate endpoint will typically require large numbers of observations coming from several trials. Moreover, meta-analytic data usually carry a degree of heterogeneity not encountered in a single trial, caused by differences in patient population, study design, treatment regimens, and so forth. We shall argue that these sources of heterogeneity increase one's confidence in the validity of a surrogate endpoint, when the relationship between the effects of treatment on the surrogate and the true endpoints tends to remain constant across such different situations.

The notion of relative effect can then be extended to a trial-level measure of association between the effects of treatment on both endpoints. The two measures of association, one at the individual level, and the other at the trial level, are proposed as an alternative way to assess the usefulness of a surrogate endpoint. This approach also naturally yields a prediction for the effect of treatment on the true endpoint, based on the observation of the effect of treatment on the surrogate endpoint.

In Section 24.3.2, Prentice's definition and criteria, as well as Freedman's proportion explained, are reviewed. Notation and motivating examples are presented in Section 24.3.3. The new concepts and an alternative validation strategy are introduced in Section 24.3.4. The examples are analyzed in Section 24.3.5. Fitting of some of the models in Section 24.3.4 by means of linear mixed-models methodology is computationally not straightforward. Section 24.3.6 examines through simulations when numerical problems are likely to occur. The emphasis is on normally distributed endpoints, for which standard linear mixed models are appropriate. The mixed-models methodology provides an easy-to-use framework that avoids many of the complexities encountered with different response types. In practice, however, endpoints are seldom normally distributed. In Section 24.3.7, a brief discussion of possible extensions to more general situations where the surrogate and true endpoints are of a different nature, such as the highly relevant situation where the surrogate endpoint is binary and the final endpoint is a survival time, possibly censored (Lin, Fleming, and DeGruttola 1997), is presented. A profound treatment of these extensions is outside the scope of the current text.

24.3.2   Validation Criteria

PRENTICE'S DEFINITION

Prentice proposed to define a surrogate endpoint as "a response variable for which a test of the null hypothesis of no relationship to the treatment groups under comparison is also a valid test of the corresponding null hypothesis based on the true endpoint" (Prentice 1989, p. 432). We adopt the following notation: T and S are random variables that denote the true and surrogate endpoints, respectively, and Z is an indicator variable for treatment. Prentice's definition can be written

$$f(S|Z) \;=\; f(S) \Leftrightarrow f(T|Z) = f(T), \tag{24.4}$$

where $f(X)$ denotes the probability distribution of random variable X and $f(X|Z)$ denotes the probability distribution of X conditional on the value of Z. As such, this definition is of limited value since a direct verification that a triplet (T, S, Z) fulfills the definition would require a large number of experiments to be conducted with information on the triplet. Even if many experiments were available, the equivalence of the statistical tests implied in (24.4) might not be true in all of them because of chance fluctuations and/or lack of statistical power. Operational criteria are therefore needed to check if definition (24.4) is fulfilled.

PRENTICE'S CRITERIA

Four operational criteria have been proposed to check if a triplet (T, S, Z) fulfills the definition. The first two verify departures from the null hypotheses implicit in (24.4):

$$f(S|Z) \;\neq\; f(S), \tag{24.5}$$
$$f(T|Z) \;\neq\; f(T). \tag{24.6}$$

Strictly speaking, (24.5) and (24.6) are not criteria since having both $f(T|Z) = f(T)$ and $f(S|Z) = f(S)$ is consistent with the definition (24.4). However, in this case, the validation is practically impossible since one may fail to detect differences due to lack of power. Thus, in practice, the validation requires Z to have an effect on both T and S. Several authors have pointed out that requiring Z to have a statistically significant effect on T may be excessively stringent, for in that case, from the limited perspective of significance testing, there would no longer be a need to establish the surrogacy of S (Fleming et al. 1994).

The other two criteria are

$$f(T|S) \;\neq\; f(T), \tag{24.7}$$
$$f(T|S, Z) \;=\; f(T|S). \tag{24.8}$$

Buyse and Molenberghs (1998) reproduce the arguments that establish the sufficiency of conditions (24.7) and (24.8) for (24.4) to hold in the case of binary responses. It is also easy to show that condition (24.7) is always necessary for (24.4), and that condition (24.8) is necessary for binary endpoints but not in general. Indeed, suppose (24.8) does not hold; then, assuming that $f(S|Z) = f(S)$,

$$f(T|Z) \;=\; \int f(T|S, Z) f(S) \, dS, \tag{24.9}$$

$$f(T) = \int f(T|S)f(S)\,dS. \tag{24.10}$$

However, (24.9) and (24.10) are, in general, not equal to one another, in which case definition (24.4) is violated. However, it is possible to construct examples where $f(T|Z) = f(T)$, in which case the definition still holds despite the fact that (24.8) does not hold. Hence, (24.8) is not a necessary condition, except for binary endpoints.

Next, assume (24.8) holds but (24.7) does not. Then,

$$f(T|Z) = \int f(T|S)f(S|Z)\,dS = \int f(T)f(S|Z)\,dS = f(T),$$

and hence $f(T|Z) = f(T)$ regardless of the relationship between S and Z. The simplest example is the situation where T is independent of the pair (S, Z). Thus, (24.7) is necessary to avoid situations where one null hypothesis is true while the other is not. However, criteria (24.5) and (24.6) already imply that both null hypotheses must be rejected, and therefore criterion (24.7) is of no additional value. In fact, criterion (24.7) indicates that the surrogate endpoint has prognostic relevance for the final endpoint, a condition which will obviously be fulfilled by any sensible surrogate endpoint. Conditions (24.5)–(24.8) are informative and will tend to be fulfilled for valid surrogate endpoints, but they should not be regarded as strict criteria. Condition (24.8) captures the essential notion of surrogacy by requiring that the treatment is irrelevant for predicting the true outcome, given the surrogate. In the next section, we discuss how Freedman, Graubard, and Schatzkin (1992) used this concept in estimation rather than in testing. Our meta-analytic development, laid out in Section 24.3.4, also emphasizes estimation and prediction rather than hypothesis testing.

FREEDMAN'S PROPORTION EXPLAINED

Freedman, Graubard, and Schatzkin (1992) argued that criterion (24.8) raises a conceptual difficulty in that it requires the statistical test for treatment effect on the true endpoint to be nonsignificant after adjustment for the surrogate. The nonsignificance of this test does not prove that the effect of treatment upon the true endpoint is fully captured by the surrogate, and therefore Freedman, Graubard, and Schatzkin (1992) proposed to calculate the proportion of the treatment effect explained by the surrogate. In this paradigm, a good surrogate is one for which this proportion explained (PE) is close to unity (Prentice's criterion (24.8) would require that $PE = 1$). Buyse and Molenberghs (1998) and Molenberghs, Buyse, et al. (1999) outlined some conceptual difficulties with the PE, in particular that it is not a proportion: PE can be estimated to be anywhere on the real line, which

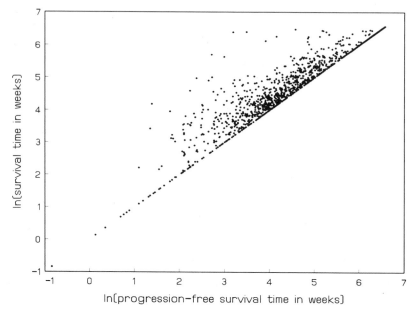

FIGURE 24.5. *Advanced Ovarian Cancer. Scatter plot of progression free survival versus survival.*

complicates its interpretation. They argued that PE can advantageously be replaced by two related quantities: the relative effect (RE), which is the ratio of the effects of treatment upon the final and the surrogate endpoint, and the treatment-adjusted association between the surrogate and the true endpoint, ρ_z. Therefore, these proposals are extended using data from several experiments. Motivating examples are introduced in the next section, and our approach in Section 24.3.4.

24.3.3   Notation and Motivating Examples

Suppose we have data from N trials, in the ith of which N_i subjects are enrolled. Let T_{ij} and S_{ij} be random variables that denote the true and surrogate endpoints, respectively, for the jth subject in the ith trial, and let Z_{ij} be an indicator variable for treatment. Although the main focus of this work is on binary treatment indicators, the methods proposed generalize without difficulty to multiple category indicators for treatment, as well as to situations where covariate information is used in addition to the treatment indicators.

AN EXAMPLE IN OVARIAN CANCER

Our methods will first be illustrated using data from a meta-analysis of four randomized multicenter trials in advanced ovarian cancer (Ovarian Cancer Meta-Analysis Project 1991). Individual patient data are available in these four trials for the comparison of two treatment modalities: cyclophosphamide plus cisplatin (CP) versus cyclophosphamide plus adriamycin plus cisplatin (CAP). The binary indicator for treatment (Z_{ij}) will be set to 0 for CP and to 1 for CAP. The surrogate endpoint S_{ij} will be the logarithm of time to progression, defined as the time (in weeks) from randomization to clinical progression of the disease or death due to the disease, and the final endpoint T_{ij} will be the logarithm of survival, defined as the time (in weeks) from randomization to death from any cause. The full results of this meta-analysis were published with a minimum follow-up of 5 years in all trials (Ovarian Cancer Meta-Analysis Project 1991). The data set was subsequently updated to include a minimum follow-up of 10 years in all trials (Ovarian Cancer Meta-Analysis Project 1998). After such long follow-up, most patients have had a disease progression or have died (952 of 1194 patients, i.e., 80%), so censoring will be ignored in our analyses. Methods that account for censoring would admittedly be preferable, but we ignore it here for the purposes of illustrating the case where the surrogate and final endpoints are both normally distributed.

The ovarian cancer data set contains only four trials. This will turn out to be insufficient to apply the meta-analytic methods of Section 24.3.4. In the two larger trials, information is also available on the centers in which the patients had been treated. We can then use center as the unit of analysis for the two larger trials, and the trial as the unit of analysis for the two smaller trials. A total of 50 units are thus available for analysis, with a number of individual patients per unit ranging from 2 to 274. To assess sensitivity, all analyses will be performed with and without the two smaller trials in which center is unknown. A scatter plot of the surrogate and true endpoints for all individuals in the trials included is presented in Figure 24.5.

The first three Prentice criteria (24.5)–(24.7) are provided by tests of significance of parameters α, β, and γ in the following models:

$$S_{ij}|Z_{ij} = \mu_S + \alpha Z_{ij} + \varepsilon_{sij}, \tag{24.11}$$

$$T_{ij}|Z_{ij} = \mu_T + \beta Z_{ij} + \varepsilon_{Tij}, \tag{24.12}$$

$$T_{ij}|S_{ij} = \mu + \gamma S_{ij} + \varepsilon_{ij}, \tag{24.13}$$

where ε_{sij}, ε_{Tij}, and ε_{ij} are independent Gaussian errors with mean zero. If the analysis is restricted to the two large trials in which center is known, $\alpha = 0.228$ [standard error (s.e.) 0.091, $p = 0.013$], $\beta = 0.149$ (s.e. 0.085, $p = 0.079$), and $\gamma = 0.874$ (s.e. 0.011, $p < 0.0001$). Strictly speaking, the criteria are not fulfilled because β fails to reach statistical significance. This will often be the case, since a surrogate endpoint is needed when there is no convincing evidence of a treatment effect upon the true endpoint.

As emphasized in Section 24.3.2, we cannot strictly show that the last criterion (24.8) is fulfilled. Instead, we can calculate Freedman's proportion explained:

$$PE \;=\; 1 - \frac{\beta_S}{\beta}, \tag{24.14}$$

where β is the estimate of the effect of Z on T as in (24.12) and β_S is the estimate of the effect of Z on T after adjustment for S:

$$T_{ij} | Z_{ij}, S_{ij} \;=\; \tilde{\mu}_T + \beta_S Z_{ij} + \gamma_Z S_{ij} + \tilde{\varepsilon}_{Tij}, \tag{24.15}$$

Here, $\beta_S = -0.051$ (s.e. 0.028) and $PE = 1.34$ (95% delta confidence limits $[0.73; 1.95]$). The proportion explained is larger than 100%, because the direction of the effect of Z on T is reversed after adjustment for S. Another problem would arise if there were a strong interaction between Z and S, which would require the following model to be fitted instead of (24.15):

$$T_{ij} | Z_{ij}, S_{ij} \;=\; \breve{\mu}_T + \breve{\beta}_S Z_{ij} + \breve{\rho}_Z S_{ij} + \delta Z_{ij} S_{ij} + \breve{\varepsilon}_{Tij}, \tag{24.16}$$

With this model, PE would cease to be captured by a single number and the validation process would have to stop (Freedman, Graubard, and Schatzkin 1992). In the two large ovarian cancer trials, the interaction term is not statistically significant ($\delta = 0.014$, s.e. 0.022), and therefore model (24.15) may be used.

Buyse and Molenberghs (1998) suggested to replace the PE by two quantities: the relative effect

$$RE \;=\; \frac{\beta}{\alpha} \tag{24.17}$$

and the association ρ_Z between T and S, adjusted for Z, which can be calculated from jointly modeling (24.11) and (24.12). To this end, the error terms of (24.11) and (24.12) are assumed to follow a bivariate Gaussian distribution with zero mean and general 2×2 covariance matrix. In this case, $RE = 0.65$ (95% confidence limits $[0.36; 0.95]$) and $\rho_Z = 0.944$ (95% confidence limits $[0.94; 0.95]$). Thus, the adjusted correlation is very close to 1 and estimated with high precision. The relative effect is determined with reasonable precision and enables calculation of the predicted effect of treatment upon survival based on the observed effect upon time to progression in a new trial. However, this prediction is based on the strong assumption of a regression through the origin based on a single pair $(\hat{\alpha}, \hat{\beta})$.

When the two smaller trials are included in the analysis, the results change very little, providing evidence for the validity of considering each of the smaller trials as a single center. The p-values for α, β, and γ become 0.003, 0.054, and < 0.0001, respectively, and $PE = 1.46$ (95% confidence limits $[0.80; 2.13]$), $RE = 0.60$ (95% confidence limits $[0.32; 0.87]$), and $\rho_z = 0.942$ (95% confidence limits $[0.94; 0.95]$). By including both trials, the precision is somewhat improved. However, in this case, the interaction term in model (24.16) is statistically significant ($\delta = 0.037$, s.e. 0.018), further complicating the interpretation of PE.

AN EXAMPLE IN OPHTHALMOLOGY

The second example concerns a clinical trial for patients with age-related macular degeneration, a condition in which patients progressively lose vision (Pharmacological Therapy for Macular Degeneration Study Group 1997). In this example, the binary indicator for treatment (Z_{ij}) is set to 0 for placebo and to 1 for interferon-α. The surrogate endpoint S_{ij} is the change in the visual acuity (which we assume to be normally distributed) at 6 months after starting treatment, and the final endpoint T_{ij} is the change in the visual acuity at 1 year. The data are presented in Figure 24.6. The first three Prentice criteria (24.5)–(24.7) are again provided by tests of significance of parameters α, β, and γ. Here, $\alpha = -1.90$ (s.e. 1.87, $p = 0.312$), $\beta = -2.88$ (s.e. 2.32, $p = 0.216$), and $\gamma = 0.92$ (s.e. 0.06, $p < 0.001$). Only γ is statistically significant and therefore the validation procedure has to stop inconclusively. Note, however, that the lack of statistical significance of α and β could merely be due to the insufficient number of observations available in this trial. Also note that α and β are negative, hinting at a negative effect of interferon-α upon visual acuity. Freedman's proportion explained is calculated as $PE = 0.61$ (95% confidence limits $[-0.19; 1.41]$). The relative effect is $RE = 1.51$ (95% confidence limits $[-0.46; 3.49]$), and the adjusted association $\rho_z = 0.74$ (95% confidence limits $[0.68; 0.81]$). The adjusted association is determined rather precisely, but the confidence limits of PE and RE are too wide to convey any useful information. Even so, as we will see in Section 24.3.5, some conclusions can be reached in this example that are in sharp contrast to those reached in the ovarian cancer example.

AN EXAMPLE IN COLORECTAL CANCER

As a third example, we will use data from two randomized multicenter trials in advanced colorectal cancer (Corfu-A Study Group 1995; Greco et $al.$ 1996). In one trial, treatment with fluorouracil and interferon (5FU/IFN) was compared to treatment with 5FU plus folinic acid (5FU/LV) (Corfu-A

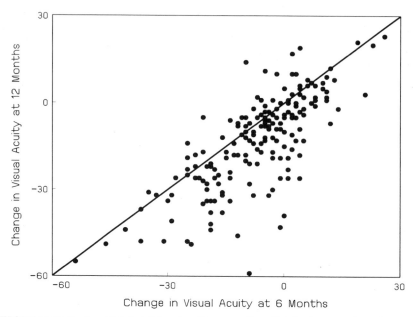

FIGURE 24.6. *Age-Related Macular Degeneration Trial. Scatter plot of visual acuity at 6 months (surrogate) versus visual acuity at 12 months (true endpoint).*

Study Group 1995). In the other trial, treatment with 5FU plus interferon (5FU/IFN) was compared to treatment with 5FU alone (Greco *et al.* 1996). The binary indicator for treatment (Z_{ij}) will be set to 0 for 5FU/IN and to 1 for 5FU/LV or 5FU alone. The surrogate endpoint S_{ij} will be progression-free survival time, defined as the time (in years) from randomization to clinical progression of the disease or death, and the final endpoint T_{ij} will be survival time, defined as the time (in years) from randomization to death from any cause. Most patients in the two trials have had a disease progression or have died (694 of 736 patients, i.e., 94.3%).

Similarly to the ovarian cancer example, we will use center as the unit of analysis. A total of 76 units are thus available for analysis. However, in eight centers, one of the treatment arms accrued no patients. These eight centers were therefore excluded from the analysis. As a result, the data used for illustration contained 68 units, with a number of individual patients per unit ranging from 2 to 38. The data are graphically depicted in Figure 24.7. Fitting (24.11)–(24.13) yields $\alpha = 0.021$ (standard error, s.e. 0.066, $p = 0.749$), $\beta = 0.002$ (s.e. 0.075, $p = 0.794$), and $\gamma = 0.917$ (s.e. 0.031, $p < 0.0001$). As with the ovarian cancer case, the criteria are not fulfilled because both β and α fail to reach statistical significance. The proportion explained is estimated as 0.985 (95% delta confidence limits $[-3.44; 5.41]$). Further, $RE = 0.931$ (95% confidence limits $[-3.23; 5.10]$). Both quantities have estimated confidence limits that are too wide to be

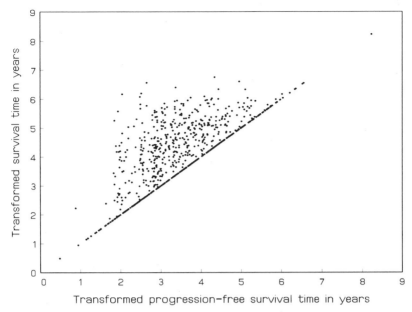

FIGURE 24.7. *Advanced Colorectal Cancer. Scatter plot of progression free survival versus survival.*

practically useful. Finally, $\rho_z = 0.802$ (95% confidence limits $[0.77; 0.83]$). This estimate of the adjusted association indicates that there is a substantial correlation between an individual's two endpoints.

24.3.4   A Meta-Analytic Approach

We focus on surrogate and true endpoints which are assumed to be jointly normally distributed. Two distinct modeling strategies will be followed, based on a two-stage fixed-effects representation on the one hand (see also Section 3.2) and the linear mixed model on the other hand.

Let us describe the two-stage model first. The first stage is based upon a fixed-effects model:

$$S_{ij}|Z_{ij} = \mu_{si} + \alpha_i Z_{ij} + \varepsilon_{sij}, \qquad (24.18)$$
$$T_{ij}|Z_{ij} = \mu_{Ti} + \beta_i Z_{ij} + \varepsilon_{Tij}, \qquad (24.19)$$

where μ_{si} and μ_{Ti} are trial-specific intercepts, α_i and β_i are trial-specific effects of treatment Z on the endpoints in trial i and ε_{si} and ε_{Ti} are correlated error terms, assumed to be zero-mean normally distributed with covariance matrix

$$\Sigma = \begin{pmatrix} \sigma_{SS} & \sigma_{ST} \\ & \sigma_{TT} \end{pmatrix}. \qquad (24.20)$$

At the second stage, we assume

$$
\begin{pmatrix} \mu_{si} \\ \mu_{Ti} \\ \alpha_i \\ \beta_i \end{pmatrix} = \begin{pmatrix} \mu_s \\ \mu_T \\ \alpha \\ \beta \end{pmatrix} + \begin{pmatrix} m_{si} \\ m_{Ti} \\ a_i \\ b_i \end{pmatrix}, \tag{24.21}
$$

where the second term on the right-hand side of (24.21) is assumed to follow a mean-zero normal distribution with dispersion matrix

$$
D = \begin{pmatrix} d_{ss} & d_{sT} & d_{sa} & d_{sb} \\ & d_{TT} & d_{Ta} & d_{Tb} \\ & & d_{aa} & d_{ab} \\ & & & d_{bb} \end{pmatrix}. \tag{24.22}
$$

Next, the random-effects representation is based upon combining both steps:

$$
S_{ij}|Z_{ij} = \mu_s + m_{si} + \alpha Z_{ij} + a_i Z_{ij} + \varepsilon_{sij}, \tag{24.23}
$$
$$
T_{ij}|Z_{ij} = \mu_T + m_{Ti} + \beta Z_{ij} + b_i Z_{ij} + \varepsilon_{Tij}, \tag{24.24}
$$

where now μ_s and μ_T are fixed intercepts, α and β are the fixed effects of treatment Z on the endpoints, m_{si} and m_{Ti} are random intercepts, and a_i and b_i are the random effects of treatment Z on the endpoints in trial i. The vector of random effects $(m_{si}, m_{Ti}, a_i, b_i)$ is assumed to be zero-mean normally distributed with covariance matrix (24.22). The error terms ε_{si} and ε_{Ti} follow the same assumptions as in fixed-effects model (24.18)–(24.19), with covariance matrix (24.20), thereby completing the specification of the linear mixed model. Section 24.3.10 provides sample SAS code to fit this particular model.

Much debate has been devoted to the relative merits of fixed versus random effects, especially in the context of meta-analysis (Thompson and Pocock 1991, Thompson 1993, Fleiss 1993, Senn 1998). Although the underlying models rest on different assumptions about the nature of the experiments being analyzed, the two approaches yield discrepant results only in pathological situations, or in very small samples where a fixed-effects analysis can yield artificially precise results if the experimental units truly constitute a random sample from a larger population. In our setting, both approaches are very similar, and the two-stage procedure can be used to introduce random effects (Section 3.2; see also Laird and Ware 1982). As the data analyses in Section 24.3.5 will illustrate, the choice between random and fixed effects can also be guided by pragmatic arguments.

TRIAL-LEVEL SURROGACY

The key motivation for validating a surrogate endpoint is to be able to predict the effect of treatment on the true endpoint based on the observed effect of treatment on the surrogate endpoint. It is essential, therefore, to explore the quality of the prediction of the treatment effect on the true endpoint in trial i by (a) information obtained in the validation process based on trials $i = 1, \ldots, N$ and (b) the estimate of the effect of Z on S in a new trial $i = 0$. Fitting either the fixed-effects model (24.18)–(24.19) or the mixed-effects model (24.23)–(24.24) to data from a meta-analysis provides estimates for the parameters and the variance components. Suppose then the new trial $i = 0$ is considered for which data are available on the surrogate endpoint but not on the true endpoint. We then fit the following linear model to the surrogate outcomes S_{0j}:

$$S_{0j} \quad = \quad \mu_{s0} + \alpha_0 Z_{0j} + \varepsilon_{s0j}. \tag{24.25}$$

Estimates for m_{s0} and a_0 are

$$\widehat{m}_{s0} \quad = \quad \widehat{\mu}_{s0} - \widehat{\mu}_s,$$
$$\widehat{a}_0 \quad = \quad \widehat{\alpha}_0 - \widehat{\alpha}.$$

We are interested in the estimated effect of Z on T, given the effect of Z on S. To this end, observe that $(\beta + b_0 | m_{s0}, a_0)$ follows a normal distribution with mean and variance:

$$E(\beta + b_0 | m_{s0}, a_0)$$
$$= \quad \beta + \left(\begin{array}{c} d_{sb} \\ d_{ab} \end{array} \right)' \left(\begin{array}{cc} d_{ss} & d_{sa} \\ d_{sa} & d_{aa} \end{array} \right)^{-1} \left(\begin{array}{c} \mu_{s0} - \mu_s \\ \alpha_0 - \alpha \end{array} \right), \tag{24.26}$$

$$\text{var}(\beta + b_0 | m_{s0}, a_0)$$
$$= \quad d_{bb} - \left(\begin{array}{c} d_{sb} \\ d_{ab} \end{array} \right)' \left(\begin{array}{cc} d_{ss} & d_{sa} \\ d_{sa} & d_{aa} \end{array} \right)^{-1} \left(\begin{array}{c} d_{sb} \\ d_{ab} \end{array} \right). \tag{24.27}$$

This suggests calling a surrogate *perfect at the trial level* if the conditional variance (24.27) is equal to zero. A measure to assess the quality of the surrogate at the trial level is the coefficient of determination

$$R^2_{\text{trial(f)}} = R^2_{b_i | m_{Si}, a_i} = \frac{\left(\begin{array}{c} d_{sb} \\ d_{ab} \end{array} \right)' \left(\begin{array}{cc} d_{ss} & d_{sa} \\ d_{sa} & d_{aa} \end{array} \right)^{-1} \left(\begin{array}{c} d_{sb} \\ d_{ab} \end{array} \right)}{d_{bb}}. \tag{24.28}$$

Coefficient (24.28) is unitless and ranges in the unit interval if the corresponding variance-covariance matrix is positive definite, two desirable features, facilitating interpretation. Intuition can be gained by considering the special case where the prediction of b_0 can be done independently of the random intercept m_{s0}. Expressions (24.26) and (24.27) then reduce to

$$
\begin{aligned}
E(\beta + b_0|a_0) &= \beta + \frac{d_{ab}}{d_{aa}}(\alpha_0 - \alpha), \\
\text{var}(\beta + b_0|a_0) &= d_{bb} - \frac{d_{ab}^2}{d_{aa}}
\end{aligned}
$$

with corresponding

$$
R^2_{\text{trial}(r)} = R^2_{b_i|a_i} = \frac{d_{ab}^2}{d_{aa}d_{bb}}. \tag{24.29}
$$

Now, $R^2_{\text{trial}(r)} = 1$ if the trial-level treatment effects are simply multiples of each other. We will refer to this simplified version as the reduced random-effects model, and the original expression (24.28) will be said to derive from the full random-effects model.

An estimate for $\beta + b_0$ is obtained by replacing the right-hand side of (24.26) with the corresponding parameter estimates. A confidence interval is obtained by applying the delta method to (24.26). The covariance matrix of the parameters involved is obtained from the meta-analysis, except for μ_{s0} and α_0, which are obtained from fitting (24.25) to the data of the new trial. The corresponding prediction interval is found by adding (24.27) to the variance obtained for the confidence interval. Details are given in Section 24.3.9.

There is a close connection between the prediction approach followed here and empirical Bayes estimation (Chapter 7). To see this, consider a similar but nonidentical approach where all data are analyzed together. This means that a meta-analysis is performed of the surrogate data on trials $i = 0, \ldots, N$ and of the true endpoint data on trials $i = 1, \ldots, N$. The estimate of b_0 will be based only on the surrogate data, since the true endpoint is unknown for trial $i = 0$, and on the parameter estimates. The expression for the empirical Bayes estimate of b_0 is identical to (24.26), but the numerical value will be slightly different since the parameters of the linear mixed model are determined on a larger set of data. For example, with the MIXED procedure in SAS, obtaining the empirical Bayes estimate of b_0 is immediate, and the conditional variance follows from the estimated standard error of prediction (Littell et al. 1996).

INDIVIDUAL-LEVEL SURROGACY

To validate a surrogate endpoint, Buyse and Molenberghs (1998) suggested considering the association between the surrogate and the final endpoints after adjustment for the treatment effect. To this end, we need to construct the conditional distribution of T, given S and Z. From (24.18)–(24.19), we derive

$$
\begin{aligned}
T_{ij}|Z_{ij}, S_{ij} \quad \sim \quad & N\left\{\mu_{Ti} - \sigma_{TS}\sigma_{SS}^{-1}\mu_{Si} + (\beta_i - \sigma_{TS}\sigma_{SS}^{-1}\alpha_i)Z_{ij} \; ; \right. \\
& \left. +\sigma_{TS}\sigma_{SS}^{-1}S_{ij}\sigma_{TT} - \sigma_{TS}^2\sigma_{SS}^{-1}\right\}.
\end{aligned}
\tag{24.30}
$$

Similarly, the random-effects model (24.23)–(24.24) yields

$$
\begin{aligned}
T_{ij}|Z_{ij}, S_{ij} \quad \sim \quad & N\left\{\mu_T + m_{Ti} - \sigma_{TS}\sigma_{SS}^{-1}(\mu_S + m_{Si}) \; ; \right. \\
& + [\beta + b_i - \sigma_{TS}\sigma_{SS}^{-1}(\alpha + a_i)]Z_{ij} \\
& \left. + \sigma_{TS}\sigma_{SS}^{-1}S_{ij}; \sigma_{TT} - \sigma_{TS}^2\sigma_{SS}^{-1}\right\},
\end{aligned}
\tag{24.31}
$$

where conditioning is also on the random effects. The association between both endpoints after adjustment for the treatment effect is in both (24.30) and (24.31) captured by

$$
R^2_{\text{indiv}} = R^2_{\varepsilon_{Ti}|\varepsilon_{Si}} = \frac{\sigma^2_{ST}}{\sigma_{SS}\sigma_{TT}},
\tag{24.32}
$$

the squared correlation between S and T after adjustment for both the trial effects and the treatment effect. Note that $R_{\varepsilon_{Ti}|\varepsilon_{Si}}$ generalizes the adjusted association ρ_Z of Section 24.3.3 to the case of several trials.

A NEW APPROACH TO SURROGATE EVALUATION

The above developments suggest to term a surrogate *trial-level valid* if $R^2_{\text{trial(f)}}$ (or $R^2_{\text{trial(r)}}$) is sufficiently close to one, and to call it *individual-level valid* if R^2_{indiv} is sufficiently close to one. Finally, a surrogate is termed *valid* if it is both trial-level and individual-level valid. In order to replace the words "valid" with "perfect," the corresponding R-squared values are required to equal one.

To be useful in practice, a valid surrogate must be able to predict the effect of treatment upon the true endpoint with sufficient precision to distinguish safely between effects that are clinically worthwhile and effects that are not. This requires both that the estimate of $\beta + b_0$ be sufficiently large and that the prediction interval of this quantity be sufficiently narrow.

It should be noted that the validation criteria proposed here do not require the treatment to have a significant effect on either endpoint. In particular, it is possible to have $\alpha \equiv 0$ and yet have a perfect surrogate. Indeed, even though the treatment may not have any effect on the surrogate endpoint as a whole, the fluctuations around zero in individual trials (or other experimental units) can be very strongly predictive of the effect on the true endpoint. However, such a situation is unlikely to occur since the heterogeneity between the trials is generally small compared to that between individual patients.

VALIDATION IN A SINGLE TRIAL

If data are available on a single trial (or, more generally, on a single experimental unit), the above developments are only partially possible. While the individual-level reasoning (producing ρ_z as in (24.32)) carries over by virtue of the within-trial replication, the trial-level reasoning breaks down and one cannot go beyond the relative effect (RE) as suggested in Buyse and Molenberghs (1998). Recall that the RE is defined as the ratio of the effects of Z on S and T, respectively, as expressed in (24.17). The confidence limits of RE can be used to assess the uncertainty about the value of β predicted from that of α, but in contrast to the above developments, no sensible prediction interval can be calculated for β.

24.3.5 Data Analysis

ADVANCED OVARIAN CANCER

As in Section 24.3.3, all analyses have been performed with and without the two smaller trials. Excluding the two smaller trials has very little impact on the estimates of interest, and therefore the results reported are those obtained with all four trials. Two-stage fixed-effects models (24.18)–(24.19) could be fitted, as well as a reduced version of the mixed-effects model (24.23)–(24.24), with random treatment effects but no random intercepts. Point estimates for the two types of model are in close agreement, although standard errors are smaller by roughly 35% in the random-effects model. Figure 24.8 shows a plot of the treatment effects on the true endpoint (logarithm of survival) by the treatment effects on the surrogate endpoint (logarithm of time to progression). These effects are highly correlated. Similarly to the random-effects situation, we refer to the models with and without the intercept used for determining R^2 as the reduced and full fixed-effects models. The reduced fixed-effects model provides $R^2_{\text{trial(r)}} = 0.939$ (s.e. 0.017). When the sample sizes of the experimental units are used to weigh the pairs (a_i, b_i), then $R^2_{\text{trial(r)}} = 0.916$ (s.e. 0.023). The full fixed-effects model yields $R^2_{\text{trial(f)}} = 0.940$ (s.e. 0.017). In the reduced random-effects model, $R^2_{\text{trial(r)}} = 0.951$ (s.e. 0.098).

Predictions of the effect of treatment on log(survival) based on the observed effect of treatment on log(time to progression) are of interest. Table 24.6

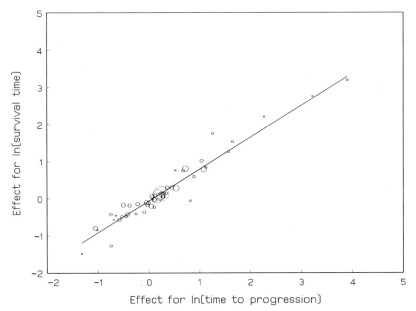

FIGURE 24.8. *Advanced Ovarian Cancer. Treatment effects on the true endpoint (logarithm of survival time) versus treatment effects on the surrogate endpoint (logarithm of time to progression) for all units of analysis. The size of each point is proportional to the number of patients in the corresponding unit.*

reports prediction intervals for several experimental units: six centers taken at random from the two large trials, and the two small trials in which center is unknown. Note that none of the predictions is significantly different from zero. The predicted values for $\beta + b_0$ agree reasonably well with the effects estimated from the data. The ratio $\widehat{\beta}_0/\widehat{\alpha}_0$ ranges from 0.69 to 0.73, which is close to the RE estimated in Section 24.3.3.

At the individual level, $R^2_{\text{indiv}} = 0.886$ (s.e. 0.006) in the fixed-effects model, and $R^2_{\text{indiv}} = 0.888$ (s.e. 0.006) in the reduced random-effects model. The square roots of these quantities are respectively 0.941 and 0.942, very close to the value of ρ_Z estimated in Section 24.3.3. Figure 24.9 displays a scatter plot of the residuals on both endpoints. It exhibits the close relationship which exists between both endpoints at the individual level.

Thus, we conclude that time to progression can be used as a surrogate for survival in advanced ovarian cancer. The effect of treatment can be observed earlier if time to progression is used instead of survival, and it is also more pronounced, as shown by the overall Kaplan-Meier estimates of Figure 24.10. Hence, a trial that used time to progression would require less follow-up time and less patients to establish the statistical significance

TABLE 24.6. *Advanced Ovarian Cancer. Predictions. Standard errors are shown in parenthesis. The number of patients is reported for each unit, as well as which sample is used for the estimation (only two trials or all four). $\widehat{\alpha}_0$ and $\widehat{\beta}_0$ are values estimated from the data; $E(\beta + b_0|a_0)$ is the predicted effect of treatment on survival (β_0), given its effect upon time to progression $(\widehat{\alpha}_0)$. The DACOVA and GONO trials are the two smaller studies, for which predictions are based on parameter estimates from the centers in the two larger studies.*

Unit	n_i	Trials	$\widehat{\alpha}_0$	$E(\beta + b_0|a_0)$	$\widehat{\beta}_0$
6	17	2	−0.58 (0.33)	−0.45 (0.29)	−0.56 (0.32)
		4		−0.45 (0.29)	
8	10	2	0.67 (0.76)	0.49 (0.57)	0.76 (0.39)
		4		0.47 (0.56)	
37	12	2	1.02 (0.61)	0.76 (0.54)	1.04 (0.70)
		4		0.73 (0.53)	
49	40	2	0.54 (0.34)	0.39 (0.26)	0.28 (0.28)
		4		0.37 (0.25)	
55	31	2	1.08 (0.56)	0.80 (0.44)	0.79 (0.45)
		4		0.77 (0.44)	
BB	21	2	−1.05 (0.55)	−0.80 (0.46)	−0.79 (0.51)
		4		−0.79 (0.46)	
DACOVA	274	2	0.25 (0.15)	0.17 (0.13)	0.14 (0.14)
GONO	125	2	0.15 (0.25)	0.10 (0.20)	0.03 (0.22)

of a truly superior treatment than a trial that used survival (Chen *et al.* 1998).

The results derived here are considerably more useful than the conclusions in Section 24.3.3. Indeed, the first three Prentice criteria provide only marginal evidence and PE cannot be estimated on the full data set, since there is a three-way interaction between Z, S, and T. RE is meaningful and estimated with precision, but it is derived from a regression through the origin based on a single data point. In contrast, the approach used here combines evidence from several experimental units and allows prediction intervals to be calculated for the effect of treatment on the true endpoint.

AGE-RELATED MACULAR DEGENERATION

The age-related macular degeneration data come from a single multicenter trial. Therefore, it is natural to consider the center in which the patients

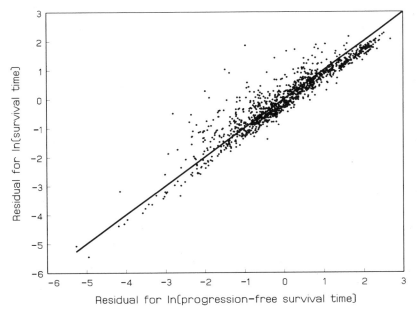

FIGURE 24.9. *Advanced Ovarian Cancer. Residuals of progression free survival versus survival, after correction for treatment and center effect.*

were treated as the unit of analysis. A total of 36 centers were thus available for analysis, with a number of individual patients per center ranging from 2 to 18.

Figure 24.11(a) shows a plot of the raw data (true endpoint versus surrogate endpoint for all individual patients). Irrespective of the software tool used (SAS, SPlus, MLwiN), the random effects are difficult to obtain. Therefore, we report only the result of a two-stage fixed-effects model and explore the computational issues further in Section 24.3.6. Figure 24.11(b) shows a plot of the treatment effects on the true endpoint by the treatment effects on the surrogate endpoint. These effects are moderately correlated, with $R^2_{trial(f)} = 0.692$ (s.e. 0.087). The estimates based on the reduced model are virtually identical. At the individual level, $R^2_{indiv} = 0.483$ (s.e. 0.053). Note that $R_{indiv} = 0.69$ is close to $\rho_Z = 0.74$, as estimated in Section 24.3.3. The coefficients of determination $R^2_{trial(r)}$ and R^2_{indiv} are both too low to make visual acuity at 6 months a reliable surrogate for visual acuity at 12 months. Figure 24.11(c) shows that the correlation of the measurements at 6 months and at 1 year is indeed rather poor at the individual level. Therefore, even with the limited data available, it is clear that the assessment of visual acuity at 6 months is not a good surrogate for the same assessment at 1 year. This is in contrast with the inconclusive analysis in Section 24.3.3.

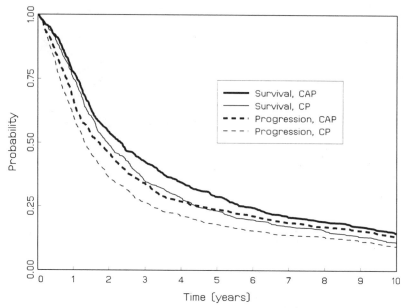

FIGURE 24.10. *Advanced Ovarian Cancer. Survival curves. Kaplan-Meier esti-mates of survival (S) and time to progression (TTP) for the two treatment groups: cyclophosphamide plus cisplatin (CP) and cyclophosphamide plus adriamycin plus cisplatin (CAP).*

ADVANCED COLORECTAL CANCER

Figure 24.12 shows a plot of the treatment effects on the true endpoint (logarithm of survival) by the treatment effects on the surrogate endpoint (logarithm of time to progression). Clearly, the correlation between both is considerably weaker than in the advanced ovarian cancer case. The corresponding $R^2_{\text{trial(r)}} = 0.454$ (95% confidence limits $[0.23; 0.68]$).

At the individual level (Figure 24.13), $R^2_{\text{indiv}} = 0.665$ (95% confidence limits $[0.62; 0.71]$). The square roots of this quantity is 0.815, very close to the estimate of ρ_Z, which is 0.805.

In contrast to the results obtained for advanced ovarian cancer, time to progression seems less qualified as a surrogate for survival in the context of advanced colorectal cancer. Both the trial level as well as the individual level R^2 are relatively low.

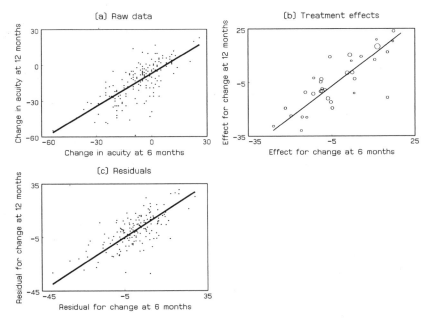

FIGURE 24.11. *Age-Related Macular Degeneration Trial. (a) True endpoint (change in visual acuity at 1 year) versus surrogate endpoint (change in visual acuity at 6 months) for all individual patients, raw data. (b) Treatment effects on the true endpoint versus treatment effects on the surrogate endpoint in all centers. The size of each point is proportional to the number of patients in the corresponding center. (c) True endpoint versus surrogate endpoint for all individual patients, after correction for treatment effect.*

24.3.6   Computational Issues

In this section, we investigate convergence properties of the random-effects approach as proposed in Section 24.3.4. The need for such an investigation arises from the observation that in many practical instances, convergence of the Newton-Raphson algorithm yielding (restricted) maximum likelihood solutions could hardly be achieved (see Section 24.3.5). Therefore, it is worth exploring what features of the problem at hand may be of influence in easing convergence of the algorithm, since this may be an additional factor to decide between a two-stage or a random-effects model.

We explored the following factors: number of trials, size of the between-trial variability (compared to residual variability), number of patients per trial, normality assumption, and strength of the correlation between random treatment effects. Since only the first two factors were found significantly to affect convergence of the algorithm, we do not report on the others in the remainder of this paragraph.

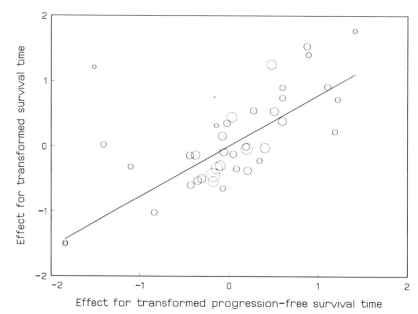

FIGURE 24.12. *Advanced Colorectal Cancer. Treatment effects on the true end-point (logarithm of survival time) versus treatment effects on the surrogate end-point (logarithm of time to progression) for all units of analysis. The size of each point is proportional to the number of patients in the corresponding unit.*

Table 24.7 shows the number of runs for which convergence could be achieved within 20 iterations. In each case, 500 runs were performed, assuming a model of the following form:

$$
\begin{aligned}
S_{ij} \mid Z_{ij} &= 45 + m_{S_i} + (3 + a_i)Z_{ij} + \varepsilon_{Sij}, \\
T_{ij} \mid Z_{ij} &= 50 + m_{Ti} + (5 + b_i)Z_{ij} + \varepsilon_{Tij},
\end{aligned}
$$

where $(m_{Si}, m_{Ti}, a_i, b_i) \sim N(0, D)$ with

$$
D = \delta^2 \begin{pmatrix} 1 & 0.8 & 0 & 0 \\ & 1 & 0 & 0 \\ & & 1 & 0.9 \\ & & & 1 \end{pmatrix}
$$

and $(\varepsilon_{Sij}, \varepsilon_{Tij}) \sim N(0, \Sigma)$ with

$$
\Sigma = 3 \begin{pmatrix} 1 & 0.8 \\ & 1 \end{pmatrix}.
$$

The number of trials was fixed to either 10, 20, or 50, each trial involving 10 subjects randomly assigned to treatment groups. The δ^2 parameter was set to 0.1 or 1.

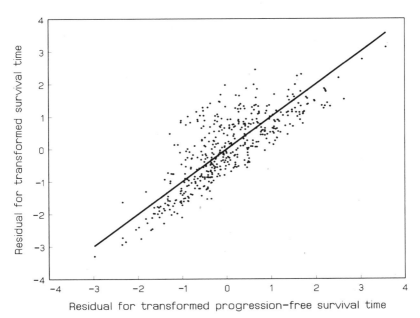

FIGURE 24.13. *Advanced Colorectal Cancer. Residuals of progression free survival versus survival, after correction for treatment and center effect.*

TABLE 24.7. *Simulation results. Number of runs for which convergence was achieved within 20 iterations. Total number of runs: 500; percentages are given in parentheses.*

	Number of trials		
δ^2	50	20	10
1	500 (100%)	498 (100%)	412 (82%)
0.1	491 (98%)	417 (83%)	218 (44%)

From Table 24.7, we see that when the between-trial variability is large ($\delta^2 = 1$), no convergence problems occur, except when the number of trials gets very small. When the between-trial variability gets smaller, convergence problems do arise and worsen as the number of trials decreases.

These simulation results indicate that there should be enough variability at the trial level, and a sufficient number of trials, to obtain convergence of the Newton-Raphson algorithm for fitting mixed-effects models. When these requirements are not fulfilled, one must rely on simpler fixed-effects models, or mixed-effects models with random treatment effects but no random intercepts. The investigator should reflect carefully on whether such simplifications are allowable. It may well be that a two-stage model is the more sensible choice.

24.3.7 Extensions

In Section 24.3.4, we focused on the methodologically appealing case of normally distributed endpoints. In practice, situations abound with binary and time-to-event endpoints, and in addition with surrogate and final endpoints of a different type (Molenberghs, Geys, and Buyse 1998). Whereas the linear mixed model provides a unified and flexible framework to analyze multivariate and/or repeated measurements that are normally distributed, similar tools for non-normal outcomes are unfortunately less well developed.

For binary outcomes, there are both marginal models such as generalized estimating equations (Liang and Zeger 1986) or full likelihood approaches (Fitzmaurice and Laird 1993, Molenberghs and Lesaffre 1994, Lang and Agresti 1994, Glonek and McCullagh 1995) and random-effects models (Stiratelli, Laird, and Ware 1984, Zeger, Liang, and Albert 1988, Breslow and Clayton 1993, Wolfinger and O'Connell, 1993; Lee and Nelder 1996). Reviews are given in Diggle, Liang, and Zeger (1994) and Fahrmeir and Tutz (1994). For additional references, see also Section 24.2.1 (p. 414).

Since our validation measures make use not only of main-effect parameters, such as treatment effects, but prominently of association (random-effects structure and residual covariance structure), standard generalized estimating equations are less suitable to extend ideas toward noncontinuous data. Possible approaches are second-order generalized estimating equations (Liang, Zeger, and Qaqish 1992, Molenberghs and Ritter 1996) and random-effects models. Since the latter are computationally involved, the likelihood-based approaches need to be supplemented with alternative methods of estimation such as quasi-likelihood. Also, pseudo-likelihood is a viable alternative.

Burzykowski *et al.* (1999) studied the situation where S_{ij} and T_{ij} are failure-time endpoints. In order to extend the approach used in the case of two normally distributed endpoints described in Section 24.3.4, they replaced model (24.18)–(24.19) by a model for two correlated failure-time random variables. This is based on a copula model (Shih and Louis 1995). More specifically, they assume that the joint survivor function of (S_{ij}, T_{ij}) can be written as

$$F(s, t) = P(S_{ij} \geq s, T_{ij} \geq t) = C_\delta\{F_{sij}(s), F_{Tij}(t)\} \qquad (24.33)$$

$(s, t \geq 0)$, where (F_{sij}, F_{Tij}) denotes marginal survivor functions and C_δ is a distribution function on $[0, 1]^2$ with $\delta \in \mathbb{R}^1$. C_δ is called a *copula function*. It describes the association between S_{ij} and T_{ij}. An attractive feature of model (24.33) is that the margins do not depend on the choice of the copula function. Specifically, the copulas of Clayton (1978) and Hougaard (1987) are studied in detail. Burzykowski *et al.* replace the R^2 at the individual level by Kendall's τ, while maintaining R^2 as a measure for trial-level surrogacy.

This area is currently in full development. For example, the important situation of a binary (or categorical) response, such as tumor response, as a surrogate for survival time, is being studied.

24.3.8   Reflections on Surrogacy

The validation of surrogate endpoints is a controversial issue. Difficulties have arisen on several fronts. First, some endpoints used as surrogates have been shown to provide wholly misleading predictions of the treatment effect upon the important clinical endpoints: The case of encainide and flecainide, two harmful drugs that were approved by the Food and Drug Administration based on their anti-arrhythmic effects, will remain a painful illustration of such an unfortunate circumstance (Fleming 1992). Second, some endpoints that have not been so catastrophically misleading have still failed to explain the totality of the treatment effect upon the final endpoints: The case of the CD4+ lymphocyte counts in patients with AIDS is an example (Lin, Fischl, and Schoenfeld 1993, DeGruttola *et al.* 1993, Choi *et al.* 1993, DeGruttola and Tu 1995). Many of these problems were mentioned in Prentice (1989). All these reasons have led some authors to express reservations about attempts to validate surrogate endpoints statistically (Fleming and DeMets 1996, DeGruttola *et al.* 1997). Their reservations rest to a large extent on biological considerations: A good surrogate must be shown to be causally linked to the true endpoint, and even so, it is implausible that the surrogate will ever capture the whole effect of treatment upon the true endpoint. These reservations are well taken, but biologically complex situations lend themselves to statistical evaluations that may shed light on the underlying mechanisms involved (Chuang-Stein and DeMasi 1998). The approach proposed here indirectly addresses these issues: A large individual-level coefficient of determination (R^2_{indiv} close to 1) indicates that the endpoints are likely to be causally linked to each other, whereas a large trial-level coefficient of determination ($R^2_{\text{trial(r)}}$ close to 1) indicates that a large proportion of the treatment effect is captured by the surrogate.

We obtain a quantitative assessment of the value of a surrogate, as well as predictions of the expected effect of treatment upon the true endpoint (Boissel *et al.* 1992, Chen *et al.* 1998). It evaluates the "validity" of a surrogate in terms of coefficients of determination, which are intuitively appealing quantities in the unit interval and enables the construction of prediction intervals for the effect of treatment on the true endpoint based on its action upon the surrogate endpoint. Such an approach is more informative than a mere dichotomization of surrogate endpoints as being "valid" or "invalid." Moreover, the validation procedure no longer requires statistical tests to be statistically significant; for instance, an endpoint with a low individual-level coefficient of determination $(R^2_{\text{indiv}} \ll 1)$ is unlikely to be a good surrogate (even if $R^2_{\text{trial(f)}} = 1$), a conclusion that may be reached with a limited number of observations.

The need for validated surrogate endpoints is as acute as ever, particularly in diseases where an accelerated approval process is deemed necessary (Cocchetto and Jones 1998, Weihrauch and Demol 1998). Some surrogate endpoints or combinations of endpoints, such as viral load measures combined with CD4+ lymphocyte counts, have, in fact, already replaced assessment of clinical outcomes in AIDS clinical trials (O'Brien *et al.* 1996, Mellors *et al.* 1997). The approach presented here may offer a better understanding of the worth of a surrogate endpoint, provided that large enough sets of data from multiple randomized experiments are available to estimate the required parameters (Daniels and Hughes 1997). Large numbers of observations are needed for the estimates to be sufficiently precise, and multiple studies are needed to distinguish individual-level from trial-level associations between the endpoints and effects of interest. However, it has to be emphasized that, even if the results of a surrogate evaluation seem encouraging based on several trials, applying these results to a new trial requires a certain amount of extrapolation that may or may not be deemed acceptable. In particular, when a new treatment is under investigation, is it reasonable to assume that the quantitative relationship between its effects on the surrogate and true endpoints will be the same as with other treatments ? The leap of faith involved in making that assumption rests primarily on biological considerations, although the type of statistical information presented above may provide essential supporting evidence.

24.3.9 Prediction Intervals

Denote $f = E(\beta + b_0 | m_{s0}, a_0) = \beta + D_1 D_2^{-1} D_3$, where D_1, D_2, and D_3 refer to the corresponding matrices in (24.26). Let f_d be the derivative of f w.r.t. the parameter vector

$$(\beta, \mu_s, \alpha, d_{sb}, d_{ab}, d_{ss}, d_{sa}, d_{aa}, \mu_{s0}, \alpha_0)'.$$

The components of f_d are

$$\frac{\partial f}{\partial \beta} = 1,$$

$$\frac{\partial f}{\partial \mu_{s0}} = -\frac{\partial f}{\partial \mu_s} = D_1 D_2^{-1} \begin{pmatrix} 1 \\ 0 \end{pmatrix},$$

$$\frac{\partial f}{\partial \alpha_0} = -\frac{\partial f}{\partial \alpha} = D_1 D_2^{-1} \begin{pmatrix} 0 \\ 1 \end{pmatrix},$$

$$\frac{\partial f}{\partial d_{sb}} = \begin{pmatrix} 1 \\ 0 \end{pmatrix}' D_2^{-1} D_3,$$

$$\frac{\partial f}{\partial d_{ab}} = \begin{pmatrix} 0 \\ 1 \end{pmatrix}' D_2^{-1} D_3,$$

$$\frac{\partial f}{\partial d_{ss}} = -D_1 D_2^{-1} \begin{pmatrix} 1 & 0 \\ 0 & 0 \end{pmatrix} D_2^{-1} D_3,$$

$$\frac{\partial f}{\partial d_{sa}} = -D_1 D_2^{-1} \begin{pmatrix} 0 & 1 \\ 1 & 0 \end{pmatrix} D_2^{-1} D_3,$$

$$\frac{\partial f}{\partial d_{aa}} = -D_1 D_2^{-1} \begin{pmatrix} 0 & 0 \\ 0 & 1 \end{pmatrix} D_2^{-1} D_3.$$

Denoting the asymptotic covariance matrix of the estimated parameter vector by V, the asymptotic variance of f is given by $f_d V f_d$, producing a confidence interval in the usual way. For a prediction interval, the variance to be used is $f_d V f_d + \text{var}(\beta + b_0 | m_{s0}, a_0)$.

24.3.10 SAS Code for Random-Effects Model

We describe how to use the SAS system to fit the random-effects model proposed in Section 24.3.4. Note that other packages such as MLwiN (see also Section A.2) are also particularly well suited for fitting this type of multivariate multilevel models and could therefore be utilized instead.

The SAS code to fit model (24.23)–(24.24) may be written as follows:

```
proc mixed data = data set covtest;
class endpoint subject trial treat;
model outcome = endpoint|treat / s noint;
random endpoint*treat / sub = trial type = un;
```

```
repeated endpoint / subject = subject(trial) type = un;
run;
```

The above syntax presumes that there are two records per subject in the input data set, one corresponding to the surrogate endpoint and the other to the true endpoint. The variable *endpoint* is an indicator for the kind of endpoint (coded 0 for the surrogate and 1 for the true endpoint) and the variable *outcome* contains measurements obtained from each endpoint.

The RANDOM statement defines the covariance matrix D in (24.22) of random effects at the trial level, and the REPEATED statement builds up the residual covariance matrix Σ in (24.20). Note that the nesting notation in the 'subject=' option enables SAS to recognize the nested structure of the data (subjects are clustered within trials). Acknowledgment of the hierarchical nature of the data prompts SAS to build a block-diagonal covariance matrix, with diagonal blocks corresponding to the different trials, which speeds up computations considerably.

24.4   The Milk Protein Content Trial

24.4.1   Introduction

Diggle (1990) and Diggle and Kenward (1994) analyzed data taken from Verbyla and Cullis (1990), who, in turn, had discovered the data at a workshop at Adelaide University in 1989. The data consist of assayed protein content of milk samples taken weekly for 19 weeks from 79 Australian cows. The cows entered the experiment after calving and were randomly allocated to 1 of 3 diets: barley, mixed barley-lupins, and lupins alone, with 25, 27, and 27 animals in the 3 groups, respectively. The time profiles for all 79 cows are plotted in Figure 24.19. All cows remained on study during the first 14 weeks, whereafter the sample reduced to 59, 50, 46, 46, and 41, respectively, due to dropout. This means that dropout is as high as 48% by the end of the study. Table 24.8 shows the number of cows per arm and per dropout pattern.

The primary objective of the milk protein experiment was to describe the effects of diet on the mean response profile of milk protein content over time. Previous analyses of the same data are reported by Diggle (1990, Chapter 5), Verbyla and Cullis (1990), Diggle, Liang, and Zeger (1994), and Diggle and Kenward (1994), under different assumptions and with different modeling of the dropout process. Diggle (1990) assumed random dropout, whereas Diggle and Kenward (1994) concluded that dropout was

TABLE 24.8. *Milk Protein Content Trial. Number of cows per arm and per dropout pattern.*

	Diet		
Dropout week	Barley	Mixed	Lupins
Week 15	6	7	7
Week 16	2	3	4
Week 17	2	1	1
Week 18			
Week 19	2	2	1
Completers	13	14	14
Total	25	27	27

nonrandom, based on their selection model. Of course, it has been noted in Chapters 17 and 19 that appropriate care should be taken with nonrandom selection models for their reliance on unverifiable assumptions.

In addition to the usual problems with this type of models, serious doubts have been raised about even the appropriateness of the "dropout" concept in this study. Cullis (1994) warned that the conclusions inferred from the statistical model are very unlikely since usually there is no relationship between dropout and a relatively low level of milk protein content. In the discussion of the Diggle and Kenward (1994) paper, one is informed by Cullis that Valentine, who originally conducted the experiment, had previously revealed the real reasons for dropout. The explanation elucidates that the experiment terminated when feed availability declined in the paddock in which animals were grazing. Thus, this would imply that a nonrandom dropout mechanism is very implausible. A nonrandom dropout mechanism would wrongly relate dropout to response, whereas, to the contrary, dropout depends on food availability only. Thus, there are actually no dropouts but rather five cohorts representing the different starting times. Together with Cullis (1994), and in agreement with our pleas for sensitivity analysis (Chapters 19 and 20), we conclude that especially with incomplete data, a statistical analysis should not proceed without a thorough discussion with the experimenters.

The complex and somewhat vague history of the data set probably is the main cause for so many conflicting issues related to the analysis of the milk data. At the same time, it becomes a perfect candidate for sensitivity analysis. Modeling will be based upon the linear mixed-effects model with serial correlation (3.11), introduced in Section 3.3.4. In Section 24.4.2, we examine the validity of the conclusions made in Diggle and Kenward (1994) by incorporating subject matter information into the method of analysis. As dropout was due to design, the method of analysis should reflect this. We will investigate two approaches. The first approach involves restructuring the data set and then analyzing the resulting data set using a selection modeling framework, whereas the second method involves fitting pattern-mixture models taking the missingness pattern into account. Both analyses consider the sequences as *unbalanced in length* rather than a formal instance of dropout. Local influence diagnostics, as introduced in Section 19.3 and supplemented with global influence measures, are presented in Section 24.4.3.

24.4.2 Informal Sensitivity Analysis

Since there has been some confusion about the actual design employed, we cannot avoid making subjective assumptions such as the following: Several matched paddocks are randomly assigned to either of three diets: barley, lupins, or a mixture of the two. The experiment starts as the first cow experiences calving. As the first 5 weeks have passed, all 79 cows have entered their randomly assigned, randomly cultivated paddock. By week 19, all paddocks appear to approach the point of exhausting their food availability (in a synchronous fashion) and the experiment is terminated for all animals simultaneously.

All previous analyses assumed a fixed date for entry into the trial and the crucial issue then becomes how the dropout process should be handled and analyzed. However, it seems intuitive that since entry into the study was at random time points (i.e., after calving) and since the experiment was terminated at a fixed time point, that this time point should be the reference for all other time points. It is therefore also appealing to reverse the time axis and to analyze the data in reverse, starting from time of dropout. Under the aforementioned assumptions, we have found a partial solution to the problem of potentially nonrandom dropout since dropout has been replaced by ragged entry. Note, however, that a crucial simplification arises: Since entry into the trial depends solely on calving and gestation, it can be thought of as totally independent from the unobserved responses.

A problem with the alignment lies in the fact that virtually all cows showed a very steep decrease in milk protein content immediately after calving, lasting until the third week into the experiment. This behavior could be due to a special hormone regulation of milk composition following calving, which lasts only for a few weeks. Such a process is most likely totally independent of diet and, probably, can also be observed in the absence of food, to the expense of the animal's natural reserves. Since entry is now

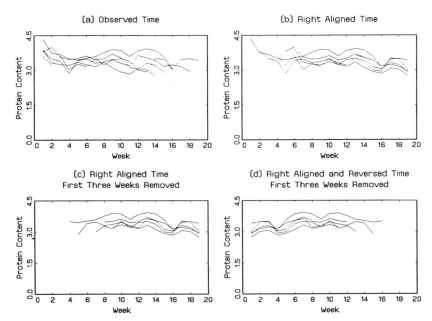

FIGURE 24.14. *Milk Protein Content Trial. Data manipulations on five selected cows: (a) raw profiles; (b) right aligned profiles; (c) deletion of the first three observations; (d) profiles with in addition time reversal.*

ragged, the process is spread and influences the mean response level during the first 8 weeks. Of course, one might construct an appropriate model for the first 3 weeks with a separate model specification, in analogy to the one used in Diggle and Kenward (1994). Instead, we prefer to ignore the first 3 weeks, analogous in spirit to the approach taken in Verbyla and Cullis (1990). Hence, we have time series of length 16, with some observations missing at the beginning. Figure 24.14 displays the data manipulations for five selected cows. In Figure 24.14(a), the raw profiles are shown. In Figure 24.14(b), the plots are right aligned. Figure 24.14(c) illustrates the protein content levels for the five cows with the first three observations deleted and Figure 24.14(d) presents these profiles when time is reversed.

In order to explore the patterns after transformation, we plotted the newly obtained mean profiles. Figures 24.15(a) and 24.15(b) display the mean profiles before and after the transformation, respectively. Notice that the mean profiles have become parallel in Figure 24.15(b). To address the issue of correlation, we shall compare the two variograms (see also Section 10.4). The two graphs shown in Figure 24.16 are very similar although slight differences can be noted in the estimated process variance, which is slightly lower after transformation. Complete decay of serial correlation appears to happen between time lags 9 and 10 in both variograms. There is virtually

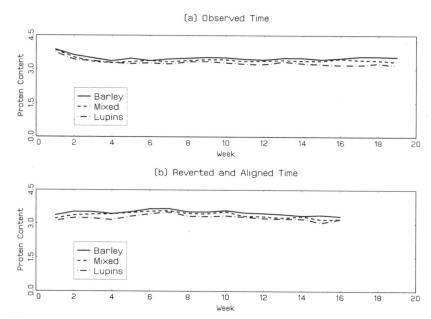

FIGURE 24.15. *Milk Protein Content Trial. Mean response profiles on the original data and after aligning and reverting.*

no evidence for random effects as the serial correlation levels of toward the process variance.

Table 24.10 presents maximum likelihood estimates for the original data, similar to the analysis by Diggle and Kenward (1994). The corresponding parameters after aligning and reverting are 3.45 (s.e. 0.06) for the barley group, with differences for the lupins and mixed groups estimated to be -0.21 (s.e. 0.08) and -0.12 (s.e. 0.08), respectively. The variance parameters roughly retain their relative magnitude, although even more weight is given to the serial process (90% of the total variance).

The analysis using aligned and reverted data shows little difference if compared to the original analysis by Diggle and Kenward (1994). It would be interesting to acquire knowledge about what mechanisms determined the systematic increase and decrease observed for the three parallel profiles illustrated in Figure 24.15(b). It is difficult to envisage that the parallelism of the profiles and their systematic peaks and troughs shown in Figure 24.15(b) is due entirely to chance. Indeed, many of the previous analyses had debated the influence on variability for those factors, common to the paddocks cultivated with the three different diets, such as meteorological factors, that had not been reported by the experimenter. These factors may account for a large amount of variability in the data. Hence, the data exploration performed in this analysis may be shown to be a useful tool

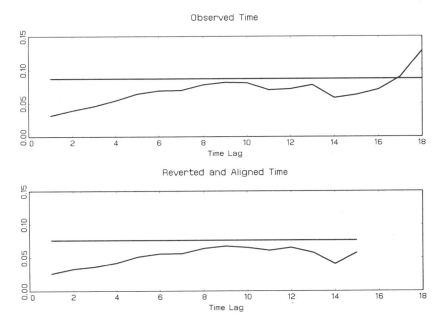

FIGURE 24.16. *Milk Protein Content Trial. Variogram for the original data and after aligning and reverting.*

in gaining insight about the response process. For example, we note that after transformation, the inexplicable trend toward an increase in milk protein content, as the paddocks approach exhaustion has, in fact, vanished or even reverted to a possible decrease. This was also confirmed in the stratified analysis where the protein level content tended to decrease prior to termination of the experiment (see Figure 24.17).

An alternative method of analysis is based on the premise that the protein content levels form distinct homogeneous subgroups for cows based on their dropout pattern. This leads very naturally to pattern-mixture models (Chapters 18 and 20). Parameters in (3.11) are now made to depend on pattern. In its general form, the fixed effects as well as the covariance parameters are allowed to vary unconstrained according to the dropout pattern. Alternatively, simplifications can be sought. For example, diet effect can vary linearly with pattern or can be pattern independent. In the latter case, this effect becomes marginal. When the diet effect is pattern dependent, an extra calculation is necessary to obtain the marginal diet effect, as was illustrated on the toenail data in Section 18.3 and formalized in Section 20.6.2. Precisely, the marginal effect can be computed as in (20.45), whereas the delta method variance expression is given by (20.46).

Denoting the parameter for diet effect $\ell = 1, 2$ (difference with the barley group) in pattern $t = 1, 2, 3$ by $\beta_{\ell t}$ and letting π_t be the proportion of cows in pattern t, then the matrix A in (20.48) assumes the form

$$A = \frac{\partial(\beta_1, \beta_2)}{\partial(\beta_{11}, \beta_{12}, \beta_{13}, \beta_{21}, \beta_{22}, \beta_{23}, \pi_1, \pi_2, \pi_3)}$$

$$= \begin{pmatrix} \pi_1 & \pi_2 & \pi_3 & 0 & 0 & 0 & \beta_{11} & \beta_{12} & \beta_{13} \\ 0 & 0 & 0 & \pi_1 & \pi_2 & \pi_3 & \beta_{21} & \beta_{22} & \beta_{23} \end{pmatrix}.$$

Note that the simple multinomial model for the dropout probabilities could be extended when additional information concerning the dropout mechanism. For example, if covariates are known or believed to influence dropout, the simple multinomial model can be replaced by logistic regression or time-to-event methods (Hogan and Laird 1997).

Recall that Table 24.8 presents the dropout pattern by time in each of the three diet groups. As few dropouts occurred in weeks 16, 17, and 19, these three dropout patterns were collapsed into a single pattern. Thus, three patterns remain with 20, 18, and 41 cows, respectively. The model fitting results are presented in Table 24.9. The most complex model for the mean structure assumes a separate mean for each diet by time by dropout pattern combination. As the variogram indicated no random effects, the covariance matrix was taken as first-order autoregressive with a residual variance term $\sigma_{jk} = \sigma^2 \rho^{|j-k|}$. Also the variance-covariance parameters are allowed to vary according to the dropout pattern. This model is equivalent to including time and diet as covariates in the model and stratifying for dropout pattern and it provides a starting point for model simplification through backward selection. We refer to the description of Strategy 3 in Section 20.5.1. The protein content levels over time are presented by pattern and diet in Figure 24.17. Note that the protein content profiles appear to vary considerably according to missingness pattern and time. Additionally, Diggle and Kenward (1994) suggested an increase in protein content level toward the end of the experiment. This observation is not consistent for the three plots in Figure 24.17. In fact, there is a tendency for a decrease in all diet by pattern subgroups prior to dropout.

To simplify the covariance structure presented in Model 1, Model 2 assumes the residual covariance parameter is equal in the three patterns. The likelihood ratio test indicates that Model 2 compares favorably with Model 1, suggesting a common residual variance (measurement error component) parameter (2 for the three groups; see Table 24.9 for details). However, comparing Model 3 with Model 2, we reject a common variance-covariance structure in the three groups.

Next, we investigate the mean structure. In Model 4, the three-way interaction among pattern, time, and diet is removed. This simplified model is acceptable when contrasted to Model 2, based on $p = 0.987$. Models 5, 6,

TABLE 24.9. *Milk Protein Content Trial. Model fit summary for pattern-mixture models.*

	Mean	Covar
1	Full interaction	AR1(t), meas(t)
2	Full interaction	AR1(t), meas
3	Full interaction	AR(1), meas
4	Two-way interactions	AR1(t), meas
5	Diet, time, pattern, diet∗time, diet∗pattern	AR1(t), meas
6	Diet, time, pattern, diet∗time, time∗pattern	AR1(t), meas
7	Diet, time, pattern, diet∗pattern, time∗pattern	AR1(t), meas
8	Time, pattern, time∗pattern	AR1(t), meas
9	Time, diet(time)	AR(1), meas
10	Time, diet	AR(1), meas

	# par	-2ℓ	Ref	G^2	df	p
1	162	−474.93				
2	160	−470.79	1	4.44	2	0.111
3	156	−428.26	2	42.23	4	<0.001
4	100	−439.96	2	30.53	50	0.987
5	70	−202.40	4	237.56	30	<0.001
6	96	−430.55	4	9.41	4	0.052
7	64	−404.04	4	35.92	36	0.472
8	58	−378.22	7	25.82	6	<0.001
			6	52.33	38	0.061
9	60					
10	24					

and 7 are fitted to investigate the pairwise interaction terms. Comparing Models 5 and 4 suggests a strong interaction between dropout pattern and time. Model 6 results in a borderline decrease in goodness-of-fit ($p = 0.052$). From Table 24.9, we observe that Model 7 is a plausible simplification of Model 4. Moreover, there is an apparent lack of fit for Model 8, which includes only one interaction term, time by pattern, when compared to Model 7. In conclusion, among the models presented, Model 7 is the preferred one to summarize the data, as it is the simplest model consistent with the data. However, Model 6 should be given some attention as well. In analogy to Diggle and Kenward (1994), we attempted to include time as a separate linear factor for the first 3 weeks and the subsequent 16 weeks. These models did not improve the fit (results not shown).

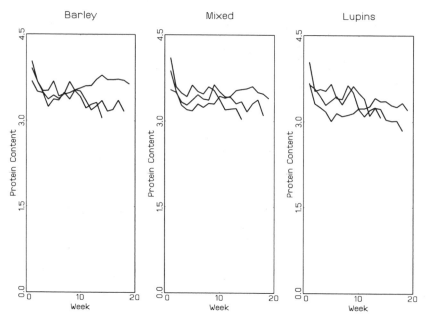

FIGURE 24.17. *Milk Protein Content Trial. Mean response level per diet and per dropout pattern.*

Recall that the objective of the experiment was to assess the influence of diet on protein content level. With selection models, the corresponding null hypothesis of no effect can be tested using, for example, the standard F-tests on 2 numerator degrees of freedom as provided by the SAS procedure MIXED or similar software. In the pattern-mixture framework, such a standard test can be used only if the treatment effects do not interact with pattern. Otherwise, the marginal treatment (diet) effect has to be determined as in (20.45) and the delta method can be used to test the hypothesis of no effect. In Model 6, the diet effect is independent of pattern and the reverse holds for Model 7. Reparameterizing Model 6 by including the diet effect and diet by time interaction as one effect in the model provides us with an appropriate F-test for the three diet profiles. The F-test rejects the null hypothesis of no diet effect ($F = 1.57$ on 38 degrees of freedom, $p = 0.015$). In the corresponding selection model, Model 9, we remove all the terms from Model 6 which include pattern. In that case, the F-test is *not* significant ($F = 1.26$ on 38 degrees of freedom, $p = 0.133$). The difference in the tests may be explained by the variance parameters which were larger in the selection model in the absence of stratification for pattern, thereby effectively diluting the strength of the difference. Additionally, the standard errors for the estimates of the fixed effects were slightly smaller in the pattern-mixture model. This is not surprising, as in the model fitting we found that the means and variance parameters were dependent on

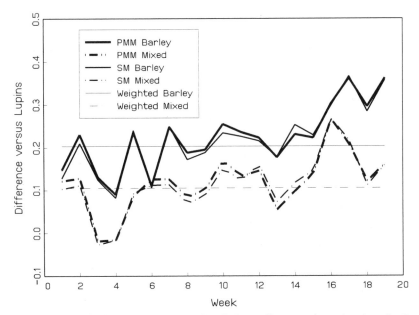

FIGURE 24.18. *Milk Protein Content Trial. Diet effect over time for the selection model (SM), the corresponding pattern-mixture model (PMM), and the estimate obtained after weighting the PMM contributions (weighted).*

pattern. Thus, stratifying for pattern results in more homogeneous subgroups of cows, reducing the variance within each group and subsequently providing more precise estimates for the diet effect.

Using Model 7, we test the global null hypothesis of no diet effect in any of the patterns. This analysis can be seen as a stratified analysis where a diet effect is estimated separately within each pattern. This model results in a significant F-test for the diet effect ($F = 6.05$, on 6 degrees of freedom, $p < 0.001$). Alternatively, we can consider the pooled estimate for the diet effect, provided by equation (20.45), and calculate the test statistic using the delta method. This test also indicates a significant diet effect ($F = 17.82$ on 2 degrees of freedom, $p < 0.001$), as does the corresponding selection model, Model 10 ($F = 8.51$ on 2 degrees of freedom, $p < 0.001$).

Figure 24.18 presents the diet by time parameter estimates for selection Model 10, for the corresponding pattern-mixture Model 7 and the weighted average estimates used in the delta method. The estimates for the selection model and the pattern-mixture model appear to differ only slightly. Since the model building within both families is done separately, this is a very reassuring sensitivity analysis outcome.

In conclusion, including pattern in the model improves the model fit significantly. In particular, the time by pattern and diet by pattern interactions are maintained in Model 7, which is considered to be the most parsimonious model consistent with the data (Figure 24.17). In addition, the covariance parameters are also dependent on the missingness pattern. Dividing cows into more homogeneous groups based on their missingness patterns reduces the unexplained variation in the data and subsequently provides more precise parameter estimates.

This example and, in particular, the absence of genuine dropout illustrate once more that care has to be taken when analyzing longitudinal outcomes with a nonrectangular structure.

The analyses discussed here provide an alternative to those obtained by Diggle and Kenward (1994) but generally do not contradict them. Rather, they convey the message that the use of sensitivity analysis should become standard practice when dropout occurs. We strongly stress the importance of careful data verification to be undertaken prior to any statistical analysis. To this end, we might add that the effect of an erroneous initial description of a data set should not be underestimated, as it can lead to subsequent mismodeling of the data, thus adding confusion to an already complex undertaking of analyzing longitudinally measured observations.

Our analysis of the correlation structure appears to agree with the general conclusions retained in the Diggle and Kenward (1994) analysis. Particularly, it is interesting to notice the absence of random effects. We do not completely share the surprise expressed by Diggle and Kenward (1994) since it should be noted that the study animals are highly selected through centuries of cow eugenics and race selection. Had we dealt with wild animals, the role played by random effects would most likely have been much more substantial. To explain the absence of random effects, we may assume that there were additional eligibility criteria for the trial (e.g., a specific breed of cow), which made random effects even more unlikely.

Analyzing a data set using various approaches to answer a particular question is seen as a simple and informal way of sensitivity analysis, as is supplementing the main analysis with additional ones to gain extra insight. Each method used requires certain assumptions about the measurement process and the dropout process. In particular, pattern-mixture models and selection models approach the issue of dropout in different ways. It may also be useful to investigate the fundamental assumptions concerning the design of the experiment since dropout may be design driven.

24.4.3 Formal Sensitivity Analysis

Supplementing the results of Section 24.4.2 with a more formal look at sensitivity, in the selection-model spirit of Chapter 19, can be done using local influence methods, described in Section 19.3 and applied in Sections 19.4 and 19.5 to the rats and mastitis data sets, respectively. In addition, we will describe and apply global influence techniques as well.

Cook (1986) suggests that more confidence can be put in a model which is relatively stable under small modifications. The best known perturbation schemes are based on case deletion (Cook and Weisberg 1982, Chatterjee and Hadi 1988), in which the effect of completely removing cases from the analysis is studied. They were introduced by Cook (1977a, 1979) for the linear regression context. Denote the likelihood function, corresponding to measurement model (3.8) and dropout model (17.17) and (17.4), by

$$\ell(\gamma) = \sum_{i=1}^{N} \ell_i(\gamma), \tag{24.34}$$

in which $\ell_i(\gamma)$ is the contribution of the ith individual to the log-likelihood and where $\gamma = (\theta, \psi, \omega)$ is the s-dimensional vector, grouping the parameters of the measurement model and the dropout model. Further, we denote by

$$\ell_{(-i)}(\gamma) \tag{24.35}$$

the log-likelihood function, where the contribution of the ith subject has been removed. Cook's distances (CD) are based on measuring the discrepancy between either the maximized likelihoods (24.34) and (24.35) or (subsets of) the estimated parameter vectors $\widehat{\gamma}$ and $\widehat{\gamma}_{(-i)}$, with obvious notation. Precisely, we will consider both

$$\mathrm{CD}_{1i} = 2(\widehat{\ell} - \widehat{\ell}_{(-i)}) \tag{24.36}$$

as well as

$$\mathrm{CD}_{2i}(\gamma) = 2\,(\widehat{\gamma} - \widehat{\gamma}_{(-i)})' \ddot{L}^{-1}\,(\widehat{\gamma} - \widehat{\gamma}_{(-i)}). \tag{24.37}$$

Formulation (24.37) easily allows to consider the global influence in a subvector of γ, such as the dropout parameters ψ, or the nonrandom parameter ω. This will be indicated using notation of the form $\mathrm{CD}_{2i}(\psi)$, $\mathrm{CD}_{2i}(\omega)$, and so forth.

Recall that Diggle and Kenward (1994) considered model (3.11), where the mean model includes separate intercepts for the barley, mixed, and lupins groups, and a common time effect which is linear during the first 3 weeks and constant thereafter. The covariance structure is described by a random intercept, an exponential serial process, and measurement error.

TABLE 24.10. *Milk Protein Content Trial. Maximum likelihood estimates (standard errors) of random and nonrandom dropout models. Dropout starts from week 15 onward.*

Effect	Par.	MAR	MNAR
Measurement model:			
Barley	μ_1	4.147 (0.053)	4.152 (0.053)
Mixed	μ_2	4.046 (0.052)	4.050 (0.052)
Lupins	μ_3	3.935 (0.052)	3.941 (0.052)
Time effect	β	−0.226 (0.015)	−0.224 (0.015)
Random intercept variance	d	−0.001 (0.010)	0.002 (0.009)
Measurement error variance	σ^2	0.024 (0.002)	0.025 (0.002)
Serial variance	τ^2	0.073 (0.012)	0.067 (0.011)
Serial correlation	ρ	0.152 (0.037)	0.163 (0.039)
Dropout model:			
Intercept	ψ_0	17.870 (3.147)	15.642 (3.535)
Previous measurement	ψ_1	−6.024 (0.998)	−10.722 (2.015)
Current measurement	ψ_2		5.176 (1.487)
-2 log-likelihood		51.844	37.257

The dropout model includes dependence on the previous and current, possibly unobserved, measurements. Since dropout only happens from week 15 onward, Diggle and Kenward (1994) chose to set the dropout probability for earlier occasions equal to zero. Thereafter, they allowed separate intercepts per time point, but common dependencies on previous and current measurements. We will now introduce two models which use the same measurement model as Diggle and Kenward (1994) but different dropout models.

A first dropout model is closely related to the one of Diggle and Kenward (1994), who defined occasion-specific intercepts ψ_{0k} ($k = 15, 16, 17, 19$), assumed slopes common, and set the dropout probability equal to zero at other occasions. We also model dropout from week 15 onward, but we will keep the intercepts constant for occasions 15 to 19. Precisely, our first model contains three parameters (intercept ψ_0, dependence on the previous measurement ψ_1, and dependence on the current measurement ψ_2).

Parameter estimates for this model under both MAR and MNAR are listed in Table 24.10. The fitted model is qualitatively equivalent to the model used by Diggle and Kenward (1994), who concluded overwhelming evidence

TABLE 24.11. *Milk Protein Content Trial. Maximum likelihood estimates (standard errors) of random and nonrandom dropout models. Dropout starts from week 1 onwards.*

Effect	Par.	MAR	MNAR
Measurement model:			
Barley	μ_1	4.147 (0.053)	4.152 (0.053)
Mixed	μ_2	4.046 (0.052)	4.050 (0.052)
Lupins	μ_3	3.935 (0.052)	3.941 (0.052)
Time effect	β	−0.226 (0.015)	−0.224 (0.015)
Random intercept variance	d	−0.001 (0.010)	0.002 (0.009)
Measurement error variance	σ^2	0.024 (0.002)	0.025 (0.002)
Serial variance	τ^2	0.073 (0.012)	0.067 (0.011)
Serial correlation	ρ	0.152 (0.037)	0.163 (0.040)
Dropout model:			
Intercept	ψ_0	10.483 (2.010)	6.477 (2.867)
Previous measurement	ψ_1	−4.326 (0.651)	−5.917 (1.069)
Current measurement	ψ_2		2.732 (1.396)
-2 log-likelihood		194.316	190.691

for nonrandom dropout (likelihood ratio statistic 13.9). In line with these results, we also could decide in favor of a nonrandom process (likelihood ratio statistic 14.59).

In our second dropout model, we allow dropout starting from the second week. Specifically, this model contains three parameters (intercept ψ_0, dependence on the previous measurement ψ_1, and dependence on the current measurement ψ_2) which are assumed constant throughout the whole 19-week period. The fit of this model is listed in Table 24.11. A striking difference with the previous analysis is that the MAR assumption is borderline not rejected (likelihood ratio statistic 3.63). Apparently, this is a major source of sensitivity, to be explored further. As results from theory, the measurement model parameters do not change under the MAR model, compared to those displayed in Table 24.10. The measurement model obtained under MNAR has changed only slightly.

Which of the two analyses is to be preferred is debatable and depends on substantive rather than statistical considerations. The first analysis accounts for the *post hoc observation* that no dropout occurred prior to week 15. However, there is a, perhaps small, chance for the experiment to terminate in a particular field with less than 15 weeks of measurements, and our second model acknowledges this possibility. It is clear that there is an enormous sensitivity of the results due to this model choice and, therefore, an assessment of influence seems appropriate. In general though, it may be questionable that the dropout model parameters remain constant over an extended period of time. Not only can the rate chance over time, but also the dominant causes and the magnitude of their effect can change.

GLOBAL INFLUENCE

Global influence results are shown in Figures 24.19–24.21. They are based on fitting a MNAR model with each of the cows deleted in turn. The Cook's distances for the first and the second model are shown in Figure 24.20 and 24.21, respectively. The individual curves with influential subjects highlighted are plotted in Figure 24.19 where subject #38 should not be highlighted for the second model.

There is very little difference in some of the Cook's distance plots, when Figures 24.20 and 24.21 are compared. Precisely, CD_{1i}, $CD_{2i}(\gamma)$, and $CD_{2i}(\theta)$ are virtually identical. The three others are similar in the sense that there is some overlap in the subjects indicated as peaks, but with varying magnitudes. Subject #38 is influential on the dropout measures $CD_{2,38}(\psi,\omega)$, $CD_{2,38}(\psi)$, and $CD_{2,38}(\omega)$. This is not surprising since #38 is rather low in the middle portion of the measurement sequence, whereas it is very high from week 15 onward. Therefore, this sequence is picked up in the second analysis only. By looking at a plot with the evolution of the parameters separately during the deletion process (not shown here), we can conclude that subject #38 has some impact on the serial correlation parameter while #65 is rather influential for the measurement error. In view of the fairly smooth deviation from a straight line of the former and the abrupt peaks in the latter, this is not a surprise.

Based on our second model, all forms of $CD_{2i}(\cdot)$, whether based on the entire parameter vector γ, the dropout parameters (ψ_0, ψ_1, ω), or subsets of the latter, indicate that subjects #51, #59, and #68 are influential. In contrast, CD_{1i}, which is based directly on the likelihood, does not reveal these subjects, but rather subject #65 jumps out. Thus, although the former three subjects have a substantial impact on the parameter estimates, they do not change the likelihood in a noticeable fashion. From a plot of the dropout parameter estimates for each deleted case (not shown here), it is very clear that upward peaks in $\widehat{\psi}_{0(-i)}$ for subjects #51 and #59 are compensated with downward peaks in $\widehat{\omega}_{(-i)}$. An explanation for this phenomenon can be found in the variance-covariance matrix of the dropout

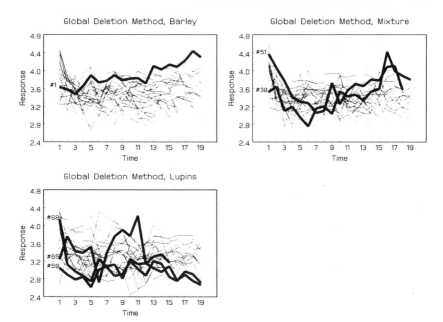

FIGURE 24.19. *Milk Protein Content Trial. Individual profiles, with globally influential subjects highlighted. Dropout modeled from week 15.*

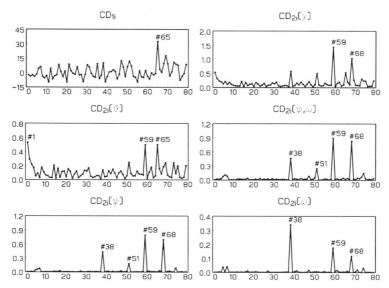

FIGURE 24.20. *Milk Protein Content Trial. Index plots of* CD_{1i}, $CD_{2i}(\gamma)$, $CD_{2i}(\theta)$, $CD_{2i}(\psi,\omega)$, $CD_{2i}(\psi)$, *and* $CD_{2i}(\omega)$. *Dropout modeled from week 15.*

parameters (correlations shown in the lower triangle):

$$\begin{pmatrix} 8.22 & 0.43 & -2.85 \\ (0.14) & 1.14 & -1.18 \\ (-0.71) & (-0.79) & 1.94 \end{pmatrix}.$$

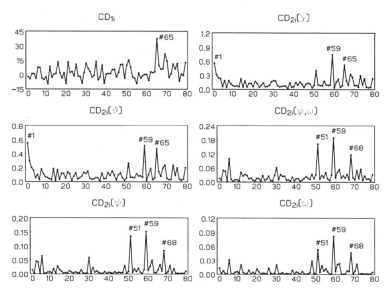

FIGURE 24.21. *Milk Protein Content Trial. Index plots of* CD_{1i}, $CD_{2i}(\gamma)$, $CD_{2i}(\theta)$, $CD_{2i}(\psi,\omega)$, $CD_{2i}(\psi)$, *and* $CD_{2i}(\omega)$. *Dropout modeled from week 1.*

From a principal components analysis, it follows that more than 90% of the variation is captured in the linear combination $0.93\psi_0 - 0.37\omega$. Hence, there is mass transfer between these two parameters, of course with sign reversal, little impact on the likelihood value, and little effect on the MAR parameter ψ_1. Note that a similar plot for the measurement model parameters can be constructed (not shown).

Let us now turn to the subjects which are globally influential. A first and common reason for those subjects to show up is the fact that they all have a rather strange profile. Remember the overall trend to be sloping downward during the first 3 weeks and constant thereafter. Subject #65 appears with large $CD_{65,1}$ and large $CD_2(\theta)$. The reason for this can be found in the fact that its profile shows extremely low and high peaks. Subjects #51, #59, and #68, on the other hand, only show large values for $CD_2(\psi,\omega)$, $CD_2(\psi)$, $CD_2(\omega)$. This means that these subjects are influential for the dropout parameters. For subject #51, this can be explained by the fact that it drops out in spite of the rather large profile. Subjects #59 and #68, on the contrary, stay in the experiment even though they both have rather low profiles.

LOCAL INFLUENCE

Local influence plots and individual profiles, with the influential subjects highlighted (bold type), for the first model for raw and incremental data,

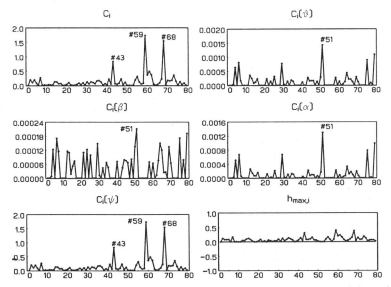

FIGURE 24.22. *Milk Protein Content Trial. Index plots of C_i, $C_i(\theta)$, $C_i(\beta)$, $C_i(\alpha)$, and $C_i(\psi)$, and of the components of the direction $\boldsymbol{h}_{\max}$ of maximal curvature. Dropout modeled from week 15.*

respectively, are depicted in Figures 24.22–24.25. Corresponding graphs for our second model are shown in Figures 24.26–24.29. It is more convenient to discuss results of the second model up front and then compare them to the first model.

Observe that the plots for C_i and $C_i(\psi)$ are virtually identical. This is due to the relative magnitudes of the ψ and θ components. Profiles #51, #59, and #66–#68 are highlighted in Figure 24.27. An explanation for the influence in ψ is found by studying (19.12). Indeed, for ψ_0 and ψ_1 as in Table 24.11, the maximum is obtained for $y = 2.51$, exactly as seen in the influential profiles, which are all in the lupins group (Figure 24.27). Further, note that there is some agreement between the locally and globally influential subjects, although there is no compelling need for the two approaches to be identical (#51 appears in different influential components in the two approaches). Indeed, although global influence lumps together all sources of influence, our local influence approach is designed to detect subjects which, due to several causes, tend to have a strong impact on ω and therefore on the conclusion about the nature of the dropout mechanism.

Observe that one factor in (19.12) is the square of the response. This is a direct consequence of our parameterization of the dropout process, the logit of which is in terms of the previous and current outcomes, to which no transformation is applied. As was argued in the mastitis case (p. 321), since two subsequent measurements are usually positively correlated, it is

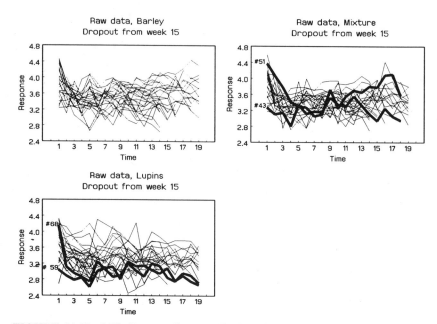

FIGURE 24.23. *Milk Protein Content Trial. Individual profiles, with locally influential subjects highlighted. Dropout modeled from week 15.*

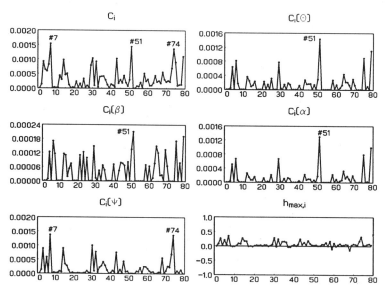

FIGURE 24.24. *Milk Protein Content Trial. Index plots of C_i, $C_i(\theta)$, $C_i(\beta)$, $C_i(\alpha)$, and $C_i(\psi)$, and of the components of the direction $\mathbf{h}_{\max}$ of maximal curvature. Dropout modeled from week 15. Incremental analysis.*

not unusual for both of them to be high. It is therefore wise to repara-

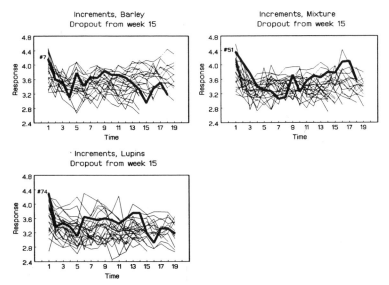

FIGURE 24.25. *Milk Protein Content Trial. Individual profiles, with locally influential subjects highlighted. Dropout modeled from week 15. Incremental analysis.*

meterize the dropout model (19.1) in terms of the *increment*; that is, y_{ij} is replaced by $y_{ij} - y_{i,j-1}$. This is related to the approach of Diggle and Kenward (1994), who reparameterized their dropout model in terms of the increment just introduced and the size (the average of both measurements). Even though a dropout model in the outcomes themselves, termed direct variables model, is equivalent to a model in the first variable Y_{i1} and the increment $Y_{i2} - Y_{i1}$, termed incremental variable representation, it was shown that they lead to different perturbation schemes of the form (19.1). Indeed, from equality (19.18) or, more generally, from

$$\psi_0 + \psi_1 y_{i,j-1} + \psi_2 y_{ij} = \lambda_0 + \lambda_1 y_{i,j-1} + \lambda_2 (y_{ij} - y_{i,j-1}), \quad (24.38)$$

it follows that $\psi_0 = \lambda_0$ and $\psi_1 = \lambda_1 - \lambda_2$. Thus, local influence is now focusing on a different set of parameters and one should not expect it to give the same answer. Therefore, it is crucial to guide the parameterization by careful substantive knowledge. In a sense, dependence on the increment is most dramatic since, at the time of dropout, there is no information about the increment, whereas size can be assessed reasonably well from $Y_{i,j-1}$, especially if the correlation is sufficiently high. The results of this analysis are presented in Figure 24.28 and the most influential profiles are highlighted in Figure 24.29. A slightly different but overlapping set of profiles is responsible for the influence now. The most important feature is that the influence is very minor. The components of the direction of maximal curvature $\boldsymbol{h}_{\max}$ shows virtually no peaks.

Finally, we will compare both models. The direct-variable results found in Figure 24.22 agree fairly well with those in Figure 24.26, the differences being the absence of #66 and #67 and the appearance of #43. The latter profile is extremely low at the end of the period, where dropout is modeled, and therefore yields a large value for (19.12). For #66 and #67, there is a logical explanation for their disappearance. Indeed, these profiles are very low during the first part of the experimental period, in spite of which they do not drop out. However, during the latter part, their profile is still low *and* they drop out, which is totally plausible behavior and, hence, their influence is marked in the second but not in this analysis.

For the incremental analysis, there is a larger discrepancy between both models as one can observe by comparing Figure 24.24 to Figure 24.28. While the direction of maximal curvature still shows no unusual subjects, C_i shows somewhat different subjects to be influential. Precisely, subjects #7, #51, and #74 are highly influential for the first model, whereas subjects #51 (again), #66, #67, and #73 are the ones detected with the second model. Although the cutoff is rather arbitrary, it is noteworthy that #51 appears in both C_i and $C_i(\boldsymbol{\theta})$ for the first model, indicating that the measurement model influence $C_i(\boldsymbol{\theta})$ is of the same order of magnitude as the dropout model influence $C_i(\boldsymbol{\psi})$, which is in contrast to the other analysis. Both #7 and #74 are *on average* not particularly low profiles, but they are among the lowest ones during the last month of the experiment and, although there are some others with the same feature, these two have a low overall level, but a *high* increment, which is very unusual.

OVERVIEW

Table 24.12 summarizes the subjects which are found to be influential in the various analyses. Although it can be argued that the various influence analyses serve different purposes, it is of some importance to distinguish between those subjects who are influential overall and others which turn up in one or a few analyses. Cow #51 is highlighted (bold type) in all six analyses and cows #59 and #68 show up four times, all others being seen three times or less. Clearly, #51 shows up unambiguously in the global influence plots and it yields the highest $C_i(\boldsymbol{\theta})$, $C_i(\boldsymbol{\beta})$, and $C_i(\boldsymbol{\alpha})$ values in the local influence analysis, even though one might argue that in some local influence plots, it is closely followed by slightly lower peaks. Inspecting its profile more closely, we conclude that it deviates from the typical profile in a number of ways. First, it is among the highest profiles during the period of initial drop, whereafter it is fairly low during the first half of the period, followed by a period of almost linear increase until the end of the study. The other two, #59 and #68, are, on average, the lowest profiles, not only within their group, but overall.

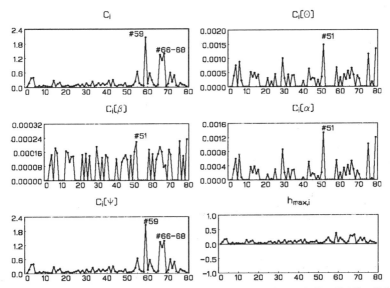

FIGURE 24.26. *Milk Protein Content Trial. Index plots of* C_i, $C_i(\boldsymbol{\theta})$, $C_i(\boldsymbol{\beta})$, $C_i(\boldsymbol{\alpha})$, *and* $C_i(\boldsymbol{\psi})$, *and of the components of the direction* $\boldsymbol{h}_{\max}$ *of maximal curvature. Dropout modeled from week 1.*

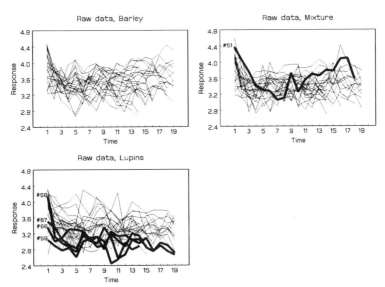

FIGURE 24.27. *Milk Protein Content Trial. Individual profiles, with locally influential subjects highlighted. Dropout modeled from week 1.*

Whereas global influence, as stated earlier, starts from deleting one subject completely, local influence only changes the dropout process for one subject from random dropout to nonrandom dropout. Because of the completely

TABLE 24.12. *Milk Protein Content Trial. Summary of influential subjects.*

Subject Subject	Drop From Week 15			Drop From Week 1		
		Local			Local	
	Glob.	Raw	Inc	Glob.	Raw	Inc
1	*			*		
7			*			
38	*					
43		*				
51	*	*	*	*	*	*
59	*	*		*	*	
65	*			*		
66					*	*
67					*	*
68	*	*		*	*	
73						*
74			*			

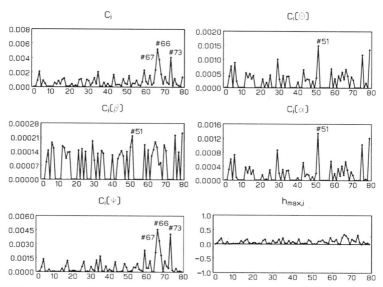

FIGURE 24.28. *Milk Protein Content Trial. Index plots of C_i, $C_i(\theta)$, $C_i(\beta)$, $C_i(\alpha)$, and $C_i(\psi)$, and of the components of the direction $\boldsymbol{h}_{\max}$ of maximal curvature. Dropout modeled from week 1. Incremental analysis.*

different approach, there is no need for both methods to yield similar results, although by looking at the influential subjects for all cases studied above, we notice some overlap.

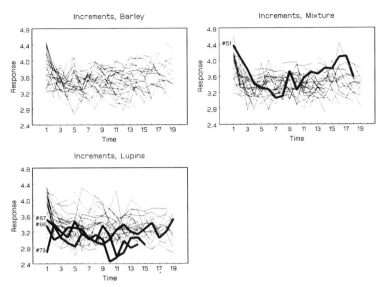

FIGURE 24.29. *Milk Protein Content Trial. Individual profiles, with locally influential subjects highlighted. Dropout modeled from week 1. Incremental analysis.*

Substantial differences are seen between the two models we formulated. The first one models dropout from week 15 onward. A second one allows dropout during the complete 19-week period. When dropout is based on the last 5 weeks, the model fitting results are, as expected, very close to those of Diggle and Kenward (1994), with a highly significant indication for nonrandom dropout. When the dropout model is based on the entire period, there is little evidence for nonrandom dropout. Moreover, influential subjects in the two approaches are entirely different. Both analyses concentrate on behavior in the period during which dropout is modeled. The latter indicates that the choice of period to which dropout applies is crucial.

In addition, we compared the direct variable analysis with an incremental one, where dropout depends on the difference between the current and previous measurement. In line with our analysis of the mastitis data set (Section 19.5.2), each analysis leads to different influential subjects, indicating that one should carefully discuss which analysis is preferable. Although both model formulations in (24.38) are equivalent, they lead to a different influence analysis, simply because the parameters at which the influence is targeted are different. Which one is chosen may depend on substantive considerations, as well as on the observation made by Molenberghs, Verbeke, *et al.* (1999) that the parameter of $Y_{i,j-1}$ is the most efficiently calculated in the incremental model, provided $\hat{\psi}_1$ and $\hat{\psi}_2$ are negatively correlated.

The latter condition is satisfied in many longitudinal applications, as was already noted by Diggle and Kenward (1994).

24.5   Hepatitis B Vaccination

The mentally handicapped residing in institutions are at high risk for hepatitis B virus (HBV) acquisition and subsequent carrier state. The higher risk of nonparental transmission in this population is due to the typical behavior of mentally retarded patients, the type of mental retardation, and the closed setting of the institutions, which all enhance spreading of the virus.

Hepatitis B vaccination of residents and staff is a general recommendation and has become part of today's hepatitis B prevention program. Data on long-term persistence of antibodies against HBV are scarce, especially in this population. Data available from other high-risk populations showed that 67% to 85% of the vaccinated individuals still had antibody levels higher than 10 international units/liter (IU/L), 9 to 12 years after the first vaccine dose (Hadler et al. 1986, Coursaget et al. 1994).

We describe the use of linear mixed-effects models for the analysis of antibodies against hepatitis B surface antigen (anti-HBs) data from a mentally handicapped population vaccinated against hepatitis B, 11 years earlier (Van Damme et al. 1989). Use of random effects in this setting was already proposed by Coursaget et al. (1994) and Gilks et al. (1993), who considered between-individuals variability in a Bayesian random-effects model. In previous studies, several factors have been described to cause a higher risk in the acquisition of hepatitis B virus infection. These factors include age, age at admission, duration of residency, type of mental retardation [Down's syndrome (DS) or other types of mental retardation (OMR)], sex, and use of anti-epileptic medication (Vellinga et al. 1999a). Sex, age, and type of mental retardation are also of influence on the response to vaccination (Vellinga et al. 1999b). The linear mixed model describes the decline in antibody titer in relation to the significant risk factors across measurement occasions. A detailed account of the medical results is reported in Vellinga et al. (1999c).

In 1985–1986, a hepatitis B vaccination program was conducted in a Belgian institution for mentally handicapped to evaluate the long-term persistence of anti-HBs after vaccination in this population. Blood samples were drawn from residents in that institution, who were then all vaccinated with three doses of hepatitis B vaccine (Engerix-B™, SmithKline Beecham Biologicals, Rixensart, Belgium) according to a month 0-1-6 schedule. Serum samples were taken after each vaccine dose, and if residents did not meet the (arbitrary) anti-HBs level of 100 IU/L at month 7, they received a booster vaccine dose at month 12. If the requirement of 100 IU/L were still not met at month 13, additional booster doses were administered (these residents were, however, not further included in the program). All residents received a booster dose after 5 years (i.e., at month 60).

Of the 196 seronegative residents originally included in the program, only 97 were included in the analysis of the follow-up after 11 years. They had blood samples taken yearly for the first 5 years and at year 11. Sixty-seven of them received 4 vaccine doses (at months 0, 1, 6, plus a booster at month 60) and 30 received 5 doses (at months 0, 1, 6, 12, plus a booster at month 60). Further details can be found in Van Damme *et al.* (1989) and Vellinga *et al.* (1999a). We will denote the group of residents vaccinated according to a month 0-1-6 (versus 0-1-6-12) schedule by G1 (versus G2).

Interest focuses on describing the evolution of the mean log(anti-HBs+1) titer over time (where log denotes the natural logarithm), while accounting for prognostic factors such as sex, body mass index, duration of residency, age at admission into the institution, type of mental retardation, use of anti-epileptic drugs, and number of vaccine doses received. It is also of importance to predict antibody level at years 11 (end of study) and 12 (1 year past the end) based on the fitted model.

Although the main epidemiological interest lies in the population-averaged prediction, the model enables one to perform individual-specific predictions as well. Both model building and prediction are complicated by the fact that individual and average profiles are highly nonlinear (Figures 24.30(a) and 24.30(b)), combined with the absence of measurements between years 5 to 11.

In order to deal with these complications, we consider two types of models. The first one is saturated in time effects and is helpful, for example, for making comparisons between groups at different time points. Although the nonlinearity of the profiles raises no particular problem there, such a model does not parsimoniously describe the temporal decline in antibody titer, nor is it truly useful for making long-term predictions. This could be achieved by restricting the analysis to post-vaccination data (i.e., data available after months 6 or 12, depending on the number of doses administered). These data are indeed somewhat "smoother" than the original data and we can employ prebooster data (until month 60) to set up a model that can then be transposed to model postbooster decline in anti-HBs. This is accomplished using fractional polynomials (Royston and Altman 1994), which provide a very flexible tool for parametric modeling. (see also Section 10.3).

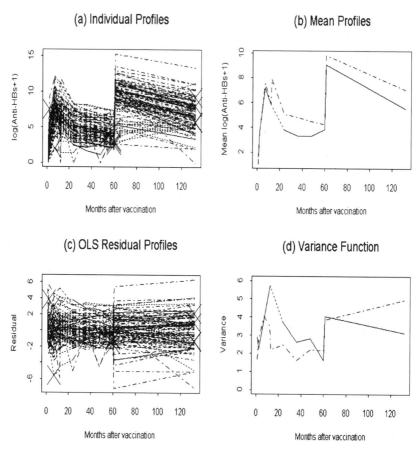

FIGURE 24.30. *Hepatitis B Vaccination Study. (a) Longitudinal trends in log(anti-HBs+1) for residents with DS (solid line) and OMR (dashed line). Cross symbols indicate missing values. (b) Average log(anti-HBs+1) over time for residents with DS (solid line) and OMR (dashed line). (c) Ordinary least squares (OLS) residual profiles obtained upon fitting a saturated mean structure to log(anti-HBs+1). (d) Variance of the OLS residuals over time for residents with DS (solid line) and OMR (dashed line).*

Section 24.5.1 presents model building for the two models we just described. Section 24.5.2 is devoted to the issue of prediction at year 12, where sensitivity to assumptions and type of model used can be assessed.

24.5.1 Time Evolution of Antibodies

In this section, we address the question of building a model that adequately describes the evolution of log antibody titer over time. Hence, we need to

Variogram

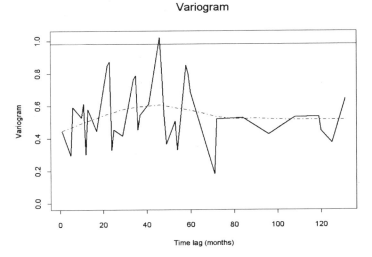

FIGURE 24.31. *Hepatitis B Vaccination Study. Sample variogram of log anti-body residuals (the horizontal line estimates the process variance; the dashed line represents a smooth estimate of the variogram).*

consider appropriate mean, variance, and covariance structures. Since the profiles are quite messy due to unequally spaced measurement occasions and booster effects, it is imperative to conduct an exploratory data analysis.

As shown in Figures 24.30(a) and 24.30(b), individual and mean profiles of log(anti-HBs+1) for DS and OMR are clearly nonlinear and show peaks after booster doses. Individual profiles follow approximately the same pattern, with the main difference between profiles residing in the vertical shift. This suggests a strong contribution of an individual random intercept. Average profiles show a difference in anti-HBs between both groups. Also, these profiles exhibit steep increases immediately after boosters, followed by a gradual decrease which appears to be nonlinear.

Figure 24.31 depicts an estimate of the empirical variogram for these data. It was constructed using standardized ordinary least squares residuals obtained upon fitting a saturated groups by times model (where group is type of mental retardation). Also shown in this figure is a loess-smoothed estimate (Cleveland 1979) of the variogram. The between-subject variance seems relatively large in these data, accounting for about one-half of the total variability. The measurement error is also substantial, accounting for the other half of the process variance. This variogram leaves little room for a serially correlated component. Note that in this context, it is essential to use standardized residuals to remove variance heterogeneity in the data, ensuring that the process variance is constant and equal to 1.

We now turn to model construction and outline the successive steps to retain a final model. Type of mental retardation, duration of residency, and number of vaccine doses (as a group variable) are allowed to have specific effects at each sample occasion. We also include time-constant effects for sex (male versus female), use of anti-epileptic drugs (yes versus no), body mass index, and age at admission in the institution, since a time trend for these covariates could not be detected.

In Chapter 9, it was suggested to select a variance-covariance structure based on the most complex mean structure one is willing to consider. Once such a structure is chosen, simplification of the mean model can proceed.

The preliminary variance model acknowledges presence of serial correlation and includes the following random effects: an intercept, a linear time slope, and number of vaccine doses (as a 0/1-coded group variable). An unstructured form is assumed for the 3×3 random-effects variance matrix D.

We first select an appropriate serial process, as shown in Table 24.13. Models with exponential (B) and Gaussian (C) serial correlation are compared to the model with no serial process (A) using the likelihood ratio test statistic (denoted G^2 in this table). These tests strongly reject the null hypothesis of no serial process. At this stage, it was decided to keep the exponential model.

Next, the random-effects structure can be simplified. Three hierarchically ordered models are presented in the second part of Table 24.13. One has to be very careful in interpreting the significance of random effects using the likelihood ratio test statistic G^2. As described in Section 6.3.4, the associated testing problem is indeed nonstandard, as the null hypothesis lies on the boundary of the parameter space of the alternative hypothesis. The reference distribution for the B–D comparison is a 50:50 mixture of χ_2^2 and χ_3^2. Similarly, for the comparison of Models D and E, we obtain a 50:50 mixture of χ_1^2 and χ_2^2 variables. These distributions have been utilized to calculate the corresponding p-values. Thus, at this stage, we select Model D, comprising a random intercept and a random time slope.

Finally, retaining the covariance structure we just selected, the mean model can be reduced. Effects kept in the final model were time, type of mental retardation, number of vaccine doses (with unstructured time effects), duration of residency (with a linear time trend), use of anti-epileptic medication, and sex (time-constant effects). Although not significant, sex was kept in the model for reasons of external comparison.

Parameter estimates for this model are shown in Tables 24.14 and 24.15. It can be seen that no covariance parameter is given for the random effects in these tables. A comparison between model-based and empirical standard er-

TABLE 24.13. *Hepatitis B Vaccination Study. Selection of a serial correlation process and a random-effects structure.*

Model	Description	# Param. Rand.	# Param. Ser.	-2ℓ	Comp.*	G^2	df	p
Selection of a serial correlation process:								
A	None	6	0	2623.72				
B	Exponential	5†	2	2575.23	A	48.49	1	< 0.0001
C	Gaussian	6	2	2574.31	A	49.41	2	< 0.0001
Selection of a random-effects structure:								
B	Intercept, time, # doses	5†	2	2575.23				
D	Intercept, time	3	2	2578.54	B	3.31		0.269††
E	Intercept	1	2	2606.22	D	23.68		< 0.0001†††

* Comparison model.

† One parameter could not be estimated due to parameter constraints.

†† From a 50-50 mixture of χ^2_2- and χ^2_3-distributions.

††† From a 50-50 mixture of χ^2_1- and χ^2_2-distributions.

rors revealed large discrepancies (relative increases more than threefold for most of the estimates) in the final model. Empirically corrected (or robust) standard errors (Section 6.2.4) counteract the effect of potential misspecification of the covariance structure (Diggle, Liang, and Zeger 1994, Liang and Zeger 1986) and disagreement between both types of standard errors might point to an inadequately specified covariance structure. Arguably, we had little reason to believe that the selected covariance function is substantially incorrect. Therefore, it is wise to attain a trade-off between model fit as reported by likelihood ratios, and differences occurring between model-based and empirical standard errors. In particular, assuming a diagonal instead of an unstructured covariance matrix for the random effects yields a much better model in this respect and was therefore retained as our final model. Most of the estimated empirical standard errors in Tables 24.14 and 24.15 do not exhibit changes of more than 25% compared to model-based standard errors.

It is worth noting that the effect of Down's syndrome on antibody titer was significant at months 24, 36, and 48, indicating a faster decline in anti-HBs in this population than in other mentally retarded. There did not seem to be

TABLE 24.14. *Hepatitis B Vaccination Study. Parameter estimates and standard errors (model based; empirically corrected) for the final model (original data). Part I.*

Effect	Time	Estimate (s.e.)
Mean Structure:		
Intercept		9.36 (0.41; 0.38)
Time	1	−6.80 (0.21; 0.24)
Time	2	−4.30 (0.21; 0.20)
Time	7	†
Time	12	−1.66 (0.18; 0.13)
Time	13	−0.54 (0.36; 0.37)
Time	24	−3.36 (0.20; 0.19)
Time	36	−3.68 (0.21; 0.19)
Time	48	−4.21 (0.28; 0.27)
Time	60	−4.69 (0.31; 0.25)
Time	61	1.62 (0.34; 0.25)
Time	132	−1.92 (0.59; 0.55)
DS/OMR	1	0.60 (0.64; 0.54)
DS/OMR	2	0.37 (0.68; 0.62)
DS/OMR	7	−0.02 (0.56; 0.56)
DS/OMR	12	−0.30 (0.61; 0.82)
DS/OMR	13	††
DS/OMR	24	−1.56 (0.53; 0.74)
DS/OMR	36	−1.75 (0.46; 0.60)
DS/OMR	48	−1.50 (0.56; 0.62)
DS/OMR	60	−0.61 (0.54; 0.44)
DS/OMR	61	−0.83 (0.69; 0.54)
DS/OMR	132	−1.18 (0.69; 0.39)
# doses	1	−1.34 (0.37; 0.24)
# doses	2	−1.78 (0.40; 0.36)
# doses	7	−2.45 (0.33; 0.36)
# doses	12	−2.71 (0.35; 0.42)
# doses	13	†††
# doses	24	−0.11 (0.31; 0.36)

DS/OMR: 1 = DS, 0 = OMR; # doses: 1 = 5 doses, 0 = 4 doses.

Sex: 1 = male, 0 = female; Anti-epileptic drugs: 1 = use, 0 = no use.

† Month 7 taken as reference point because the decision to give an extra booster dose was taken at that time.

†† No measurements were available at month 13 in DS patients.

††† No measurements were available at month 13 in the group that was administered four vaccine doses.

TABLE 24.15. *Hepatitis B Vaccination Study. Parameter estimates and standard errors (model based; empirically corrected) for the final model (original data). Part II.*

Effect	Time	Estimate (s.e.)
Mean Structure (continued):		
# doses	36	-0.25 (0.27; 0.30)
# doses	48	0.00 (0.33; 0.35)
# doses	60	0.24 (0.32; 0.34)
# doses	61	-2.01 (0.41; 0.45)
# doses	132	-2.00 (0.41; 0.41)
Residency		-0.04 (0.02; 0.01)
Residency$\star$Time		0.005 (0.002; 0.002)
Sex		-0.01 (0.23; 0.23)
Anti-epileptic		-0.63 (0.24; 0.23)
Random Effects:		
Intercept		0.66
Time		0.02
Serial Structure:		
Variance		0.58
Rate of exponential decrease $(1/\rho)$	2.32	
Measurement Error:		
Time	1	1.32
Time	2	1.37
Time	7	0.76
Time	12	0.70
Time	13	0.78
Time	24	0.48
Time	36	0.00
Time	48	0.46
Time	60	0.47
Time	61	1.31
Time	132	0.31

a difference in immediate response to vaccination between these two groups. Also, we see that the extra dose given at month 12 in G2 had sufficiently elevated anti-HBs titer so as to render it almost indistinguishable from anti-HBs titer in G1 until year 5. Yet, administration of a booster dose at that time again led to better responses in G1 and this was still visible at year 11.

A similar modeling strategy can be applied to the postvaccination data [i.e., data available after the last vaccination, at month 6 (G1) or 12 (G2)]. We need to specify different models for each of the two groups since post-vaccination times are different. We can set up a model for prebooster data (until month 60) and then transpose this model to postbooster data, using an indicator variable for the time of booster administration. For instance, a simple model, ignoring potential covariates could be written as follows:

$$E(Y_{ij}) = \begin{cases} \beta_0^{(1)} + \beta_1^{(1)} I(t_j \geq 55) + g(t_j - 55.I(t_j \geq 55)) & \text{in group G1,} \\ \beta_0^{(2)} + \beta_1^{(2)} I(t_j \geq 49) + g(t_j - 49.I(t_j \geq 49)) & \text{in group G2,} \end{cases}$$

where $g(t)$ is a fractional polynomial (Royston and Altman 1994) (i.e., a linear combination of real-valued powers of t). This strategy is discussed in detail in Section 10.3. Let us briefly recapitulate the key concepts. Royston and Altman (1994) argue that conventional low-order polynomials offer only a limited family of shapes and that high-order polynomials may fit poorly at the extreme values of the covariates. Moreover, polynomials do not have finite asymptotes and cannot fit the data where limiting behavior is expected. This is a severe limitation in many cases. As a result, Royston and Altman (1994) propose an extended family of curves, which they call fractional polynomials. Conventional polynomials are included as a subset of this family. For a given degree m and an argument $t > 0$ (e.g., time), fractional polynomials are defined as

$$\beta_0 + \sum_{j=1}^{M} \beta_j t^{p_j},$$

where the β_j are regression parameters and $t^0 \equiv \ln(t)$ and the powers $p_1 < \cdots < p_m$ are positive or negative integers or fractions (Royston and Altman 1994). They argue that polynomials with degree higher than 2 are rarely required in practice and further restrict the powers of dose to a small prede-fined set of possibly noninteger values: $\Pi = \{-2, -1, -1/2, 0, 1/2, 1, 2, \ldots, \max(3, m)\}$. For example, setting $m = 2$ generates (1) four quadratics in powers of d, represented by $(1/t^2, 1/t)$, $(1/t, 1/\sqrt{t})$, $(\sqrt{t}, t)$, and (t, t^2); (2) a quadratic in $\ln(t)$, and (3) other curves which have shapes different from those of conventional low-degree polynomials. The full definition includes possible "repeated powers" which involve powers of $\ln(t)$. For example, a fractional polynomial of degree $m = 3$ with powers $(-1, -1, 2)$ is of the form

$$\beta_0 + \beta_1 y^{-1} + \beta_2 y^{-1} \ln(y) + \beta_3 y^2$$

(Royston and Altman 1994, Sauerbrei and Royston 1999). In this case, the set of powers we have considered ranged from -3 to 3 with increments of

TABLE 24.16. *Hepatitis B Vaccination Study. Parameter estimates and standard errors (model-based; empirically corrected) for the fixed effects of the final model (postvaccination data).*

Effect	Estimate (s.e.)
Intercept G1	8.808 (0.299; 0.284)
$I(t \geq 55)$	3.501 (0.151; 0.163)
$(t - 55I(t \geq 55))^{-3}$	-0.384 (0.202; 0.178)
$\log(t - 55I(t \geq 55))$	-1.120 (0.052; 0.057)
DS/OMR G1	-0.494 (0.673; 0.702)
Sex	0.037 (0.278; 0.276)
Intercept G2	6.389 (0.503; 0.582)
$I(t \geq 49)$	1.358 (0.272; 0.346)
$(t - 49I(t \geq 49))^{-3}$	1.601 (0.455; 0.446)
$\log(t - 49I(t \geq 49))$	-0.435 (0.131; 0.154)
DS/OMR G2	-2.706 (0.688; 0.533)
Use of anti-epileptic drugs	-0.632 (0.284; 0.281)

0.5. We then searched for the best pair or triple of powers among this set. The combination $(-3, 0)$ turned out to give an adequate fit to these data. See also Section 10.3.

Selection of the covariance structure was performed similarly to what was previously done, with the difference that no spatial process was found necessary. For the selection of the mean structure, the antibody titer measurement obtained at the last vaccination could be included as a baseline covariate on top of the usual covariates, but this could not be done because no measurement were taken at month 6. Instead, the first log antibody titer measurement was used as baseline. Selected covariates were type of mental retardation (with a different effect depending on the number of vaccine doses administered), use of anti-epileptic medication, and sex. Table 24.16 presents the parameter estimates of the final model, for the fixed effects only.

In order to visually assess the fit of these two models, Figure 24.32 shows observed and predicted average profiles for combinations of number of vaccine doses and type of mental retardation. For predicted average profiles, all other covariate effects were set equal to their mean values.

On accommodating individual-specific effects, our model enables a much more precise assessment of important explanatory variables, such as number of vaccine doses received, whether or not a person has Down's syn-

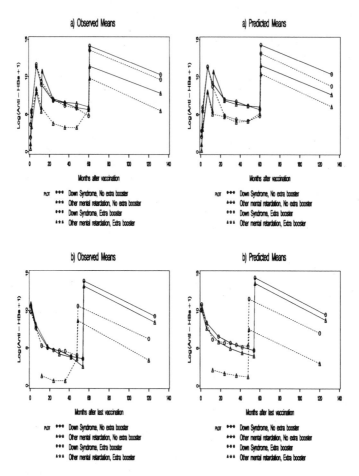

FIGURE 24.32. *Hepatitis B Vaccination Study. Observed and predicted mean profiles for combinations of number of vaccine doses and type of mental retardation: (a) original data; (b) postvaccination data.*

drome, and, of course, time effects. In particular, the strong contributions of random intercepts and serial correlation show the importance of the initial response as well as the individual trajectory for the further evolution of anti-HBs profiles. Models that restrict attention to geometric mean titers (GMT) calculation are not able to include such individual-specific effects, typically resulting in less precise inference, also for the fixed effects.

In this study, no difference can be detected between DS and OMR patients in their immediate response, but we find that DS induces an accelerated decrease in antibody titers, implying that the rate of decline in antibody titers might be different in these two populations. This might explain why some other studies attempting to demonstrate a difference between DS and OMR patients in their anti-HBs response after vaccination have failed. See Vellinga *et al.* (1999c) for further discussion. Another point concerns whether anti-epileptic medication has an influence on the immune system. However, it is hard to decide whether this is due to the medication itself, or rather an indication for the influence of epilepsy, or both (De Ponti *et al.* 1993).

Although there is some interest in modeling the complete set of data from a descriptive viewpoint, this could only be achieved with the help of a time-saturated model to account for the high nonlinearity in the profiles. If one is interested in a more parsimonious, parametric description of the temporal decline in antibody titer, to address such questions as long-term influence, one needs to resort to an alternative solution. Focusing the analysis on postvaccination data is a suitable alternative, enabling simple parametric modeling of the anti-HBs evolution over time. Obviously, the absence of intermittent measurements between years 5 and 11 weakens the long-term prediction process, and we need to make certain assumptions such that the rate of decline after booster administration at month 60 is similar to the rate of decline after time of last vaccination. It is nevertheless reassuring to see that predicted values at year 12 were all in good agreement, independently of the model or method chosen. We conclude that each of these two models may bring their own insight into the data and their combined use may better serve the purpose of a sensitivity analysis.

24.5.2   Prediction at Year 12

We now address the issue of predicting antibody titer at year 12 (i.e., 1 year after last follow-up contact). This extrapolation problem is complicated by the design feature that no measurements are available between months 61 and 132, whereas apparently we are dealing with nonlinear profiles. Although the use of a time-saturated model for the mean structure is suitable for building an acceptable model, it is less so when it comes to prediction, in particular when interest centers on future prediction. Therefore, we propose two simple methods to perform such a prediction and compare the results to the fractional-polynomial approach on postvaccination data.

The first approach merely uses a linear extrapolation based on the last two measurements (at months 61 and 132) of an individual. The resulting extrapolations are then averaged out to obtain a prediction at month 144. Obviously, this approach can be criticized as being overly simple since the profiles are clearly nonlinear over the first 5 years. A refinement of this method might consist of overlaying profiles for the month 61–132 period with profiles from the first part of the study and then extrapolating until month 144. It raises some technical difficulties though, since the starting

TABLE 24.17. *Hepatitis B Vaccination Study. Predicting log antibody titer (IU/L) at year 12: (a) Approach 1: linear interpolation (original data); (b) Approach 2: refined linear interpolation (original data); (c) Approach 3: fractional polynomials model (post-vaccination data).*

Group	Approach 1	Approach 2	Approach 3
4 vaccine doses (G1)	7.05	7.38	7.13
5 vaccine doses (G2)	5.01	5.42	5.41

point of the first period (time of the last vaccine dose) depends upon the group being considered: month 7 for group G1 and month 13 for G2. Using month 24 as a cutoff point to split the first time period into two pieces, we can linearly approximate the profiles in these two time windows, translate them to the month 61–132 period, extrapolate until month 144, and eventually average the results across the two groups.

Predictions at year 12 based on these two extrapolation methods are displayed in Table 24.17, together with the prediction inferred from the model on postvaccination data. As can be seen, all three approaches yield very similar results.

We conclude with two remarks. First, in this study, prediction takes place only 1 year upon completion of the study, which is not too distant in time compared to the duration of the study. Would prediction have to be done several years later, we would likely observe larger discrepancies. Second, in studies where more emphasis is to be put on prediction, it is a good idea to plan intermittent assessment occasions to aid in modeling long-term temporal evolution. A simple model (e.g., using fractional polynomials), might then be used straightforwardly for making long-term inference with more confidence.

24.5.3 SAS Code for Vaccination Models

We describe how to use the SAS statistical software package to fit our starting and thus the most complex model on the one hand, and the finally selected model on the other hand. The models are discussed in Section 24.5.1.

The SAS code to fit the initial model is

```
proc mixed data = hbv method = ml info noclprint;
class timecls ds_omr dosegrp sex epilepsy;
model loganti = timecls ds_omr(timecls)
                dosegrp(timecls) dur_res(timecls)
                sex bmi age epilepsy / s;
random intercept time ds_omr / subject = patid
                               type = un;
repeated timecls / subject = patid type = sp(exp)(time)
                   local = exp(timecls);
run;
```

Apart from an unstructured time trend, the fixed-effects structure includes time-varying effects of Down's syndrome versus other mental retardation (DS/OMR), dose group, and duration of residency. Effects of sex, body mass index, age, and influence of epilepsy are time invariant. Note that time is included as a class variable, implying that there is a different time effect parameter for each time point, as well as a different parameter at each time for the time-varying covariates. Apart from a random-intercepts term, there is a linear random time effect and an effect of DS/OMR. A different copy of *time* is used so that it can be left out of the CLASS statement. The serial correlation structure in time is specified by means of the REPEATED statement, where the 'type=' option is included to indicate that a spatial exponential process is required. To ensure that, in addition to random effects and serial variance, also measurement error is allowed to be present, the 'local' option is used. More precisely, 'local=' allows the measurement error to depend on covariates. In our case, 'exp(timecls)' includes a different measurement error variance component for each time point. Formally, the following contribution is added to Σ_i:

$$\sigma^2 \text{diag}\left[\exp(U_i \delta)\right]$$

where U_i is a design matrix built from the covariates included and δ are the estimated and reported parameters. Note that the exponential function ensures non-negative components of variance.

As an aside, the 'info' option in the PROC MIXED statement provides preliminary information, such as an expression for the model fitted and the dimensions of relevant matrices.

The final model is fitted using the following syntax:

```
proc mixed data = hbv method = ml info noclprint;
class timecls ds_omr dosegrp sex;
model loganti = timecls ds_omr(timecls) dosegrp(timecls)
                dur_res dur_res*time sex / s;
random intercept time / subject = patid type = vc;
repeated timecls / subject = patid type = sp(exp)(time)
                local = exp(timecls);
run;
```

Appendix A

Software

A.1 The SAS System

A.1.1 Standard Applications

Many of the analyses done in this book have been performed in SAS, in particular using the procedure MIXED. An extensive overview is to be found in Chapter 8. Here, the focus is on the Baltimore Longitudinal Study of Aging (prostate cancer data; Section 2.3.1). The main program features are discussed, as well as the output. The SAS procedure MIXED is explicitly discussed in a number of other chapters. In Chapter 9, the prostate cancer data are used to illustrate model building principles. In Chapter 17, both complete as well as incomplete versions of the growth data (Section 2.6) are analyzed using PROC MIXED. The use of SAS for pattern-mixture based sensitivity analysis is documented in Chapter 20. Finally, SAS code is included for several case studies in Chapter 24.

A.1.2 New Features in SAS Version 7.0

In this book, all analyses using the SAS System, have been carried out using Version 6.12. Although this may seem anomalous to many, given the availability of Version 7.0 and higher, it has to be noted that Version 7.0

(SAS 1999) was not available on a commercial basis in 1999, for example, in Europe. For a thorough description of PROC MIXED in SAS Version 7.0, we refer to the on-line manual (SAS 1999).

In this section, we will highlight some of the important changes that have been implemented in Version 7.0 with respect to PROC MIXED. They are ordered by statement.

PROC MIXED STATEMENT

The option 'CL', requesting confidence intervals for the covariance parameter estimates, has been modified. For those parameters with a default lower bound of zero (diagonal elements in a covariance matrix), Satterthwaite approximations are used:

$$\frac{\nu\widehat{\sigma^2}}{\chi^2_{\nu,1-\alpha/2}} \leq \sigma^2 \leq \frac{\nu\widehat{\sigma^2}}{\chi^2_{\nu,\alpha/2}}, \tag{A.1}$$

where $\nu = 2Z^2$ and Z is the classical Wald statistic $\widehat{\sigma^2}/\text{s.e.}(\widehat{\sigma^2})$. Using 'CL<=Wald>' requests a Wald version, rather than the modified limits in (A.1).

The 'method' option has been extended to include, apart from REML, ML, and MIVQUE0, also TYPE1, TYPE2, and TYPE3. The new methods request analysis of variance (ANOVA) estimation of the corresponding type, producing method-of-moment variance component estimates. Of course, these methods are available only when there is no 'subject=<effects>' option and no REPEATED statement.

Also, 'namelen=<number>' is a new option, which specifies the length to which long effect names are shortened.

MAKE STATEMENT AND OUTPUT DELIVERY

A major change to PROC MIXED (and all other procedures) is the way in which output is handled. PROC MIXED now makes use of the integrated ODS (output delivery system). This implies that the MAKE statement has become obsolete. Although still supported, it is envisaged that this will no longer be the case in later versions.

For example, a Version 6 statement of the form

```
make 'covparms' out = cp;
```

within the MIXED procedure is replaced by

```
ods output covparms = cp;
```

to be placed in front of PROC MIXED. For a thorough discussion on the capabilities of ODS, we refer to SAS (1999).

MODEL STATEMENT

In order to have a more efficient access to the covariance structure of the estimated fixed effects, two new options have been added: 'corrb', producing the asymptotic correlation matrix, and 'covbi', producing the inverse of the asymptotic covariance matrix. For the latter, there is still an alias: 'xpvix'.

The 'ddfm=satterth' option is now extensively documented.

Given the ODS structure, predicted means and predicted values are now handled in a different way. Precisely, the option 'outpredm=SAS-data set' and 'outpred=SAS-data set' have to be used. Options 'pred' and 'pred-means' have become obsolete. Aliases for the new options are 'outpm=SAS-data set' and 'outp=SAS-data set'.

PRIOR STATEMENT

This statement has undergone a major revision. New options are as follows:

'data=': allows input of user-defined prior densities for variance components.

'alg=': specifies the algorithm used for generating the posterior sample.

'bdata': allows input of the base densities used by the sampling algorithm.

'grid=': a grid of values over which to estimate the posterior density.

'gridt=': specifies a transformed grid of values over which to estimate the posterior density.

'lognote=number': writes a log note to screen each time a sample is generated of which the sequence number is a multiple of the specified *number*.

'logrbound=number': specifies the bounding constant for rejection sampling. This option has been available since Version 6.12.

'out=': output data set with the sample of the posterior density.

'outg=': output data set from the grid evaluations.

'outgt=': output data set from the transformed grid evaluations.

'psearch': displays the search used to determine the parameters for the inverted gamma densities.

'ptrans': displays the transformation of the variance components.

'seed=': starting value for the random number generation within a call of PROC MIXED.

'tdata=': allows to input the transformation of the covariance parameters used by the sampling algorithm.

'trans=': specifies the algorithm used to determine the transformation of the covariance parameters.

RANDOM STATEMENT

The option 'nofullz' eliminates the columns in Z corresponding to missing levels of random effects involving class variables.

REPEATED STATEMENT

Since Version 6.12, multivariate repeated measures can be fitted. To this end, direct-product covariance structure are available, of which the first factor specifies the correlation among the components of the multivariate outcome vector and the second component specifies the correlation over time within a component. Three structures are available: 'type=un@ar(1)', 'type=un@cs', and 'type=un@un'. An example is

```
model y = var time var*time;
repeated var time / type = un@ar(1) subject = subject;
```

Note that all outcomes for a given individual are still stacked, as with univariate repeated measures. Thus, two indicators are needed: *var* to indicate which of the multivariate outcomes is listed and *time* indicating the longitudinal structure. The 'type=' option specifies an unstructured covariance matrix among the components at a given time (a typical assumption in multivariate data, cf. PROC GLM), and further an AR(1) process for repeated measures of a specific outcome. It is then assumed that for different outcomes at unequal measurement times, the Kronecker product specifies the appropriate covariance. Note that this is an assumption. Although not implemented in the SAS procedure MIXED, the modeler could in principle consider more complex structures.

TESTING FOR VARIANCE COMPONENTS

The PROC MIXED documentation explicitly acknowledges the complexity of hypothesis testing for variance components, as discussed in Section 6.3.4.

A.2   Fitting Mixed Models Using MLwiN

The package MLwiN can be used to fit a wide variety of linear mixed models. This package is based on the *multilevel model* concept (Goldstein 1979, 1995). Using the growth data, introduced in Section 2.6 and analyzed in Section 17.4, we will briefly introduce multilevel modeling, with emphasis on longitudinal studies, and illustrate the use of the MLwiN package. The most important source of information is Goldstein (1995), which contains ample references to further reading, and the User's Guide to MLwiN. The manual and other relevant information can be obtained from the websites on multilevel models (with Australian and American mirror sites) and the website on MLwiN:

http://www.ioe.ac.uk/multilevel/
http://www.medent.umontreal/multilevel/
http://www.edfac.unimelb.edu.au/multilevel/
http://www.ioe.ac.uk/mlwin/

For the growth data, we will focus on Model 6, introduced on p. 252. MLwiN Version 1.02.0002 will be used. This model assumes separate linear profiles for boys and girls, implying four fixed-effects parameters, as well as a random intercept and a random age slope, with an unstructured 2×2 covariance matrix D. Finally, a measurement error term is added. Similar to (17.10), which assumes an unstructured covariance matrix Σ_i, we can write this model as

$$Y_{ij} = \beta_0 + b_{0i} + \beta_{01}x_i + \beta_{10}t_j(1 - x_i) + \beta_{11}t_jx_i + b_{1i}t_j + \varepsilon_{ij}, \quad \text{(A.2)}$$

where $(b_{0i}, b_{1i})'$ follows a $N(\mathbf{0}, D)$ distribution and the residual errors are uncorrelated with zero mean and variance σ^2. Recall that x_i is 1 for boys and 0 for girls and that t_j indicates measurement ages 8, 10, 12, and 14. Now, let us group the model terms per covariate effect:

$$Y_{ij} = (\beta_0 + b_{0i} + \varepsilon_{ij})1 \quad \text{(A.3)}$$

TABLE A.1. *Growth Data. Model effects grouped per covariate.*

Number	Effect	Fixed	Level 2	Level 1
0	Intercept	β_0	b_{0i}	ε_{ij}
1	Male	β_{01}		
2	Female*Age	β_{10}		
3	Male*Age	β_{11}		
4	Age		b_{1i}	

$$+\beta_{01}x_i \tag{A.4}$$

$$+\beta_{10}t_j(1 - x_i) \tag{A.5}$$

$$+\beta_{11}t_jx_i \tag{A.6}$$

$$+b_{1i}t_j. \tag{A.7}$$

Model (A.2) is equivalent to Models (A.3)–(A.7). The intercept or constant term is explicitly included in (A.3). In the latter, the coefficients are grouped per effect, as detailed in Table A.1. Fixed effects are listed in the column labeled "Fixed." By "Level 2" we indicate all random effects (i.e., those effects which are random at the level of the individual subject). "Level 1" contains the measurement error (i.e., the term which is random at the level of the individual measurement within a subject). The level can also be deduced from the subscripts i and j affixed to the parameters. Fixed effects carry no subscripts, neither i nor j. Level 2 effects carry only the first subscript i, whereas level 1 effects are subscripted by both i and j. Observe that the coefficient of the intercept consists of three parts: a fixed effect, a random intercept, and a residual error term. Alternatively, these terms can be termed fixed effect, random effect at level 2, and random effect at level 1. Covariates 1–3 have a fixed effect only, whereas age only has a random effect at level 2 (random slope). This is due to the fact that the fixed slope is allowed to differ for boys and girls, whereas a common random slope is assumed for both. In multilevel notation, this model can be rewritten as

$$Y_{ij} = \beta_{0ij}1 + \beta_1 x_j + \beta_2 t_i(1 - x_j) + \beta_3 t_i x_j + u_{4j}t_j, \tag{A.8}$$

$$\beta_{0ij} = \beta_0 + u_{0j} + e_{0ij}. \tag{A.9}$$

Using the MLwiN package, Models (A.8)–(A.9) can be implemented very easily using the drop-down window structure. The result is shown in Figure A.1. The corresponding maximum likelihood estimates are displayed in Figure A.2.

$$\text{measure}_{ij} \sim N(XB, \Omega)$$

$$\text{measure}_{ij} = \beta_{0ij}\text{intercep} + \beta_1\text{male}_j + \beta_2\text{maleage}_{ij} + \beta_3\text{femage}_{ij} + u_{4j}\text{age}_{ij}$$

$$\beta_{0ij} = \beta_0 + u_{0j} + e_{0ij}$$

$$\begin{bmatrix} u_{0j} \\ u_{4j} \end{bmatrix} \sim N(0, \Omega_u) : \Omega_u = \begin{bmatrix} \sigma_{u0}^2 & \\ \sigma_{u40} & \sigma_{u4}^2 \end{bmatrix}$$

$$\begin{bmatrix} e_{0ij} \end{bmatrix} \sim N(0, \Omega_e) : \Omega_e = \begin{bmatrix} \sigma_{e0}^2 \end{bmatrix}$$

$$-2*log(like) = 427{,}806$$

FIGURE A.1. *Growth Data. Symbolic MLwiN equation for Models (A.8)–(A.9).*

$$\text{measure}_{ij} \sim N(XB, \Omega)$$

$$\text{measure}_{ij} = \beta_{0ij}\text{intercep} + -1{,}032(1{,}535)\text{male}_j + 0{,}784(0{,}083)\text{maleage}_{ij} + $$
$$0{,}480(0{,}100)\text{femage}_{ij} + u_{4j}\text{age}_{ij}$$

$$\beta_{0ij} = 17{,}373(1{,}182) + u_{0j} + e_{0ij}$$

$$\begin{bmatrix} u_{0j} \\ u_{4j} \end{bmatrix} \sim N(0, \Omega_u) : \Omega_u = \begin{bmatrix} 4{,}557(4{,}672) & \\ -0{,}198(0{,}379) & 0{,}024(0{,}034) \end{bmatrix}$$

$$\begin{bmatrix} e_{0ij} \end{bmatrix} \sim N(0, \Omega_e) : \Omega_e = \begin{bmatrix} 1{,}716(0{,}330) \end{bmatrix}$$

$$-2*log(like) = 427{,}806$$

FIGURE A.2. *Growth Data. MLwiN estimates for Models (A.8)–(A.9).*

The coefficient of every effect is allowed to include any combination of a fixed effect and random effects at all levels. In this case, there are only two levels, but higher-order hierarchies are implemented as well and are no more difficult to construct. Estimation is done using maximum likelihood or iterative generalized least squares. Alternative estimation methods are restricted iterative generalized least squares, parametric bootstrap, Gibbs sampling, and Metropolis-Hastings.

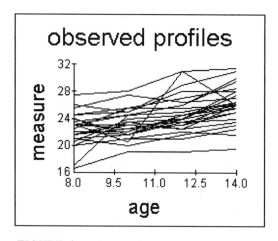

FIGURE A.3. *Growth Data. Observed profiles.*

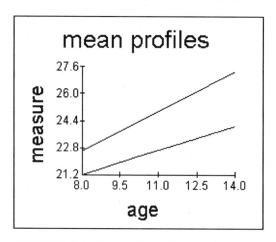

FIGURE A.4. *Growth Data. Predicted means.*

Model simplification can flexibly be conducted by clicking on terms that need to be deleted. For example, if the covariance between both random effects is judged to be redundant, it can be removed by simply clicking on this term and reestimating the model.

There are ample graphical capabilities. For example, a simple plot of the raw profiles is shown in Figure A.3. Predicted means are displayed in Figure A.4. This plot is based on the fixed-effects structure. Therefore, two fitted straight lines are shown, one for boys and one for girls. Empirical Bayes predictions are displayed in Figure A.5. Since the model assumes random intercepts and random slopes, individual linear profiles are produced. MLwiN uses data by means of a work sheet. This can easily be manipulated. Variables can be renamed and transformed, and new ones

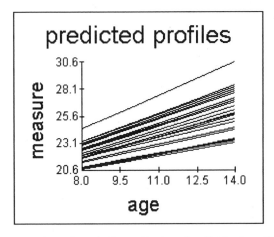

FIGURE A.5. *Growth Data. Predicted individual profiles.*

can be created. For example, if predicted values are computed, they can be added to the work sheet as a new column. Subsequently, predicted values can be used in the graphical window to generate plots such as Figures A.4 and A.5.

Multilevel modeling is not restricted to repeated measures. Rather, this modeling paradigm applies equally well to hierarchical survey data, hierarchical multivariate outcomes, time series, and so forth. Apart from random effects at the various levels, correlated errors at level 1 (i.e., serial correlation) are allowed and can be fitted using a macro provided within MLwiN. Apart from linear models, nonlinear models are allowed. Further, the methodology also applies to categorical and time-to-event outcomes.

A.3   Fitting Mixed Models Using SPlus

SPlus provides various ways to estimate mixed models. On the one hand, the built-in functions lme() for linear mixed-effects models can be used. Note that there is a companion function for nonlinear mixed-effects models, nmle(). These functions are based on work by Lindstrom and Bates (1988), Laird and Ware (1982), Box, Jenkins, and Reinsel (1994), and Davidian and Giltinan (1995). The lme() function will be discussed in Section A.3.1. A third-party suite of SPlus functions, termed OSWALD (Smith, Robertson, and Diggle 1996) has been developed to fit longitudinal models. The package is based on the methods described in Diggle, Liang, and Zeger (1994). In particular, the pcmid() function fits linear mixed models. Many of the functionalities between lme() and pcmid() are shared, but an attractive feature specific to pcmid() is its capability to jointly estimate measurement and dropout models, based on the model of Diggle and Kenward (1994). This selection model allows for MCAR, MAR, as well as MNAR dropout. It will be studied in Section A.3.2.

A.3.1 Standard SPlus Functions

As in Section A.2 on multilevel modeling and MLwiN, we will use the growth data (Section 2.6) to illustrate the built-in function mle(). SPlus Version 4.5 is used. Apart from the references mentioned earlier which give the theoretical underpinning, there is ample documentation within SPlus. The on-line manual provides a 53-page discussion of linear and nonlinear mixed-effects models. The function lme() is generic. The on-line help system of SPlus provides a brief account of the syntax of this generic function. Methods functions are being developed for specific classes of objects. The methods function lme.formula() comes with ample documentation.

Let us discuss the main arguments:

Fixed effects. The structure is specified by means of the fixed argument, using standard formulas.

Random effects. The random-effects structure is specified through random. Additional arguments to tune the random-effects model are re.block (describing the blocking structure), re.structure (specifying the form of the D matrix), and re.paramtr (specifying how the D matrix is *internally* parameterized). The latter argument is included to improve numerical stability and to ensure that the resulting D matrix is positive define. Values of this argument refer to the Cholesky decomposition, the matrix logarithm, and several others.

Serial correlation. This structure is defined by means of the argument serial.structure. In the case that a serial correlation structure depending on time is assumed, the arguments serial.covariate and serial.covariate.transformation can be used to specify this aspect of the serial process.

Residual variance. The residual variance function is defined by means of var.function. Fine-tuning can be done using var.covariate and var.estimate (indicating whether the variance parameters are to be estimated or to be kept fixed at their initial values).

Clusters. The clusters (subjects, units, etc.) are defined using cluster.

Method of estimation. Both maximum likelihood and REML are provided. The user's preference can be specified by means of the argument est.method.

Other tools include subsetting, specifying the action to be undertaken on missing data, and control over the estimation algorithm.

Let us apply the function lme.formula() to fit Model 6 to the growth data, as was done using MLwiN in Section A.2. The following program can be used.

```
my.lme <- lme.formula(
fixed = MEASURE ~ 1 + MALE + MALEAGE + FEMAGE,
random =  ~ 1 + AGE,
cluster = ~ IDNR,
data = growth5.df,
re.structure = "unstructured",
na.action = "na.omit",
est.method = "ML")
```

Printing the object my.lme produces

```
Call:
  Fixed: MEASURE ~ 1 + MALE + MALEAGE + FEMAGE
 Random:  ~ 1 + AGE
Cluster:  ~ (IDNR)
   Data: growth5.df

Variance/Covariance Components Estimate(s):

  Structure: unstructured
  Parametrization: matrixlog
  Standard Deviation(s) of Random Effect(s)
(Intercept)       AGE
   2.134752 0.1541473
Correlation of Random Effects
   (Intercept)
AGE -0.6025632

Cluster Residual Variance: 1.716206

Fixed Effects Estimate(s):
  (Intercept)    MALE MALEAGE    FEMAGE
      17.37273 -1.032102 0.784375 0.4795455
```

```
Number of Observations: 108
Number of Clusters: 27
```

Although the above output is rather brief, one can obtain a more extensive summary:

```
> my.lme.2 <- summary(my.lme)
> my.lme.2

Call:
  Fixed: MEASURE ~ 1 + MALE + MALEAGE + FEMAGE
  Random:  ~ 1 + AGE
Cluster:  ~ (IDNR)
    Data: growth5.df

Estimation Method: ML
Convergence at iteration: 6
Log-likelihood: -213.903
         AIC:  443.806
         BIC:  465.263

Variance/Covariance Components Estimate(s):
  Structure: unstructured
  Parametrization: matrixlog
  Standard Deviation(s) of Random Effect(s)
  (Intercept)       AGE
      2.134752 0.1541473
  Correlation of Random Effects
       (Intercept)
  AGE -0.6025632

Cluster Residual Variance: 1.716206

Fixed Effects Estimate(s):
                  Value Approx. Std.Error  z ratio(C)
  (Intercept) 17.3727273        1.18203467 14.6973077
         MALE -1.0321023        1.53550808 -0.6721568
      MALEAGE  0.7843750        0.08275405  9.4783886
       FEMAGE  0.4795455        0.09980513  4.8048175
```

```
Conditional Correlation(s) of Fixed Effects Estimates
           (Intercept)          MALE          MALEAGE
    MALE -7.698004e-001
 MALEAGE  6.198039e-016 -5.617972e-001
  FEMAGE -8.801671e-001  6.775530e-001 -1.691642e-016


Random Effects (Conditional Modes):
    (Intercept)            AGE
  1 -0.68278894 -0.039972872
  2 -0.45926352  0.071886460
  3 -0.03109489  0.093020178
  4  1.61182535  0.030832363
  5  0.43850471 -0.043000835
 . . . . .
 25  0.50935427 -0.055453935
 26 -0.10573027  0.083999487
 27 -0.89462307 -0.076992100



Standardized Population-Average Residuals:
        Min         Q1        Med        Q3        Max
 -3.335979 -0.4153858 0.01039114 0.4916851 3.858188

Number of Observations: 108
Number of Clusters: 27
```

The estimates and standard errors coincide with those obtained with ML-wiN (Figure A.2). This is immediately clear for the fixed-effects estimates, their standard errors, and the residual variance. The components of the D matrix have to be derived from the standard deviations and correlation of the random effects:

$$
\begin{aligned}
d_{11} &= 2.134752^2 = 4.577, \\
d_{12} &= (-0.6025632)(2.134752)(0.1541473) = -0.198, \\
d_{22} &= 0.1541473^2 = 0.024.
\end{aligned}
$$

As is the case with MLwiN, SPlus in general, and lme() in particular, have extensive graphical capabilities.

A.3.2   OSWALD for Nonrandom Nonresponse

Even though a word of care was issued in Chapters 17–20 about nonrandom dropout models restricting the model-building exercise simply to MAR mechanisms is equally dangerous, since the MAR assumption is in itself fundamentally untestable. Often, the choice to restrict model building to MAR is driven by the lack of software for fitting more general models. Many of the software packages do not allow, at present, the exploration of nonrandom dropout mechanisms.

In this section, we will illustrate how a family of nonrandom dropout models can be fitted with the OSWALD (Smith, Robertson, and Diggle 1996) software in SPlus. Several data sets have been fitted using OSWALD (the toenail data in Section 17.2 and the Vorozole study in Section 17.6). We will study another, relatively simple example in this section.

In a heart failure study, the primary efficacy endpoint is based upon the ability to do physical exercise. This ability is measured in the number of seconds a subject is able to ride the exercise bike. There are 25 subjects assigned to placebo and 25 to treatment. The treatment consisted of the administration of ACE inhibitors. Four measurements were taken at monthly intervals. Table A.2 presents outcome scores, transformed to normality. We will refer to them as the exercise bike data.   All 50 subjects are observed at the first occasion, whereas there are 44, 41, and 38 subjects seen at the second, the third, and the fourth visits, respectively.

To be able to perform a comparison between PROC MIXED and OSWALD, we will restrict attention to a set of models that can be fitted reasonably well with both packages (Diggle 1988). The measurement models belong to the general class (3.8). We consider it useful to decompose the variability explicitly in three components:

$$\text{var}(\boldsymbol{Y}_i) \;=\; Z_i D Z_i' + \tilde{H}_i + \tau^2 I_i.$$

The notation used here is chosen to reflect the OSWALD output and hence deviates slightly from earlier conventions. Recall that $\tilde{H}_i$ represents the serial correlation and τ^2 is the measurement error. For the remainder of this section, we will restrict the random effects to a random intercept, and to balanced designs with measurement occasions common to all subjects. As a result, the variance can be selected to have the following, simplified form:

$$\text{var}(\boldsymbol{Y}_i) \;=\; \nu^2 J + \tilde{H} + \tau^2 I.$$

Finally, writing $\tilde{H}$ in terms of its correlation matrix yields

$$\text{var}(\boldsymbol{Y}_i) \;=\; \nu^2 J + \sigma^2 H + \tau^2 I. \tag{A.10}$$

The symbols ν^2, σ^2, and τ^2 are chosen to reflect the names used in OSWALD.

The first model we consider is the random intercept Model 7 used for the growth data (see Section 17.4.1, p. 253). It is given by omitting the H

TABLE A.2. *Exercise Bike Data.*

Placebo				Treatment			
1	2	3	4	1	2	3	4
0.43	0.94	4.32	4.51	−2.54	−0.20	−0.15	3.53
3.10	5.82	5.59	6.32	4.33	5.57	6.86	6.87
0.56	2.21	1.18	1.54	−2.46	.	.	.
−1.18	−0.30	2.48	2.67	2.30	4.64	7.37	7.99
1.24	2.83	1.98	3.21	0.73	3.29	5.23	6.12
−1.87	−0.06	1.16	1.84	0.38	1.25	2.91	4.71
−0.28	1.30	.	.	1.51	4.00	5.98	.
2.93	.	.	.	0.38	0.94	3.28	4.05
−0.20	3.34	3.71	3.69	0.42	2.53	.	.
−0.12	2.01	2.35	2.70	2.41	4.24	4.79	8.14
−1.60	1.42	0.41	0.72	0.12	1.48	3.12	3.69
0.64	.	.	.	−3.46	−0.93	2.78	3.02
−1.14	−1.20	0.09	2.39	−0.55	.	.	.
2.24	2.12	3.00	1.52	0.74	2.40	4.04	5.61
−0.44	0.88	2.83	1.47	2.37	2.79	4.05	5.91
0.39	1.77	3.62	4.35	1.94	5.05	3.06	5.89
−4.37	−2.43	−0.43	−0.13	0.77	2.46	.	.
0.20	2.05	3.18	5.13	1.32	.	.	.
1.31	3.82	2.70	3.59	2.15	4.84	7.70	8.29
−0.38	−1.92	−0.12	−0.40	−0.09	2.02	4.68	5.29
−0.78	.	.	.	2.10	4.91	7.48	8.91
−0.48	0.32	0.66	3.03	1.36	0.62	1.87	.
−0.64	1.53	1.29	.	3.14	5.79	5.95	7.50
0.88	2.10	1.90	3.51	−0.94	−0.08	3.57	3.80
2.02	3.10	4.93	4.76	0.89	1.51	3.14	5.96

term from (A.10). The next two models are new and, to avoid confusion, they will be assigned numbers 9 and 10. The second model (Model 9) supplements a random intercept with serial correlation. We will choose the serial correlation to be of the AR(1) type. This model is found by omitting the measurement error component $\tau^2 I$ from (A.10) and choosing the elements of H to be $h_{jk} = \rho^{|j-k|} = \exp\left(-\phi|j-k|\right)$. Model 10 combines all three sources of variability. SAS code for these models is

```
proc mixed data = m.bike method = ml covtest;
title 'Exercise Bike, Dropout, Model 7';
class group id;
model y = group time*group / s;
repeated / type = cs subject = id r rcorr;
run;

proc mixed data = m.bike method = ml covtest;
title 'Exercise Bike, Dropout, Model 9';
class group id;
model y = group time*group / s;
repeated / type = ar(1) subject = id r rcorr;
random intercept / type = un subject = id g;
run;

proc mixed data = m.bike method = ml covtest;
title 'Exercise Bike, Dropout, Model 10';
class group id;
model y = group time*group / s;
repeated / type = ar(1) local subject = id r rcorr;
random intercept / type = un subject = id g;
run;
```

The mean and covariance model parameters are summarized in Table A.3. The parameters supplied by PROC MIXED are supplemented with some additional quantities in order to obtain both sets of intercepts and slopes for the two treatment groups.   Thus, intercept 0 is the sum of intercept 1 and the group 0 effect. Further, ϕ and ρ are connected by $\phi = -\ln(\rho)$. Although the mean models are straightforward to interpret from the SAS output, it is necessary to approach the covariance parameters output with some care.

As with the growth data, the random intercept Model 7 is introduced via the 'type=CS' option in the REPEATED statement. Alternatively, we could have chosen to use a random intercept, using the RANDOM statement. The fitted covariance matrix is

$$\tau^2 I + \nu^2 J \;=\; \begin{pmatrix} 3.1403 & 2.3781 & 2.3781 & 2.3781 \\ 2.3781 & 3.1403 & 2.3781 & 2.3781 \\ 2.3781 & 2.3781 & 3.1403 & 2.3781 \\ 2.3781 & 2.3781 & 2.3781 & 3.1403 \end{pmatrix}.$$

The SAS summary of the covariance parameters is

TABLE A.3. *Exercise Bike Data. SAS output on Models 7, 9, and 10.*

Parameter	Interpretation	Model 7	Model 9	Model 10
Intercept	Intercept 1	−0.8082	−0.8218	−0.8236
Group 0		0.1915	0.1608	0.1606
	Intercept 0	−0.6167	−0.6610	−0.6630
Time*group 0	Slope 0	0.9200	0.9236	0.9230
Time*group 1	Slope 1	1.6434	1.6449	1.6451
ν^2	Random interc.	2.3781	2.1585	2.0938
τ^2	Meas. error	0.7622		0.2097
σ^2	Serial variance		0.9484	0.8012
ρ	Serial corr.		0.3080	0.4440
ϕ	Serial corr. (exp.)		1.1778	0.8119
Deviance		564.27	559.69	559.68

```
Covariance Parameter Estimates (MLE)

Cov Parm    Subject      Estimate

CS          ID           2.37805794
Residual                 0.76223505
```

from which it follows that $\nu^2 = 2.3781$ and $\tau^2 = 0.7622 = 3.1403 - 2.3781$.

Model 9 replaces the independent measurement errors with serially correlated errors. This requires the use of the RANDOM and REPEATED statements simultaneously. The 'r' matrix produced by SAS is now interpreted as

$$\sigma^2 H = \begin{pmatrix} 0.9484 & 0.2921 & 0.0899 & 0.0277 \\ 0.2921 & 0.9484 & 0.2921 & 0.0899 \\ 0.0899 & 0.2921 & 0.9484 & 0.2921 \\ 0.0277 & 0.0899 & 0.2921 & 0.9484 \end{pmatrix}.$$

Observe that this 'r' matrix has a completely different interpretation in these two models, since they refer to different sources of variability. The corresponding correlation matrix is

$$H = \begin{pmatrix} 1.0000 & 0.3080 & 0.0948 & 0.0292 \\ 0.3080 & 1.0000 & 0.3080 & 0.0948 \\ 0.0948 & 0.3080 & 1.0000 & 0.3080 \\ 0.0292 & 0.0948 & 0.3080 & 1.0000 \end{pmatrix}$$

from which we deduce that $\rho = 0.3080$. In addition, the output labeled 'G Matrix' corresponds to the variance of the intercept, $\tau^2 = 2.1585$. The covariance parameters are given in

Covariance Parameter Estimates (MLE)

Cov Parm	Subject	Estimate
UN(1,1)	ID	2.15845062
AR(1)	ID	0.30795893
Residual		0.94837375

of which all components have been discussed. It is slightly misleading that $\sigma^2 = 0.9484$ is labeled "residual" since it does not refer to measurement error, but rather to the variance of the serially correlated process. This point will be clearer from Model 10. Indeed, this model combines all three components of variability. The full covariance structure output is

R Matrix for ID 1

Row	COL1	COL2	COL3	COL4
1	1.01091685	0.35573508	0.15795174	0.07013296
2	0.35573508	1.01091685	0.35573508	0.15795174
3	0.15795174	0.35573508	1.01091685	0.35573508
4	0.07013296	0.15795174	0.35573508	1.01091685

R Correlation Matrix for ID 1

Row	COL1	COL2	COL3	COL4
1	1.00000000	0.35189351	0.15624603	0.06937559
2	0.35189351	1.00000000	0.35189351	0.15624603
3	0.15624603	0.35189351	1.00000000	0.35189351
4	0.06937559	0.15624603	0.35189351	1.00000000

G Matrix

Parameter	Subject	Row	COL1
INTERCEPT	ID 1	1	2.09382622

```
           Covariance Parameter Estimates (MLE)

           Cov Parm     Subject       Estimate

           UN(1,1)      ID          2.09382622
           Variance     ID          0.80117791
           AR(1)        ID          0.44401509
           Residual                 0.20973894
```

This output is easier to understand since the parameters are nicely grouped by source of variability: (1) random intercept, with $\nu^2 = 2.0938$, (2) serial correlation, with $\sigma^2 = 0.8012$ and $\rho = 0.4440$, and (3) measurement error (residual), with $\tau^2 = 0.2097$.

Formal inspection of the deviances shows that Model 7 is too simple and that Model 10 does not improve Model 9 significantly.

Our next goal is to fit the same three models with OSWALD. For a full documentation on OSWALD we refer to Smith, Robertson, and Diggle (1996) or to the web page:

http://www.maths.lancs.ac.uk:2080/~maa036/OSWALD/

The output is presented in the form of a typical SPlus list object. For Model 7,

```
Longitudinal Data Analysis Model
assuming completely random dropout

Call:
pcmid(formula = bike.bal ~ group * time,
      vparms = c(0, 1.5, 0), correxp = 1)

Analysis Method: Maximum Likelihood (ML)
Correlation structure: exp(- phi * |u| ^ 1 )

Maximised likelihood:
[1] -482.4208

Mean Parameters:
            (Intercept)        group        time group:time
PARAMETER    -0.6166502   -0.1915053  0.92002700   0.7233457
STD.ERROR     0.3792614    0.5374845  0.08554001   0.1232477
```

```
Variance Parameters:
nu.sq sigma.sq    tau.sq phi
    0 2.378172 0.7622228    0
```

Since we want to include only a random intercept and measurement error as components of variability, we would like to omit σ^2 from the model. One way to do this is by specifying initial values for the covariance parameters using the VPARMS argument to the PCMID function. The argument of VPARMS is a vector with three components, containing initial values for ν^2, τ^2, and ϕ, respectively. Setting one or more of the initial values equal to 0 prevents these parameters from maximization. However, σ^2 is not included as a component of the initial values vector. One way to circumvent this problem is to set $\phi \equiv 0$. This implies that the serial correlation matrix H reduces to a matrix of ones, $H = J$, whence it takes on the same role as the random intercept component. As a result, ν^2 can be omitted from the model. Thus, the component we are actually interested in needs to be excluded ! Model fit for all three models is summarized in Table A.4. Comparison with Table A.3 shows that the fits are virtually identical. The only substantial difference is seen in the deviances. Adding 400.57 to the deviances in Table A.3 yields the deviances in Table A.4. Presumably, OSWALD uses a slightly different objective function.

Model 9 omits the measurement error, which is simply done by setting the initial value for $\tau^2 \equiv 0$:

```
Longitudinal Data Analysis Model
assuming completely random dropout

Call:
pcmid(formula = bike.bal ~ group * time,
        vparms = c(1, 0, 1), correxp = 1)

Analysis Method: Maximum Likelihood (ML)
Correlation structure: exp(- phi * |u| ^ 1 )

Maximised likelihood:
[1] -480.13

Mean Parameters:
          (Intercept)       group       time group:time
PARAMETER  -0.6609958 -0.1607851 0.92363457  0.7212975
STD.ERROR   0.3923346  0.5559721 0.09760774  0.1403651
```

TABLE A.4. *Exercise Bike Data. SPlus (OSWALD) output on Models 7, 9, and 10.*

Parameter	Interpretation	Model 7	Model 9	Model 10
Intercept	Intercept 0	−0.6167	−0.6610	−0.6630
Group		−0.1915	−0.1608	−0.1606
	Intercept 1	−0.8082	−0.8218	−0.8236
Time	Slope 0	0.9200	0.9236	0.9230
Group:time		0.7233	0.7213	0.7221
	Slope 1	1.6434	1.6449	1.6451
ν^2	Random interc.		2.1584	2.0944
τ^2	Measurem. error	0.7622		0.2087
σ^2	Serial variance	2.3782	0.9484	0.8017
ρ	Serial corr.		0.3080	0.4431
ϕ	Expon. param.		1.1778	0.8139
Deviance		964.84	960.26	960.24

```
Variance Parameters:
   nu.sq  sigma.sq tau.sq      phi
2.158405 0.9483919      0 1.177751
```

Finally, the unrestricted Model 10 produces the following output:

```
Longitudinal Data Analysis Model
assuming completely random dropout

Call:
pcmid(formula = bike.bal ~ group * time,
      vparms = c(2, 0.2, 1), correxp = 1,
      reqmin = 1e-012)

Analysis Method: Maximum Likelihood (ML)
Correlation structure: exp(- phi * |u| ^ 1 )

Maximised likelihood:
[1] -480.1222

Mean Parameters:
           (Intercept)       group      time group:time
PARAMETER  -0.6629645 -0.1606256 0.92304454  0.7221020
STD.ERROR   0.3932421  0.5572348 0.09842758  0.1415133
```

```
Variance Parameters:
    nu.sq  sigma.sq    tau.sq        phi
   2.094417 0.8016666 0.2087339 0.8139487
```
.

In all of the above models, CORREXP=1 ensures that an AR(1) serial correlation structure is used. OSWALD accepts any value between 1 and 2 as arguments to CORREXP (the latter corresponds to Gaussian decay). The model formula is equivalent to a SAS model:

```
model y = group time group*time
```

and follows standard model formulating conventions of SPlus. The object BIKE.BAL is a member of the BALANCED class, designed within OS-WALD for collections of time series with a common set of measurement times. For unbalanced designs, a different class (LDA.MAT) is conceived.

All models considered in this section have a simple random-effects structure. Indeed, Models 7, 9, and 10 include only a random intercept. Clearly, more elaborate random-effects models can be fitted with PROC MIXED using the RANDOM statement. In OSWALD Version 2.6, the function PCMID does not allow more complex structures. The predecessor of the function PCMID (the function REML.FIT) allows the user to specify several random effects, but constrains its variance-covariance matrix D to be diagonal.

All analyses done on the exercise bike data so far assumed ignorable nonresponse, in the spirit of Section 17.3. This means that they are valid under MAR (and not only under MCAR, in spite of the claim printed in the OSWALD output). Of great potential value is the feature of OSWALD to be able to go beyond an ignorable analysis and to fit a specific class of nonrandom models. We will exemplify this power using the exercise bike data. Illustration will be on the basis of the most general Model 10, even though the slightly simpler Model 9 fits the data equally well. The analysis featured by OSWALD couples a linear mixed-effects model for the measurements with a logistic model for dropout, with predictors given by the current outcome as well as a set of previous responses. Details of the model are to be found in Diggle and Kenward (1994). For instance, assuming that the dropout probability at occasion j depends on both the current outcome Y_{ij} and the previous one $Y_{i,j-1}$, leads to the following model:

$$\ln\left(\frac{P(R_{ij} = 0|\boldsymbol{y}_i)}{1 - P(R_{ij} = 0|\boldsymbol{y}_i)}\right) = \psi_0 + \psi_1 y_{ij} + \psi_2 y_{i,j-1}. \quad (A.11)$$

Such an analysis is done in OSWALD through the DROP.PARMS, DROP-MODEL, and DROP.COV.PARMS arguments of the PCMID function.

Explicitly, DROP.PARMS specifies starting values for a number of time points, starting from the current one, to be included in the dropout model. In model (A.11), this number is 2 (ψ_1 and ψ_2). A starting value for the intercept ψ_0 is given by means of the DROP.COV.PARMS statement. A very important feature of this argument is that it can be used to include *covariate* effects as well. In that case, model (A.11) is extended to

$$\ln\left(\frac{P(R_{ij} = 0|\boldsymbol{y}_i)}{1 - P(R_{ij} = 0|\boldsymbol{y}_i)}\right) = \psi_0 + \boldsymbol{x}_i\boldsymbol{\psi}_c + \psi_1 y_{ij} + \psi_2 y_{i,j-1}, \text{(A.12)}$$

where $\boldsymbol{x}_i$ is a vector of covariates and $\boldsymbol{\psi}_c$ is an additional vector of parameters. The actual form of the dropout model is specified in the DROP-MODEL argument, using standard SPlus model building conventions. An illustration will be given in the sequel. As before, setting one or more of the initial values equal to zero prevents their inclusion in the maximization process. This is a useful feature, since it allows the user to estimate the dropout parameters under MCAR and MAR assumptions, not only under nonrandom assumptions. For example, model (A.11) corresponds to an MAR process by setting the initial value for $\psi_1 \equiv 0$.

Let us discuss OSWALD analyses for dropout model (A.11), in the MCAR, MAR, and nonrandom contexts. The MCAR analysis output is as follows:

```
Longitudinal Data Analysis Model
assuming random dropout based on 0 previous observations

Call:
pcmid(formula = bike.bal ~ group * time,
      vparms = c(2, 0.2, 1), drop.parms = c(0),
      drop.cov.parms = c(-2), dropmodel =  ~ 1,
      correxp = 1, maxfn = 1600, reqmin = 1e-012)

Analysis Method: Maximum Likelihood (ML)
Correlation structure: exp(- phi * |u| ^ 1 )

Maximised likelihood:
[1] -520.6167

Mean Parameters:
            (Intercept)       group       time group:time
PARAMETER    -0.6629601 -0.1606322 0.9230446   0.7221012
STD.ERROR     0.3859817  0.5470223 0.1009714   0.1451370
```

```
Variance Parameters:
    nu.sq  sigma.sq    tau.sq        phi
 2.094412 0.8016692 0.2087333 0.8139474

Dropout parameters:
 (Intercept) y.d
    -2.32728    0


Iteration converged after 993 iterations.
```

The indication that the dropout model is random with a dependence on 0 previous observations effectively refers to a MCAR process. The mechanism is turned into MCAR by specifying DROP.PARMS=c(0) and by setting DROP.COV.PARMS=c(-2). We increased the maximum number of iterations MAXFN to 1600, since more complex dropout models tend to require more iterations. As is seen from the output, a bit under 1000 iterations were actually needed. In addition, the tolerance of the relative gradient was set equal to 10^{-12} by means of the REQMIN argument.

The MAR program and output are similar:

```
Longitudinal Data Analysis Model
assuming random dropout based on 1 previous observations

Call:
pcmid(formula = bike.bal ~ group * time,
      vparms = c(2, 0.2, 1), drop.parms = c(0, -0.1),
      drop.cov.parms = c(-2), dropmodel =  ~ 1,
      correxp = 1, maxfn = 3000, reqmin = 1e-012)

Analysis Method: Maximum Likelihood (ML)
Correlation structure: exp(- phi * |u| ^ 1 )

Maximised likelihood:
[1] -520.3494

Mean Parameters:
            (Intercept)       group       time group:time
PARAMETER    -0.6629684  -0.1606169  0.9230426   0.7221043
STD.ERROR     0.3859815   0.5470220  0.1009713   0.1451369
```

TABLE A.5. *Exercise Bike Data. Nonrandom models. Model 10.*

Parameter	Ign.	Dropout modeled		
		MCAR	MAR	MNAR
Intercept	−0.6630	−0.6630	−0.6630	0.6666
Group	−0.1606	−0.1606	−0.1606	0.1740
Time	0.9230	0.9230	0.9230	0.9377
Group:time	0.7221	0.7221	0.7221	0.7317
ν^2	2.0944	2.0944	2.0944	2.1067
σ^2	0.8017	0.8017	0.8017	0.8349
τ^2	0.2087	0.2087	0.2087	0.1577
ϕ	0.8139	0.8139	0.8139	0.9311
ψ_0		−2.3273	−2.1661	−2.7316
ψ_1				0.3280
ψ_2			−0.0992	−0.3869
Deviance	960.24	1041.23	1040.70	1040.46

```
Variance Parameters:
   nu.sq   sigma.sq    tau.sq        phi
 2.09441 0.8016771 0.2087185 0.8139738

Dropout parameters:
 (Intercept) y.d        y.d-1
   -2.166139    0 -0.09920587

Iteration converged after 1203 iterations.
```

Note that the number of iterations has increased somewhat, even though the extra dropout parameter, ψ_2, appears to be very small. In fact, the likelihood has increased only marginally over the MCAR analysis.

Finally, we allow for nonrandom dropout.

```
Longitudinal Data Analysis Model
assuming informative dropout
based on 1 previous observations

Call:
pcmid(formula = bike.bal ~ group * time,
      vparms = c(2, 0.4, 0.5), drop.parms = c(2, -2),
      drop.cov.parms = c(-3), dropmodel =  ~ 1,
```

```
        correxp = 1, maxfn = 10000, reqmin = 1e-012)
```

Analysis Method: Maximum Likelihood (ML)
Correlation structure: exp(- phi * |u| ^ 1)

Maximised likelihood:
[1] -520.2316

Mean Parameters:
            (Intercept)   group   time group:time
PARAMETER   -0.6666244 -0.1740098 0.9377247   0.7317099
STD.ERROR   NA   NA   NA   NA

Variance Parameters:
     nu.sq sigma.sq   tau.sq   phi
  2.106741 0.8348588 0.1577192 0.9310764

Dropout parameters:
  (Intercept)   y.d   y.d-1
    -2.731573 0.3279598 -0.3869054

Iteration converged after 5034 iterations.

The number of iterations has increased considerably, which is a typical feature of nonrandom dropout models. The likelihood has changed only marginally, and the dropout parameters are all somewhat larger, even though this is not a precise statement due to the lack of precision estimates. Note also that no standard errors for the mean parameters are provided, in contrast to the other PCMID analyses. This is a very sensible decision since, unlike with ignorable analyses, standard errors are not obtained as simple by-products of the maximization process and require in general considerable extra code. A few methods are listed on p. 376 and on p. 389.

Results for the three dropout models are summarized in Table A.5, under the headings MCAR, MAR, and MNAR, respectively. Also included is the earlier ignorable analysis. Since MCAR and MAR are ignorable, whether or not the dropout model parameters are estimated explicitly, the first three models in Table A.5 yield exactly the same values for the mean and covariance parameter estimates, as they should. Note that the deviance from the ignorable model is not comparable to the other deviances, since the dropout parameters are not estimated. Comparing the parameters in these models to the nonrandom dropout models shows some shifts, although they are very modest.

TABLE A.6. *Exercise Bike Data. Non-Random models. Model 10. Treatment assignment (group) included into the dropout model.*

Parameter	MAR	MNAR
intercept	-0.6630	-0.6214
group	-0.1606	-0.2288
time	0.9231	0.9032
group:time	0.7221	0.7252
ν^2	2.0944	2.0438
σ^2	0.8017	0.7859
τ^2	0.2087	0.2931
ϕ	0.8139	0.6289
ψ_0	-2.4059	-2.1036
ψ_1		0.3036
ψ_2	-0.1289	0.1238
group	0.5395	0.7979
Deviance	1039.96	1039.83

We may now want to compare the dropout models. The likelihood ratio test statistic to compare MAR with MCAR is 0.53 on 1 degree of freedom ($p = 0.4666$). This means that MCAR would be acceptable *provided MAR were the correct alternative hypothesis* and the actual parametric form for the MAR process were correct. In addition, a comparison between the nonrandom and random dropout models yields a likelihood ratio test statistic of 0.24 ($p = 0.6242$). Of course, for reasons outlined in Chapters 19 and 20 (see also Section 18.1.2), one should use nonrandom dropout models with caution, since they rely on assumptions that are at best only partially verifiable. These issues of sensitivity are illustrated in Sections 19.4, 19.5, and 24.4.

To conclude, let us illustrate the capability of OSWALD to incorporate covariates into the dropout model, as in model (A.12). Including the treatment assignment (group) into the MAR and nonrandom models of Table A.5 yields

```
Longitudinal Data Analysis Model
assuming random dropout based on 1 previous observations

Call:
pcmid(formula = demo2.bal ~ group * time,
        vparms = c(2, 0.2, 0.8), drop.parms = c(0, -0.1),
```

```
      drop.cov.parms = c(-2, 0.1),
      dropmodel =  ~ 1 + group,
      correxp = 1, maxfn = 5000, reqmin = 1e-012)
```

Analysis Method: Maximum Likelihood (ML)
Correlation structure: exp(- phi * |u| ^ 1)

Maximised likelihood:
[1] -519.9814

Mean Parameters:
            (Intercept)   group   time group:time
PARAMETER   -0.6629779 -0.1606298 0.9230512 0.7220979
STD.ERROR   0.3859810 0.5470213 0.1009712 0.1451367

Variance Parameters:
    nu.sq sigma.sq   tau.sq   phi
 2.094418 0.8016748 0.2087164 0.8139889

Dropout parameters:
 (Intercept)   group y.d   y.d-1
   -2.405863 0.5395115   0 -0.1289393

Iteration converged after 2983 iterations.

and

Longitudinal Data Analysis Model
assuming informative dropout
based on 1 previous observations

Call:
```
pcmid(formula = demo2.bal ~ group * time,
      vparms = c(2, 0.2, 0.8),
      drop.parms = c(0.3, -0.3),
      drop.cov.parms = c(-2, 0.1),
      dropmodel =  ~ 1 + group,
      correxp = 1, maxfn = 10000, reqmin = 1e-012)
```

Analysis Method: Maximum Likelihood (ML)
Correlation structure: exp(- phi * |u| ^ 1)

Maximised likelihood:
[1] -519.9143

```
Mean Parameters:
          (Intercept)      group      time group:time
PARAMETER  -0.6213682 -0.228779 0.9032322  0.7251818
STD.ERROR          NA        NA        NA         NA

Variance Parameters:
    nu.sq  sigma.sq    tau.sq       phi
 2.043767 0.7858604 0.2930629 0.6288833

Dropout parameters:
 (Intercept)       group       y.d     y.d-1
   -2.103619 0.7979308 -0.303576 0.1238007

Iteration converged after 5930 iterations.
```

The model fit is summarized in Table A.6.

Again, the main effect and variance parameters in the MAR column have not changed relative to their ignorable counterparts in Table A.5. In contrast, the nonrandom dropout model parameters are all different, compared to the nonrandom model in Table A.5.

Appendix B

Technical Details for Sensitivity Analysis

This appendix contains the more technical material, given for completeness but not essential to understanding, about sensitivity analysis for selection models (Section B.1) and pattern-mixture models (Section B.2).

B.1 Local Influence: Derivation of Components of $\boldsymbol{\Delta}_i$

Let us provide some computational details that lead to expression (19.2)–(19.5). We will consider complete and incomplete sequences in turn.

The log-likelihood contribution for a complete sequence is

$$\ell_{i\omega} \;=\; \ln f(\boldsymbol{y}_i) + \sum_{j=2}^{n_i} \ln[1 - g(\boldsymbol{h}_{ij}, y_{ij})],$$

where the parameter dependencies are suppressed for notational ease. The mixed derivatives are particularly easy to calculate, immediately yielding expressions (19.2) and (19.3).

The log-likelihood contribution from an incomplete sequence equals

$$
\begin{aligned}
\ell_{i\omega} &= \ln \int f(\boldsymbol{y}_i) \prod_{j=2}^{d-1} [1 - g(\boldsymbol{h}_{ij}, y_{ij})] g(\boldsymbol{h}_{id}, y_{id}) dy_{id} \\
&= \ln f(\boldsymbol{h}_{id}) + \sum_{j=2}^{d-1} \ln[1 - g(\boldsymbol{h}_{ij}, y_{ij})] + \ln \int f(y_{id}|\boldsymbol{h}_{id}) g(\boldsymbol{h}_{id}, y_{id}) dy_{id},
\end{aligned}
$$

of which the first component depends on $\boldsymbol{\theta}$ only, the second one on $\boldsymbol{\psi}$ only, and the third one contains both.

The mixed derivatives of the log-likelihood w.r.t. ω_i can be written as

$$
\begin{aligned}
\frac{\partial^2 \ell_{i\omega}}{\partial \boldsymbol{\theta} \partial \omega_i} &= \frac{\int f(y_{id}|\boldsymbol{h}_{id}) g(\boldsymbol{h}_{id}, y_{id}) dy_{id} \int \frac{\partial f(y_{id}|\boldsymbol{h}_{id})}{\partial \boldsymbol{\theta}} \frac{\partial g(\boldsymbol{h}_{id}, y_{id})}{\partial \omega_i} dy_{id}}{\left[\int f(y_{id}|\boldsymbol{h}_{id}) g(\boldsymbol{h}_{id}, y_{id}) dy_{id} \right]^2} \\
&\quad - \int f(y_{id}|\boldsymbol{h}_{id}) \frac{\partial g(\boldsymbol{h}_{id}, y_{id})}{\partial \omega_i} dy_{id} \\
&\quad \times \frac{\int \frac{\partial f(y_{id}|\boldsymbol{h}_{id})}{\partial \boldsymbol{\theta}} g(\boldsymbol{h}_{id}, y_{id}) dy_{id}}{\left[\int f(y_{id}|\boldsymbol{h}_{id}) g(\boldsymbol{h}_{id}, y_{id}) dy_{id} \right]^2},
\end{aligned} \tag{B.1}
$$

$$
\begin{aligned}
\frac{\partial^2 \ell_{i\omega}}{\partial \boldsymbol{\psi} \partial \omega_i} &= - \sum_{j=2}^{d-1} h_{ij} y_{ij} g(\boldsymbol{h}_{ij}, y_{ij}) [1 - g(\boldsymbol{h}_{ij}, y_{ij})] \\
&\quad + \frac{\int f(y_{id}|\boldsymbol{h}_{id}) g(\boldsymbol{h}_{id}, y_{id}) dy_{id} \int f(y_{id}|\boldsymbol{h}_{id}) \frac{\partial^2 g(\boldsymbol{h}_{id}, y_{id})}{\partial \boldsymbol{\psi} \partial \omega_i} dy_{id}}{\left[\int f(y_{id}|\boldsymbol{h}_{id}) g(\boldsymbol{h}_{id}, y_{id}) dy_{id} \right]^2} \\
&\quad - \int f(y_{id}|\boldsymbol{h}_{id}) \frac{\partial g(\boldsymbol{h}_{id}, y_{id})}{\partial \omega_i} dy_{id} \\
&\quad \times \frac{\int f(y_{id}|\boldsymbol{h}_{id}) \frac{\partial g(\boldsymbol{h}_{id}, y_{id})}{\partial \boldsymbol{\psi}} dy_{id}}{\left[\int f(y_{id}|\boldsymbol{h}_{id}) g(\boldsymbol{h}_{id}, y_{id}) dy_{id} \right]^2}.
\end{aligned} \tag{B.2}
$$

In order to evaluate these expressions under $\omega_i = 0$, we set $\omega_i = 0$ in the integrands and calculate the resulting simplified integrals:

$$\int f(y_{id}|\boldsymbol{h}_{id})g(\boldsymbol{h}_{id}, y_{id})dy_{id}\Big|_{\omega_i=0} = \int f(y_{id}|\boldsymbol{h}_{id})g(\boldsymbol{h}_{id})dy_{id}$$

$$= g(\boldsymbol{h}_{id}), \qquad (\text{B.3})$$

$$\int f(y_{id}|\boldsymbol{h}_{id})\frac{\partial g(\boldsymbol{h}_{id}, y_{id})}{\partial \omega_i}dy_{id}\Big|_{\omega_i=0} = g(\boldsymbol{h}_{id})[1 - g(\boldsymbol{h}_{id})]$$

$$\times \int y_{id}f(y_{id}|\boldsymbol{h}_{id})dy_{id}$$

$$= g(\boldsymbol{h}_{id})[1 - g(\boldsymbol{h}_{id})]$$

$$\times \lambda(y_{id}|\boldsymbol{h}_{id}), \qquad (\text{B.4})$$

$$\int \frac{\partial f(y_{id}|\boldsymbol{h}_{id})}{\partial \boldsymbol{\theta}}g(\boldsymbol{h}_{id}, y_{id})dy_{id}\Big|_{\omega_i=0} = g(\boldsymbol{h}_{id}) \int \frac{\partial f(y_{id}|\boldsymbol{h}_{id})}{\partial \boldsymbol{\theta}}dy_{id}$$

$$= 0 \qquad (\text{B.5})$$

$$\int \frac{\partial f(y_{id}|\boldsymbol{h}_{id})}{\partial \boldsymbol{\theta}}\frac{\partial g(\boldsymbol{h}_{id}, y_{id})}{\partial \omega_i}dy_{id}\Big|_{\omega_i=0} = g(\boldsymbol{h}_{id})[1 - g(\boldsymbol{h}_{id})]$$

$$\times \int y_{id}\frac{\partial f(y_{id}|\boldsymbol{h}_{id})}{\partial \boldsymbol{\theta}}dy_{id}$$

$$= g(\boldsymbol{h}_{id})[1 - g(\boldsymbol{h}_{id})]$$

$$\times \frac{\partial \lambda(y_{id}|\boldsymbol{h}_{id})}{\partial \boldsymbol{\theta}} \qquad (\text{B.6})$$

$$\int f(y_{id}|\boldsymbol{h}_{id})\frac{\partial g(\boldsymbol{h}_{id}, y_{id})}{\partial \boldsymbol{\psi}}dy_{id}\Big|_{\omega_i=0} = g(\boldsymbol{h}_{id})[1 - g(\boldsymbol{h}_{id})]\boldsymbol{h}_{id} \quad (\text{B.7})$$

$$\int f(y_{id}|\boldsymbol{h}_{id})\frac{\partial^2 g(\boldsymbol{h}_{id}, y_{id})}{\partial \boldsymbol{\psi}\partial \omega_i}dy_{id}\Big|_{\omega_i=0} = g(\boldsymbol{h}_{id})[1 - g(\boldsymbol{h}_{id})]$$

$$\times [1 - 2g(\boldsymbol{h}_{id})]\boldsymbol{h}_{id}$$

$$\times \lambda(y_{id}|\boldsymbol{h}_{id}). \qquad (\text{B.8})$$

Combining (B.3)–(B.8) with (B.1)–(B.2) immediately yields expressions (19.4) and (19.5).

B.2 Proof of Theorem 20.1

The MAR assumption states that

$$f(d = t + 1 | y_1, \ldots, y_n) \quad = \quad f(d = t + 1 | y_1, \ldots, y_t) \qquad \text{(B.9)}$$

and the ACMV assumption that for all $t \geq 2, \forall j < t$,

$$f(y_t | y_1, \ldots, y_{t-1}, d = j + 1) \quad = \quad f(y_t | y_1, \ldots, y_{t-1}, d > t). \quad \text{(B.10)}$$

First, a lemma will be established.

Lemma B.1 *In a longitudinal setting with dropout, ACMV* $\Longleftrightarrow \forall t \geq 2, \forall j < t : f(y_t | y_1, \ldots, y_{t-1}, d = j + 1) = f(y_t | y_1, \ldots, y_{t-1}).$

Proof. Take $t \geq 2, j < t$, then ACMV leads to

$$f(y_t | y_1, \ldots, y_{t-1})$$

$$= \quad \sum_{i=1}^{t-1} f(y_t | y_1, \ldots, y_{t-1}, d = i + 1) f(d = i + 1)$$
$$+ f(y_t | y_1, \ldots, y_{t-1}, d > t) f(d > t)$$

$$= \quad \sum_{i=1}^{t-1} f(y_t | y_1, \ldots, y_{t-1}, d = j + 1) f(d = i + 1)$$
$$+ f(y_t | y_1, \ldots, y_{t-1}, d = j + 1) f(d > t)$$

$$= \quad f(y_t | y_1, \ldots, y_{t-1}, d = j + 1) \left[\sum_{i=1}^{t-1} f(d = i + 1) + f(d > t) \right]$$

$$= \quad f(y_t | y_1, \ldots, y_{t-1}, d = j + 1).$$

To show the reverse direction, take again $t \geq 2, j < t$:

$$f(y_t | y_1, \ldots, y_{t-1}, d > t) f(d > t)$$

$$= \quad f(y_t | y_1, \ldots, y_{t-1}) - \sum_{i=1}^{t-1} f(y_t | y_1, \ldots, y_{t-1}, d = i + 1) f(d = i + 1)$$

$$= \quad f(y_t | y_1, \ldots, y_{t-1}) - \sum_{i=1}^{t-1} f(y_t | y_1, \ldots, y_{t-1}) f(d = i + 1)$$

$$= \quad f(y_t | y_1, \ldots, y_{t-1}) \left[1 - \sum_{i=1}^{t-1} f(d = i + 1) \right]$$

$$\begin{aligned} &= f(y_t|y_1,\ldots,y_{t-1},d=j+1)\left[1-\sum_{i=1}^{t-1}f(d=i+1)\right] \\ &= f(y_t|y_1,\ldots,y_{t-1},d=j+1)f(d>t). \end{aligned}$$

This completes the proof. We are now able to prove Theorem 20.1.

MAR $\Rightarrow$ ACMV

Consider the ratio Q of the complete data likelihood to the observed data likelihood. This gives, under the MAR assumption,

$$\begin{aligned} Q &= \frac{f(y_1,\ldots,y_n)f(d=i+1|y_1,\ldots,y_i)}{f(y_1,\ldots,y_i)f(d=i+1|y_1,\ldots,y_i)} \\ &= f(y_{i+1},\ldots,y_n|y_1,\ldots,y_i). \end{aligned} \tag{B.11}$$

Further, one can always write,

$$\begin{aligned} Q &= f(y_{i+1},\ldots,y_n|y_1,\ldots,y_i,d=i+1) \\ &\quad\times \frac{f(y_1,\ldots,y_i|d=i+1)f(d=i+1)}{f(y_1,\ldots,y_i|d=i+1)f(d=i+1)} \\ &= f(y_{i+1},\ldots,y_n|y_1,\ldots,y_i,d=i+1). \end{aligned} \tag{B.12}$$

Equating expressions (B.11) and (B.12) for Q, we see that

$$f(y_{i+1},\ldots,y_n|y_1,\ldots,y_i,d=i+1)=f(y_{i+1},\ldots,y_n|y_1,\ldots,y_i). \tag{B.13}$$

To show that (B.13) implies the ACMV conditions (B.10), we will use the induction principle on t. First, consider the case $t=2$. Using (B.13) for $i=1$, and integrating over $y_3,\ldots,y_n$, we obtain

$$f(y_2|y_1,d=2) = f(y_2|y_1),$$

leading to, using Lemma B.1,

$$f(y_2|y_1,d=2) = f(y_2|y_1,d>2).$$

Suppose, by induction, ACMV holds for all $t\leq i$, We will now prove the hypothesis for $t=i+1$. Choose $j\leq i$. Then, from the induction hypothesis and Lemma B.1, it follows that for all $j<t\leq i$:

$$\begin{aligned} f(y_t|y_1,\ldots,y_{t-1},d=j+1) &= f(y_t|y_1,\ldots,y_{t-1},d>t) \\ &= f(y_t|y_1,\ldots,y_{t-1}). \end{aligned}$$

Taking the product over $t=j+1,\ldots,i$ then gives

$$f(y_{j+1},\ldots,y_i|y_1,\ldots,y_j,d=j+1)=f(y_{j+1},\ldots,y_i|y_1,\ldots,y_j). \tag{B.14}$$

After integration over $y_{i+2}, \ldots, y_n$, (B.13) leads to

$$
\begin{aligned}
& f(y_{j+1}, \ldots, y_{i+1} | y_1, \ldots, y_j, d = j + 1) \\
& = \quad f(y_{j+1}, \ldots, y_{i+1} | y_1, \ldots, y_j).
\end{aligned} \tag{B.15}
$$

Dividing (B.15) by (B.14) and equating the left- and right-hand sides, we find that

$$
f(y_{i+1} | y_1, \ldots, y_i, d = j + 1) \quad = \quad f(y_{i+1} | y_1, \ldots, y_i).
$$

This holds for all $j \le i$, and Lemma B.1 shows this is equivalent to ACMV.

ACMV $\Rightarrow$ MAR

Starting from the ACMV assumption and Lemma 1, we have

$$
\forall t \ge 2, \forall j < t : f(y_t | y_1, \ldots, y_{t-1}, d = j + 1) = f(y_t | y_1, \ldots, y_{t-1}). \tag{B.16}
$$

We now factorize the full data density as

$$
\begin{aligned}
& f(y_1, \ldots, y_n, d = i + 1) \\
& = \quad f(y_1, \ldots, y_i, d = i + 1) f(y_{i+1}, \ldots, y_n | y_1, \ldots, y_i, d = i + 1) \\
& = \quad f(y_1, \ldots, y_i, d = i + 1) \prod_{t=i+1}^{T} f(y_t | y_1, \ldots, y_{t-1}, d = i + 1).
\end{aligned}
$$

Using (B.16), it follows that

$$
\begin{aligned}
& f(y_1, \ldots, y_n, d = i + 1) \\
& = \quad f(y_1, \ldots, y_i | d = i + 1) f(d = i + 1) \prod_{t=i+1}^{T} f(y_t | y_1, \ldots, y_{t-1}) \\
& = \quad f(y_1, \ldots, y_i | d = i + 1) f(d = i + 1) f(y_{i+1}, \ldots, y_n | y_1, \ldots, y_i) \\
& = \quad \frac{f(y_1, \ldots, y_i | d = i + 1) f(d = i + 1)}{f(y_1, \ldots, y_i)} \\
& \qquad \times f(y_1, \ldots, y_i) f(y_{i+1}, \ldots, y_n | y_1, \ldots, y_i) \\
& = \quad \frac{f(y_1, \ldots, y_i | d = i + 1) f(d = i + 1)}{f(y_1, \ldots, y_i)} f(y_1, \ldots, y_n) \\
& = \quad f(d = i + 1 | y_1, \ldots, y_i) f(y_1, \ldots, y_n).
\end{aligned} \tag{B.17}
$$

An alternative factorization of $f(\boldsymbol{y}, d)$ gives

$$
f(y_1, \ldots, y_n, d = i + 1) = f(d = i + 1 | y_1, \ldots, y_n) f(y_1, \ldots, y_n). \tag{B.18}
$$

It follows from (B.17) and (B.18) that

$$f(d = i + 1 | y_1, \ldots, y_n) \quad = \quad f(d = i + 1 | y_1, \ldots, y_i),$$

completing the proof of Theorem 20.1.

References

Afifi, A. and Elashoff, R. (1966) Missing observations in multivariate statistics I: Review of the literature. *Journal of the American Statistical Association*, **61**, 595–604.

Agresti, A. (1990) *Categorical Data Analysis*. New York: John Wiley & Sons.

Aitkin, M. (1999) A general maximum likelihood analysis of variance components in generalized linear models. *Biometrics*, **55**, 218–234.

Aitkin, M. and Francis, B. (1995) Fitting overdispersed generalized linear models by nonparametric maximum likelihood. *The GLIM Newsletter*, **25**, 37–45.

Aitkin, M. and Rubin, D.B. (1985) Estimation and hypothesis testing in finite mixture models. *Journal of the Royal Statistical Society, Series B*, **47**, 67–75.

Akaike, H. (1974) A new look at the statistical model identification. *IEEE Transactions on Automatic Control*, **19**, 716–723.

Allison, P.D. (1987) Estimation of linear models with incomplete data. *Sociology Methodology*, 71–103.

Altham, P.M.E. (1984) Improving the precision of estimation by fitting a model. *Journal of the Royal Statistical Society, Series B*, **46**, 118–119.

Amemiya, T. (1984) Tobit models: a survey. *Journal of Econometrics*, **24**, 3–61.

Ashford, J.R. and Sowden, R.R. (1970) Multivariate probit analysis. *Biometrics*, **26**, 535–546.

Bahadur, R.R. (1961) A representation of the joint distribution of responses to n dichotomous items. In: *Studies in Item Analysis and Prediction,,* H. Solomon (Ed.). Stanford Mathematical Studies in the Social Sciences VI. Stanford, CA: Stanford University Press.

Baker, S.G. (1992) A simple method for computing the observed information matrix when using the EM algorithm with categorical data. *Journal of Computational and Graphical Statistics*, **1**, 63–76.

Baker, S.G. (1994) Regression analysis of grouped survival data with incomplete covariates: non-ignorable missing-data and censoring mechanisms. *Biometrics*, **50**, 821–826.

Baker, S.G. and Laird, N.M. (1988) Regression analysis for categorical variables with outcome subject to non-ignorable non-response. *Journal of the American Statistical Association*, **83**, 62–69.

Baker, S.G., Rosenberger, W.F., and DerSimonian, R. (1992) Closed-form estimates for missing counts in two-way contingency tables. *Statistics in Medicine*, **11**, 643–657.

Bartlett, M.S. (1937) Some examples of statistical methods of research in agriculture and applied botany. *Journal of the Royal Statistical Society, Series B*, **4**, 137–170.

Beckman, R.J., Nachtsheim, C.J., and Cook, R.D. (1987) Diagnostics for mixed-model analysis of variance. *Technometrics*, **29**, 413–426.

Bickel, P.J. and Doksum, K.A. (1977) *Mathematical Statistics*. Englewood Cliffs, NJ: Prentice-Hall.

Birnbaum, Z.W. (1952) Numerical tabulation of the distribution of Kolmogorov's statistic for finite sample size. *Journal of the American Statistical Association*, **47**, 425–441.

Böhning, D. and Lindsay, B.G. (1988) Monotonicity of quadratic approximation algorithms. *The Annals of the Institute of Statistical Mathematics*, **40**, 641–663.

Boissel, J.P., Collet, J.P., Moleur, P., and Haugh, M. (1992) Surrogate endpoints: a basis for a rational approach. *European Journal of Clinical Pharmacology*, **43**, 235–244.

Box, G.E.P., Jenkins, G.M., and Reinsel, G.C. (1994) *Time Series Analysis: Forecasting and Control* (3rd ed.). London: Holden-Day.

Box, G.E.P. and Tiao, G.C. (1992) *Bayesian Inference in Statistical Analysis.* Wiley Classics Library edition. New York: John Wiley & Sons.

Bozdogan, H. (1987) Model selection and Akaike's Information Criterion (AIC): The general theory and its analytical extensions. *Psychometrika*, **52**, 345–370.

Brant, L.J. and Fozard, J.L. (1990) Age changes in pure-tone hearing thresholds in a longitudinal study of normal human aging. *Journal of the Acoustic Society of America*, **88**, 813–820.

Brant, L.J. and Pearson, J.D. (1994) Modeling the variability in longitudinal patterns of aging. In: *Biological Anthropology and Aging: Perspectives on Human Variation over the Life Span*, Ch. 14, D.E. Crews and R.M.Garruto (Eds.). New York: Oxford University Press, pp. 373–393.

Brant, L.J., Pearson, J.D., Morrell, C.H., and Verbeke, G. (1992) Statistical methods for studying individual change during aging. *Collegium Antropologicum*, **16**, 359–369.

Brant, L.J. and Verbeke, G. (1997a) Describing the natural heterogeneity of aging using multilevel regression models. *International Journal of Sports Medicine*, **18**, S225–S231.

Brant, L.J. and Verbeke, G. (1997b) Modelling longitudinal studies of aging. In: *Proceedings of the 12th International Workshop on Statistical Modelling, Schriftenreihe der Osterreichischen Statistischen Gesellschaft*, Vol. 5, C.E. Minder and H. Friedl (Eds.). Biel/Bienne, Switzerland, pp. 19–30.

Breslow, N.E. and Clayton, D.G. (1993) Approximate inference in generalized linear mixed models. *Journal of the American Statistical Association*, **88**, 9–25.

Brown, N.A. and Fabro, S. (1981) Quantitation of rat embryonic development in vitro: a morphological scoring system. *Teratology*, **24**, 65–78.

Buck, S.F. (1960) A method of estimation of missing values in multivariate data suitable for use with an electronic computer. *Journal of the Royal Statistical Society, Series B*, **22**, 302–306.

Burzykowski, T., Molenberghs, G., Buyse, M., Geys, H., and Renard, D. (1999) Validation of surrogate endpoints in multiple randomized clinical trials with failure-time endpoints. *Submitted for publication.*

Butler, S.M. and Louis, T.A. (1992) Random effects models with nonparametric priors. *Statistics in Medicine*, **11**, 1981–2000.

Buyse, M. and Molenberghs, G. (1998) The validation of surrogate endpoints in randomized experiments. *Biometrics*, **54**, 1014–1029.

Buyse, M., Molenberghs, G., Burzykowski, T., Renard, D., and Geys, H. (2000) The validation of surrogate endpoints in meta-analyses of randomized experiments, *Biostatistics*, **1**, 000–000.

Carlin, B.P. and Louis, T.A. (1996) *Bayes and Empirical Bayes Methods for Data Analysis.* London: Chapman & Hall.

Carter, H.B. and Coffey, D.S. (1990) The prostate: an increasing medical problem. *The Prostate*, **16**, 39–48.

Carter, H.B., Morrell, C.H., Pearson, J.D., Brant, L.J., Plato, C.C., Metter, E.J., Chan, D.W., Fozard, J.L., and Walsh, P.C. (1992a) Estimation of prostate growth using serial prostate-specific antigen measurements in men with and without prostate disease. *Cancer Research*, **52**, 3323–3328.

Carter, H.B., Pearson, J.D., Metter, E.J., Brant, L.J., Chan, D.W., Andres, R., Fozard, J.L., and Walsh, P.C. (1992b) Longitudinal evaluation of prostate-specific antigen levels in men with and without prostate disease. *Journal of the American Medical Association*, **267**, 2215–2220.

Catalano, P.J. and Ryan, L.M. (1992) Bivariate latent variable models for clustered discrete and continuous outcomes. *Journal of the American Statistical Association*, **87**, 651–658.

Catalano, P.J., Scharfstein, D.O., Ryan, L.M., Kimmel, C.A., and Kimmel, G.L. (1993) Statistical model for fetal death, fetal weight, and malformation in developmental toxicity studies. *Teratology*, **47**, 281–290.

Chatterjee, S. and Hadi, A.S. (1988) *Sensitivity Analysis in Linear Regression.* New York: John Wiley & Sons.

Chen, J. (1993) A malformation incidence dose-response model incorporating fetal weight and/or litter size as covariates. *Risk Analysis*, **13**, 559–564.

Chen, T.T., Simon, R.M., Korn, E.L., Anderson, S.J., Lindblad, A.D., Wieand, H.S., Douglass Jr., H.O., Fisher, B., Hamilton, J.M., and Friedman, M.A. (1998) Investigation of disease-free survival as a surrogate endpoint for survival in cancer clinical trials. *Communications in Statistics*, A, **27**, 1363–1378.

Chi, E.M. and Reinsel, G.C. (1989) Models for longitudinal data with random effects and AR(1) errors. *Journal of the American Statistical Association*, **84**, 452–459.

Choi, S., Lagakos, S., Schooley, R.T., and Volberding, P.A. (1993) CD4+ lymphocytes are an incomplete surrogate marker for clinical progression in persons with asymptomatic HIV infection taking zidovudine. *Annals of Internal Medicine*, **118**, 674–680.

Christensen, R., Pearson, L.M., and Johnson, W. (1992) Case-deletion diagnostics for mixed models. *Technometrics*, **34**, 38–45.

Chuang-Stein, C. and DeMasi, R. (1998) Surrogate endpoints in AIDS drug development: current status (with discussion). *Drug Information Journal*, **32**, 439–448.

Clayton, D.G. (1978) A model for association in bivariate life tables and its application in epidemiological studies of familial tendency in chronic disease incidence. *Biometrika*, **65**, 141–151.

Clayton, D. and Hills, M. (1993) *Statistical Methods in Epidemiology*. Oxford: Oxford University Press.

Cleveland, W.S. (1979) Robust locally-weighted regression and smoothing scatterplots. *Journal of the American Statistical Association*, **74**, 829–836.

Cocchetto, D.M. and Jones, D.R. (1998) Faster access to drugs for serious or life-threatening illnesses through use of the accelerated approval regulation in the United States. *Drug Information Journal*, **32**, 27–35.

Cohen, J. and Cohen, P. (1983) *Applied multiple regression/correlation analysis for the behavioral sciences* (2nd ed.). Hillsdale, NJ: Erlbaum.

Conaway, M.R. (1992) The analysis of repeated categorical measurements subject to nonignorable nonresponse. *Journal of the American Statistical Association*, **87**, 817–824.

Conaway, M.R. (1993) Non-ignorable non-response models for time-ordered categorical variables. *Applied Statistics*, **42**, 105–115.

Cook, R.D. (1977a) Detection of influential observations in linear regression. *Technometrics*, **19**, 15–18.

Cook, R.D. (1977b) Letter to the editor. *Technometrics*, **19**, 348.

Cook, R.D. (1979) Influential observations in linear regression. *Journal of the American Statistical Association*, **74**, 169–174.

Cook, R.D. (1986) Assessment of local influence. *Journal of the Royal Statistical Society, Series B,* **48**, 133–169.

Cook, R.D. and Weisberg, S. (1982) *Residuals and Influence in Regression.* London: Chapman & Hall.

Copas, J.B. and Li, H.G. (1997) Inference from non-random samples (with discussion). *Journal of the Royal Statistical Society, Series B,* **59**, 55–96.

Corfu-A Study Group (1995) Phase III randomized study of two fluorouracil combinations with either interferon alfa-2a or leucovorin for advanced colorectal cancer. *Journal of Clinical Oncology,* **13**, 921–928.

Coursaget, P., Leboulleux, D., Soumare, M., le Cann P., Yvonnet, B., Chiron, J.P., and Collseck A.M. (1994) Twelve-year follow-up study of hepatitis immunization of Senegalese infants. *Journal of Hepatology,* **21**, 250–254.

Cowles, M.K., Carlin, B.P., and Connett, J.E. (1996) Bayesian tobit modeling of longitudinal ordinal clinical trial compliance data with nonignorable missingness. *Journal of the American Statistical Association,* **91**, 86–98.

Cox, D.R. (1972) The analysis of multivariate binary data. *Applied Statistics,* **21**, 113–120.

Cox, D.R. and Hinkley, D.V. (1974) *Theoretical Statistics.* London: Chapman & Hall.

Cox, D.R. and Hinkley, D.V. (1990) *Theoretical Statistics.* London: Chapman & Hall.

Cox, D.R. and Wermuth, N. (1992) Response models for mixed binary and quantitative variables. *Biometrika,* **79**, 441–461.

Cox, D.R. and Wermuth, N. (1994) *Multivariate Dependencies: Models, Analysis and Interpretation.* London: Chapman & Hall.

Crépeau, H., Koziol, J., Reid, N., and Yuh, Y.S. (1985) Analysis of incomplete multivariate data from repeated measurements experiments. *Biometrics,* **41**, 505–514.

Cressie, N.A.C. (1991) *Statistics for Spatial Data.* New York: John Wiley & Sons.

Crowder, M.J. and Hand, D.J. (1990) *Analysis of Repeated Measures.* London: Chapman & Hall.

Cullis, B.R. (1994) Discussion to Diggle, P.J. and Kenward, M.G.: Informative dropout in longitudinal data analysis. *Applied Statistics*, **43**, 79–80.

Curran, D., Pignatti, F., and Molenberghs, G. (1997) Milk protein trial: informative dropout versus random drop-in. *Submitted for publication*.

D'Agostino, R.B. (1971) An omnibus test of normality for moderate and large size samples. *Biometrika*, **58**, 341–348.

Dale, J.R. (1986) Global cross-ratio models for bivariate, discrete, ordered responses. *Biometrics*, **42**, 909–917.

Daniels, M.J. and Hughes, M.D. (1997) Meta-analysis for the evaluation of potential surrogate markers. *Statistics in Medicine*, **16**, 1515–1527.

Davidian, M. and Giltinan, D.M. (1995) *Nonlinear Models for Repeated Measurement Data*. London: Chapman & Hall.

De Backer, M., De Keyser, P., De Vroey, C., and Lesaffre, E. (1996) A 12-week treatment for dermatophyte toe onychomycosis: terbinafine 250mg/day vs. itraconazole 200mg/day–a double-blind comparative trial. *British Journal of Dermatology*, **134**, 16–17.

DeGruttola, V., Fleming, T.R., Lin, D.Y., and Coombs, R. (1997) Validating surrogate markers - are we being naive ? *Journal of Infectious Diseases*, **175**, 237–246.

DeGruttola, V., Lange, N., and Dafni, U. (1991) Modeling the progression of HIV infection. *Journal of the American Statistical Association*, **86**, 569–577.

DeGruttola, V. and Tu, X.M. (1994) Modelling progression of CD4 lymphocyte count and its relationship to survival time. *Biometrics*, **50**, 1003–1014.

DeGruttola, V. and Tu, X.M. (1995) Modelling progression of CD-4 lymphocyte count and its relationship to survival time. *Biometrics*, **50**, 1003–1014.

DeGruttola, V., Ware, J.H., and Louis, T.A. (1987) Influence analysis of generalized least squares estimators. *Journal of the American Statistical Association*, **82**, 911–917.

DeGruttola, V., Wulfsohn, M., Fischl, M.A., and Tsiatis, A. (1993) Modelling the relationship between survival and CD4 lymphocytes in patients with AIDS and AIDS-related complex. *Journal of Acquired Immune Deficiency Syndrome*, **6**, 359–365.

Dempster, A.P., Laird, N.M., and Rubin, D. B. (1977) Maximum likelihood from incomplete data via the EM algorithm (with discussion). *Journal of the Royal Statistical Society, Series B*, **39**, 1–38.

Dempster, A.P. and Rubin, D.B. (1983) Overview. In: *Incomplete Data in Sample Surveys, Vol. II: Theory and Annotated Bibliography*, W.G. Madow, I. Olkin, and D.B. Rubin (Eds.). New York: Academic Press, pp. 3–10.

Dempster, A.P., Rubin, R.B., and Tsutakawa, R.K. (1981) Estimation in covariance components models. *Journal of the American Statistical Association*, **76**, 341–353.

De Ponti, F., Lecchini, S., Cosentino, M., Castelletti, C.M., Malesci, A., and Frigo, G.M. (1993) Immunological adverse effects of anticonvulsants. What is their clinical relevance ? *Drug Safety*, **8**, 235–250.

Diem, J.E. and Liukkonen, J.R. (1988) A comparative study of three methods for analysing longitudinal pulmonary function data. *Statistics in Medicine*, **7**, 19–28.

Diggle, P.J. (1983) *Statistical Analysis of Spatial Point Patterns*. Mathematics in Biology. London: Academic Press.

Diggle, P.J. (1988) An approach to the analysis of repeated measures. *Biometrics*, **44**, 959–971.

Diggle, P.J. (1989) Testing for random dropouts in repeated measurement data. *Biometrics*, **45**, 1255–1258.

Diggle, P.J. (1990) *Time Series: A Biostatistical Introduction*. Oxford: Oxford University Press.

Diggle, P.J. (1992) On informative and random dropouts in longitudinal studies. Letter to the Editor. *Biometrics*, **48**, 947.

Diggle, P.J. (1993) Estimation with missing data. Reply to a Letter to the Editor. *Biometrics*, **49**, 580.

Diggle, P.J. and Kenward, M.G. (1994) Informative drop-out in longitudinal data analysis (with discussion). *Applied Statistics*, **43**, 49–93.

Diggle, P.J., Liang, K.-Y., and Zeger, S.L. (1994) *Analysis of Longitudinal Data*. Oxford Science Publications. Oxford: Clarendon Press.

Draper, D. (1995) Assessment and propagation of model uncertainty (with discussion). *Journal of the Royal Statistical Society, Series B*, **57**, 45–97.

Dyer, A.R. (1974) Comparison of tests for normality with a quationary note. *Biometrika*, **61**, 185–189.

Edlefsen, L.E. and Jones, S.D. *GAUSS*. Aptech Systems Inc., Kent, WA.

Edwards, A.W.F. (1972) *Likelihood*. Cambridge: Cambridge University Press.

Efron, B. (1994) Missing data, imputation, and the bootstrap (with discussion). *Journal of the American Statistical Association*, **89**, 463–479.

Efron, B. and Hinkley, D.V. (1978) Assessing the accuracy of the maximum likelihood estimator: observed versus expected Fisher information. *Biometrika*, **65**, 457–487.

Ekholm, A. and Skinner, C. (1998) The muscatine children's obesity data reanalysed using pattern mixture models. *Applied Statistics*, **47**, 251–263.

Ellenberg, S.S. and Hamilton, J.M. (1989) Surrogate endpoints in clinical trials: cancer. *Statistics in Medicine*, **8**, 405–413.

Fahrmeir, L. and Tutz, G. (1994) *Multivariate Statistical Modelling Based on Generalized Linear Models*. Heidelberg: Springer-Verlag.

Fitzmaurice, G.M. and Laird, N.M. (1993) A Likelihood-based method for analysing longitudinal binary responses. *Biometrika*, **80**, 141–151.

Fitzmaurice, G.M. and Laird, N.M. (1995) Regression models for a bivariate discrete and continuous outcome with clustering. *Journal of the American Statistical Association*, **90**, 845–852.

Fitzmaurice, G.M. and Laird, N.M. (1997) Regression models for mixed discrete and continuous responses with potentially missing values. *Biometrics*, **53**, 110–122.

Fitzmaurice, G.M., Laird, N.M., and Lipsitz, S.R. (1994) Analysing incomplete longitudinal binary responses: a likelihood-based approach. *Biometrics*, **50**, 601–612.

Fitzmaurice, G.M., Laird, N.M., and Rotnitzky, A. (1993) Regression models for discrete longitudinal responses. *Statistical Science*, **8**, 284–309.

Fitzmaurice, G.M., Molenberghs, G., and Lipsitz, S.R. (1995) Regression models for longitudinal binary responses with informative dropouts. *Journal of the Royal Statistical Society, Series B*, **57**, 691–704.

Fleiss, J.L. (1993) The statistical basis of meta-analysis. *Statistical Methods in Medical Research*, **2**, 121–145.

Fleming, T.R. (1992) Evaluating therapeutic interventions: some issues and experiences (with discussion) *Statistical Science*, **7**, 428–456.

Fleming, T.R. and DeMets, D.L. (1996) Surrogate endpoints in clinical trials: are we being misled ? *Annals of Internal Medicine*, **125**, 605–613.

Fleming, T.R., Prentice, R.L., Pepe, M.S., and Glidden, D. (1994) Surrogate and auxiliary endpoints in clinical trials, with potential applications in cancer and AIDS research. *Statistics in Medicine*, **13**, 955–968.

Follman, D. and Wu, M. (1995) An approximate generalized linear model with random effects for informative missing data. *Biometrics*, **51**, 151–168.

Fowler, F. J. (1988) *Survey Research Methods*. Newbury Park, CA: Sage.

Freedman, L.S., Graubard, B.I., and Schatzkin, A. (1992) Statistical validation of intermediate endpoints for chronic diseases. *Statistics in Medicine*, **11**, 167–178.

Friedman, L.M., Furberg, C.D., and DeMets, D.L. (1998) *Fundamentals of Clinical Trials*. New York: Springer-Verlag.

Gaylor, D.W. (1989) Quantitative risk analysis for quantal reproductive and developmental effects. *Environmental Health Perspectives*, **79**, 243–246.

Gelman, A., Carlin, J.B., Stern, H.S., and Rubin, D.B. (1995) *Bayesian Data Analysis*, Texts in Statistical Science. London: Chapman & Hall.

Geys, H., Molenberghs, G., and Ryan, L.M. (1997) Pseudo-likelihood inference for clustered binary data. *Communications in Statistics: Theory and Methods*, **26**, 2743–2767.

Geys, H., Molenberghs, G., and Ryan, L. (1999) Pseudolikelihood modeling of multivariate outcomes in developmental toxicology. *Journal of the American Statistical Association*, **94**, 734–745.

Geys, H., Molenberghs, G., and Williams, P. (1999) Analysis of clustered binary data with covariates specific to each observation. *Submitted for publication*.

Geys, H., Regan, M., Catalano, P., and Molenberghs, G. (1999) Two latent variable risk assessment approaches for combined continuous and discrete outcomes from developmental toxicity data. *Submitted for publication*.

Ghosh, J.K. and Sen, P.K. (1985) On the asymptotic performance of the log likelihood ratio statistic for the mixture model and related results. In: *Proceedings of the Berekely Conference in Honor or Jerzy Neyman and Jack Kiefer*, Vol. 2, L.M. Le Cam and R.A. Olshen (Eds.). Monterey: Wadsworth, Inc., pp. 789–806.

Gilks, W.R., Wang, C.C., Yvonnet, B., and Coursaget, P. (1993) Random-effects models for longitudinal data using Gibbs sampling. *Biometrics*, **49**, 441–453.

Glonek, G.F.V. and McCullagh, P. (1995) Multivariate logistic models. *Journal of the Royal Statistical Society, Series B*, **81**, 477–482.

Glynn, R.J., Laird, N.M., and Rubin, D.B. (1986) Selection modelling versus mixture modelling with non-ignorable nonresponse. In: *Drawing Inferences from Self Selected Samples*, H. Wainer (Ed.). New York: Springer-Verlag, pp. 115–142.

Goldstein, H. (1979) *The Design and Analysis of Longitudinal Studies*. London: Academic Press.

Goldstein, H. (1995) *Multilevel Statistical Models*. Kendall's Libary of Statistics 3. London: Arnold.

Golub, G.H. and Van Loan, C.F. (1989) *Matrix Computations*. (2nd ed.). Baltimore: The Johns Hopkins University Press.

Goss, P.E., Winer, E.P., Tannock, I.F., and Schwartz, L.H. (1999) Breast cancer: randomized phase III trial comparing the new potent and selective third-generation aromatase inhibitor vorozole with megestrol acetate in postmenopausal advanced breast cancer patients. *Journal of Clinical Oncology*, **17**, 52–63.

Gould, A.L. (1980) A new approach to the analysis of clinical drug trials with withdrawals. *Biometrics*, **36**, 721–727.

Greco, F.A., Figlin, R., York, M., Einhorn, L., Schilsky, R., Marshall, E.M., Buys, S.S., Froimtchuk, M.J., Schuller, J., Buyse, M., Ritter, L., Man, A., and Yap, A.K.L. (1996) Phase III randomized study to compare interferon alfa-2a in combination with fluorouracil versus fluorouracil alone in patients with advanced colorectal cancer. *Journal of Clinical Oncology*, **14**, 2674–2681.

Green, S., Benedetti, J., and Crowley, J. (1997) *Clinical Trials in Oncology*. London: Chapman & Hall.

Greenlees, W.S., Reece, J.S., and Zieschang, K.D. (1982) Imputation of missing values when the probability of response depends on the variable being imputed. *Journal of the American Statistical Association*, **77**, 251–261.

Gregoire, T., Brillinger, D.R., Diggle, P.J., Russek-Cohen, E., Warren, W.G., and Wolfinger, R.D. (1997) *Modelling Longitudinal and Spatially Correlated Data.* Lecture Notes in Statistics 122. New York: Springer-Verlag.

Haber, F. (1924) Zur Geschichte des Gaskrieges (On the history of gas warfare). In: *Funf Vortrage aus den Jahren 1920-1923 (Five Lectures from the Years 1920-1923)*, Berlin: Springer-Verlag, pp. 76–92.

Hadler, S.C., Francis, D.P., Maynard, J.E., Thompson, S.E., Judson, F.N., Echenberg, D.F., Ostrow, D.G., O'Malley, P.M., Penley, K.A., Altman, N.L., *et al.* (1986) Long-term immunogenicity and efficacy of hepatitis B vaccine in homosexual men. *New England Journal of Medicine*, **315**, 209–214.

Hand, D.J., Daly, F., Lunn, A.D., McConway, K.J., and Ostrowski, E. (1994) *A Handbook of Small Data Sets* (1st ed.). London: Chapman & Hall.

Hand, D.J. and Taylor, C.C. (1987) *Multivariate Analysis of Variance and Repeated Measures.* London: Chapman & Hall.

Hannan, E.J. and Quinn, B.G. (1979) The determination of the order of an autoregression. *Journal of the Royal Statistical Society, Series B*, **41**, 190–195.

Hartley, H.O. and Hocking, R. (1971) The analysis of incomplete data. *Biometrics*, **27**, 7783–808.

Harville, D.A. (1974) Bayesian inference for variance components using only error contrasts. *Biometrika*, **61**, 383–385.

Harville, D.A. (1976) Extension of the Gauss-Markov theorem to include the estimation of random effects. *The Annals of Statistics*, **4**, 384–395.

Harville, D.A. (1977) Maximum likelihood approaches to variance component estimation and to related problems. *Journal of the American Statistical Association*, **72**, 320–340.

Heckman, J.J. (1976) The common structure of statistical models of truncation, sample selection and limited dependent variables and a simple estimator for such models. *Annals of Economic and Social Measurement*, **5**, 475–492.

Hedeker, D. and Gibbons, R.D. (1994) A random-effects ordinal regression model for multilevel analysis. *Biometrics*, **50**, 933–944.

Hedeker, D. and Gibbons, R.D. (1996) MIXOR: A computer program for mixed-effects ordinal regression analysis. *Computer Methods and Programs in Biomedicine*, **49**, 157–176.

Hedeker, D. and Gibbons, R.D. (1997) Application of random-effects pattern-mixture models for missing data in longitudinal studies. *Psychological Methods*, **2**, 64–78.

Heitjan, D.F. (1993) Estimation with missing data. Letter to the Editor. *Biometrics*, **49**, 580.

Heitjan, D.F. (1994) Ignorability in general incomplete-data models. *Biometrika*, **81**, 701–708.

Helms, R.W. (1992) Intentionally incomplete longitudinal designs: Methodology and comparison of some full span designs. *Statistics in Medicine*, **11**, 1889–1913.

Henderson, C.R. (1984) *Applications of Linear Models in Animal Breeding*. Guelph, Canada: University of Guelph Press.

Henderson, C.R., Kempthorne, O., Searle, S.R., and Von Krosig, C.N. (1959) Estimation of environmental and genetic trends from records subject to culling. *Biometrics*, **15**, 192–218.

Heyting, A., Tolboom, J.T.B.M., and Essers, J.G.A. (1992) Statistical handling of drop-outs in longitudinal clinical trials. *Statistics in Medicine*, **11**, 2043–2061.

Hogan, J.W. and Laird, N.M. (1997) Mixture models for the joint distribution of repeated measures and event times. *Statistics in Medicine*, **16**, 239–258.

Holmes, L.B. (1988) Human teratogens: delineating the phenotypic effects, the period of greatest sensitivity, and the dose-response relationship and mechanisms of action. In: *Transplacental Effects on Fetal Health*. New York: Alan R. Liss, Inc., pp. 171–191.

Hosmer, D.W. and Lemeshow, S. (1989) *Applied Logistic Regression*. New York: John Wiley & Sons.

Hougaard, P. (1987) Modelling multivariate survival. *Scandinavian Journal of Statistics*, **14**, 291–304.

Jennrich, R.I. and Schluchter, M.D. (1986) Unbalanced repeated measures models with structured covariance matrices. *Biometrics*, **42**, 805–820.

Johnson, R.A. and Wichern, D.W. (1992) *Applied Multivariate Statistical Analysis* (3rd ed.). Englewood Cliffs, NJ: Prentice-Hall.

Kahn, H. and Sempos, C.T. (1989) *Statistical Methods in Epidemiology*. New York: Oxford University Press.

Kenward, M.G. (1998) Selection models for repeated measurements with nonrandom dropout: an illustration of sensitivity. *Statistics in Medicine*, **17**, 2723–2732.

Kenward, M.G. and Molenberghs, G. (1998) Likelihood based frequentist inference when data are missing at random. *Statistical Science*, **12**, 236–247.

Kenward, M.G. and Molenberghs, G. (1999) Parametric models for incomplete continuous and categorical longitudinal studies data. *Statistical Methods in Medical Research*, **8**, 51–83.

Kenward, M.G., Molenberghs, G. and Lesaffre, E. (1994) An application of maximum likelihood and estimating equations to the analysis of ordinal data from a longitudinal study with cases missing at random. *Biometrics*, **50**, 945–953.

Kenward, M.G. and Roger, J.H. (1997) Small sample inference for fixed effects from restricted maximum likelihood. *Biometrics*, **53**, 983–997.

Khoury, M.J., Adams, M.M., Rhodes, P., and Erickson, J.D. (1987) Monitoring multiple malformations in the detection of epidemics of birth defects. *Teratology*, **36**, 345–354.

Kimmel, G.L., Cuff, J.M., Kimmel, C.A., Heredia, D.J., Tudor, N., and Silverman, P.M. (1993) Embryonic development in vitro following short-duration exposure to heat. *Teratology*, **47**, 243–251.

Kimmel, G.L., Williams, P.L., Kimmel, C.A., Claggett, T.W., and Tudor, N. (1994) The effects of temperature and duration of exposure on in vitro development and response-surface modelling of their interaction. *Teratology*, **49**, 366–367.

Krzanowski, W.J. (1988) *Principles of Multivariate Analysis.* Oxford: Clarendon Press.

Laird, N.M. (1978) Nonparametric maximum likelihood estimation of a mixing distribution. *Journal of the American Statistical Association*, **73**, 805–811.

Laird, N.M. (1988) Missing data in longitudinal studies. *Statistics in Medicine*, **7**, 305–315.

Laird, N.M. (1994) Discussion to Diggle, P.J. and Kenward, M.G.: Informative dropout in longitudinal data analysis. *Applied Statistics*, **43**, 84.

Laird, N.M., Lange, N., and Stram, D. (1987) Maximum likelihood computations with repeated meausres: application of the EM algorithm. *Journal of the American Statistical Association*, **82**, 97–105.

Laird, N.M. and Ware, J.H. (1982) Random effects models for longitudinal data. *Biometrics*, **38**, 963–974.

Lang, J.B. and Agresti, A. (1994) Simultaneously modeling joint and marginal distributions of multivariate categorical responses. *Journal of the American Statistical Association*, **89**, 625–632.

Lange, N. and Ryan, L. (1989) Assessing normality in random effects models. *The Annals of Statistics*, **17**, 624–642.

Lee, Y. and Nelder, J.A. (1996) Hierarchical generalized linear models (with discussion). *Journal of the Royal Statistical Society, Series B*, **58**, 619–678.

Lehmann, E.L. and D'Abrera, H.J.M. (1975) *Nonparametrics. Statistical Methods Based on Ranks*. San Francisco: Holden-Day.

Lesaffre, E., Asefa, M., and Verbeke, G. (1999) Assessing the goodness-of-fit of the Laird and Ware model: an example: the Jimma Infant Survival Differential Longitudinal Study. *Statistics in Medicine*, **18**, 835–854.

Lesaffre, E. and Verbeke, G. (1998) Local influence in linear mixed models. *Biometrics*, **54**, 570–582.

Leslie, J.R., Stephens, M.A., and Fotopoulos, S. (1986) Asymptotic distribution of the Shapiro-Wilk W for testing for normality. *The Annals of Statistics*, **14**, 1497–1506.

Li, K.H., Raghunathan, T.E., and Rubin, D.B. (1991) Large-sample significance levels from multiply imputed data using moment-based statistics and an F reference distributions. *Journal of the American Statistical Association*, **86**, 1065–1073.

Liang, K.-Y. and Zeger, S.L. (1986) Longitudinal data analysis using generalized linear models. *Biometrika*, **73**, 13–22.

Liang, K.-Y. and Zeger, S.L. (1989) A class of logistic regression models for multivariate binary time series. *Journal of the American Statistical Association*, **84**, 447–451.

Liang, K.-Y., Zeger, S.L., and Qaqish, B. (1992) Multivariate regression analyses for categorical data. *Journal of the Royal Statistical Society, Series B*, **54**, 3–40.

Lilienfeld, D.E. and Stolley, P.D. (1994) *Foundations of Epidemiology*. New York: Oxford University Press.

Lin, D.Y., Fischl, M.A., and Schoenfeld, D.A. (1993) Evaluating the role of CD4-lymphocyte change as a surrogate endpoint in HIV clinical trials. *Statistics in Medicine*, **12**, 835–842.

Lin, D.Y., Fleming T.R., and De Gruttola, V. (1997) Estimating the proportion of treatment effect explained by a surrogate marker. *Statistics in Medicine*, **16**, 1515–1527.

Lin, X., Raz, J., and Harlow, S. (1997) Linear mixed models with heterogeneous within-cluster variances. *Biometrics*, **53**, 910–923.

Lindley, D.V. and Smith, A.F.M. (1972) Bayes estimates for the linear model. *Journal of the Royal Statistical Society, Series B*, **34**, 1–41.

Lindstrom, M.J. and Bates, D.M. (1988) Newton-Raphson and EM algorithms for linear mixed-effects models for repeated-measures data. *Journal of the American Statistical Association*, **83**, 1014–1022.

Littell, R.C., Milliken, G.A., Stroup, W.W., and Wolfinger, R.D. (1996) *SAS System for Mixed Models*. Cary, NC: SAS Institute Inc.

Little, R.J.A. (1976) Inference about means for incomplete multivariate data. *Biometrika*, **63**, 593–604.

Little, R.J.A. (1986) A note about models for selectivity bias. *Econometrika*, **53**, 1469–1474.

Little, R.J.A. (1993) Pattern-mixture models for multivariate incomplete data. *Journal of the American Statistical Association*, **88**, 125–134.

Little, R.J.A. (1994a) A class of pattern-mixture models for normal incomplete data. *Biometrika*, **81**, 471–483.

Little, R.J.A. (1994b) Discussion to Diggle, P.J. and Kenward, M.G.: Informative dropout in longitudinal data analysis. *Applied Statistics*, **43**, 78.

Little, R.J.A. (1995) Modeling the drop-out mechanism in repeated measures studies. *Journal of the American Statistical Association*, **90**, 1112–1121.

Little, R.J.A. and Rubin, D.B. (1987) *Statistical Analysis with Missing Data*. New York: John Wiley & Sons.

Little, R.J.A. and Wang, Y. (1996) Pattern-mixture models for multivariate incomplete data with covariates. *Biometrics*, **52**, 98–111.

Little, R.J.A. and Yau, L. (1996) Intent-to-treat analysis for longitudinal studies with drop-outs. *Biometrics*, **52**, 1324–1333.

Liu, C. and Rubin, D.B. (1994) The ECME algorithm: a simple extension of EM and ECM with faster monotone convergence. *Biometrika*, **81**, 633–648.

Liu, C. and Rubin, D.B. (1995) MI estimation of the *t* distribution using EM and its extensions, ECM and ECME. *Statistica Sinica*, **5**, 19–39.

Liu, C., Rubin, D.B., and Wu, Y.N. (1998) Parameter expansion to accelerate EM: the PX-EM algorithm. *Biometrika*, **85**, 755–770.

Liu, G. and Liang, K.-Y. (1997) Sample size calculations for studies with correlated observations. *Biometrics*, **53**, 937–947.

Longford, N.T. (1993) *Random Coefficient Models*. Oxford: Oxford University Press.

Louis, T.A. (1982) Finding the observed information matrix when using the EM algorithm. *Journal of the Royal Statistical Society, Series B*, **44**, 226-233.

Louis, T.A. (1984) Estimating a population of parameter values using bayes and empirical Bayes methods. *Journal of the American Statistical Association*, **79**, 393–398.

Magder, L.S. and Zeger, S.L. (1996) A smooth nonparametric estimated of a mixing distribution using mixtures of Gaussians. *Journal of the American Statistical Association*, **91**, 1141–1152.

Mansour, H., Nordheim, E.V., and Rutledge, J.J. (1985) Maximum likelihood estimation of variance components in repeated measures designs assuming autoregressive errors. *Biometrics*, **41**, 287–294.

McArdle, J.J. and Hamagami, F. (1992) Modeling incomplete longitudinal and cross-sectional data using latent growth structural models. *Experimental Aging Research*, **18**, 145–166.

McCullagh, P. and Nelder, J.A. (1989) *Generalized Linear Models*. London: Chapman & Hall.

McLachlan, G.J. and Basford, K.E. (1988) *Mixture models. Inference and Applications to Clustering*. New York: Marcel Dekker.

McLachlan, G.J. and Krishnan, T. (1997) *The EM Algorithm and Extensions*. New York: John Wiley & Sons.

McLean, R.A., Sanders, W.L., and Stroup, W.W. (1991) A unified approach to mixed linear models. *The American Statistician*, **45**, 54–64.

Meilijson, I. (1989) A fast improvement to the EM algorithm on its own terms. *Journal of the Royal Statistical Society, Series B*, **51**, 127–138.

Mellors, J.W., Munoz, A., Giorgi, J.V., Margolich, J.B., Tassoni, C.J., Gupta, P., Kingsley, L.A., Todd, J.A., Saah, A.J., Phair, J.P., and Rinaldo, C.R. (1997) Plasma viral load and CD4+ lymphocytes as prognostic markers of HIV-1 infection. *Annals of Internal Medicine*, **126**, 946–954.

Meng, X.-L. (1997) The EM algorithm and medical studies: a historical link. *Statistical Methods in Medical Research*, **6**, 3–23.

Meng, X.-L. and Rubin, D.B. (1991) Using EM to obtain asymptotic variance covariance matrices: the SEM algorithm. *Journal of the American Statistical Association*, **86**, 899–909.

Meng, X.-L. and Rubin, D.B. (1993) Maximum likelihood estimation via the ECM algorithm: a general framework. *Biometrika*, **80**, 267–278.

Meng, X.-L. and van Dyk, D. (1997) The EM algorithm–an old folk-song sung to a fast new tune. *Journal of the Royal Statistical Society, Series B*, **3**, 511–567.

Meng, X.-L. and van Dyk, D. (1998) Fast EM-type implementation for mixed effects models. *Journal of the Royal Statistical Society, Series B*, **3**, 559–578.

Mentré, F., Mallet, A., and Baccar, D. (1997) Optimal design in random-effects regression models. *Biometrika*, **84**, 429–442.

Michiels, B., Molenberghs, G., Bijnens, L., and Vangeneugden, T. (1999) Selection models and pattern-mixture models to analyze longitudinal quality of life data subject to dropout. *Submitted for publication*.

Michiels, B., Molenberghs, G., and Lipsitz, S.R. (1999). Selection models and pattern-mixture models for incomplete categorical data with covariates. *Biometrics*, **55**, 978–983.

Miller, J.J. (1977) Asymptotic properties of maximum likelihood estimates in the mixed model of the analysis of variance. *The Annals of Statistics*, **5**, 746–762.

Molenberghs, G., Buyse, M., Geys, H., Renard, D., and Burzykowski, T. (1999) Statistical challenges in the evaluation of surrogate endpoints in randomized trials. *Submitted for publication*.

Molenberghs, G., Geys, H., and Buyse, M. (1998) Validation of surrogate endpoints in randomized experiments with mixed discrete and continuous outcomes. *Submitted for publication*.

Molenberghs, G., Goetghebeur, E.J.T., Lipsitz, S.R., Kenward, M.G. (1999) Non-random missingness in categorical data: strengths and limitations. *The American Statistician*, **53**, 110–118.

Molenberghs, G., Kenward, M. G., and Lesaffre, E. (1997) The analysis of longitudinal ordinal data with non-random dropout. *Biometrika*, **84**, 33–44.

Molenberghs, G. and Lesaffre, E. (1994) Marginal modelling of correlated ordinal data using a multivariate Plackett distribution. *Journal of the American Statistical Association*, **89**, 633–644.

Molenberghs, G., Michiels, B., and Kenward, M.G. (1998) Pseudo-like-lihood for combined selection and pattern-mixture models for missing data problems. *Biometrical Journal*, **40**, 557–572.

Molenberghs, G., Michiels, B., Kenward, M.G., and Diggle, P.J. (1998) Missing data mechanisms and pattern-mixture models. *Statistica Neerlandica*, **52**, 153–161.

Molenberghs, G., Michiels, B., and Lipsitz, S.R. (1999) A pattern-mixture odds ratio model for incomplete categorical data. *Communications in Statistics: Theory and Methods*, **28**, 000–000.

Molenberghs, G. and Ritter, L. (1996) Likelihood and quasi-likelihood based methods for analysing multivariate categorical data, with the association between outcomes of interest. *Biometrics*, **52**, 1121–1133.

Molenberghs, G. and Ryan, L.M. (1999) Likelihood inference for clustered multivariate binary data. *Environmetrics*, **10**, 279–300.

Molenberghs, G., Verbeke, G., Thijs, H., Lesaffre, E., and Kenward, M.G. (1999) Mastitis in dairy cattle: influence analysis to assess sensitivity of the dropout process. *Submitted for publication*.

Morrell, C.H. (1998) Likelihood ratio testing of variance components in the linear mixed-effects model using restricted maximum likelihood. *Biometrics*, **54**, 1560–1568.

Morrell, C.H. and Brant, L.J. (1991) Modelling hearing thresholds in the elderly. *Statistics in Medicine*, **10**, 1453–1464.

Morrell, C.H., Pearson, J.D., Ballentine Carter, H., and Brant, L.J. (1995) Estimating unknown transition times using a piecewise non-linear mixed-effects model in men with prostate cancer. *Journal of the American Statistical Association*, **90**, 45–53.

Morrell, C.H., Pearson, J.D., and Brant, L.J. (1997) Linear transformations of linear mixed-effects models. *The American Statistician*, **51**, 338–343.

Murray, G.D. and Findlay, J.G. (1988) Correcting for the bias caused by drop-outs in hypertension trials. *Statististics in Medicine*, **7**, 941-946.

Muthén, B., Kaplan, D., and Hollis, M. (1987) On structural equation modeling with data that are not missing completely at random. *Psychometrika*, **52**, 431–462.

Neave, H.R. (1986) *Statistics Tables for Mathematicians, Engineers, Economists and the Behavioural and Management Sciences*. London: George Allen & Unwin.

Nelder, J.A. (1954) The interpretation of negative components of variance. *Biometrika*, **41**, 544–548.

Nelder, J.A. and Mead, R. (1965) A simplex method for function minimisation. *The Computer Journal*, **7**, 303–313.

Neter, J., Wasserman, W., and Kutner, M.H. (1990) *Applied Linear Statistical Models. Regression, Analysis of Variance and Experimental Designs* (3rd ed.). Homewood, IL: Richard D. Irwin, Inc.

Neuhaus, J.M. and Kalbfleisch, J.D. (1998) Between- and within-cluster covariate effects in the analysis of clustered data. *Biometrics*, **54**, 638–645.

Nordheim, E.V. (1984) Inference from nonrandomly missing categorical data: an example from a genetic study on Turner's syndrome. *Journal of the American Statistical Association*, **79**, 772–780.

Núñez-Antón, V. and Woodworth, G.G. (1994) Analysis of longitudinal data with unequally spaced observations and time-dependent correlated errors. *Biometrics*, **50**, 445–456.

O'Brien, W.A., Hartigan, P.M., Martin, D., Eisnhart, J., Hill, A., Benoit, S., Rubin, M., Simberkoff, M.S., and Hamilton, J.D. (1996) Changes in plasma HIV-1 RNA and CD4+ lymphocyte counts and the risk of progression to AIDS. *New England Journal of Medicine*, **334**, 426–431.

Olkin, I. and Tate, R.F. (1961) Multivariate correlation models with mixed discrete and continuous variables. *Annals of Mathematical Statistics*, **32**, 448–465 (with correction in **36**, 343–344).

O'Neill, B. (1966) *Elementary Differential Geometry*. New York: Academic Press.

Ovarian Cancer Meta-Analysis Project (1991) Cyclophosphamide plus cisplatin versus cyclophosphamide, doxorubicin, and cisplatin chemotherapy of ovarian carcinoma: a meta-analysis. *Journal of Clinical Oncology*, **9**, 1668–1674.

Ovarian Cancer Meta-Analysis Project (1998) Cyclophosphamide plus cisplatin versus cyclophosphamide, doxorubicin, and cisplatin chemotherapy of ovarian carcinoma: a meta-analysis. *Classic Papers and Current Comments*, **3**, 237–43.

Pan, H. and Goldstein, H. (1998) Multi-level repeated measures growth modelling using extended spline functions. *Statistics in Medicine*, **17**, 2755–2770.

Park, T. and Brown, M.B. (1994) Models for categorical data with nonignorable nonresponse. *Journal of the American Statistical Association*, **89**, 44–52.

Park, T. and Lee, S.-L. (1999) Simple pattern-mixture models for longitudinal data with missing observations: analysis of urinary incontinence data. *Statistics in Medicine*, **18**, 2933–2941.

Patel, H.I. (1991) Analysis of incomplete data from clinical trials with repeated measurements. *Biometrika*, **78**, 609-619.

Patterson, H.D. and Thompson, R. (1971) Recovery of inter-block information when block sizes are unequal. *Biometrika*, **58**, 545–554.

Pearson, E.S., D'Agostino, R.B., and Bowman, KO. (1977) Tests for departure from normality: comparison of powers. *Biometrika*, **64**, 231–246.

Pearson, J.D., Kaminski, P., Metter, E.J., Fozard, J.L., Brant, L.J., Morrell, C.H., and Carter, H.B. (1991) Modeling longitudinal rates of change in prostate specific antigen during aging. *Proceedings of the Social Statistics Section of the American Statistical Assciation, Washington, DC*, pp. 580–585.

Pearson, J.D., Morrell, C.H., Gordon-Salant, S., Brant, L.J., Metter, E.J., Klein, L.L., and Fozard J.L. (1995) Gender differences in a longitudinal study of age-associated hearing loss. *Journal of the Acoustical Society of America*, **97**, 1196–1205.

Pearson, J.D., Morrell, C.H., Landis, P.K., Carter, H.B., and Brant, L.J. (1994) Mixed-effects regression models for studying the natural history of prostate disease. *Statistics in Medicine*, **13**, 587–601.

Peixoto, J.L. (1987) Hierarchical variable selection in polynomial regression models. *The American Statistician*, **41**, 311–313.

Peixoto, J.L. (1990) A property of well-formulated polynomial regression models. *The American Statistician*, **44**, 26–30.

Pendergast, J.F., Gange, S.J., Newton, M.A., Lindstrom, M.J., Palta, M., and Fisher, M.R. (1996) A survey of methods for analyzing clustered binary response data. *International Statistical Review*, **64**, 89–118.

Pharmacological Therapy for Macular Degeneration Study Group (1997) Interferon α-IIA is ineffective for patients with choroidal neovascularization secondary to age-related macular degeneration. Results of a prospective randomized placebo-controlled clinical trial. *Archives of Ophthalmology*, **115**, 865–872.

Piantadosi, S. (1997) *Clinical Trials: A Methodologic Perspective.* New York: John Wiley & Sons.

Potthoff, R.F. and Roy, S.N. (1964) A generalized multivariate analysis of variance model useful especially for growth curve problems. *Biometrika*, **51**, 313–326.

Prasad, N.G.N. and Rao, J.N.K. (1990) The estimation of mean squared error of small-area estimators. *Journal of the American Statistical Association*, **85**, 163–171.

Pregibon, D. (1979) *Data analytic methods for generalized linear models.* Ph.D. Thesis, University of Toronto.

Pregibon, D. (1981) Logistic regression diagnostics. *The Annals of Statistics*, **9**, 705–724.

Prentice, R.L. (1988) Correlated binary regression with covariates specific to each binary observation. *Biometrics*, **44**, 1033–1048.

Prentice, R.L. (1989) Surrogate endpoints in clinical trials: definitions and operational criteria. *Statistics in Medicine*, **8**, 431–440.

Rang, H.P. and Dale, M.M. (1990) *Pharmacology.* Edinburgh: Churchill Livingstone.

Rao, C.R. (1973) *Linear Statistical Inference and Its Applications* (2nd ed.). New York: John Wiley & Sons.

Regan, M.M. and Catalano, P.J. (1999a) Likelihood models for clustered binary and continuous outcomes: Application to developmental toxicology. *Biometrics*, **55**, 760–768.

Regan, M.M. and Catalano, P.J. (1999b) Bivariate dose-response modeling and risk estimation in developmental toxicology. *Journal of Agricultural, Biological and Environmental Statistics*, **4**, 217–237.

Ripley, B.D. (1981) *Spatial Statistics.* New York: John Wiley & Sons.

Roberts, D.T. (1992) Prevalence of dermatophyte onychomycosis in the United Kingdom: results of an omnibus survey. *British Journal of Dermatology*, **126**, 23–27.

Robins, J.M. (1997) Non-respone models for the analysis of non-monotone non-ignorable missing data. *Statistics in Medicine*, **16**, 21–38.

Robins, J.M. and Gill, R. (1997) Non-respone models for the analysis of non-monotone ignorable missing data. *Statistics in Medicine*, **16**, 39–56.

Robins, J.M. and Rotnitzky, A. (1995) Semiparametric efficiency in multivariate regression models with missing data. *Journal of the American Statistical Association*, **90**, 122–129.

Robins, J.M., Rotnitzky, A., and Scharfstein, D.O. (1998) Semiparametric regression for repeated outcomes with non-ignorable non-response. *Journal of the American Statistical Association*, **93**, 1321–1339.

Robins, J.M., Rotnitzky, A., and Zhao, L.P. (1995) Analysis of semiparametric regression models for repeated outcomes in the presence of missing data. *Journal of the American Statistical Association*, **90**, 106–121.

Robinson, G.K. (1991) That BLUP is a good thing: the estimation of random effects. *Statistical Science*, **1**, 15–51.

Rochon, J. (1992) ARMA covariance structures with time heteroscedasticity for repeated measures experiments. *Journal of the American Statistical Association*, **87**, 777–784.

Roger, J.H. (1993) A new look at the facilities in PROC MIXED. *Proceedings SEUGI*, **93**, 521–532.

Roger, J.H. and Kenward, M.G. (1993) Repeated measures using proc mixed instead of proc glm. In: *Proceedings of the First Annual South-East SAS Users Group Conference, Cary, NC, U.S.A.* Cary, NC: SAS Institute Inc. pp. 199–208.

Rosner, B. (1984) Multivariate methods in ophtalmology with applications to other paired-data situations. *Biometrics*, **40**, 1025–1035.

Rotnitzky, A. and Robins, J.M. (1995) Semi-parametric estimation of models for means and covariances in the presence of missing data. *Scandinavian Journal of Statistics: Theory and Applications*, **22**, 323–334.

Rotnitzky, A. and Robins, J.M. (1997) Analysis of semiparametric regression models with non-ignorable non-response. *Statistics in Medicine*, **16**, 81–102.

Royston, P. and Altman, D.G. (1994) Regression using fractional polynomials of continuous covariates: parsimonious parametric modelling. *Applied Statistics*, **43**, 429–468.

Rubin, D.B. (1976) Inference and missing data. *Biometrika*, **63**, 581–592.

Rubin, D.B. (1978) Multiple imputations in sample surveys – a phenomenological Bayesian approach to nonresponse. In: *Imputation and Editing of Faulty or Missing Survey Data*. Washington, DC: U.S. Department of Commerce, pp. 1–23.

Rubin, D.B. (1987) *Multiple Imputation for Nonresponse in Surveys*. New York: John Wiley & Sons.

Rubin, D.B. (1994) Discussion to Diggle, P.J. and Kenward, M.G.: Informative dropout in longitudinal data analysis. *Applied Statistics*, **43**, 80–82.

Rubin, D.B. (1996) Multiple imputation after 18+ years. *Journal of the American Statistical Association*, **91**, 473–489.

Rubin, D.B. and Schenker, N. (1986) Multiple imputation for interval estimation from simple random samples with ignorable nonresponse. *Journal of the American Statistical Association*, **81**, 366–374.

Rubin, D.B., Stern H.S., and Vehovar V. (1995) Handling "don't know" survey responses: the case of the Slovenian plebiscite. *Journal of the American Statistical Association*, **90**, 822–828.

Ryan, L.M. (1992a) Quantitative risk assessment for developmental toxicity. *Biometrics*, **48**, 163–174.

Ryan, L.M. (1992b) The use of generalized estimating equations for risk assessment in developmental toxicity. *Risk Analysis*, **12**, 439–447.

Ryan, L.M. and Molenberghs, G. (1999) Statistical methods for developmental toxicity: analysis of clustered multivariate binary data. *Annals of the New York Academy of Sciences*, **00**, 000–000.

SAS Institute Inc. (1989) *SAS/STAT User's guide, Version 6, Volume 1* (4th ed.). Cary, NC: SAS Institute Inc.

SAS Institute Inc. (1991) *SAS System for Linear Models* (3rd ed.). Cary, NC: SAS Institute Inc.

SAS Institute Inc. (1992) *SAS Technical Report P-229, SAS/STAT Software: Changes and Enhancements, Release 6.07*. Cary, NC: SAS Institute Inc.

SAS Institute Inc. (1996) *SAS/STAT Software: Changes and Enhancements through Release 6.11*. Cary, NC: SAS Institute Inc.

SAS Institute Inc. (1997) *SAS/STAT Software : Changes and Enhancements through Release 6.12*. Cary, NC: SAS Institute Inc.

SAS Institute Inc. (1999) *SAS/STAT User's guide, Version 7*. Cary, NC: SAS Institute Inc.

Satterthwaite, F.E. (1941) Synthesis of variance. *Psychometrika*, **6**, 309–316.

Sauerbrei, W. and Royston, P. (1999) Building multivariable prognostic and diagnostic models: transformation of the predictors by using fractional polynomials. *Journal of the Royal Statistical Society, Series A*, **162**, 71–94.

Schafer J.L. (1997) *Analysis of Incomplete Multivariate Data*. London: Chapman & Hall.

Schafer J.L., Khare M., and Ezatti-Rice T.M. (1993) Multiple imputation of missing data in NHANES III. In: *Proceedings of the Annual Research Conference* Washington, DC: Bureau of the Census. pp. 459–487.

Schipper, H., Clinch, J., and McMurray, A. (1984) Measuring the quality of life of cancer patients: the Functional-Living Index-Cancer: development and validation. *Journal of Clinical Oncology*, **2**, 472–483.

Schluchter, M.D. (1992) Methods for the analysis of informatively censored longitudinal data. *Statistics in Medicine*, **11**, 1861–1870.

Schwarz, G. (1978) Estimating the dimension of a model. *The Annals of Statistics*, **6**, 461–464.

Schwetz, B.A. and Harris, M.W. (1993) Developmental toxicology: status of the field and contribution of the National Toxicology Program. *Environmental Health Perspectives*, **100**, 269–282.

Searle, S.R. (1987) *Linear Models for Unbalanced Data*. New York: John Wiley & Sons.

Searle, S.R., Casella, G., and McCulloch, C.E. (1992) *Variance Components*. New York: John Wiley & Sons.

Seber, G.A.F. (1977) *Linear Regression Analysis*. New York: John Wiley & Sons.

Seber, G.A.F. (1984) *Multivariate Observations*. New York: John Wiley & Sons.

Self, S.G. and Liang, K.Y. (1987) Asymptotic properties of maximum likelihood estimators and likelihood ratio tests under nonstandard conditions. *Journal of the American Statistical Association*, **82**, 605–610.

Selvin, S. (1996) *Statistical Analysis of Epidemiologic Data*. New York: Oxford University Press.

Sen, A. and Srivastava, M. (1990) *Regression Analysis. Theory, Methods and Applications*. New York: Springer-Verlag.

Senn, S. (1998) Some controversies in planning and analysing multi-centre trials. *Statistics in Medicine*, **17**, 1753–1765.

Shapiro, S.S. and Wilk, M.B. (1965) An analysis of variance test for normality (complete samples). *Biometrika*, **52**, 591–611.

Shapiro, S.S. and Wilk, M.B. (1968) Approximations for the null distribution of the W statistic. *Technometrics*, **10**, 861–866.

Sharples, K. and Breslow, N.E. (1992) Regression analysis of correlated binary data: some small sample results for the estimating equation approach. *Journal of Statistical Computation and Simulation*, **42**, 1–20.

Sheiner, L.B., Beal, S.L., and Dunne, A. (1997) Analysis of nonrandomly censored ordered categorical longitudinal data from analgesic trials. *Journal of the American Statistical Association*, **92**, 1235–1244.

Shih, J.H. and Louis, T.A. (1995) Inferences on association parameter in copula models for bivariate survival data. *Biometrics*, **51**, 1384–1399.

Shih, W.J. and Quan, H. (1997) Testing for treatment differences with dropouts in clinical trials—a composite approach. *Statistics in Medicine*, **16**, 1225–1239.

Shock, N.W., Greullich, R.C., Andres, R., Arenberg, D., Costa, P.T., Lakatta, E.G., and Tobin, J.D. (1984) *Normal Human Aging: The Baltimore Longitudinal Study of Aging*. National Institutes of Health Publication 84–2450. Washington, DC: National Institutes of Health.

Siddiqui, O. and Ali, M.W. (1998) A comparison of the random-effects pattern-mixture model with last-observation-carried-forward (LOCF) analysis in longitudinal clinical trials with dropouts. *Journal of Biopharmaceutical Statistics*, **8**, 545–563.

Smith, A.F.M. (1973) A general Bayesian linear model. *Journal of the Royal Statistical Society, Series B*, **35**, 67–75.

Smith, D.M., Robertson, B., and Diggle, P.J. (1996) *Object-oriented Software for the Analysis of Longitudinal Data in S*. Technical Report MA 96/192. Department of Mathematics and Statistics, University of Lancaster, LA1 4YF, United Kingdom.

Sprott, D.A. (1975) Marginal and conditional sufficiency. *Biometrika*, **62**, 599–605.

Stasny, E.A. (1986) Estimating gross flows using panel data with nonresponse: an example from the Canadian Labour Force Survey. *Journal of the American Statistical Association*, **81**, 42–47.

Stiratelli, R., Laird, N., and Ware, J. (1984) Random effects models for serial observations with dichotomous response. *Biometrics*, **40**, 961–972.

Stram, D.O. and Lee, J.W. (1994) Variance components testing in the longitudinal mixed effects model. *Biometrics*, **50**, 1171–1177.

Stram, D.A. and Lee, J.W. (1995) Correction to: Variance components testing in the longitudinal mixed effects model. *Biometrics*, **51**, 1196.

Strenio, J.F., Weisberg, H.J., and Bryk, A.S. (1983) Empirical Bayes estimation of individual growth-curve parameters and their relationship to covariates. *Biometrics*, **39**, 71–86.

Tanner, M.A. and Wong, W.H. (1987) The calculation of posterior distributions by data augmentation. *Journal of the American Statistical Association*, **82**, 528–550.

Thijs, H., Molenberghs, G., and Verbeke, G. (2000) The milk protein trial: influence analysis of the dropout process. *Biometrical Journal*, **00**, 000–000.

Thompson, S.G. (1993) Controversies in meta-analysis: the case of the trials of serum cholesterol reduction. *Statistical Methods in Medical Research*, **2**, 173–192.

Thompson, S.G. and Pocock, S.J. (1991) Can meta-analyses be trusted ? *Lancet*, **338**, 1127–1130.

Thompson, W.A., Jr. (1962) The problem of negative estimates of variance components. *Annals of Mathematical Statistics*, **33**, 273–289.

Titterington, D.M., Smith, A.F.M., and Makov, U.E. (1985) *Statistical Analysis of Finite Mixture Distributions*. New York: John Wiley & Sons.

Tomasko, L., Helms, R.W., and Snapinn, S.M. (1999) A discriminant analysis extension to mixed models. *Statistics in Medicine*, **18**, 1249–1260.

Troxel, A.B., Harrington, D.P., and Lipsitz, S.R. (1998) Analysis of longitudinal data with non-ignorable non-monotone missing values. *Applied Statistics*, **47**, 425–438.

U.S. Environmental Protection Agency (1991) Guidelines for developmental toxicity risk assessment. *Federal Register*, **56**, 63,798–63,826.

Vach, W. and Blettner, M. (1995) Logistic regresion with incompletely observed categorical covariates–investigating the sensitivity against violation of the missing at random assumption. *Statistics in Medicine*, **12**, 1315–1330.

Van Damme, P., Vranckx, R., Safary, A., Andre F.E., and Meheus, A. (1989) Protective efficacy of a recombinant desoxyribonucleic acid hepatitis B vaccine in institutionalized mentally handicapped clients. *American Journal of Medicine*, **87**, 26S–29S.

van Dyk, D. Meng, X.-L., and Rubin, D.B. (1995) Maximum likelihood estimation via the ECM algorithm: computing the asymptotic variance. *Statistica Sinica*, **5**, 55-75.

Vellinga, A., Van Damme, P., and Meheus, A. (1999a) Hepatitis B and C in institutions for individuals with mental retardation: a review. *Journal of Intellectual Disability Research*, 00, 000–000.

Vellinga, A., Van Damme, P., Weyler, J.J., Vranckx, R., Meheus, A. (1999b) Hepatitis B vaccination in mentally retarded: effectiveness after 11 years. *Vaccine*, **17**, 602–606.

Vellinga, A., Van Damme, P., Bruckers, L., Weyler, J.J., Molenberghs, G., and Meheus, A. (1999c) Modelling long term persistence of hepatitis B antibodies after vacination. *Journal of Medical Virology*, **57**, 100–103.

Verbeke, G. (1995) *The linear mixed model. A critical investigation in the context of longitudinal data analysis.* Ph.D. thesis, Catholic University of Leuven, Faculty of Science, Department of Mathematics.

Verbeke, G. and Lesaffre, E. (1994) The effect of misspecifying the random effects distribution in a linear mixed effects model. In: *Proceedings of the 9th International Workshop on Statistical Modelling*, Exeter, U.K.

Verbeke, G. and Lesaffre, E. (1996a) A linear mixed-effects model with heterogeneity in the random-effects population. *Journal of the American Statistical Association*, **91**, 217–221.

Verbeke, G. and Lesaffre, E. (1996b) *Large Sample Properties of the Maximum Likelihood Estimators in Linear Mixed Models with Misspecified Random-Effects Distributions.* Technical report 1996.1, Biostatistical Centre for Clinical Trials, Catholic University of Leuven, Belgium.

Verbeke, G. and Lesaffre, E. (1997a) The effect of misspecifying the random effects distribution in linear mixed models for longitudinal data. *Computational Statistics and Data Analysis*, **23**, 541–556.

Verbeke, G. and Lesaffre, E. (1997b) The linear mixed model. A critical investigation in the context of longitudinal data. In: *Proceedings of the Nantucket conference on Modelling Longitudinal and Spatially Correlated Data: Methods, Applications, and Future Directions*, T. Gregoire (Ed.), Lecture Notes in Statistics 122. New York: Springer-Verlag, pp. 89–99.

Verbeke, G. and Lesaffre, E. (1999) The effect of drop-out on the efficiency of longitudinal experiments. *Applied Statistics*, **48**, 363–375.

Verbeke, G., Lesaffre, E., and Brant L.J. (1998) The detection of residual serial correlation in linear mixed models. *Statistics in Medicine*, **17**, 1391–1402.

Verbeke, G., Lesaffre, E., and Spiessens, B. (2000) The practical use of different strategies to handle dropout in longitudinal studies. *Submitted for publication.*

Verbeke, G. and Molenberghs, G. (1997) *Linear Mixed Models in Practice: A SAS-Oriented Approach.* Lecture Notes in Statistics 126. New York: Springer-Verlag.

Verbeke, G., Spiessens, B., and Lesaffre, E. (2000) Conditional linear mixed models. *American Statistician*, **00**, 000–000.

Verbeke, G., Spiessens, B., Lesaffre, E., and Brant, L.J. (1999) Conditional linear mixed models. In: *Proceedings of the 14th International Workshop on Statistical Modelling*, H. Friedl, A. Berghold, and G. Kauermann (Eds), Graz, Austria, pp. 386–393.

Verbyla, A.P. and Cullis, B.R. (1990) Modelling in repeated measures experiments. *Applied Statistics*, **39**, 341–356.

Verdonck, A., De Ridder, L., Verbeke, G., Bourguignon, J.P., Carels, C., Kuhn, E.R., Darras, V., and de Zegher, F. (1998) Comparative effects of neonatal and prepubertal castration on craniofacial growth in rats. *Archives of Oral Biology*, **43**, 861–871.

Wang-Clow, F., Lange, N., Laird, N.M., and Ware, J.H. (1995) A simulation study of estimators for rate of change in longitudinal studies with attrition. *Statistics in Medicine*, **14**, 283–297.

Waternaux, C., Laird, N.M., and Ware, J.H. (1989) Methods for analysis of longitudinal data: bloodlead concentrations and cognitive development. *Journal of the American Statistical Association*, **84**, 33–41.

Weihrauch, T.R. and Demol, P. (1998) The value of surrogate endpoints for evaluation of therapeutic efficacy. *Drug Information Journal*, **32**, 737–43.

Weiner, D.L. (1981) Design and analysis of bioavailability studies. In: *Statistics in the pharmaceutical industry*, C.R. Buncher and J.-Y. Tsay (Eds.). New York: Marcel Dekker, pp. 205–229.

Welham, S.J. and Thompson, R. (1997) Likelihood ratio tests for fixed model terms using residual maximum likelihood. *Journal of the Royal Statistical Society, Series B*, **59**, 701–714.

White, H. (1980) Nonlinear regression on cross-section data. *Econometrica*, **48**, 721–746.

White, H. (1982) Maximum likelihood estimation of misspecified models. *Econometrica*, **50**, 1–25.

Williams, P.L., Molenberghs, G., and Lipsitz, S.R. (1996) Analysis of multiple ordinal outcomes in developmental toxicity studies. *Journal of Agricultural, Biological, and Environmental Statistics*, **1**, 250–274.

Wolfinger, R. and O'Connell, M. (1993) Generalized linear mixed models: a pseudo-likelihood approach. *Journal of Statistical Computation and Simulation*, **48**, 233–243.

Wu, M.C. and Bailey, K.R. (1988) Analysing changes in the presence of informative right censoring caused by death and withdrawal. *Statistics in Medicine*, **7**, 337–346.

Wu, M.C. and Bailey, K.R. (1989) Estimation and comparison of changes in the presence of informative right censoring: conditional linear model. *Biometrics*, **45**, 939–955.

Wu, M.C. and Carroll, R.J. (1988) Estimation and comparison of changes in the presence of informative right censoring by modeling the censoring process. *Biometrics*, **44**, 175–188.

Yates, F. (1933) The analysis of replicated experiments when the field results are incomplete. *Empirical Journal of Experimental Agriculture*, **1**, 129–142.

Zeger, S.C., Liang, K.-Y., and Albert, P.S. (1988) Models for longitudinal data: a generalized estimating equation approach. *Biometrics*, **44**, 1049–1060.

Index

Springer Series in Statistics   *(continued from p. ii)*